Molecular Biology and Biotechnology

Third Edition

Molecular Biology and Biotechnology

Third Edition

Edited by
J. M. Walker
University of Hertfordshire, Hatfield, UK
E. B. Gingold
South Bank University, London, UK

A catalogue record for this book is available from the British Library.

ISBN 0-85186-794-4

First published 1985.
Second edition published 1988.
Third edition published 1993.

Published by The Royal Society of Chemistry,
Thomas Graham House, The Science Park, Cambridge CB4 4WF

Typeset by Computape (Pickering) Ltd, North Yorkshire
Printed and bound by The Bath Press, Lower Bristol Road, Bath

Preface

Had one been asked twenty five years ago to select the most theoretical and least applied area of the biological sciences, there is no doubt that molecular biology would have been near the top of most people's list. Yet today this same area is at the centre of an unparalleled expansion of industrial biology, for out of those earlier theoretical studies has come genetic engineering, and with it the ability to manipulate organisms to produce large quantities of products previously only obtainable by expensive and difficult routes. So dramatic has been the increase in possibilities that a new term, 'biotechnology', has been introduced.

It should never be forgotten, however, that although biotechnology may be a new term, there is a long history of the use of biological processes in the manufacture of products, ranging from the ancient process of alcohol fermentation to the somewhat more recent production of antibiotics. Biotechnology could thus be thought of as the traditional fields such as industrial microbiology and process biochemistry united under a new name. But it is undoubtedly the vast increase in potential arising from the developments of molecular biology that has been responsible for the public recognition that biotechnology now enjoys. Scientific workers from fields well outside the traditional fields of biology have seen the importance of these developments and the need to become involved in the area. It is in this context that the Royal Society of Chemistry organized a series of residential courses at the University of Hertfordshire which form the basis of this book.

It is very much our belief that the importance of the long-standing fields of industrial biology should not be lost in the excitement of the recent developments. It is one thing to manipulate a micro-organism into producing an exotic product, and quite another thing to produce it commercially. For this reason we have included in this volume subjects such as fermentation technology, enzyme technology, and downstream processing as well as the methodology of genetic engineering, and in discussions of the new genetics we have emphasized the aspects relevant to industrial processes. Since the publication of the first edition of this book, there has been a rapid increase in interest in moving beyond bacteria as hosts for industrial processes, and the increased coverage of cloning in yeasts, animal cells, and plant genetic engineering is a reflection of this.

Many of the chapters in this book relate directly to applications of the new technology. These include discussions on its impact on medical care and on the pharmaceutical industry, on monoclonal antibodies, and on the food industry. The exciting field of biosensors is reviewed as is the development of enzyme engineering with its promise of genuinely new protein products.

In producing the third edition of this volume we have aimed to reflect the rapid developments in the field. Probably the most significant has been the development of the polymerase chain reaction (PCR) as a day-to-day tool. The introduction of this technique is already seen as a landmark in the development of molecular biology technology and has allowed molecular biologists to carry out experiments unthinkable just a few years ago. We have thus committed an entire new chapter to introducing this technology. The influence of PCR can also be seen in the updated versions of other molecular biology chapters. The chapter on molecular diagnosis of human genetic disease has been entirely rewritten as a result of these and other advances. In addition, a new chapter describes the impact of molecular biology in the area of forensic science.

Over the last few years, the promise of biotechnology as a commercial success has begun to be realized. There is no doubt that it is in the field of health care that the major successes have emerged. New recombinant therapeutic peptides and proteins, monoclonal antibodies, and vaccines have all found successful markets. Gene therapy and novel approaches to disease treatment and drug discovery, such as antisense oligonucleotides and aptamers, hold great hope for the future. In addition, transgenic animals offer the possibility of cheaper polypeptide pharmaceuticals as well as providing models for disease research. These advances are reflected in this third edition in new chapters on vaccines and transgenesis as well as extensively expanded chapters on the genetic engineering in the pharmaceutical industry and on monoclonal antibodies.

The RSC course was aimed not at expert biotechnologists but at scientific workers whose experience was in entirely different fields. The contributions in this book should thus be seen as primarily having a teaching function. The book should prove of interest both to undergraduates studying for biological or chemical qualifications and to scientific workers from other fields who need a basic introduction to this rapidly expanding area.

J. M. Walker
E. B. Gingold

Contents

Contributors

S. Abbs, *Division of Medical and Molecular Genetics, Paediatric Research Unit, UMDS Guy's and St Thomas's Hospitals, London SE1 9RT, UK*

T. Atkinson, *Division of Biotechnology, PHLS Centre for Applied Microbiology and Research, Porton Down, Salisbury, Wiltshire SP4 0JG, UK*

C. Buffery, *Metropolitan Police Forensic Science Laboratories, 109 Lambeth Road, London SE1 7LP, UK*

V. C. Bugeja, *Division of Biosciences, University of Hertfordshire, College Lane, Hatfield, Hertfordshire AL10 9AB, UK*

M. F. Chaplin, *School of Applied Science, South Bank University, Borough Road, London SE1 0AA, UK*

E. S. Chukhrai, *Faculty of Chemistry, The Lomonosov Moscow State University, 119899, Moscow, Russia*

B. P. G. Curran, *School of Biological Sciences, Queen Mary and Westfield College, University of London, Mile End Road, London E1 4NS, UK*

C. W. Dykes, *Biochemical Targets Department, Glaxo Group Research Ltd., Greenford Road, Greenford, Middlesex UB6 0PG, UK*

E. B. Gingold, *School of Applied Science, South Bank University, Borough Road, London SE1 0AA, UK*

M. J. Greenhalgh, *Metropolitan Police Forensic Science Laboratory, 109 Lambeth Road, London SE1 7LP, UK*

P. M. Hammond, *Division of Biotechnology, PHLS Centre for Applied Microbiology and Research, Porton Down, Salisbury, Wiltshire SP4 0JG, UK*

M. G. K. Jones, *Plant Sciences, School of Biological and Environmental Sciences, Murdoch University, Perth, Western Australia, 6150*

K. Lindsey, *Leicester Biocentre, University of Leicester, University Road, Leicester LE1 7RH, UK*

M. Mackett, *Cancer Research Campaign Department of Molecular Biology, Paterson Institute for Cancer Research, Christie Hospital and Holt Radium Institute, Wilmslow Road, Withington, Manchester M20 9BX, UK*

C. G. P. Mathew, *Division of Medical and Molecular Genetics, Paediatric Research Unit, UMDS Guy's and St Thomas's Hospitals, London SE1 9RT, UK*

J. J. Mullins, *AFRC Centre for Genome Research, King's Buildings, Edinburgh University, West Mains Road, Edinburgh EH9 3JQ, UK*

L. J. Mullins, *AFRC Centre for Genome Research, King's Buildings, Edinburgh University, West Mains Road, Edinburgh EH9 3JQ, UK*
E. J. Murray, *Roche Research Centre, PO Box 8, Welwyn Garden City, Hertfordshire AL7 3AY, UK*
R. K. Pawsey, *Food Science Division, School of Applied Sciences, South Bank University, Borough Road, London SE1 0AA, UK*
O. M. Poltorak, *Faculty of Chemistry, The Lomonosov Moscow State University, 119899, Moscow, Russia*
A. Rosevear, *Biotechnology Department, AEA Environment and Energy, Harwell Laboratories, Didcot, Oxfordshire OX11 0RA, UK*
M. D. Scawen, *Division of Biotechnology, PHLS Centre for Applied Microbiology and Research, Porton Down, Salisbury, Wiltshire SP4 0JT, UK*
R. F. Sherwood, *Division of Biotechnology, PHLS Centre for Applied Microbiology and Research, Porton Down, Salisbury, Wiltshire SP4 0JG, UK*
R. J. Slater, *Division of Biosciences, University of Hertfordshire, College Lane, Hatfield, Hertfordshire AL10 9AB, UK*
P. F. Stanbury, *Division of Biosciences, University of Hertfordshire, College Lane, Hatfield, Hertfordshire AL10 9AB, UK*
M. D. Trevan, *Faculty of Science, Technology, Health, and Society, South Bank University, Borough Road, London SE1 0AA, UK*
J. M. Walker, *Division of Biosciences, University of Hertfordshire, College Lane, Hatfield, Hertfordshire AL10 9AB, UK*
M. Webb, *Sandoz Institute for Medical Research, 5 Gower Place, London WC1E 6BN, UK*

CHAPTER 1

Fermentation Technology

P. F. STANBURY

1 INTRODUCTION

Micro-organisms are capable of growing on a wide range of substrates and can produce a remarkable spectrum of products. The relatively recent advent of *in vitro* genetic manipulation has extended the range of products that may be produced by micro-organisms and has provided new methods for increasing the yields of existing ones. The commercial exploitation of the biochemical diversity of micro-organisms has resulted in the development of the fermentation industry and the techniques of genetic manipulation have given this well established industry the opportunity to develop new processes and to improve existing ones. The term 'fermentation' is derived from the Latin verb *fervere*, to boil, which describes the appearance of the action of yeast on extracts of fruit or malted grain during the production of alcoholic beverages. However, 'fermentation' is interpreted differently by microbiologists and biochemists. To a microbiologist the word means 'any process for the production of a product by the mass culture of micro-organisms'. To a biochemist, however, the word means 'an energy-generating process in which organic compounds act as both electron donors and acceptors', that is, an anaerobic process where energy is produced without the participation of oxygen or other inorganic electron acceptors. In this chapter 'fermentation' is used in its broader, microbiological context.

2 MICROBIAL GROWTH

The growth of a micro-organism may result in the production of a range of metabolites but to produce a particular metabolite the desired organism must be grown under precise cultural conditions at a particular growth rate. If a micro-organism is introduced into a nutrient medium which supports its growth, the inoculated culture will pass through a number of stages and the system is termed batch culture. Initially, growth does not occur and this period is referred to as the lag phase and may be considered a period of adaptation. Following an interval during which the growth rate of the cells gradually increases, the cells grow at a constant, maximum rate and this period is referred to as the log or exponential phase, which may be described by the equation

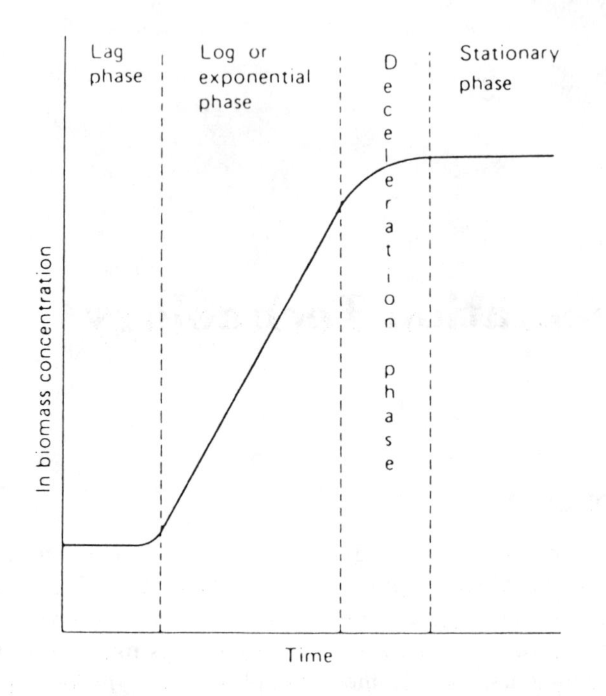

Figure 1 *Growth of a 'typical' micro-organism under batch culture conditions*
(Reproduced with permission from P. F. Stanbury and A. Whitaker, 'Principles of Fermentation Technology', Pergamon Press, Oxford, 1984)

$$dx/dt = \mu x \quad (1)$$

where x is the cell concentration (mg cm^{-3}), t is the time of incubation (h), and μ the specific growth rate (h^{-1}). On integration equation (1) gives

$$x_t = x_0 e^{\mu t} \quad (2)$$

where x_0 is the cell concentration at time zero and x_t is the cell concentration after a time interval, t h.

Thus, a plot of the natural logarithm of the cell concentration against time gives a straight line, the slope of which equals the specific growth rate which is the maximum for the prevailing conditions and is thus described as the maximum specific growth rate, or μ_{max}. Equations (1) and (2) ignore the facts that growth results in the depletion of nutrients and the accumulatoin of toxic by-products and thus predict that growth continues indefinitely. However, in reality, as substrate (nutrient) is exhausted and toxic products accumulate, the growth rate of the cells deviates from the maximum and eventually growth ceases and the culture enters the stationary phase. After a further period of time, the culture enters the death phase and the number of viable cells declines. This classic representation of microbial growth is illustrated in Figure 1. It should be remembered that this description refers to the behaviour of both unicellular and mycelial (filamentous) organisms in batch culture, the growth of the latter resulting in the exponential addition of viable biomass to the mycelial body rather than the production of separate, discrete unicells.

As already stated, the cessation of growth in a batch culture may be due to the exhaustion of a nutrient component or the accumulation of a toxic product. However, provided that the growth medium is designed such that growth is limited by the availability of a medium component, growth may be extended by addition of an aliquot of fresh medium to the vessel. If the fresh medium is added continuously, at an appropriate rate, and the culture vessel is fitted with an overflow device, such that culture is displaced by the incoming fresh medium, a continuous culture may be established. The growth of the cells in a continuous culture of this type is controlled by the availability of the growth limiting chemical component of the medium and, thus, the system is described as a chemostat. In this system a steady-state is eventually achieved and the loss of biomass via the overflow is replaced by cell growth. The flow of medium through the system is described by the term dilution rate, D, which is equal to the rate of addition of medium divided by the working volume of the culture vessel. The balance between growth of cells and their loss from the system may be described as

$$\mathrm{d}x/\mathrm{d}t = \text{growth} - \text{output}$$

or

$$\mathrm{d}x/\mathrm{d}t = \mu x - Dx$$

Under steady-state conditions,

$$\mathrm{d}x/\mathrm{d}t = 0$$

and, therefore, $\mu x = Dx$ and $\mu = D$.

Hence, the growth rate of the organisms is controlled by the dilution rate, which is an experimental variable. It will be recalled that under batch culture conditions an organism will grow at its maximum specific growth rate and, therefore, it is obvious that a continuous culture may be operated only at dilution rates below the maximum specific growth rate. Thus, within certain limits, the dilution rate may be used to control the growth rate of a chemostat culture.

The mechanism underlying the controlling effect of the dilution rate is essentially the relationship between μ, specific growth rate, and s, the limiting substrate concentration in the chemostat, demonstrated by Monod[1] in 1942:

$$\mu = \mu_{max} s/(K_s + s) \tag{3}$$

where K_s is the utilization or saturation constant, which is numerically equal to the substrate concentration when μ is half μ_{max}. At steady-state, $\mu = D$, and, therefore,

$$D = \mu_{max} \bar{s}/(K_s + \bar{s})$$

where $\bar{s}$ is the steady-state concentration of substrate in the chemostat, and

$$\bar{s} = K_s D/(\mu_{max} - D) \tag{4}$$

[1] J. Monod, 'Recherches sur les Croissances des Cultures Bacteriennes', Herman and Cie, Paris, 1942.

Equation (4) predicts that the substrate concentration is determined by the dilution rate. In effect, this occurs by growth of the cells depleting the substrate to a concentration that supports that growth rate equal to the dilution rate. If substrate is depleted below the level that supports the growth rate dictated by the dilution rate the following sequence of events takes place:

(i) The growth rate of the cells will be less than the dilution rate and they will be washed out of the vessel at a rate greater than they are being produced, resulting in a decrease in biomass concentration.
(ii) The substrate concentration in the vessel will rise because fewer cells are left in the vessel to consume it.
(iii) The increased substrate concentration in the vessel will result in the cells growing at a rate greater than the dilution rate and biomass concentration will increase.
(iv) The steady-state will be re-established.

Thus, a chemostat is a nutrient-limited self-balancing culture system which may be maintained in a steady-state over a wide range of sub-maximum specific growth rates.

Fed-batch culture is a system which may be considered to be intermediate between batch and continuous processes. The term fed-batch is used to describe batch cultures which are fed continuously, or sequentially, with fresh medium without the removal of culture fluid. Thus, the volume of a fed-batch culture increases with time. Pirt[2] described the kinetics of such a system as follows. If the growth of an organism were limited by the concentration of one substrate in the medium the biomass at stationary phase, x_{max}, would be described by the equation

$$x_{max} = YS_R$$

where Y is the yield factor and is equal to the mass of cells produced per gram of substrate consumed and S_R is the initial concentration of the growth limiting substrate. If fresh medium were to be added to the vessel at a dilution rate less than μ_{max} then virtually all the substrate would be consumed as it entered the system:

$$FS_R = \mu(X/Y)$$

where F is the flow rate and X is the total biomass in the vessel, *i.e.* the cell concentration multiplied by the culture volume.

Although the total biomass (X) in the vessel increases with time the concentration of cells, x, remains virtually constant; thus $dx/dt = 0$, $\mu = D$. Such a system is then described as quasi-steady-state. As time progresses and the volume of culture increases, the dilution rate decreases. Thus, the value of D is given by the expression

$$D = F/(V_o + Ft)$$

[2] S. J. Pirt, 'Principles of Microbe and Cell Cultivation', Blackwell, Oxford, 1975.

where F is the flow rate, V_0 is the initial culture volume, and t is time. Monod[1] kinetics predict that as D falls residual substrate should also decrease, resulting in an increase in biomass. However, over the range of growth rates operating the increase in biomass should be insignificant. The major difference between the steady-state of the chemostat and the quasi-steady-state of a fed-batch culture is that in a chemostat D (hence, μ) is constant whereas in a fed-batch system D (hence, μ) decreases with time. The dilution rate in a fed-batch system may be kept constant by increasing, exponentially, the flow rate using a computer control system.

3 APPLICATIONS OF FERMENTATION

Microbial fermentations may be classified into the following major groups:[3]

(i) Those that produce microbial cells (biomass) as the product.
(ii) Those that produce microbial metabolites.
(iii) Those that produce microbial enzymes.
(iv) Those that modify a compound which is added to the fermentation – the transformation processes.
(v) Those that produce recombinant products.

3.1 Microbial Biomass

Microbial biomass is produced commercially as single cell protein (SCP) for human food or animal feed and as viable yeast cells to be used in the baking industry. The industrial production of bakers' yeast started in the early 1900s and yeast biomass was used as human food in Germany during the First World War. However, the development of large-scale processes for the production of microbial biomass as a source of commercial protein began in earnest in the late 1960s. Several of the processes investigated did not come to fruition owing to political and economic problems but the establishment of the ICI Pruteen process for the production of bacterial SCP for animal feed was a milestone in the development of the fermentation industry.[4] This process utilized continuous culture on an enormous scale (1500 m^3) and is an excellent example of the application of good engineering to the design of a microbiological process. However, the economics of the production of SCP as animal feed were marginal which eventually led to the discontinuation of the Pruteen process. The technical expertise gained from the Pruteen process assisted ICI in collaborating with Rank Hovis MacDougall on a process for the production of fungal biomass to be used as human food.[5] A continuous fermentation process for the production of *Fusarium graminearum* biomass (marketed as Quorn ®) was developed

[3] P. F. Stanbury and A. Whitaker, 'Principles of Fermentation Technology', Pergamon Press, Oxford, 1984.
[4] D. H. Sharp, 'Bioprotein Manufacture – A Critical Assessment', Ellis Horwood, Chichester, 1989, Chapter 4, p. 53.
[5] A. P. J. Trinci, *Mycol. Res.*, 1992, **96**, 1.

Table 1 *Some examples of microbial primary metabolites and their commercial significance*

Primary metabolite	*Producing organism*	*Commercial significance*
Ethanol	*Saccharomyces cerevisiae*	'Active ingredient' in alcoholic beverages
Citric acid	*Aspergillus niger*	Various uses in the food industry
Acetone and butanol	*Clostridium acetobutyricum*	Solvents
Glutamic acid	*Corynebacterium glutamicum*	Flavour enhancer
Lysine	*Corynebacterium glutamicum*	Feed additive
Polysaccharides	*Xanthomonas* spp.	Applications in the food industry; enhanced oil recovery
Fe^{3+}	*Thiobacillus* and *Sulfolobus*	Ore leaching

utilizing a 40 m^3 air-lift fermenter and a larger scale plant is planned for the near future.

3.2 Microbial Metabolites

The kinetic description of batch culture may be rather misleading when considering the product-forming capacity of the culture during the various phases, for, although the metabolism of stationary phase cells is considerably different from that of logarithmic ones, it is by no means stationary. Bu'Lock *et al.*[6] proposed a descriptive terminology of the behaviour of microbial cells which considered the type of metabolism rather than the kinetics of growth. The term 'trophophase' was suggested to describe the log or exponential phase of a culture during which the sole products of metabolism are either essential to growth, such as amino acids, nucleotides, proteins, nucleic acids, lipids, carbohydrates, *etc.* or are the by-products of energy-yielding metabolism such as ethanol, acetone, and butanol. The metabolites produced during the trophophase are referred to as primary metabolites. Some examples of primary metabolites of commercial importance are listed in Table 1.

Bu'Lock *et al.* suggested the term 'idiophase' to describe the phase of a culture during which products other than primary metabolites are synthesized, products which do not have an obvious role in cell metabolism. The metabolites produced during the idiophase are referred to as the secondary metabolites. The interrelationships between primary and secondary metabolism are illustrated in Figure 2, from which it may be seen that secondary metabolites tend to be synthesized from the intermediates and end products of primary metabolism. Although the primary metabolic routes shown in Figure 2 are common to the vast majority of micro-organisms, each secondary metabolite would be synthesized by very few microbial taxa. Also, not all microbial taxa undergo secondary metabolism; it is a common feature of the filamentous fungi and bacteria and the

[6] J. D. Bu'Lock, D. Hamilton, M. A. Hulme, A. J. Powell, D. Shepherd, H. M. Smalley, and G. N. Smith, *Can. J. Microbiol.*, 1965, **11**, 765.

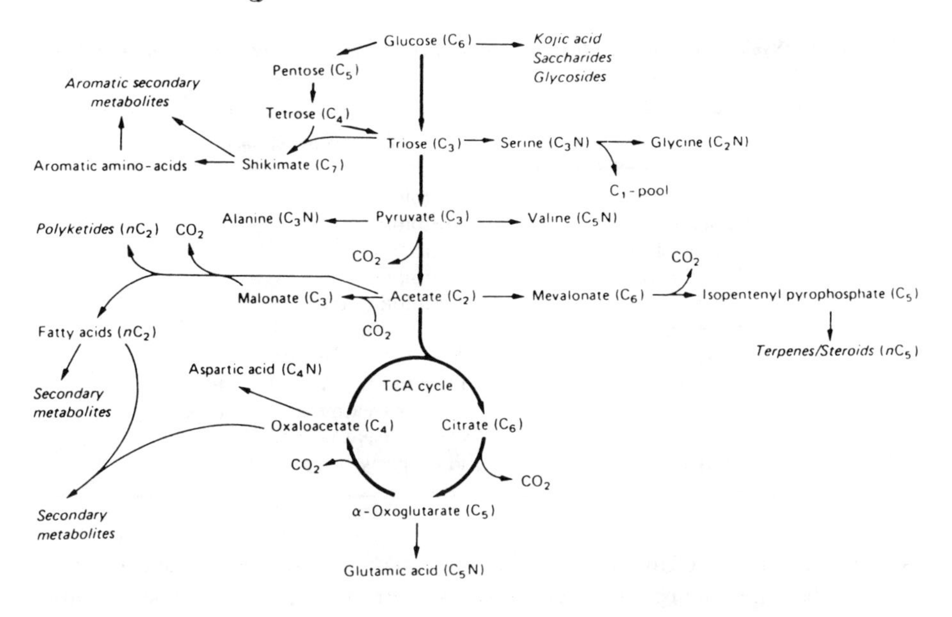

Figure 2 *The inter-relationships between primary and secondary metabolism*
(Reproduced with permission from W. B. Turner, 'Fungal Metabolites', Academic Press, 1971)

sporing bacteria but it is not, for example, a feature of the Enterobacteriaceae. Thus, although the taxonomic distribution of secondary metabolism is far more limited than that of primary metabolism, the range of secondary products produced is enormous. The classification of microbial products into secondary and primary metabolites should be considered as a convenient, but in some cases, artificial system. To quote Bushell,[7] the classification 'should not be allowed to act as a conceptual straightjacket, forcing the reader to consider all products as either primary or secondary metabolites'. It is sometimes difficult to categorize a product as primary or secondary, and the kinetics of production of certain compounds may change, depending on the growth conditions employed.

At first sight it may seem anomalous that micro-organisms produce compounds which do not appear to have any metabolic function and are certainly not by-products of catabolism as are, for example, ethanol and acetone. However, many secondary metabolites exhibit antimicrobial properties and, therefore, may be involved in competition in the natural environment;[8] others have, since their discovery in idiophase cultures, been demonstrated to be produced during the trophophase where, it has been claimed, they act in some form of metabolic control.[9] Although the physiological role of secondary metabolism continues to be the subject of considerable debate its relevance to the fermentation industry is the commercial significance of the secondary metabolites. Table 2 summarizes some of the industrially important groups of secondary metabolites.

[7] M. E. Bushell, in 'Principles of Biotechnology', ed. A. Wiseman, Chapman and Hall, New York, 1988, p. 5.
[8] A. L. Demain, *Search*, 1980, **11**, 148.
[9] I. M. Campbell, *Adv. Microb. Physiol.*, 1984, **25**, 2.

Table 2 *Some examples of microbial secondary metabolites and their commercial significance*

Secondary metabolite	*Commercial significance*
Penicillin	Antibiotic
Cephalosporin	Antibiotic
Tetracyclines	Antibiotic
Streptomycin	Antibiotic
Griseofulvin	Antibiotic (anti-fungal)
Actinomycin	Antitumour
Pepstatin	Treatment of ulcers
Cyclosporin A	Immunosuppressant
Krestin	Cancer treatment
Bestatin	Cancer treatment
Gibberellin	Plant growth regulator

The production of microbial metabolites may be achieved in continuous, as well as batch, systems. The chronological separation of trophophase and idiophase in batch culture may be studied in continuous culture in terms of dilution rate.[10–12] Secondary metabolism will occur at relatively low dilution rates (growth rates) and, therefore, it should be remembered that secondary metabolism is a property of slow-growing, as well as stationary, cells. The fact that secondary metabolites are produced by slow-growing organisms in continuous culture indicates that primary metabolism is continuing in idiophase type cells. Thus, secondary metabolism is not switched on to remove an accumulation of metabolites synthesized entirely in a different phase; synthesis of the primary metabolic precursors continues through the period of secondary biosynthesis.

The control of the onset of secondary metabolism has been studied extensively in batch culture and, to a lesser extent, in continuous culture. The outcome of this work is that a considerable amount of information is available on the interrelationships between the changes occurring in the medium and the cells at the onset of secondary metabolism and the control of the process. Primary metabolic precursors of secondary metabolites have been demonstrated to induce secondary metabolism, for example tryptophan in alkaloid[13] biosynthesis and methionine in cephalosporin biosynthesis.[14] On the other hand, medium components have been demonstrated to repress secondary metabolism, the earliest observation being that of Saltero and Johnson[15] in 1953 of the repressing effect of glucose on benzyl penicillin formation. Carbon sources which support high growth rates tend to support poor secondary metabolism and Table 3 cites some examples of this situation. Phosphate and nitrogen sources have also been implicated in the repression of secondary metabolism, as exemplified in Table 3.

[10] S. J. Pirt, *Chem. Ind. (London)*, May 1968, 601.
[11] S. J. Pirt and D. S. Callow, *J. Appl. Bacteriol.*, 1960, **23**, 87.
[12] S. J. Pirt and R. C. Righelato, *Appl. Microbiol.*, 1967, **15**, 1284.
[13] J. F. Robers and H. G. Floss, *J. Pharmacol. Sci.*, 1970, **59**, 702.
[14] K. Komatsu, M. Mizumo, and R. Kodaira, *J. Antibiot.*, **28**, 881.
[15] F. V. Saltero and M. I. Johnson, *Appl. Microbiol.*, 1953, **1**, 2.

Table 3 *Some examples of the repression of secondary metabolism by medium components*

Medium component	*Repressed secondary metabolite*
Glucose	Penicillin[a]
Glucose	Actinomycin[b]
Glucose	Neomycin[c]
Glucose	Streptomycin[d]
Phosphate	Candicidin[e]
Phosphate	Streptomycin[f]
Phosphate	Tetracycline[g]
Nitrogen source	Penicillin[h]

[a] F. V. Saltero and M. I. Johnson, *Appl. Microbiol.*, 1953, **1**, 2. [b] R. C. Vining and D. S. Westlake, 'Biotechnology of Industrial Antibiotics', ed. E. J. Vandamme, Marcel Dekker, New York, 1984, p. 387. [c] M. J. Majumbar and S. K. Majumbar, *Biochem. J.*, 1972, **122**, 397. [d] E. Inamine, D. B. Lago, and A. L. Demain, 'Fermentation Advances', ed. D. Perlman, Academic Press, New York, 1969, p. 199. [e] G. Naharrop, J. A. Gill, J. R. Villanueva, and J. F. Martin, 'Advances in Biotechnology', ed. E. Vezina and K. Singh, Pergamon Press, Toronto, 1981, Vol. 3, p. 147. [f] A. L. Demain, ref. a, p. 33. [g] V. Behal, Z. Hostalek, and Z. Vanek, *Biotechnol. Lett.*, 1979, **1**, 177. [h] S. Sanchez, L. Paniagua, R. C. Mateos, F. Lara, and J. More, ref. e, p. 147.

Therefore, it is essential that repressing nutrients should be avoided in media to be used for the industrial production of secondary metabolites or that the mode of operation of the fermentation maintains the potentially repressing components at sub-repressing levels, as discussed in a later section of this chapter.

3.3 Microbial Enzymes

The major commercial utilization of microbial enzymes is in the food and beverage industries[16] although enzymes do have considerable application in clinical and analytical situations, as well as their use in washing powders. Most enzymes are synthesized in the logarithmic phase of batch culture and may, therefore, be considered as primary metabolites. However, some, for example the amylases of *Bacillus stearothermophilus*,[17] are produced by idiophase-type cultures and may be considered as equivalent to secondary metabolites. Enzymes may be produced from animals and plants as well as microbial sources but the production by microbial fermentation is the most economic and convenient method. Furthermore, it is now possible to engineer microbial cells to produce animal or plant enzymes, as discussed in Section 3.5.

3.4 Transformation Processes

As well as the use of micro-organisms to produce biomass and microbial products, microbial cells may be used to catalyse the conversion of a compound into

[16] D. J. Jeenes, D. A. MacKenzie, I. N. Roberts, and D. B. Archer, in 'Biotechnology and Genetic Engineering Reviews', ed. M. P. Tombs, Intercept, Andover, 1991, Vol. 9, Chapter 9, p. 327.
[17] A. B. Manning and L. L. Campbell, *J. Biol. Chem.*, 1961, **236**, 2951.

a structurally similar, but financially more valuable, compound. Such fermentations are termed transformation processes, biotransformations, or bioconversions. Although the production of vinegar is the oldest and most well established transformation process (the conversion of ethanol into acetic acid), the majority of these processes involve the production of high-value compounds. Because micro-organisms can behave as chiral catalysts with high regio- and stereospecificity, microbial processes are more specific than purely chemical ones and make possible the addition, removal, or modification of functional groups at specific sites on a complex molecule without the use of chemical protection. The reactions which may be catalysed include oxidation, dehydrogenation, hydroxylation, dehydration and condensation, decarboxylation, deamination, amination, and isomerization. The anomaly of the transformation process is that a large biomass has to be produced to catalyse, perhaps, a single reaction. The logical development of these processes is to perform the reaction using the purified enzyme or the enzyme attached to an immobile support. However, enzymes work more effectively within their microbial cells, especially if co-factors such as reduced pyridine nucleotide need to be regenerated. A compromise is to employ resting cells as catalysts, which may be suspended in a medium not supporting growth or attached to an immobile support. The reader is referred to Goodhue *et al.*[18] for a detailed review of transformation processes.

3.5 Recombinant Products

The advent of recombinant DNA technology has extended the range of potential microbial fermentation products. It is possible to introduce genes from higher organisms into microbial cells such that the recipient cells are capable of synthesizing 'foreign' (or heterologous) proteins. Examples of the 'hosts' for such foreign genes include *Escherichia coli*, *Saccharomyces cerevisiae*, and other yeasts as well as filamentous fungi. Products produced in such genetically manipulated organisms include interferon, insulin, human serum albumin, factor VIII and factor IX, epidermal growth factor, and bovine somatostatin. Important factors in the design of these processes include the secretion of the product, minimization of the degradation of the product, and the control of the onset of synthesis during the fermentation, as well as maximizing the expression of the foreign gene. These aspects are considered in detail in references 19 and 20.

4 THE FERMENTATION PROCESS

Figure 3 illustrates the component parts of a generalized fermentation process. Although the central component of the system is obviously the fermenter itself, in which the organism is grown under conditions optimum for product formation,

[18] C. T. Goodhue, J. P. Rosazza, and G. P. Peruzzutti, in 'Manual of Industrial Microbiology and Biotechnology', ed. A. L. Demain and A. Solomons, American Society for Microbiology, Washington, DC, 1986, p. 97.

[19] J. R. Harris, 'Protein Production by Biotechnology', Elsevier, London, 1990.

[20] A. Wiseman, 'Genetically-engineered Proteins and Enzymes from Yeast: Production and Control', Ellis Horwood, Chichester, 1991.

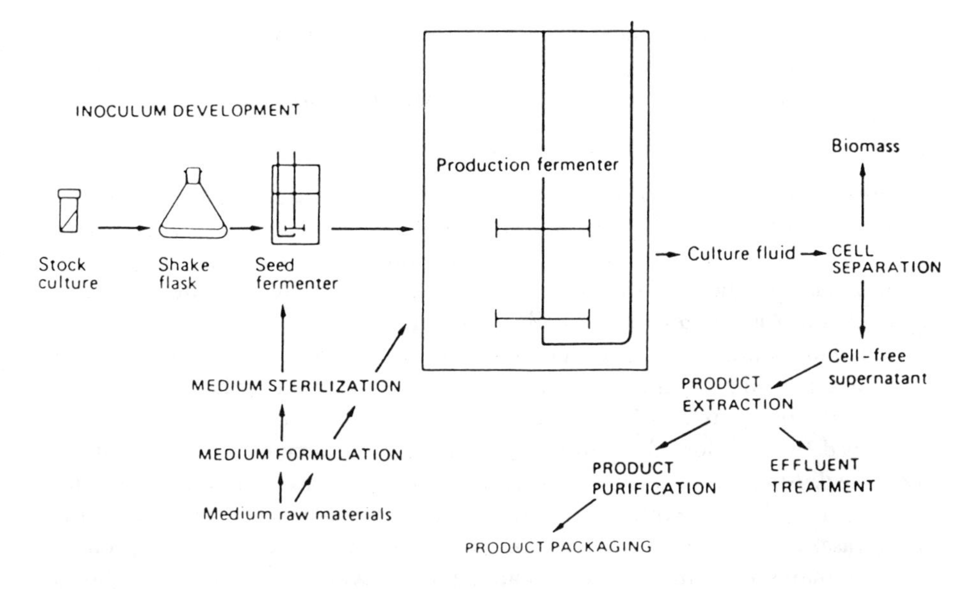

Figure 3 *A generalized, schematic representation of a fermentation process*
(Reproduced with permission from P. F. Stanbury and A. Whitaker, 'Principles of Fermentation Technology', Pergamon Press, Oxford, 1984)

one must not lose sight of operations upstream and downstream of the fermenter. Before the fermentation is started the medium must be formulated and sterilized, the fermenter sterilized, and a starter culture must be available in sufficient quantity and in the correct physiological state to inoculate the production fermenter. Downstream of the fermenter the product has to be purified and further processed and the effluents produced by the process have to be treated.

4.1 The Mode of Operation of Fermentation Processes

As discussed earlier, micro-organisms may be grown in batch, fed-batch, or continuous culture, and continuous culture offers the most control over the growth of the cells. However, the commercial adoption of continuous culture is confined to the production of biomass and, to a limited extent, the production of potable and industrial alcohol. The superiority of continuous culture for biomass production is overwhelming, as may be seen from the following account, but for other microbial products the disadvantages of the system outweigh the improved process control which the technique offers.

Productivity in batch culture may be described by the equation[3]

$$R_{batch} = (x_{max} - x_0)/(t_i + t_{ii}) \qquad (5)$$

where R_{batch} is the output of the culture in terms of biomass concentration per hour, x_{max} is the maximum cell concentration achieved at stationary phase, x_0 is the initial cell concentration at inoculation, t_i is the time during which the organism grows at μ_{max}, and t_{ii} is the time during which the organism is not growing at μ_{max} and includes the lag phase, the deceleration phase, and the periods of batching, sterilizing, and harvesting.

The productivity[3] of a continuous culture may be represented as

$$R_{cont} = D\bar{x}(1 - t_{iii}/T) \quad (6)$$

where R_{cont} is the output of the culture in terms of cell concentration per hour, t_{iii} is the time period prior to the establishment of a steady-state and includes time for vessel preparation, sterilization, and operation in batch culture prior to continuous operation, T is the time period during which steady-state conditions prevail, and $\bar{x}$ is the steady-state cell concentration.

Maximum output of biomass per unit time (*i.e.* productivity) in a chemostat may be achieved by operating at the dilution rate giving the highest value of $D\bar{x}$, this value being referred to as D_{max}. Batch fermentation productivity, as described by equation (5) is an average for the total time of the fermentation. Because $dx/dt = \mu x$ the productivity of the culture increases with time and, thus, the vast majority of the biomass in a batch process is produced near the end of the log phase. In a steady-state chemostat, operating at or near D_{max}, the productivity remains constant, and maximum, for the whole fermentation. Also, a continuous process may be operated for a very long time so that the non-productive period, t_{iii} in equation (6), may be insignificant. However, the non-productive time element for a batch culture is a very significant period, especially as the fermentation would have to be re-established many times during the running time of a comparable continuous process and, therefore, t_{ii} would be recurrent.

The steady-state nature of a continuous process is also advantageous in that the system should be far easier to control than a comparable batch one. During a batch fermentation, heat output, acid or alkali production, and oxygen consumption will range from very low rates at the start of the fermentation to very high rates during the late logarithmic phase. Thus, the control of the environment of such a system is far more difficult than that of a continuous process where, at steady-state, production and consumption rates are constant. Furthermore, a continuous process should result in a more constant labour demand than a comparable batch one.

A frequently quoted disadvantage of continuous systems is their susceptibility to contamination by 'foreign' organisms. The prevention of contamination is essentially a problem of fermenter design, construction, and operation and should be overcome by good engineering and microbiological practice. ICI recognized the overwhelming advantages of a continuous biomass process and overcame the problems of contamination by building a secure fermenter capable of very long periods of aseptic operation, as described by Smith.[21]

The production of growth-associated by-products, such as ethanol, should also be more efficient in continuous culture. However, continuous brewing has met with only limited success and UK breweries have abandoned such systems owing to problems of flavour and lack of flexibility.[22] The production of industrial

[21] S. R. L. Smith, *Philos. Trans. R. Soc. London., Ser. B*, 1980, **290**, 341.

[22] B. H. Kirsop, in 'Topics in Enzyme and Fermentation Biotechnology', ed. A. Wiseman, Ellis Horwood, Chichester, 1982, p. 79.

alcohol, on the other hand, should not be limited by the problems encountered by the brewing industry and continuous culture should be the method of choice for such a process. The adoption of continuous culture for the production of biosynthetic (as opposed to catabolic) microbial products has been extremely limited. Although, theoretically, it is possible to optimize a continuous system such that optimum productivity of a metabolite should be achieved, the long-term stability of such systems is precarious, owing to the problem of strain degeneration. A consideration of the kinetics of continuous culture reveals that the system is highly selective and will favour the propagation of the best adapted organism in a culture. 'Best adapted' in this context refers to the affinity of the organism for the limiting substrate at the operating dilution rate. A commercial organism is usually highly mutated such that it will produce very high amounts of the desired product. Therefore, in physiological terms, such commercial organisms are extremely inefficient and a revertant strain, producing less of the desired product, may be better adapted to the cultural conditions than the superior producer and will come to dominate the culture. This phenomenon, termed by Calcott[23] as contamination from within, is the major reason for the lack of use of continuous culture for the production of microbial metabolites.

Although the fermentation industry has been reluctant to adopt continuous culture for the production of microbial metabolites, very considerable progress has been made in the development of fed-batch systems.[24,25] Fed-batch culture may be used to achieve a considerable degree of process control and to extend the productive period of a traditional batch process without the inherent disadvantages of continuous culture described previously. The major advantage of feeding a medium component to a culture, rather than incorporating it entirely in the initial batch, is that the nutrient may be maintained at a very low concentration during the fermentation. A low (but constantly replenished) nutrient level may be advantageous in

(i) Maintaining conditions in the culture within the aeration capacity of the fermenter.
(ii) Removing the repressive effects of medium components such as rapidly used carbon and nitrogen sources and phosphate.
(iii) Avoiding the toxic effects of a medium component.
(iv) Providing a limiting level of a required nutrient for an auxotrophic strain.

The earliest example of the commercial use of fed-batch culture is the production of bakers' yeast. It was recognized as early as 1915 that an excess of malt in the production medium would result in a high rate of biomass production and an oxygen demand which could not be met by the fermenter.[26] This resulted in the development of anaerobic conditions and the formation of ethanol at the expense of biomass. The solution to this problem was to grow the yeast initially in a weak medium and then add additional medium at a rate less than the organism

[23] P. H. Calcott, 'Continuous Culture of Cells', CRC Press, Boca Raton, Fl., 1981, Vol. 1, p. 13.
[24] A. Whitaker, *Process Biochem.*, 1980, **15**(4), 10.
[25] T. Yamane and S. Shimizu, *Adv. Biochem. Eng./Biotechnol.*, 1984, **30**, 147.
[26] G. Reed and H. J. Peppler, 'Yeast Technology', Avi, Westport, 1973, p. 664.

could use it. It is now appreciated that a high glucose concentration represses respiratory activity, and in modern yeast production plants the feed of molasses is under strict control based on the automatic measurement of traces of ethanol in the exhaust gas of the fermenter. As soon as ethanol is detected the feed rate is reduced. Although such systems may result in low growth rates the biomass yield is near that theoretically obtainable.[27]

The penicillin fermentation provides a very good example of the use of fed-batch culture for the production of a secondary metabolite.[28] The penicillin process is a 'two-stage' fermentation; an initial growth phase is followed by the production phase or idiophase. During the production phase glucose is fed to the fermentation at a rate which allows a relatively high growth rate (and therefore rapid accumulation of biomass) yet maintains the oxygen demand of the culture within the aeration capacity of the equipment. If the oxygen demand of the biomass were to exceed the aeration capacity of the fermenter anaerobic conditions would result and the carbon source would be used inefficiently. During the production phase the biomass must be maintained at a relatively low growth rate and, thus, the glucose is fed at a low dilution rate. Phenylacetic acid is a precursor of the penicillin molecule but it is also toxic to the producer organism above a threshold concentration. Thus, the precursor is also fed into the fermentation continuously, thereby maintaining its concentration below the inhibitory level.

5 THE GENETIC IMPROVEMENT OF PRODUCT FORMATION

Owing to their inherent control systems, micro-organisms usually produce commercially important metabolites in very low concentrations and, although the yield may be increased by optimizing the cultural conditions, productivity is controlled ultimately by the organism's genome. Thus, to improve the potential productivity, the organism's genome must be modified and this may be achieved in two ways, by mutation or by recombination.

5.1 Mutation

Each time a microbial cell divides there is a small probability of an inheritable change occurring. A strain exhibiting such a changed characteristic is termed a mutant and the process giving rise to it, a mutation. The probability of a mutation occurring may be increased by exposing the culture to a mutagenic agent such as UV light, ionizing radiation, and various chemicals, for example nitrosoguanidine and nitrous acid. Such an exposure usually involves subjecting the population to a mutagen dose which results in the death of the vast majority of the cells. The survivors of the mutagen exposure may then contain some mutants, a very small proportion of which may be improved producers. Thus, it

[27] A. Fiechter, in 'Advances in Biotechnology, 1, Scientific and Engineering Principles', ed. M. Moo-Young, C. W. Robinson, and C. Vezina, Pergamon Press, Toronto, 1981, p. 261.

[28] J. M. Hersbach, C. P. Van der Beek, and P. W. M. Van Vijek, in 'Biotechnology of Industrial Antibiotics', ed. E. J. Vandamme, Marcel Dekker, New York, 1984, p. 387.

is the task of the industrial geneticist to separate the desirable mutants (the superior producers) from the very many inferior types. This approach is easier for strains producing primary metabolites than it is for those producing secondary metabolites, as may be seen from the following examples.

The synthesis of a primary microbial metabolite (such as an amino acid) is controlled such that it is only produced at a level required by the organism. The control mechanisms involved are the inhibition of enzyme activity and the repression of enzyme synthesis by the end product when it is present in the cell at a sufficient concentration. Thus, these mechanisms are referred to as feedback control. It is obvious that a good 'commercial' mutant should lack the control systems so that 'overproduction' of the end product will result. The isolation of mutants of *Corynebacterium glutamicum* capable of producing lysine will be used to illustrate the approaches which have been adopted to remove the control systems.

The control of lysine synthesis in *Corynebacterium glutamicum* is illustrated in Figure 4 from which it may be seen that the first enzyme in the pathway, aspartokinase, is inhibited only when both lysine and threonine are synthesized above a threshold level. This type of control is referred to as concerted feedback control. A mutant which could not catalyse the conversion of aspartic semi-aldehyde into homoserine would be capable of growth only in a homoserine-supplemented medium and the organism would be described as a homoserine auxotroph. If such an organism were grown in the presence of very low concentrations of homoserine the endogenous level of threonine would not reach the inhibitory level for aspartokinase control and, thus, aspartate would be converted into lysine which would accumulate in the medium. Thus, a knowledge of the control of the biosynthetic pathway allows a 'blueprint' of the desirable mutant to be constructed and makes easier the task of designing the procedure to isolate the desired type from the other survivors of a mutation treatment.

The isolation of bacterial auxotrophs may be achieved using the penicillin enrichment technique developed by Davis.[29] Under normal culture conditions an auxotroph is at a disadvantage compared with the parental (wild type) cells. However, penicillin only kills growing cells and, therefore, if the survivors of a mutation treatment were cultured in a medium containing penicillin and lacking the growth requirement of the desired mutant only those cells unable to grow would survive, *i.e.* the desired auxotrophs. If the cells were removed from the penicillin broth, washed, and resuspended in a medium containing the requirement of the desired auxotroph then the resulting culture should be rich in the required type. Nakayama *et al.*[30] used this technique to isolate a homoserine auxotroph of *C. glutamicum* which produced 44 g dm^{-3} lysine.

An alternative approach to the isolation of mutants which do not produce controlling end products (*i.e.* auxotrophs) is to isolate mutants which do not recognize the presence of controlling compounds. Such mutants may be isolated from the survivors of a mutation treatment by exploiting their capacity to grow in the presence of certain compounds which are inhibitory to the parental types.

[29] B. D. Davis, *Proc. Natl. Acad. Sci. USA*, 1949, **35**, 1.
[30] K. Nakayama, S. Kituda, and S. Kinoshita, *J. Gen. Appl. Microbiol.*, 1961, **7**, 41.

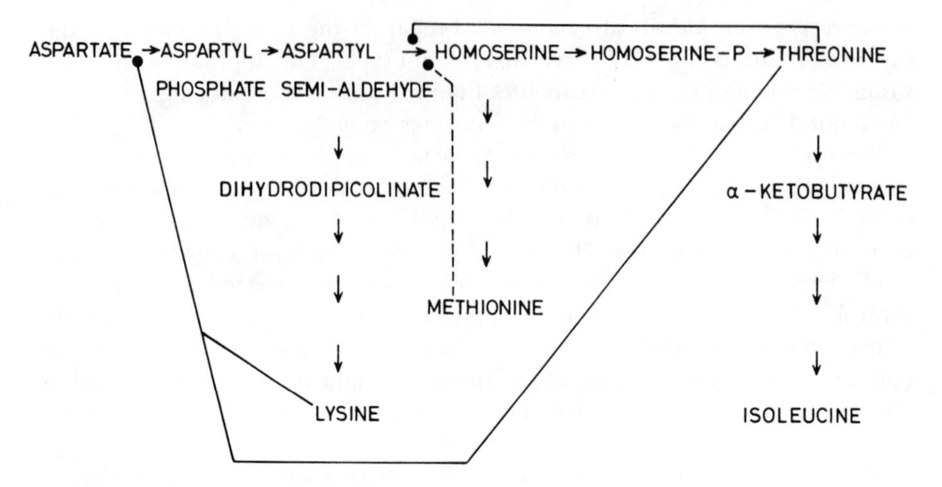

Figure 4 *The control of biosynthesis of lysine in* Corynebacterium glutamicum: *Biosynthetic route* →; *Feedback inhibition* —●; *Feedback repression* ----●

An analogue is a compound which is similar in structure to another compound and analogues of primary metabolites are frequently inhibitory to microbial cells. The toxicity of the analogue may be due to any of a number of possible mechanisms; for example, the analogue may be incorporated into a macromolecule in place of the natural product, resulting in the production of a defective compound, or the analogue may act as a competitive inhibitor of an enzyme for which the natural product is a substrate. Also, the analogue may mimic the control characteristics of the natural product and inhibit product formation despite the fact that the natural product concentration is inadequate to support growth. A mutant which is capable of growing in the presence of an analogue inhibitory to the parent may owe its resistance to any of a number of mechanisms. However, if the toxicity were due to the analogue mimicking the control characteristics of the normal end product, then the resistance may be due to the control system being unable to recognize the analogue as a control factor. Such analogue-resistant mutants may also not recognize the natural product and may, therefore, overproduce it. Thus, there is a reasonable probability that mutants resistant to the inhibitory effects of an analogue may overproduce the compound to which the analogue is analogous. Sano and Shiio[31] made use of this approach in attempting to isolate lysine producing mutants of *Brevibacterium flavum*. The control of lysine formation in *B. flavum* is the same as that illustrated in Figure 4 for *C. glutamicum*. Sano and Shiio demonstrated that the lysine analogue *S*-(2-aminoethyl)cysteine (AEC) only inhibited growth completely in the presence of threonine, which suggests that AEC combined with threonine in the concerted inhibition of aspartokinase and deprived the organism of lysine and methionine. Mutants were isolated by plating the survivors of a mutation treatment on to agar plates containing both AEC and threonine. A relatively high proportion of the resulting colonies were lysine overproducers, the best of

[31] K. Sano and I. Shiio, *J. Gen. Appl. Microbiol.*, 1970, **16**, 373.

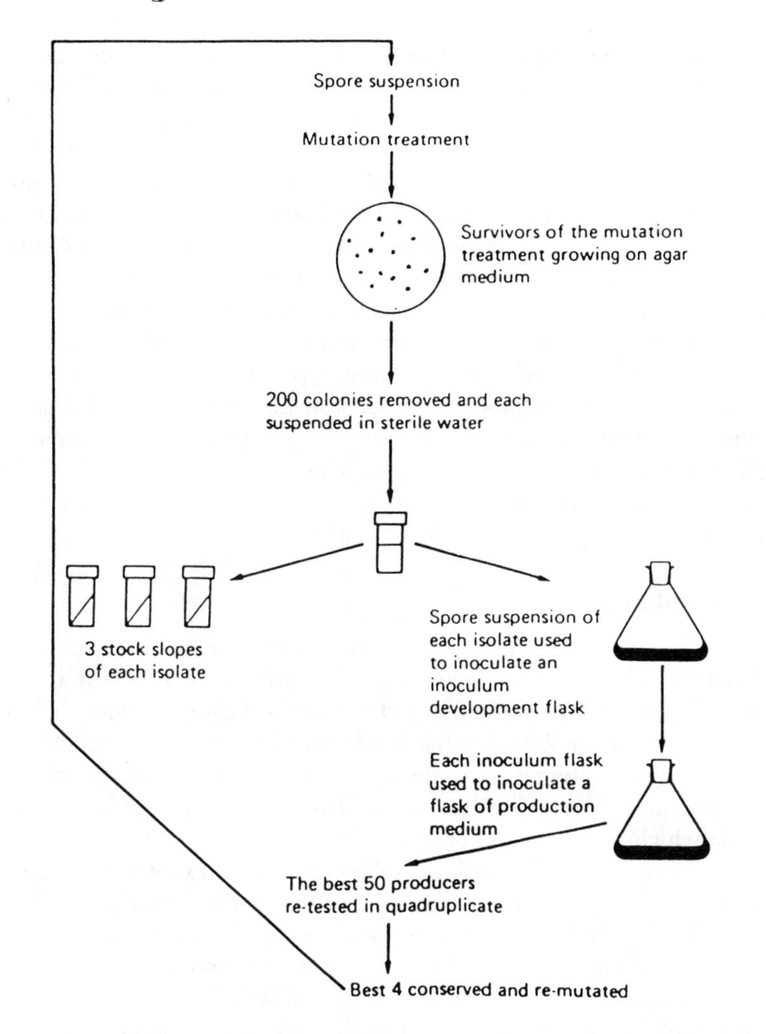

Figure 5 *A strain improvement programme for a secondary metabolite producing culture* (Reproduced with permission from P. F. Stanbury and A. Whitaker, 'Principles of Fermentation Technology', Pergamon Press, Oxford, 1984)

which produce more than 30 g dm^{-3}. Fuller accounts of the isolation of amino acid and nucleotide producing strains may be found in references 3, 32, and 33.

Thus, a knowledge of the control systems may assist in the design of procedures for the isolation of mutants overproducing primary metabolites. The design of procedures for the isolation of mutants overproducing secondary metabolites has been more difficult owing to the fact that far less information was available on the control of production and, also, that the end products of secondary metabolism are not required for growth. Thus, many current industrial strains have been

[32] S. Kinoshita and K. Nakayama, 'Primary Products of Metabolism. Economic Microbiology 2', Academic Press, London, 1978, p. 210.

[33] H. Enei and Y. Hirose, in 'Biotechnology and Genetic Engineering Research Reviews 2', ed. G. E. Russell, Intercept, Newcastle upon Tyne, UK, 1984, p. 100.

selected using direct, empirical, screens of the survivors of a mutation treatment for productivity rather than cultural systems which give an advantage to producing types. A typical programme is illustrated in Figure 5[34] although such systems have been miniaturized[3] to increase their throughput. However, attempts have been, and are being, made to adopt a more rational approach to selection techniques and to reduce the empirical nature of the screens. Elander *et al.*[35] adopted the analogue resistance approach in the isolation of mutants of *Pseudomonas aureofaciens* overproducing the antibiotic pyrrolnitrin. Tryptophan is a precursor of pyrrolnitrin and a limiting factor of productivity. These workers isolated tryptophan analogue-resistant mutants, one of which produced 2–3 times more antibiotic than the parental type. Martin *et al.*[36] removed the inhibitory effect of tryptophan on candicidin production by isolating mutants resistant to tryptophan analogues. Relief of carbon repression has been achieved by selection of mutants resistant to 2-deoxyglucose, a glucose analogue.[37] Further examples of selection methods for the isolation of improved secondary metabolite producers are given in references 3, 38, and 39.

5.2 Recombination

Hopwood[40] defined recombination as any process which helps to generate new combinations of genes that were originally present in different individuals. Compared with the use of mutation techniques for the improvement of industrial strains the use of recombination has been fairly limited. This is probably due to the success, and relative ease, of mutation techniques and to the lack of basic genetic information on industrial strains. However, techniques are now widely available which allow the use of recombination as a system of strain improvement. *In vivo* recombination may be achieved in the asexual fungi (for example *Penicillium chrysogenum*, used for the commercial production of penicillin) using the parasexual cycle.[41] The technique of protoplast fusion has increased greatly the prospects of combining together characteristics found in different production strains. Protoplasts are cells devoid of their cell walls and may be prepared by subjecting cells to the action of wall-degrading enzymes in isotonic solutions. Cell fusion, followed by nuclear fusion, may occur between protoplasts of strains which would not otherwise fuse and the resulting fused protoplast may regenerate a cell wall and grow as a normal cell. Protoplast fusion has been achieved with the filamentous fungi, yeasts, streptomycetes, and bacteria and is an increasingly used technique. For example, Tosaka *et al.*[42] improved the rate of

[34] O. L. Davies, *Biometrics*, 1964, **20**, 576.
[35] R. P. Elander, J. A. Mabe, R. L. Hamill, and M. Gorman, *Fol. Microbiol.*, 1971, **16**, 157.
[36] J. F. Martin, J. A. Gill, G. Naharro, P. Liras, and J. R. Villanueva, in 'Genetics of Industrial Micro-organisms', ed. O. K. Sebek and A. I. Laskin, American Society for Microbiology, Washington, DC, 1979, p .205.
[37] D. A. Hodgson, *J. Gen. Microbiol.*, 1982, **128**, 2417.
[38] J. P. DeWitt, J. V. Jackson, and T. J. Paulus, in 'Fermentation Process Development of Industrial Organisms', ed. J. O. Neway, Marcel Dekker, New York, 1989, Chapter 1, p. 1.
[39] R. P. Elander, ref. 38, Chapter 4, p. 169. [40] D. A. Hopwood, ref. 36, p. 1.
[41] K. D. Macdonald and G. Holt, *Sci. Progr.*, 1976, **63**, 547.
[42] O. Tosaka, M. Karasawa, S. Ikeda, and H. Yoshii, Abstracts of 4th International Symposium on Genetics of Industrial Micro-organisms, 1982, p. 61.

glucose consumption (and therefore lysine production) of a high lysine producing strain of *B. flavum* by fusing it with another *B. flavum* strain which was a non-lysine producer but consumed glucose at a high rate. Among the fusants one strain exhibited high lysine production with rapid glucose utilization. Chang *et al.*[43] used protoplast fusion to combine the desirable properties of two strains of *P. chrysogenum* producing penicillin V into one producer strain. Lein[44] has described the use of protoplast fusion for the improvement of penicillin production in the procedures used by Panlabs Inc. and DeWitt *et al.*[38] reviewed the technique for the improvement of actinomycete processes.

In vitro recombination has been achieved by the techniques of *in vitro* recombinant DNA technology discussed elsewhere in this book. Although the most well publicized recombinants achieved by these techniques are those organisms which synthesize foreign products (see section 3.5), very considerable achievements have been made in the improvement of strains producing conventional products. The efficiency of *Methylomonas methylotrophus*, the organism used in the ICI Pruteen process, was improved by the incorporation of a plasmid containing a glutamate dehydrogenase gene from *E. coli*.[45] The manipulated organism was capable of more efficient ammonia metabolism which resulted in a 5% improvement in carbon conversion. However, the strain was not used on the large scale plant due to problems of scale-up.

In vitro DNA technology has been used to amplify the number of copies in a critical pathway gene (or operon) in a process organism. Although gene amplification is not an example of recombination it is best considered in the context of DNA manipulative techniques. Threonine production by *E. coli* has been improved by incorporating the entire threonine operon of a threonine analogue-resistant mutant into a plasmid which was then introduced back into the bacterium. The plasmid copy number in the cell was approximately 20 and the activity of the threonine operon enzymes was increased 40–50 times. The organism produced 30 g dm^{-3} threonine compared with the 2–3 g dm^{-3} of the non-manipulated strain.[46] Miwa *et al.*[47] utilized similar techniques in constructing an *E. coli* strain capable of synthesizing 65 g dm^{-3} threonine.

The application of the techniques of genetic manipulation to the improvement of *Corynebacterium glutamicum* (see section 5.1) was hindered by the availability of a suitable vector. However, vectors have been constructed and considerable progress has been made in the improvement of amino acid fermentations.[48,49] Threonine, histidine, and phenylalanine production have been improved using

[43] L. T. Chang, D. T. Terasaka, and R. P. Elander, *Dev. Ind. Microbiol.*, 1982, **23**, 21.

[44] J. Lein, in 'Overproduction of Microbial Metabolites', ed. Z. Vanek and Z. Hostalek, Butterworths, Boston, 1986, p. 105.

[45] J. D. Windon, M. J. Worsey, E. M. Pioli, D. Pioli, P. T. Barth, K. T. Atherton, E. C. Dart, D. Byrom, K. Powell, and P. J. Senior, *Nature (London)*, 1980, **287**, 396.

[46] V. G. Debabov, in 'Overproduction of Microbial Products', ed. V. Krumphanzl, B. Sikyta, and Z. Vanek, Academic Press, London, 1982, p. 345.

[47] K. Miwa, S. Nakamori, and H. Momose, Abstracts of 13th International Congress of Microbiology, Boston, USA, 1982, p. 96.

[48] I. Shiio and S. Nakamori, ref. 38, Chapter 3, p. 133.

[49] J. F. Martin, in 'Microbial Products: New Approaches', ed. S. Baumberg, I. S. Hunter, and P. M. Rhodes, Society for General Microbiology Symposium, Cambridge University Press, Cambridge, 1989, Vol. 44, p. 25.

gene amplification techniques.[48] In these examples the cloned genes were mutant forms which were resistant to feedback control and had been obtained using the conventional mutagenesis/screening systems described in section 5.1. Thus, the *in vitro* DNA techniques have built upon the achievements of conventional strain improvement.

Phenylalanine has become a very important fermentation product because it is a precursor in the manufacture of the sweetener, aspartame. Backman *et al.*[50] have described the rationale used in the construction of an *E. coli* strain capable of synthesizing commercial levels of phenylalanine. *E. coli* was chosen as the producer because of its rapid growth, the availability of DNA manipulative techniques, and the extensive genetic database. Several of the phenylalanine genes are subject to control by the repressor protein of the *tyr* R gene. *In vitro* techniques were used to generate *tyr* R mutations and introduce them into the production strain. The promoter of the *phe* A gene, was replaced to remove repression and attenuation control. As an alternative to the traditional technique of generating a tyrosine auxotroph an excision vector carrying the *tyr* A gene was incorporated into the chromosome. The vector is excised from the chromosome at a slightly increased temperature. Thus, auxotrophy may be induced *during* the fermentation by careful temperature manipulation thus allowing tyrosine limitation to be imposed after the growth phase and at the beginning of the production phase. However, the final step in the genetic manipulation of the organism was the traditional step of isolating an analogue resistant mutant to relieve the feedback inhibition of DAHP synthase by phenylalanine.

The application of *in vitro* recombinant DNA technology to the improvement of secondary metabolite formation is not as developed as it is in the primary metabolite field. However, considerable advances have been made in the genetic manipulation of the streptomycetes[51] and the filamentous fungi[52] and a number of different strategies have been devised for cloning secondary metabolite genes.[53] The first such genes which were cloned were those coding for resistance of the producer organism to its own antibiotic.[54,55] Complete streptomycete antibiotic synthesizing pathways have now been cloned[56,57] and these have been shown to be clustered together on the chromosome along with a resistance gene. The penicillin biosynthetic gene cluster has been cloned from *Penicillum chrysogenum*.[58] Transcription patterns of secondary metabolism genes are also being

[50] K. Backman, M. J. O'Connor, A. Maruya, E. Rudd, D. McKay, R. Balakrishnan, M. Radjai, V. DiPasquantonio, D. Shoda, R. Hatch, and K. Venkatasubramanian, *Ann. N.Y. Acad. Sci.*, 1990, **589**, 16.

[51] D. A. Hopwood, M. J. Bibb, K. F. Chater, T. Kieser, C. J. Bruton, H. M. Kieser, J. Lydiate, C. P. Smith, J. M. Ward, and H. Schrempf, 'Genetic Manipulation in Streptomycetes', John Innes Foundation, Norwich, 1985.

[52] C. A. M. J. J. van den Hondel and P. J. Punt, in 'Applied Molecular Genetics of Fungi', ed. J. F. Peberdy, C. E. Caten, J. E. Ogden, and J. W. Bennett, Cambridge University Press, Cambridge, 1991, p. 1.

[53] I. S. Hunter and S. Baumberg, ref. 49, p. 121.

[54] M. J. Bibb, J. L. Schotel, and S. N. Cohen, *Nature (London)*, 1980, **284**, 284.

[55] C. J. Thompson, T. Kieser, J. M. Ward, and D. A. Hopwood, *Gene*, 1982, **20**, 51.

[56] F. M. Malpartida and D. A. Hopwood, *Nature (London)*, 1984, **309**, 462.

[57] H. Motamedi and C. R. Hutchinson, *Proc. Natl. Acad. Sci. USA*, 1987, **84**, 4445.

[58] D. J. Smith, M. K. R. Burnham, J. Edwards, A. J. Earl, and G. Turner, *Biotechnology*, 1990, **8**, 39.

unravelled.[53] An example of the application of recombinant DNA technology to strain improvement in secondary metabolism is given by the work of the Lilly Research Laboratories group on cephalosporin production by *Cephalosporium acremonium*. Transformation of a high producing cephalosporin strain with a plasmid containing the REXH gene resulted in significantly increased titres.[59] This work demonstrates that the REXH product is at least partially limiting in cephalosporin synthesis and that recombinant DNA methods can alleviate such bottlenecks in commercial strains. The key to such applications is the identification of the bottlenecks.

6 CONCLUSIONS

Thus, micro-organisms are capable of producing a wide range of products – a range which has been increased by the techniques of *in vitro* DNA recombination to include mammalian products. Improved productivity may be achieved by the optimization of cultural conditions and the genetic modification of the producer cells. However, a successful commercial process for the production of a microbial metabolite depends as much upon chemical engineering expertise as it does on that of microbiology and genetics.

[59] P. L. Skatrud, T. D. Ingolia, D. J. Fisher, J. L. Chapman, A. Tietz, and S. W. Queener, *Biotechnology*, 1989, **7**, 477.

CHAPTER 2

An Introduction to Recombinant DNA Technology

E. B. GINGOLD

1 INTRODUCTION

Over the last 15 years there has been a revolution in biology. Although developments initially centred around studies at the molecular level, few areas of pure and applied biology have remained unaffected. The cause of all this excitement is a series of techniques variously referred to as genetic engineering, gene cloning, *in vitro* genetic manipulation, or recombinant DNA technology.

The basis of this new technology is the ability to introduce genetic material into cells in such a way as to enable that DNA to replicate and be passed on to the progeny of the cells. In other words, to alter the genetic make-up of cells by adding DNA from the outside. The DNA can come from virtually any source: animal, plant, microbial, or synthetic. No matter what its source, it can be introduced into bacterial species, fungi, plant cells, animal cells, or even whole plants and animals. The possibilities seem limitless.

There are many reasons why such experiments are done and most of these are relevant to biotechnology. The most obvious applications are those in which a gene coding for a commercially useful product is transferred into a species in which that product can be produced cheaply and efficiently. Well known examples include the transfer into the bacterium *Escherichia coli* of the human insulin gene,[1,2] the interferon genes,[3] and the message for human growth hormone.[4] By linking the foreign genes to bacterial control sequences it was possible to arrange expression of the messages encoded and thereby create bacterial strains with the ability to manufacture these products.

But many of the breakthroughs that have been made come from the ability

[1] L. Villa-Komaroff, A. Efstratiadas, S. Broome, P. Lomedico, R. Tizard, S. P. Naber, W. L. Chick, and W. Gilbert, *Proc. Natl. Acad. Sci. USA*, 1978, **75**, 3727.

[2] D. V. Goeddel, D. G. Kleid, F. Bolivar, H. L. Heyneker, D. G. Yansura, R. Crea, T. Hirose, A. Kraszewski, K. Itakura, and A. D. Riggs, *Proc. Natl. Acad. Sci. USA*, 1979, **76**, 106.

[3] S. Nagata, H. Taira, A. Hall, L. Johnsrud, M. Sreuli, J. Ecsodi, W. Boll, K. Cantell, and C. Weissman, *Nature (London)*, 1980, **284**, 316.

[4] D. V. Goeddel, H. L. Heyneker, T. Hozumi, R. Arentzen, K. Itakura, D. G. Yansura, M. J. Ross, G. Miozzari, R. Crea, and P. H. Seeburg, *Nature (London)*, 1979, **281**, 544.

that the new technologies give us to study the structure and working of genes at a level that was undreamed of 20 years ago. As will be seen in this and following chapters, a gene can now be amplified many fold to enable detailed studies such as gene sequencing, and the analysis of the functions of the sequences thus discovered. And from this new knowledge previously unforeseen applications are emerging.

It is the aim of this chapter to introduce the fundamental processes involved in the new DNA technologies. The description will concentrate on work involving *E. coli*, the species in which the technologies were first developed. But, as will be seen in later chapters, the same basic principles apply whatever host is being used for the work.

2 GENE CLONING – THE BASIC STEPS

It has been long realized that it is possible to add DNA to cells of certain species of bacteria and change their genetic make-up.[5] This process, called *transformation*, helped to clarify the role of DNA as the genetic material, but there was one severe limitation to this approach: the DNA had to come from the same species and integrate into the chromosome to be passed on.

So what if we had, say, the human insulin gene and attempted to add it to bacterial cells? There would be little difficulty in getting the DNA into the cells. Once inside, however, it would not replicate. Even if it were not destroyed by the cell's nucleases (which as a linear fragment it probably would be) it would remain as a single copy. When the initial cell had multiplied to give a million progeny, only one would have the insulin gene!

If we want the gene to be able to replicate it must be attached to a molecule that is capable of replicating in a bacterial cell. Such a molecule must have a bacterial origin of replication, that is, the sequence of DNA recognized by the enzymes of the host cell as a point to initiate replication. A bacterial species such as *E. coli* will never recognize a eukaryotic origin; hence it is necessary to attach the foreign DNA to a molecule carrying an *E. coli* origin. Such carrier molecules are called *vectors*.

An outline of the steps involved in gene cloning can be seen in Figure 1. From this outline, it is clear that insertion of the DNA into a vector is the central feature. Although the nature of vectors used will be covered in more detail in section 6 of this chapter, a few preliminary words are required here. In theory, it would be possible to remove the main chromosome from the bacterial cell, insert the foreign DNA, and return it to a live cell. In practice, this is inconceivable. The chromosome is far too large; it could neither be isolated intact, handled in a test tube, nor returned to a cell. What is needed is a far smaller molecule that nonetheless has the ability to replicate autonomously (that is, outside the chromosome).

As it happens, two classes of suitable molecules exist. The first, *plasmids*, are extrachromosomal molecules of DNA found in many bacterial species.[6] Like the

[5] O. T. Avery, C. M. MacLeod, and M. MacCarty, *J. Exp. Med.*, 1944, **79**, 137.
[6] P. Broda, 'Plasmids', W. H. Freeman and Company, San Francisco, 1979.

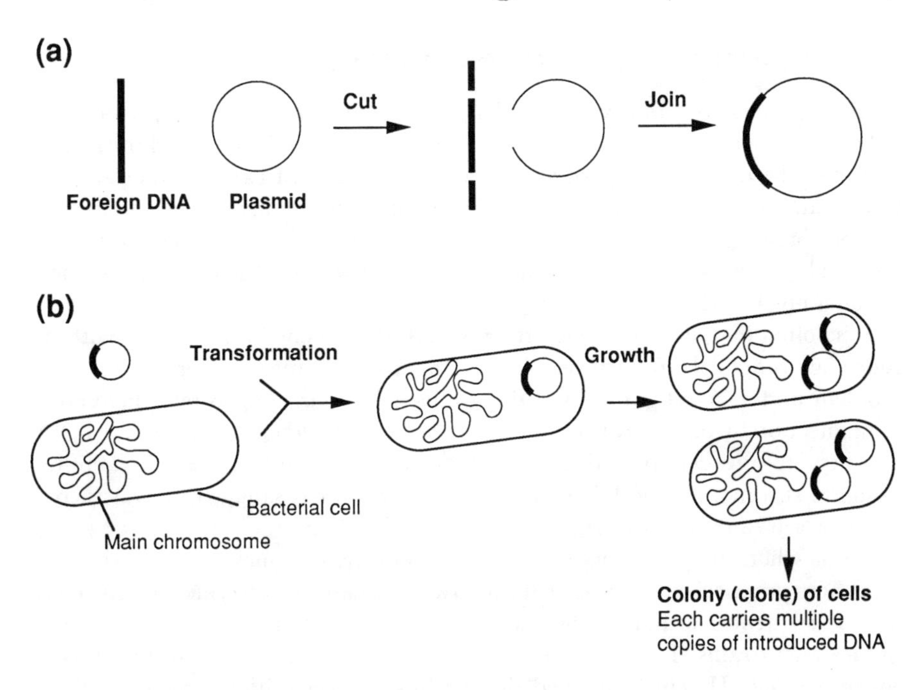

Figure 1 *The basic steps in cloning: (*a*) insertion of foreign DNA into the vector; (*b*) incorporation of the recombinant vector into bacterial cells*

chromosome, they are closed circular molecules, although of smaller size. Unlike the chromosome, they are not necessary for the general viability of the cell although they may carry genes, such as those for resistance to antibiotics, that can help survival in special conditions. In addition, some classes of plasmid have what is known as relaxed replication, that is, they are present in many copies per cell. Such multi-copy plasmids may be conveniently isolated and purified, an essential feature for a cloning vector. And they have the additional advantage of providing a large number of copies of the cloned sequence in each cell.

The second class of vectors is based on bacterial viruses or *phages*. Phage DNA can clearly also replicate inside bacteria and give rise to many progeny molecules; this is the basis of the action of phage as an infective agent.

From a consideration of the outline in Figure 1, it is clear that to enable the process to be carried out a number of techniques had to be developed. Essentially these were:

(1) A method of reproducibly cutting DNA in the required places.
(2) A method of joining (or ligating) DNA fragments.
(3) The ability to return the DNA to cells, *i.e.* transformation.
(4) A way of selecting for the transformants carrying the required gene.
(5) Methods of analysing the cloned DNA.

The development and application of these techniques is the subject of the rest of this chapter.

3 CUTTING DNA – RESTRICTION ENZYMES

At first glance, the ability to cut DNA into smaller fragments presents no difficulty. Anyone who has attempted to isolate DNA will be aware of the fragile nature of the molecule. Pipetting and sonication are just two examples of how DNA can be fragmented, but genetic engineering requires more than this random breakage. It is essential to be able to cut molecules reproducibly and predictably, that is at the same points each time. It was the discovery of a method of doing just this that provided the breakthrough.

As so often happens in science, the research that opened up this field with its enormous practical applications was, in fact, of the most esoteric kind. The problem under investigation was the ability of bacteria to protect themselves from infection by phage. It was found that bacteria could cut up the phage DNA on entry to the cell provided the phage had previously grown in another bacterial strain.[7,8] Phage DNA previously grown in the same strain was recognized as 'self' and avoided degradation.

It was when the properties of these DNA cutting enzymes were investigated that their practical value was realized. In each case the enzyme recognized a specific sequence, although the sequence differed for enzymes from different species and strains. Furthermore, it was found that the major group of such enzymes (Class II) cut at defined sites within the recognition sequence.[9] Such restriction endonucleases or restriction enzymes are thus an invaluable tool for cutting DNA at defined points.

The recognition sequences for six of the most commonly used enzymes are shown in Figure 2. In fact hundreds of such enzymes have now been identified and an ever increasing number of these are available commercially in a purified form.[10] Those illustrated will, however, serve to demonstrate the general properties. In these examples the recognition sites are between 4 and 8 base pairs in length. As a particular sequence of 8 bases will occur by chance far less frequently than a sequence of 4 bases, it follows that enzymes that have only 4 bases in their recognition sequences will cut a given DNA molecule more often than those with 8. Hence, while *Hae*III or *Sau*3A is used to cut DNA into many small fragments with average sizes of some hundreds of base pairs in length, *Not*I is used to cut human chromosomes into a smaller number of large (but manageable) fragments of average length in excess of a million base pairs.

Another feature obvious from Figure 2 is that the sequences are symmetrical, that is they have the same series of bases in each strand, though running in opposite directions. This too is a general feature of such enzymes.

The points of cutting are also shown and from this it should be clear that not all enzymes cut at the centre of the sequence. *Eco*RI, for example, makes staggered cuts and leaves single-stranded tails of four bases. Such tails are called cohesive or 'sticky' ends and are of great use in rejoining fragments. The

[7] D. M. Dussoix and W. Arber, *J. Mol. Biol.*, 1962, **5**, 37.
[8] S. Lederberg and M. Meselson, *J. Mol. Biol.*, 1964, **8**, 623.
[9] T. J. Kelly and H. O. Smith, *J. Mol. Biol.*, 1970, **51**, 393.
[10] R. J. Roberts, *Nucleic Acids Res.*, 1989, **17**, r347.

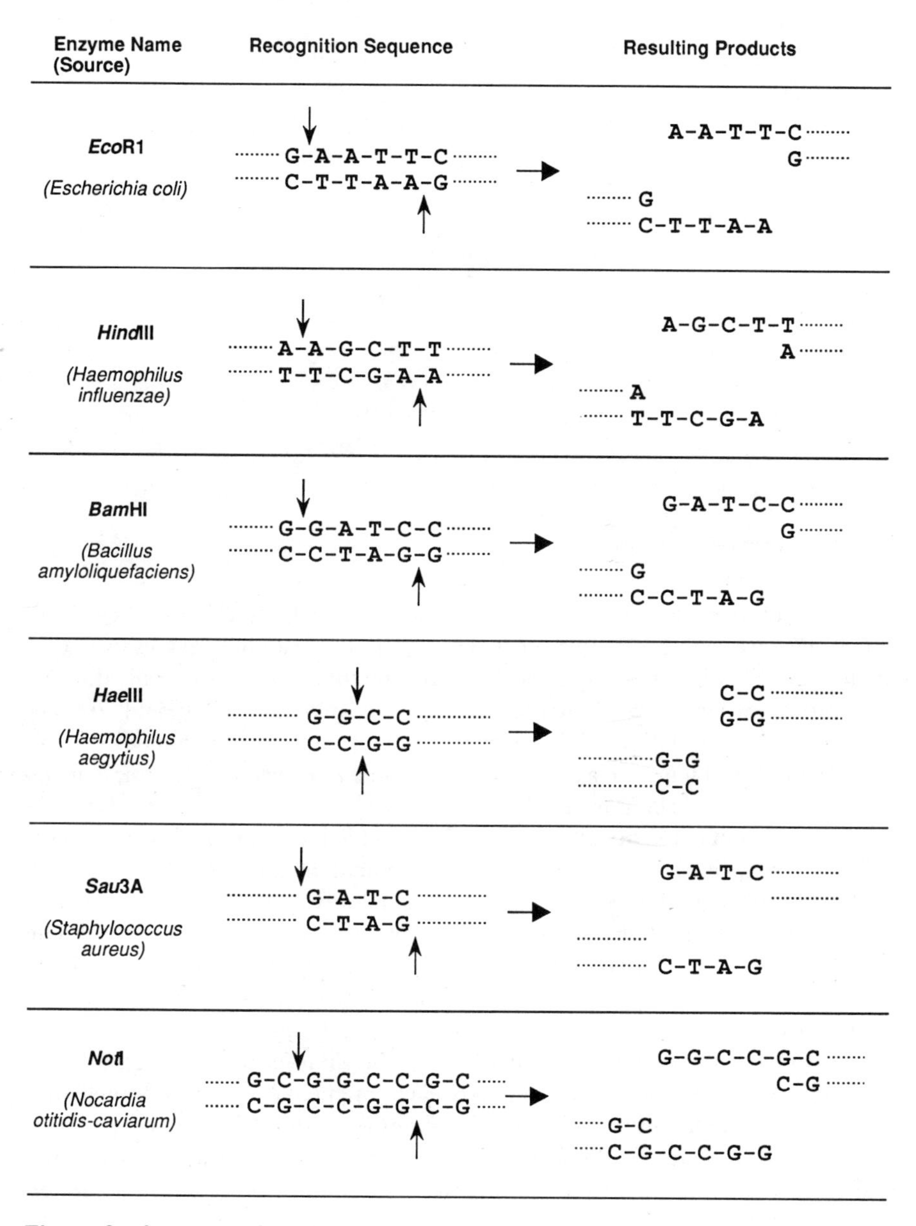

Figure 2 *Some commonly used restriction enzymes*

complete range of enzymes include those that leave 5′ and 3′ tails of lengths from two to five bases. Other enzymes, such as *Hae*III, cut centrally and leave blunt ends.

It should be noted that different enzymes may have different recognition sites yet leave the same single-stranded tails. An example of this is seen in Figure 2 with *Bam*HI and *Sau*3A. The application of this property to the later rejoining of fragments will be seen.

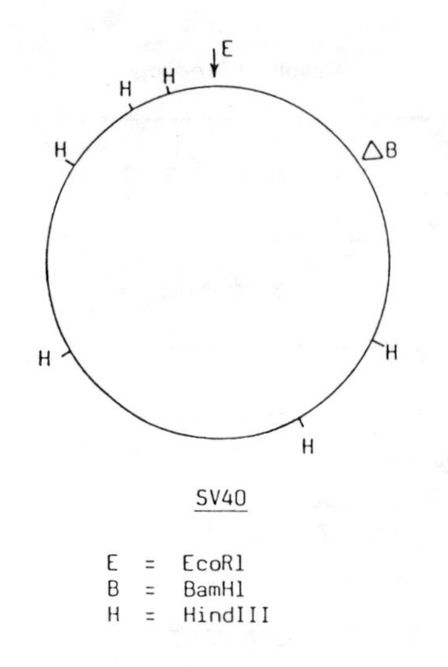

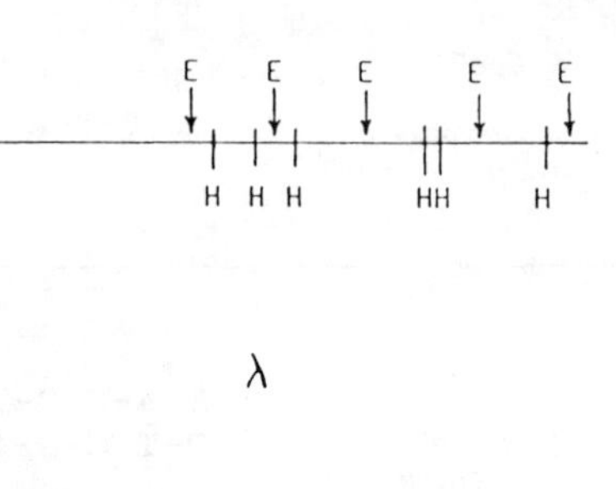

Figure 3 *Some restriction maps*

A point of interest is why such enzymes do not attack the DNA of the cells in which they are found. The answer is that a parallel series of enzymes exist that modify the DNA in such cells by adding extra methyl groups to bases within the recognition sequence. Such sequences are thus marked as self and are not subjected to attack. Foreign DNA would, of course, not have such modification, but potential problems are avoided in most cloning experiments by using as the hosts mutant cells lacking in restriction enzymes.

The benefits of being able to cut a DNA molecule into reproducible fragments would be reduced if there were not a convenient method of separating and analysing the fragments. Such separation is generally performed by gel electrophoresis, using agarose gels, or, for small fragments, polyacrylamide gels.[11] The cut DNA is simply loaded into the wells and allowed to migrate toward the anode. As all molecules should have similar charge/mass ratio the rate of migration depends simply on the molecular weight. Bands can be visualized with the help of the intercalating agent ethidium bromide under UV light. Using known standards the molecular weights of each fragment can be determined and, by a variety of methods, including double digestion using two enzymes together, restriction maps (such as those in Figure 3) can be obtained which reveal the positions of each fragment on the original molecule and the *restriction sites* on which the enzymes act.

4 JOINING DNA MOLECULES

Early in the development of procedures for gene cloning it was realized that a DNA fragment could be inserted into an open vector if both molecules had complementary single-stranded DNA tails. In the initial work a poly-A tail was

[11] E. M. Southern, *Methods Enzymol.*, 1979, **68**, 152.

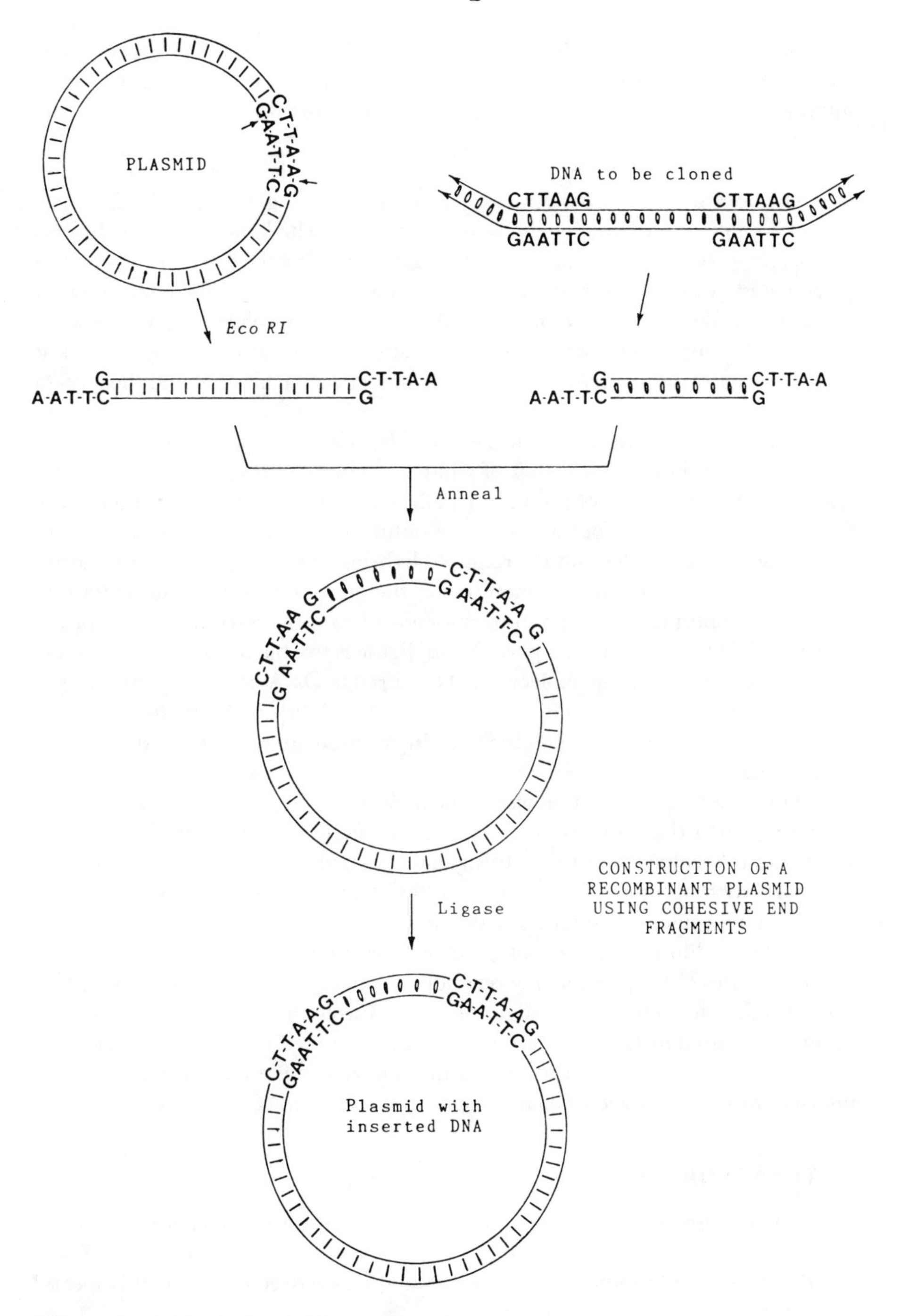

Figure 4 *Joining foreign DNA into a plasmid*

artificially added to one molecule and a poly-T tail to the other.[12] Later, it was realized that the cohesive ends left by restriction enzymes would also provide suitable complementary regions.[13] This method, which is the basis of practically all current cloning procedures, is illustrated in Figure 4.

The vector and the DNA to be inserted are cut with the same enzyme (or with different enzymes which nonetheless leave identical single-stranded tails). The cut molecules are then mixed under conditions which favour annealing of complementary strands. In fact, the hydrogen bonds involved in four sets of base pairs would not be enough to form a stable hybrid. However, the enzyme DNA ligase is included in the mix and this will complete the covalent bonds between the molecules in the otherwise transient structure. Thus, as shown in Figure 4, it is possible to form a covalently closed circle including the plasmid and the foreign DNA.

On further consideration it will become obvious that the desired structure is just one of the possible outcomes of what is a random joining operation. Generally a simple resealing of the vector molecule without inserts is an even more likely outcome. Hybrid formation can be optimized by adjusting concentrations of the reactant molecules, but the required hybrids generally still form a minority of the products. It is thus common to use the enzyme alkaline phosphatase to remove the phosphate groups from the ends of the single-stranded tails of the vector molecules. Without the phosphates, ligase is unable to catalyse the simple recircularization of these molecules. The foreign DNA to be inserted is not treated with the enzyme and hence retains the ability to be ligated into the vector. As a consequence the vector must incorporate an insert to be successfully recircularized.

There are many situations in which the molecule being inserted into the vector has blunt rather than cohesive ends (that is, no single-stranded tails). In fact, the DNA ligase from phage T4 which is generally used in this work is able to join molecules with blunt ends, but only at a low efficiency. This property is used to add small molecules called *linkers* to the blunt ends as shown in Figure 5. Linkers are synthetic double-stranded oligonucleotides that incorporate one or more restriction sites.[14] Despite the low reaction efficiency, the high molar concentration of linker free-ends in the ligation mix ensures that each molecule of the blunt-ended foreign DNA has at least one linker joined to each end. Following treatment with the appropriate restriction enzyme the molecule may now be inserted into an open vector as in the previous method.

5 TRANSFORMATION

Once the foreign DNA has been inserted into the vector it becomes necessary to return the DNA into a living cell. For many years it was believed that *E. coli* could not be transformed by added DNA and it is indeed true that this species

[12] D. A. Jackson, R. H. Symons, and P. Berg, *Proc. Natl. Acad. Sci. USA*, 1972, **69**, 2904.
[13] J. E. Mertz and R. W. Davis, *Proc. Natl. Acad. Sci. USA*, 1972, **69**, 3370.
[14] R. H. Scheller, R. E. Dickerson, H. W. Boyer, A. D. Riggs, and K. Itakura, *Science*, 1977, **196**, 177.

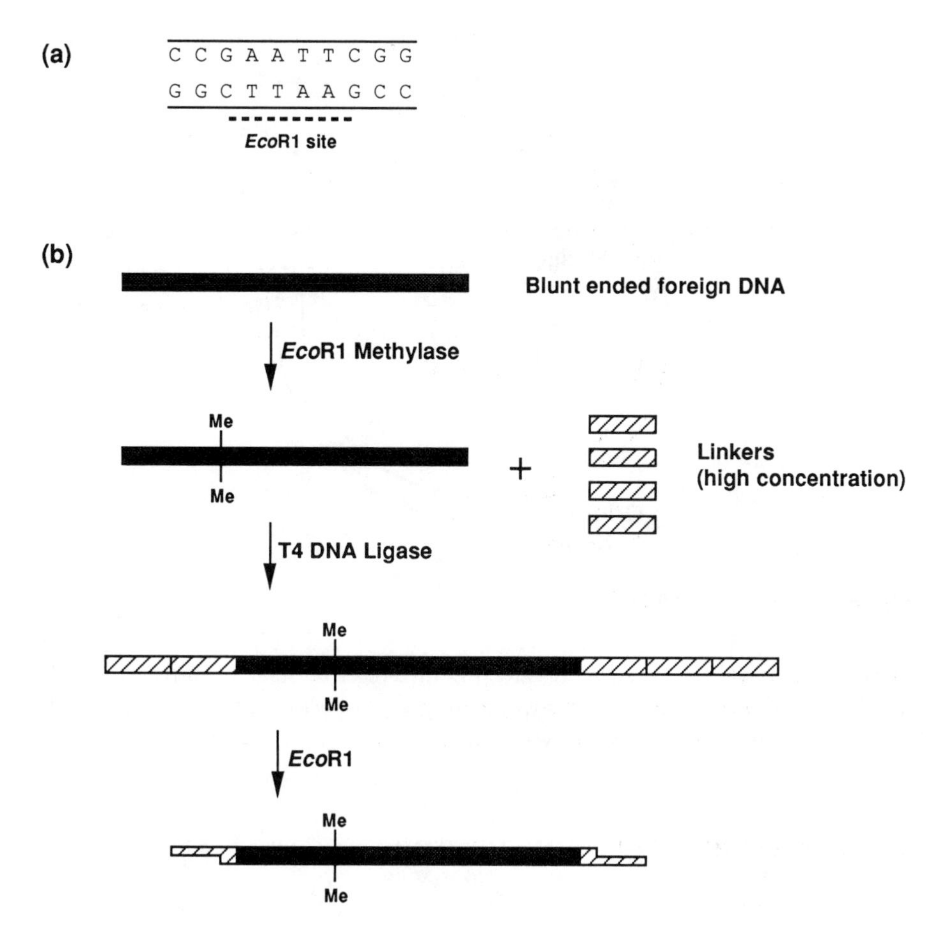

Figure 5 *The use of linkers: (a) an example of a chemically synthesized linker; (b) the addition of cohesive ends by use of this linker. Note that any* Eco*R1 sites within the DNA fragment are first protected by the use of the modification enzyme* Eco*R1 Methylase*

lacks a natural system for uptake of DNA. Nonetheless, in 1970 it was demonstrated that it is possible to force DNA uptake by a set of extreme conditions.[15] Actively growing cells are harvested and left in hypotonic $CaCl_2$ at 4°C. After 30 minutes of such treatment changes occur in the cell membrane and the cells are now said to be competent. The DNA is then added to the cell suspension and it is left at 4°C for a similar period. A brief heat shock (42°C) allows DNA uptake by the cells. A short incubation in growth medium then allows cell recovery and plasmid establishment.

A number of improvements have been made to this procedure so as to increase the efficiency of the transformation.[16] Nonetheless, even with the best techniques only a small proportion of cells actually take up the DNA. For this reason it is

[15] M. Mandel and A. Higa, *J. Mol. Biol.*, 1970, **53**, 159.
[16] D. Hanahan, *J. Mol. Biol.*, 1983, **166**, 557.

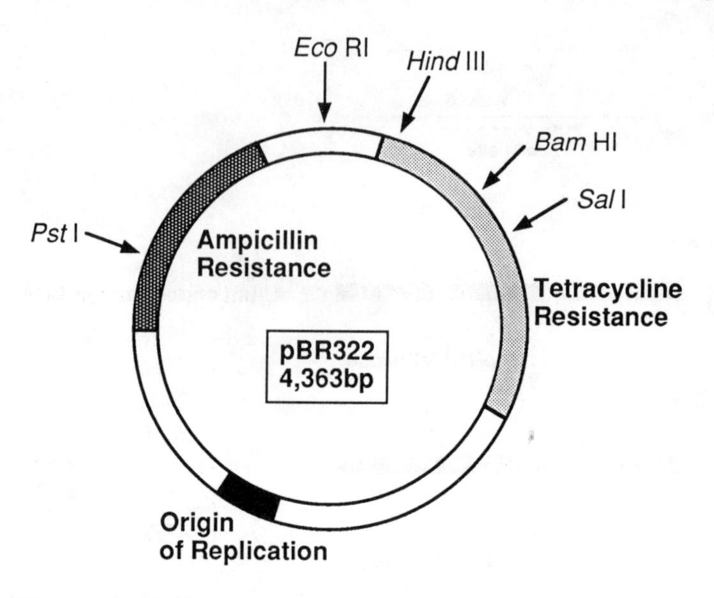

Figure 6 *The plasmid pBR322*

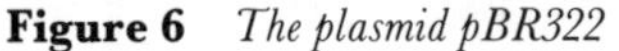

necessary to have some method of selecting the cells that have been transformed. To this end all vectors used in this work carry *selection markers*, that is genes which confer easily identifiable phenotypes (characteristics) on cells that have taken them up.

6 THE NATURE OF CLONING VECTORS

6.1 Plasmids

An ideal plasmid to be used as a cloning vector, as well as being capable of autonomous replication, would be as small as possible so as to enable easy isolation and handling. It would contain single restriction sites (so that it can be opened but not destroyed by the enzymes) and simple to detect selection markers. An examination of the natural plasmids that were available revealed that they were far from this ideal. Most plasmids with useful markers were far too large and contained multiple restriction sites. Some smaller plasmids were known, but these carried genes difficult to use for selection. An early task of the genetic engineers was thus to construct artificial plasmids combining useful features from a number of the natural plasmids into a small molecule.

The plasmid pBR322, shown in Figure 6, has undoubtedly been the most successful of the constructed plasmids,[17] and is the 'parent' or 'grandparent' of most of the other vectors used today. An examination of its properties reveals why this is so. The origin of replication incorporated into this plasmid is of the 'relaxed' type, its replication not being under direct control of the main chromo-

[17] F. Bolivar, R. L. Rodriguez, P. J. Greene, M. W. Betlach, H. L. Heynecker, H. W. Boyer, J. H. Crossa, and S. Falkow, *Gene*, 1977, **2**, 95.

some, and thus multiple copies of the plasmid are present per cell. This number can be greatly increased by inhibiting the replication of the main chromosome with the addition of chloramphenicol (so-called amplification). The high copy number is a major factor in the ease of isolation and purification of this plasmid from cell culture. The small size of 2.6 MDa or 4.3 thousand base pairs (kb) is also a major help here.

Two genes for antibiotic resistance provide easy-to-use selection markers. Cells that have taken up this plasmid can be identified by plating the culture on media containing either ampicillin or tetracycline. The position of several of the unique restriction sites within these genes is also a major benefit. Take the case, for example, in which the plasmid has been opened at the *Bam*HI site in a cloning experiment. Insertion of foreign DNA at this point would disrupt the tetracycline resistance gene whereas simple recircularization of the plasmid would not. All transformants, whether carrying inserts or not, would still be ampicillin resistant and hence this marker could be used for selection of transformants. The plasmid with inserts, however, would no longer confer tetracycline resistance. Thus a simple test for growth on tetracycline would distinguish the clones carrying inserts (they would not grow) and those that carry the original plasmid. Such a phenomenon is called 'insertional inactivation'.

Derivatives of pBR322 include pAT153, a still smaller vector with an even higher copy number.[18] Other modifications have involved the addition of alternative selection markers. The vectors pBR325, pBR328, and pBR329, for example, include a chloramphenicol resistance gene with a site for the restriction enzyme *Eco*RI and hence allow the application of insertional inactivation in conjunction with this much used enzyme.

The system for detecting inserts described to this point uses so-called negative selection – testing for the *loss* of the ability to grow on antibiotic-containing plates. This is time consuming, especially when dealing with large numbers of colonies. A more convenient system to use is employed by the plasmids in the pUC series[19] which carry a section of the *lacZ* gene. This marker allows β-galactosidase activity to be demonstrated in suitable host cells deficient in this enzyme. The activity of this enzyme is easily assayed on the plate itself as colonies expressing enzyme will hydrolyse the compound Xgal resulting in a blue derivative. It is this enzyme that is inactivated by the insertion of foreign DNA at a restriction site within the gene. Looking for colonies carrying inserts thus simply becomes a matter of visually examining the Xgal-containing plates and picking out the white colonies in which the enzyme is inactivated from the blue colonies carrying the uninterrupted marker gene.

Other plasmid derivatives of pBR322 have added sequences making them suitable for special tasks. Examples that will be encountered in later chapters include vectors tailored for cloning in yeast, plant and animal cells, and the expression vectors, with their control sequences that allow the cloned gene to produce its protein product in the host cell.

[18] A. J. Twigg and D. Sherratt, *Nature (London)*, 1980, **283**, 216.
[19] J. Vierira and J. Messing, *Gene*, 1982, **19**, 259.

6.2 Phage Vectors

Although plasmids make good general purpose cloning vectors, there are limitations to their use. The major problem relates to the size of insert which can be efficiently carried into the cell. Transformation frequencies sharply decrease as the size of the DNA circle increases. In addition, plasmids with large inserts are unstable and give rise to smaller segregants. Such limitations make plasmid vectors unsuitable for experiments of the type in which, for example, the entire human genome is cut up and cloned. A second class of vectors, based on the bacteriophage λ, are the system of choice for such applications.

The phage λ has two modes by which it replicates itself.[20] The first is the so-called lytic cycle, which involves attachment of the phage to the bacterial cell, injection of the phage DNA, and replication of this DNA inside the cell. After the packaging of the DNA molecules in protein coats the cell will lyse, liberating the progeny phage. In experimental work phage are normally detected as plaques, or holes in a bacterial lawn. The alternative mode of replication, known as lysogeny, follows the same initial path of infection, but the phage genome, rather than independently replicating, becomes integrated into the bacterial chromosome. In this case the cell does not lyse and the phage genome becomes a silent passenger.

The phage λ genome is a 49 kb molecule, packaged in a protein head attached to a tail that promotes the injection of the phage DNA into the cell. The head size limits the amount of DNA that can be packaged to a range of 75–108% of the size of the normal genome. This might appear to severely limit the amount of extra DNA that can be added to the phage genome. However, the central region of the phage genome is unnecessary for the lytic growth cycle of the phage and is concerned only with lysogeny. As most cloning methods use only the lytic cycle this DNA can be deleted and replaced with the cloned DNA.

The λ-based vectors are modifications of the natural phage with a reduced number of restriction sites.[21,22] Some vectors have only one site for the enzyme in use; such *insertion vectors* are simply opened and the foreign DNA ligated in. Sufficient phage DNA has been deleted from the vector to allow space for inserts. The site of the insertion is often into a gene, for example, *lacZ* as described above for pUC plasmids, thereby allowing for the selection of plaques with inserts. The size of inserts clonable in such vectors is limited, however, by the fact that the vector itself must be 75% the size of the wild type genome.

A way around this minimum size limit has been developed with the *replacement vectors*. The vectors carry a region of non-essential DNA, the so-called 'stuffer' region, between two restriction sites. This region serves only to make the vector large enough to be viable. After the vector is prepared this central region can be

[20] N. E. Murray, 'The Bacteriophage Lambda', Cold Spring Harbor Laboratory, New York, 1983, Vol. 2.

[21] K. Kaiser and N. E. Murray, in 'DNA Cloning', ed. D. M. Glover, IRL Press, Oxford, 1985, Vol. 1, p. 1.

[22] J. W. Dale, in 'Techniques in Molecular Biology', ed. J. M. Walker and W. Gaastra, Croom-Helm, London, 1987, Vol. 2, p. 159.

removed and replaced by a large insert. With vectors of this type it is possible to accommodate inserts of up to 24 kb and propagate them in a stable fashion.

The critical advantage of using phage vectors is that after the foreign DNA has been added the products can be packaged in the test tube into a phage coat. This process, called *in vitro packaging*, allows the recombinant DNA to be added to the cell by the normal phage injection system instead of requiring transformation, thus increasing entry efficiency by many orders of magnitude. A third class of vector, the cosmids,[23] also makes use of this feature. Such vectors are actually plasmids, but include the DNA sequence recognized by the phage packaging system. On insertion of a large (*ca.* 40 kb) fragment, the recombinant DNA can be packaged into phage coats and hence injected into the cell. Once inside it behaves as a very large plasmid. However, although such vectors solve the problem of getting large plasmids inside the cell, they do not help with the problem of plasmid instability.

Finally, it should be noted that there is another class of vector based on the single-stranded phages such as M13. The use of these vectors is discussed in the final section of this chapter.

7 CLONING ACTUAL GENES – THE PROBLEM

This discussion has so far concentrated on the techniques involved in DNA manipulation. Thus it should now be clear how a piece of foreign DNA can be inserted into a vector and then added to the genetic material of a bacterial cell. What has been avoided is any discussion of how the actual foreign gene is obtained in the first place. This is, in fact, the most difficult part of any cloning experiment. The problem is best illustrated with an example. Suppose you wanted to clone the human growth hormone gene. You could take total human DNA, cut it with an enzyme like *Eco*RI, and insert all the resulting fragments into plasmids and then into *E. coli* cells. You would obtain many colonies, each carrying inserts. Amongst them would be some carrying the human growth hormone gene (or possibly fragments if it is cut internally by the enzyme!). But *Eco*RI cuts human DNA into 700 000 or so fragments. This means that the clones you want would be outnumbered by a factor of 700 000 to 1!

Such collections of the total genetic material from a particular species, cut up into fragments and cloned into an enormous number of microbial cells, are known as *Genome Libraries* and play a central role in most approaches to obtaining genes. In practice, the human DNA would probably be cloned in a phage or cosmid vector as the larger fragment size that can be accommodated in such vectors serves to reduce the total number of clones needed to get a representative library. But the key problem remains, how do you pick out the clone you want from the vast number of others? It is most unlikely that the desired clone would have a recognizable phenotype – the presence of the human growth hormone gene will not make the bacterial colony any larger! In fact, it is most unlikely that the presence of a human chromosomal gene would lead to any protein product at all.

[23] J. Collins and B. Hohn, *Proc. Natl. Acad. Sci. USA*, 1979, **75**, 4242.

What is needed is some way of 'fishing' the required clone out from the rest. To do this some kind of specific hook or *probe* must be first obtained. Such a probe must be able to recognize and identify the required clone. To see how such probes can be obtained we must leave the genome library and consider alternative approaches to the problem of obtaining a copy of the nucleic acid sequence corresponding to the protein of interest.

8 LABORATORY SYNTHESIZED GENES – COMPLEMENTARY DNA

The genetic material is the same in all cells of the body. Nerve cells, skin cells, and liver cells all contain the same genes in the same numbers. What does differ is which genes are active and thus which products are being made.

Developing red blood cells produce great amounts of α- and β-globin, insulin is produced in the pancreas, and it is to the pituitary that one must look for growth hormone production. So although pancreas cells, for example, contain no more copies of the insulin gene than any other cell type, there is a large amount of insulin mRNA in this tissue. In the case of red blood cells, the enrichment of α- and β-globin mRNA is so great that it proved possible to isolate these molecules in pure form by classical biochemical techniques.

The mRNA cannot itself be added to a vector and cloned, but a viral enzyme is available which breaks the classical 'central dogma' of genetics and allows us to copy the mRNA message into a complementary DNA sequence. This enzyme, reverse transcriptase, originates from retroviruses which travel as RNA but convert their genomes into DNA once inside the host cell. The use of this enzyme in converting a molecule such as the β-globin mRNA into a DNA copy is illustrated in Figure 7.

All DNA synthesizing enzymes requires a template, normally a single strand of DNA, but RNA for reverse transcriptase, and the four deoxynucleotide triphosphates used as building blocks. But even provided with these requirements, no DNA synthesizing enzyme is able to start making a new strand of DNA *de novo*. For DNA synthesis a primer is required, that is, a short length of oligonucleotide that can anneal to the template and form the start of the new molecule. In the present case, we can take advantage of the fact that almost all eukaryotic mRNA molecules have a poly-A tail at their 3′ end. Hence, it is possible to use a short oligo-dT molecule for the primer. This can anneal to the complementary poly-A tail of the mRNA and allow reverse transcription to proceed. As a result of this first stage, a hybrid molecule consisting of one DNA strand and one RNA strand is formed. The following stages convert this to double-stranded DNA.

In one method, the mRNA template is hydrolysed away by alkali leaving a single-stranded DNA molecule.[24] As we have no prior knowledge of the sequence at the 5′ end of the original mRNA, a suitable primer cannot be easily designed for the synthesis of the second strand. But one can take advantage of 'self-priming'. This relies on the original strand forming a hair-pin loop and hence providing its own primer. How this looping comes about is not clear, but it

[24] A. Efstratiadis, F. C. Kafotos, A. M. Maxam, and T. Maniatis, *Cell*, 1976, **7**, 279.

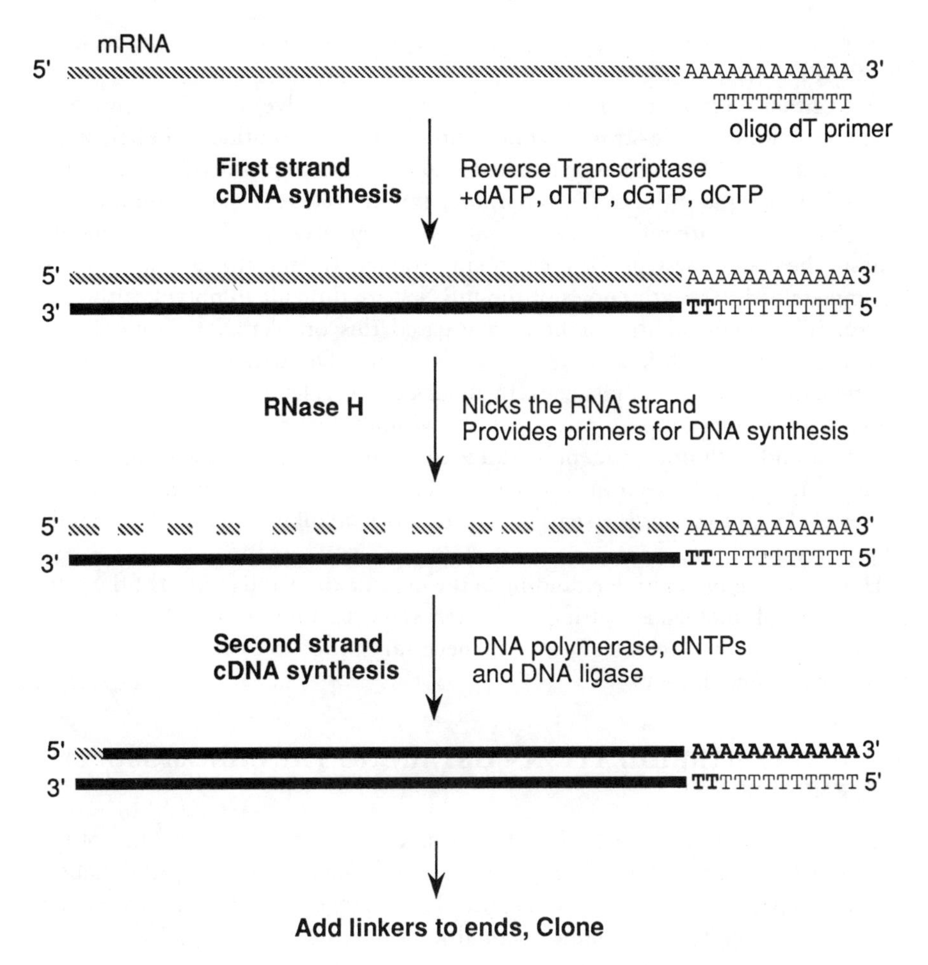

Figure 7 *Production of cDNA using RNaseH produced nicks as a method of priming second strand synthesis*

probably relies on fortuitous base pairing between regions within the original strand with some degree of homology. The loop itself must then be removed with an enzyme such as Sl nuclease that degrades single-stranded regions of DNA. A variable amount of the DNA sequence is thus lost at this stage. Hence this method tends to produce only fragments of the original sequence and has generally been replaced by improved techniques.

One such method[25] is shown in Figure 7. Rather than removing the RNA strand, the enzyme RNaseH is used to nick this strand in numerous places. The RNA fragments can then provide primers for second strand DNA synthesis. By itself, this method would yield numerous short segments of DNA. However, the addition of DNA ligase ensures that they are joined together. As can be seen from

[25] U. Gubler and B. J. Hoffman, *Gene*, 1983, **25**, 263.

Figure 7, all the RNA except a small fragment at the 5′ end is replaced by a strand of DNA.

With either method, the final stage of the process involves the trimming away of any remaining single-stranded tails followed by the addition of linkers as in Figure 5. The DNA can then be inserted into a vector. Following transformation of *E. coli*, large quantities of this laboratory synthesized gene can be obtained.

Such molecules are referred to as complementary DNA, or cDNA, and should contain the coded message for the protein product. In fact, the message is often incomplete as full length copies of the mRNA are difficult to obtain, although several refinements of the method have eased this problem.[26,27] Nonetheless, even if it were a perfect copy of the mRNA, the cDNA molecule will not be identical to the chromosomal gene. It will lack control elements such as promoter sequences and, most importantly, it will be without the 'introns' or interruptions of code found within most mammalian genes. Introns are discussed in detail in Chapter 4. As will be seen in this chapter, for production of foreign proteins in bacteria a lack of introns is essential. In such work, a full length cDNA is exactly what is required.

However, to gain an understanding of the organization and control of the gene in its normal biological setting it is necessary to obtain the chromosomal sequence. The cDNA clone gives us the necessary probe to extract this sequence from the genome library.

9 COLONY HYBRIDIZATION – OBTAINING THE CHROMOSOMAL GENE

If DNA is heated or treated with alkali it denatures, that is the strands of the double helix separate. On cooling or returning to neutrality the double helices can reform provided complementary strands come together. The complementary sequences can be from the same original molecule, or different molecules with similar sequences. Provided homology exists, hybrids can form between molecules from diverse sources. The homology between the coding regions of the chromosomal gene and its cDNA copy allows such hybridization. Using this property, a cDNA clone can be used to identify a clone carrying the corresponding chromosomal gene.

Let us return to the problem of picking out the required clone from the genome library. In theory, it would be possible to isolate DNA from each of the up to one million odd colonies and test each sample for hybridization with a given cDNA preparation. Clearly, however, the numbers involved make this impractical. Using the Grunstein–Hogness method of colony hybridization, however, many colonies can be tested at the one time.[28] The colonies are either grown on or, as shown in Figure 8, transferred to a nitrocellulose (or nylon) filter membrane. As the colonies will be destroyed during the experiment it is essential to retain the

[26] H. Okayama and P. Berg, *Mol. Cell Biol.*, 1982, **2**, 161.
[27] J. H. Han and W. J. Rutter, in 'Genetic Engineering: Principles and Methods', ed. J. K. Selkow, Plenum Publishing, New York, 1988, Vol. 10, p. 195.
[28] M. Grunstein and D. S. Hogness, *Proc. Natl. Acad. Sci. USA*, 1975, **72**, 3961.

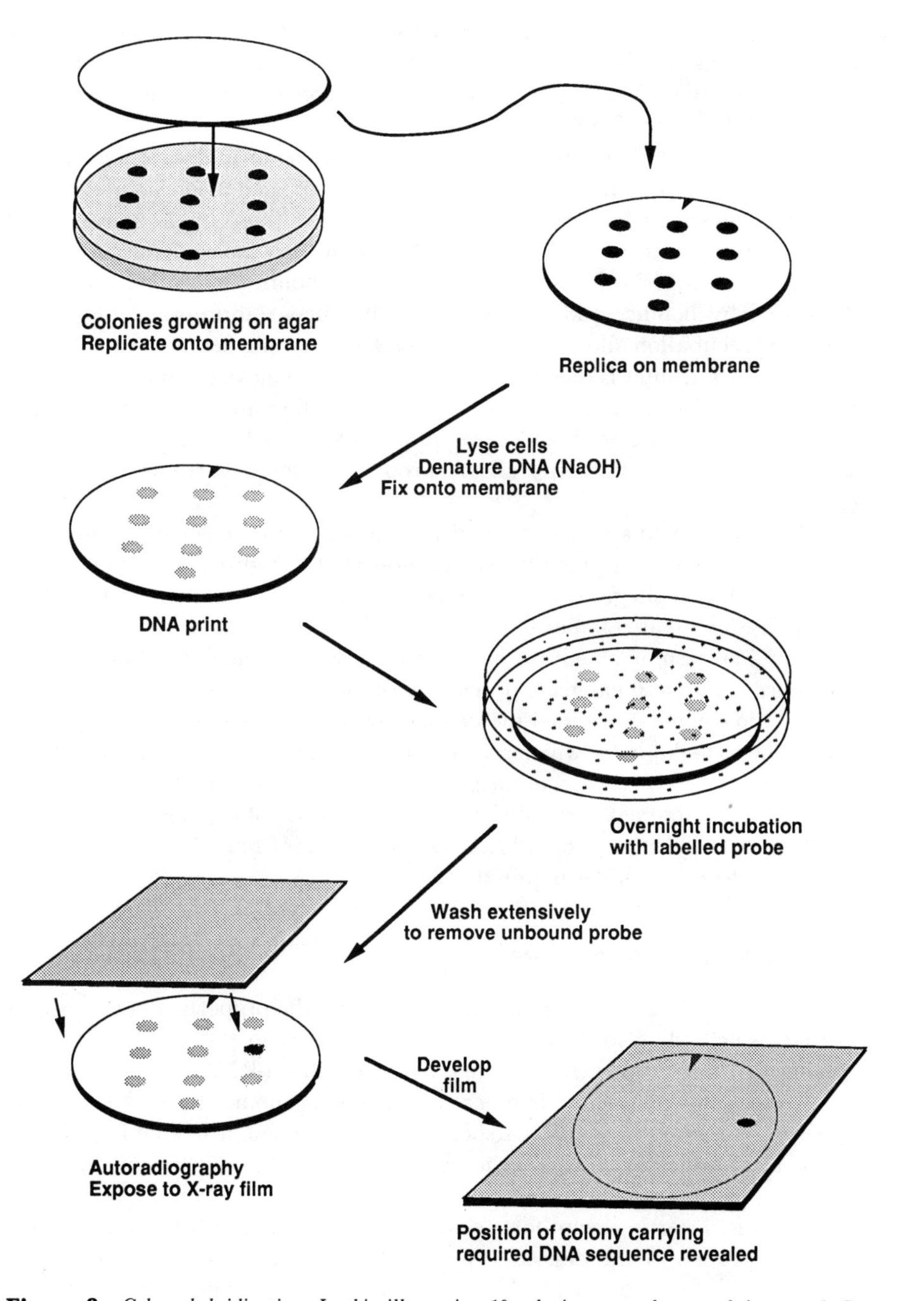

Figure 8 *Colony hybridization. In this illustration 10 colonies on a plate are being tested. In practice, many thousands can be tested on a single plate.*

initial plate or a duplicate. The test filter is then treated with NaOH to lyse the cells and denature the DNA. When the filters are subsequently baked in a vacuum oven the DNA from each colony will become bound to the filter at the position of the original colony.

The cDNA probe must be in some way labelled so as to enable detection of hybridization. This can be done *in vitro* by methods such as nick translation,[29] in which radioactively labelled nucleotides replace those in the molecule, or random primer extension,[30] in which the non-labelled molecules act as a template for the synthesis of labelled copies. Having obtained a labelled probe, it is denatured by boiling, and added to the filter in an appropriate buffer. An overnight incubation allows the probe to find any complementary sequence and anneal to it. The filter is then extensively washed to remove unbound probe and the positions of bound label are revealed by autoradiography. By reference back to the original plate it should now be possible to isolate any colonies with sequences homologous to the cDNA probe and hence carrying the desired chromosomal gene.

With developments in this method it is now possible to screen hundreds of thousands of colonies (rather than the 10 shown in the Figure) on a small number of plates.[31] When phage vectors are in use a simple variant of this method allows plaques rather than colonies to be screened.[32] It is thus clear that there is no difficulty in screening an entire genome library and pulling out a clone carrying the desired chromosomal gene, provided a suitable cDNA probe is available.

This discussion has outlined the method by which it proved possible to clone genes such as β-globin for which a source of purified mRNA was available. From the mRNA, a cDNA version of the gene could be prepared, and with the cDNA clone the chromosomal version of the gene picked out of a genome library. For most genes, however, a pure mRNA cannot be obtained and hence a somewhat more complex approach is required.

10 CLONING STRATEGIES

As a general rule, if a tissue is producing a protein, the proportion of the mRNA of that tissue coding for that protein is far higher than the proportion of the genome carrying the protein code. In other words, the mRNA is a better place to look for the coding sequence than in a chromosomal genome library. Very rarely, however, is the required mRNA species abundant enough to enable consideration of its purification. Even in the case of insulin, it was not possible to obtain pure mRNA from pancreatic extracts.

The normal starting point for a cloning exercise is an mRNA fraction with a small proportion of the desired sequence. When this is used to direct cDNA synthesis and the products are cloned, a *cDNA library* is obtained. But, while the

[29] P. W. J. Rigby, M. Dieckmann, C. Rhodes, and P. Berg, *J. Mol. Biol.*, 1977, **113**, 237.
[30] A. P. Feinberg and B. Vogelstein, *Anal. Biochem.*, 1983, **132**, 6.
[31] D. Hanahan and M. Meselson, *Gene*, 1980, **10**, 63.
[32] W. D. Benton and R. W. Davis, *Science*, 1977, **196**, 180.

Amino acid sequence	- Phen	- Trp	- Pro	- His	- Met
Nucleotides	5' T T T/C	T G G	C C T/C/A/G	C A T/C	A T C

Only Trp and Met are unique codons, therefore 16 possible sequences for the run of amino acids.

Figure 9 *An example of an oligonucleotide probe*

odds might be better than they were for the genome library, the problem of how you select the required clone from the rest still remains.

A number of approaches have been used to solve this problem. Which approach is taken depends on what is available at the start of the investigation. If the sequence of all or part of the required protein is available, and this can be obtained from a very small sample of the protein, a favoured approach involves the use of chemically synthesized oligonucleotides as probes. Using the genetic code a synthetic oligonucleotide corresponding to a run of amino acids is designed and synthesized. Although this may seem surprising, a simple calculation reveals that a particular sequence of as few as 18 bases is likely to occur only once in the human genome.

The design of an oligonucleotide probe is, however, slightly less straightforward than it might seem owing to the degeneracy of the code. As shown in Figure 9 there are a number of possible coding sequences for a given stretch of amino acids. Although there is no way of knowing which is used in the gene itself, the problem can be overcome by using a mixture of nucleotides in the synthesis reactions at the points of degeneracy. In this way the correct complementary sequence will be amongst those synthesized. An alternative approach, suitable if a longer stretch of the protein sequence is available, is to produce a 'guessmer' based on the most likely codons for each amino acid. While this will be unlikely to be totally correct, enough complementarity should exist in a run of say 50 bases to enable hybridization with the required clone.

The probe, once prepared, is used to screen a cDNA library by colony hybridization and thus pick out the clone with the required message.

A somewhat different approach is provided by the use of expression vectors. In this case the aim is to place the cDNA molecules into a vector that has the control sequences for gene expression and test the resulting clones for expression of the desired protein. The phage vector system λgt11 has proved particularly useful in this respect.[33,34] The fragments making up the cDNA library are cloned into an *Eco*RI site within a *lacZ* gene carried on the phage. Owing to the incomplete nature of most cDNA molecules and hence their variable endpoints, some of these molecules will end up fused in phase with the reading frame of the *lacZ* gene. When the *lac* system is induced, the products of the fusions will be

[33] R. A. Young and R. W. Davis, *Science*, 1983, **222**, 778.
[34] T. V. Huynh, R. A. Young, and R. W. Davis, ref. 21, Vol. 12, p. 49.

synthesized and accumulate in the plaques. An immunological screen is used to identify the clones producing the protein of interest and hence carrying the required cDNA sequence. In this case it is thus an antibody to the protein that is required to initiate the process of identification of the cDNA.

The range of methods available thus makes it possible to consider cloning practically any gene. In some cases the starting point might be an antibody to the protein product, in others the sequence of at least a small segment of the protein. Variants on the methods described here allow cloning of genes on the basis of as little information as their selective expression in different tissue types,[35] or at different points in the cell cycle.[36] Of course, once even a fragment of the cDNA has been isolated this can be used as a probe to obtain the chromosomal gene from the genome library.

It is obviously an easier matter to clone a gene such as that for β-globin which is expressed in abundance in a particular tissue than to clone genes coding for proteins found only at low levels. Nonetheless, when the desire for a sequence is sufficient these difficulties can be overcome and even the genes for proteins such as membrane receptors have been successfully cloned.[37]

Finally, it is worth emphasizing that this discussion has only covered a small subset of the cloning strategies that have been successfully used. It is by no means always the case that a cDNA clone is first found, followed by its use in obtaining a chromosomal copy of the gene. For example, many of the genes responsible for human genetic disease have been obtained directly from studies on the chromosomal DNA. It is interesting to note that the genes for both cystic fibrosis and Duchene muscular dystrophy (DMD) were first obtained as clones of chromosomal DNA from mapped chromosomal regions and, in the reverse of the above procedure, the chromosomal DNA clones then used as a probe to select the required cDNA clones from cDNA libraries.[38,39]

A full discussion of the many strategies for cloning genes is beyond the scope of this text. For more information the reader is referred to one of the excellent books dealing specifically with genetic engineering.[40–44]

[35] S. M. Hedrick, D. I. Cohen, E. A. Neilson, and M. M. Davis, *Nature (London)*, 1984, **308**, 149.

[36] L. F. Lau and D. Nathans, *EMBO J.*, 1985, **4**, 3145.

[37] A. Ullrich, J. R. Bell, E. Y. Chen, R. Herrera, L. M. Petruzzelli, T. J. Dull, A Gray, L. Coussens, Y. J. Liao, M. Tsubokawa, A. Mason, P. H. Seeburg, C. Grunfeld, O. M. Rosen, and J. Ramachandran, *Nature (London)*, 1985, **313**, 756.

[38] J. R. Riordan, J. M. Rommens, B. S. Kerem, N. Alon, R. Rozmahel, Z. Grzelczak, J. Zielenski, S. Lok, N. Plavsic, J. L. Chou, M. L. Drumm, M. C. Iannuzzi, F. S. Collins, and L. C. Tsui, *Science*, 1989, **245**, 1066.

[39] M. Koenig, E. P. Hoffman, C. J. Bertelson, A. P. Monaco, C. Feener, and L. N. Kunkel, *Cell*, 1987, **50**, 509.

[40] R. W. Old and S. B. Primrose, 'Principles of Genetic Manipulation. An Introduction to Genetic Engineering', 3rd Edn., Blackwell, Oxford, 1985.

[41] S. M. Kingsman and A. J. Kingsman, 'Genetic Engineering: An Introduction to Gene Analysis and Exploitation in Eukaryotes', Blackwell, Oxford, 1988.

[42] T. A. Brown, 'Gene Cloning – An Introduction', 2nd Edn., Chapman and Hall, London, 1990.

[43] J. D. Watson, J. Witkowski, M. Gilman, and M. Zoller, 'Recombinant DNA', 2nd Edn., Scientific American Books, W. H. Freeman and Company, New York, 1992.

[44] J. Sambrook, E. F. Fritzsch, and T. Maniatis, 'Molecular Cloning. A Laboratory Manual', 2nd Edn., Cold Spring Harbor Laboratory, New York, 1989, Vols. 1–3.

11 ANALYSIS OF CLONED DNA

11.1 Hybridization Techniques

Having obtained a cloned copy of the region of the chromosome carrying a gene of interest, the next step is to locate the position of the gene itself on the cloned segment. In most cases the cloned segment will be at least 20 kb in length (or > 40 kb if a cosmid library was used). The actual coding region will normally account for less than 10% of this length. While restriction analysis will provide a map of restriction sites in the cloned segment, how can the location on this map of the coding regions be established?

Once again, a hybridization technique using the cDNA version of the gene as a probe provides the solution. The cDNA, being derived from mRNA, is only complementary to restriction fragments corresponding to transcribed regions of the gene. It is not complementary to, and hence will not hybridize with, regions of the cloned chromosomal segment that lie outside the boundaries of the gene. In fact, as the cDNA contains no introns, it will not hybridize with regions of the segment consisting of these interruptions to the coding sequence. So, in order to locate the gene (or at least its coding sequence) onto the restriction map of the cloned chromosomal segment, it is only necessary to establish which restriction fragments hybridize to the cDNA.

In practice, it is not convenient to perform hybridization studies directly on fragments in a gel. The fragments are first transferred to a membrane (nitrocellulose or nylon) which provides a suitable support for this work. The method, known as *Southern Blotting*[45] after the originator, is outlined in Figure 10. It can be seen to share many features with the procedure used for colony hybridization. The DNA is cut with suitable restriction enzymes and run on a gel. Treatment with NaOH then denatures the DNA. The transfer to the membrane is performed using capillary action: the gel, with the membrane on top, is placed above buffer saturated filter paper and covered with dry filter paper (or other absorbent material). A flow of buffer occurs through the gel and membrane into the drop top layer. The flow carries the DNA fragments with it, but these cannot pass through the membrane and are hence trapped. The result is that the DNA is transferred from the gel to the membrane. The membrane, with the DNA fixed to it, is then exposed overnight to a solution containing the radiolabelled (and denatured) cDNA probe. After extensive washing, the bands which have bound probe (and hence correspond to coding regions of the gene) can be detected by autoradiography.

The technique of Southern Blotting is one of the most central methods used in molecular biology. In the example described above, it was used to distinguish between discrete restriction fragments arising from a cloned chromosomal segment. Very frequently, it is performed on restriction digests of total chromosomal DNA. In these cases, the initial gel does not show discrete bands corresponding to individual restriction fragments, the large numbers of which ensure that all

[45] E. M. Southern, *J. Mol. Biol.*, 1975, **98**, 503.

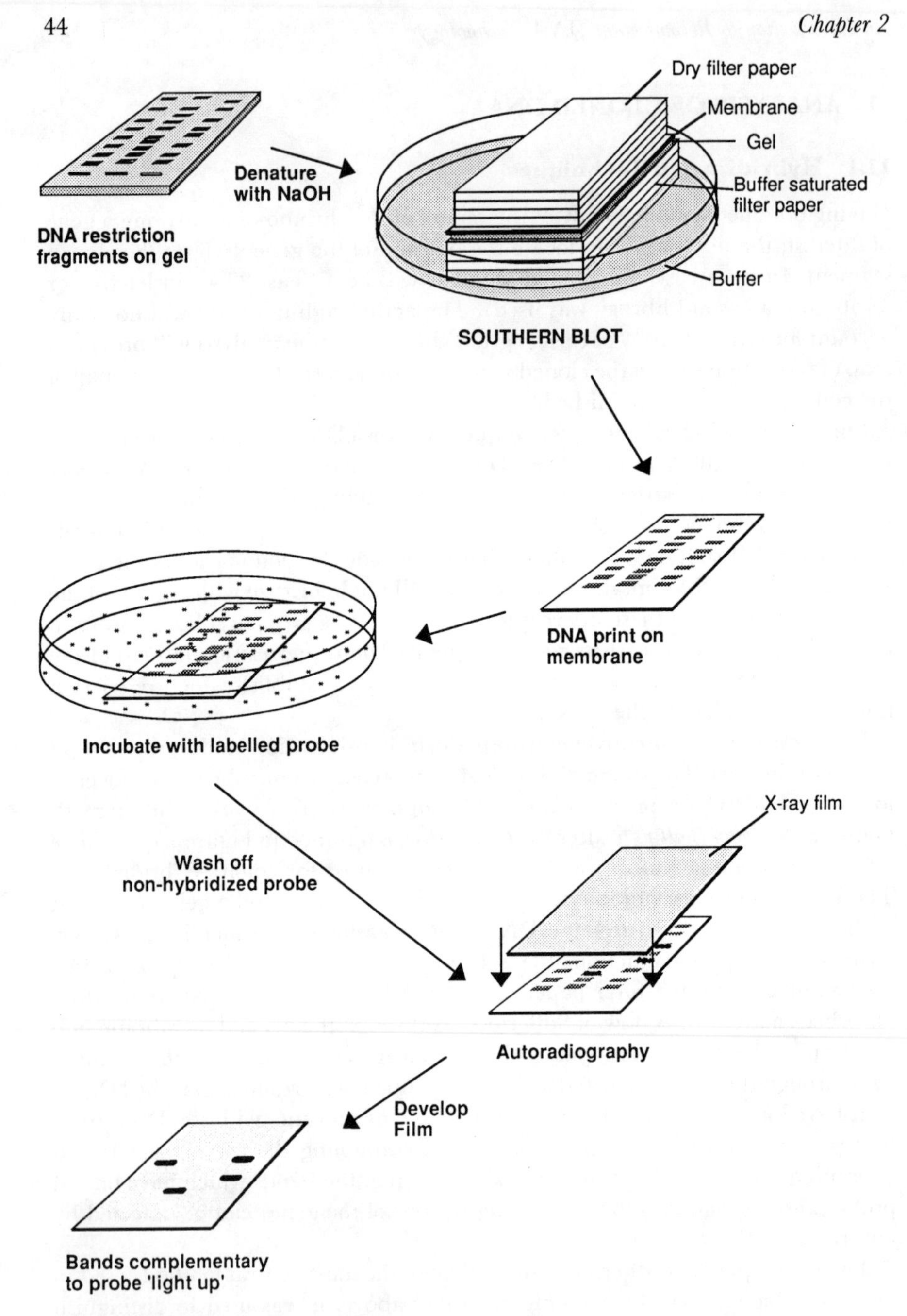

Figure 10 *Southern blotting: hybridization analysis of restriction fragments*

that is visible is a continuous smear. But following blotting and hybridization, the position of the gel of the band(s) corresponding to a given probe can be determined. And as a consequence of genetic variation, the pattern obtained using the same probe will frequently differ between individuals. This is the basis of the process known as genetic fingerprinting (discussed in Chapter 10) and also some methods of prenatal diagnosis (discussed in Chapter 11). Use of non-radioactive methods of labelling the probe make these methods even more convenient to use, and one such method is described in Chapter 10.

Other variations on this method include *Northern Blotting*,[46] (not named after its originator), in which RNA is transferred from a gel to a membrane for hybridization analysis. With a suitable cDNA probe, the activity of genes in particular tissues can be assayed by this approach.

To this point, we have concentrated on locating sequences within a cloned segment. Very frequently, however, it is necessary to look beyond the cloned segment to neighbouring DNA. Many genes and their control sequences are, in fact, spread out far beyond the confines of a single 20–40 kb segment. The DMD gene, mentioned in the last section, is over 1 million base pairs in length! Hence it is often necessary to build up a collection of clones carrying DNA from adjoining chromosomal regions.

The way this is done takes advantage of a feature in the construction of most genome libraries. So as to obtain as near as possible to a complete library, the fragments are prepared using an enzyme, such as *Sau*3A, under conditions in which it only cuts at a small proportion of its many possible sites. This results in an almost random collection of fragments, with potential for overlap between different fragments. Hence, to obtain a neighbouring segment from the library, one looks for a clone that has overlap with the original segment. This is done by using a small fragment from one end of the original segment as a probe and rescreening the genome library (colony or plaque hybridization). Once the (overlapping) neighbouring clone has been isolated, it can be used in the same way to move even further down the chromosome, a process known as *chromosome walking*.

11.2 DNA Sequencing

One of the greatest prizes of the new DNA technologies is the ability to produce base sequences of genes and other chromosomal elements. As has already been seen, cloning technology enables amplification of individual genes to the point that their study becomes feasible. DNA sequencing techniques provide a major tool in that study.

There are a number of different approaches to DNA sequencing, the two most common being the chemical degradation technique of Maxam–Gilbert,[47] and the chain termination technique of Sanger.[48] As the latter has become the more popular, it will be described here.

The technique is based on DNA synthesis. A single-stranded version of the DNA to be sequenced is used as a template, and a short complementary oligonucleotide used to prime the synthesis. The key element is the introduction of dideoxynucleotides. These molecules are nucleotides with a modified sugar

[46] J. C. Alwine, D. J. Kemp, and G. R. Stark, *Proc. Natl. Acad. Sci. USA*, 1977, **74**, 5350.
[47] A. M. Maxam and W. Gilbert, *Proc. Natl. Acad. Sci. USA*, 1977, **74**, 560.
[48] F. Sanger, S. Nicklen, and A. R. Coulson, *Proc. Natl. Acad. Sci. USA*, 1977, **74**, 5463.

THE REACTION MIX (The 'G' Tube)

The template (the DNA to be sequenced)

3' A-T-T-G-C-T-A-T-G-G-T-C-A-C-C-G-A-T-G-A-C-T-T-G-C-G-T 5'

A primer (Complementary to a section of the template)

5' T-A-A-C-G-A-T-A-C 3'

The building blocks for new DNA

dATP, dTTP, dGTP, dCTP (Either one of these or the primer is labelled)

The chain terminator

ddGTP
(dideoxy GTP)

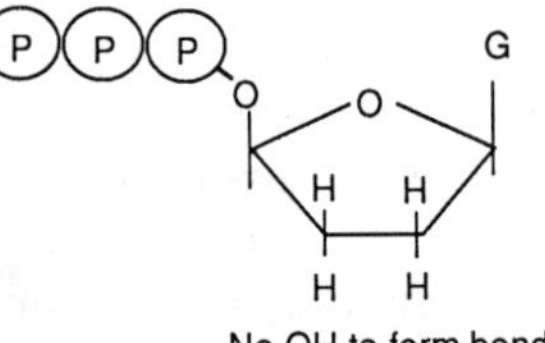

(P) = phosphate

TYPICAL REACTION PRODUCTS

A-T-T-G-C-T-A-T-G-G-T-C-A-C-C-G-A-T-G-A-C-T-T-G-C-G-T
• • • • • • • • • • • • •
T-A-A-C-G-A-T-A-C-**C-A-G▌**

A-T-T-G-C-T-A-T-G-G-T-C-A-C-C-G-A-T-G-A-C-T-T-G-C-G-T
• • • • • • • • • • • • • • •
T-A-A-C-G-A-T-A-C-**C-A-G-T-G▌**

A-T-T-G-C-T-A-T-G-G-T-C-A-C-C-G-A-T-G-A-C-T-T-G-C-G-T
• • • • • • • • • • • • • • • •
T-A-A-C-G-A-T-A-C-**C-A-G-T-G-G▌**

A-T-T-G-C-T-A-T-G-G-T-C-A-C-C-G-A-T-G-A-C-T-T-G-C-G-T
• •
T-A-A-C-G-A-T-A-C-**C-A-G-T-G-G-C-T-A-C-T-G▌**

A-T-T-G-C-T-A-T-G-G-T-C-A-C-C-G-A-T-G-A-C-T-T-G-C-G-T
• •
T-A-A-C-G-A-T-A-C-**C-A-G-T-G-G-C-T-A-C-T-G-A-A-C-G▌**

Figure 11 *Sanger DNA sequencing: the reactions taking place in the G tube. Three additional tubes, with ddATP, ddTTP, and ddCTP as chain terminators are used in addition to the one shown*

element such that the normal OH at the 3′ position is replaced by an H (ddGTP is shown in Figure 11). Such modified nucleotides are able to join a DNA chain in the normal way. Once incorporated into a DNA chain, however, the chain is unable to be extended, as it will now not have an OH group at the 3′ position for the next nucleotide to be joined to. Hence incorporation of a dideoxynucleotide terminates the chain.

Four separate tubes are used for the reaction. Each tube contains the four building blocks of DNA (dATP, dTTP, dGTP, dCTP). One of these, or the primer, is radiolabelled. Each tube also contains one of the ddNTP chain terminators. The G tube, as shown in Figure 11, contains ddGTP. Each chain that is made in that tube will continue until it incorporates such a ddGTP. Hence each chain finishes at the position of a G. Which G is random and depends on when a ddGTP happens to be incorporated into the chain. The ratio of ddGTP:dGTP is critical and must be low enough to ensure that the average chain length is not too short.

In a similar fashion, the chains in the A tube (ddATP) finish at random A positions, the chains in the T finish at random T positions, and chains in the C tube finish at random C positions. On completion of the reaction the DNA in each tube is denatured to free the newly synthesized (and radiolabelled) strand. The samples from each tube are then run on a polyacrylamide gel to separate fragments on the basis of size, and the positions of the bands revealed by autoradiography (Figure 12). By examining the sequences in Figure 11, it can be seen that the first new base added would be a **C**. Hence the shortest new fragment (and thus fastest running) would be in the C lane and correspond to termination by the addition of a ddCTP at this point. The second base added would be an **A**, and hence the second shortest fragment would be in the A lane. In fact, the entire sequence can be read directly off the autoradiograph, as shown in Figure 12.

Two factors might be considered as limiting the usefulness of this method: the need for a single strand template (isolated from its partner strand) and the need for a primer, complementary to a presumably unknown sequence. In fact, both of these problems are solved by the use the M13 cloning system.[49,50] The life cycle of this phage is shown in Figure 13. It is a single-stranded DNA phage, which infects F^+ *E. coli* through the sex pilus. Once inside the cell, however, it directs the synthesis of a complementary strand and converts to a double-stranded form (the RF or Replicative Form). RF DNA replicates until it reaches a copy number of about 100. It then changes its mode of replication and produces copies of just one strand. These form the genomes of new phage particles which are extruded through the cell wall.

When used as a vector, the double-stranded RF form is isolated and the DNA to be cloned inserted. As can be seen from the Figure, M13 vectors have been modified to include convenient restriction sites. The vector plus insert is then returned to the cell as if a plasmid. Note that the *lacZ* marker enables screening for the uptake of phages with inserts. From the resulting phage particles, single-stranded versions of the cloned sequence are recovered. Hence M13 is a

[49] J. Messing, R. Crea, and P. H. Seeburg, *Nucleic Acids Res.*, 1981, **9**, 309.
[50] J. Norrander, T. Kempe, and J. Messing, *Gene*, 1983, **27**, 101.

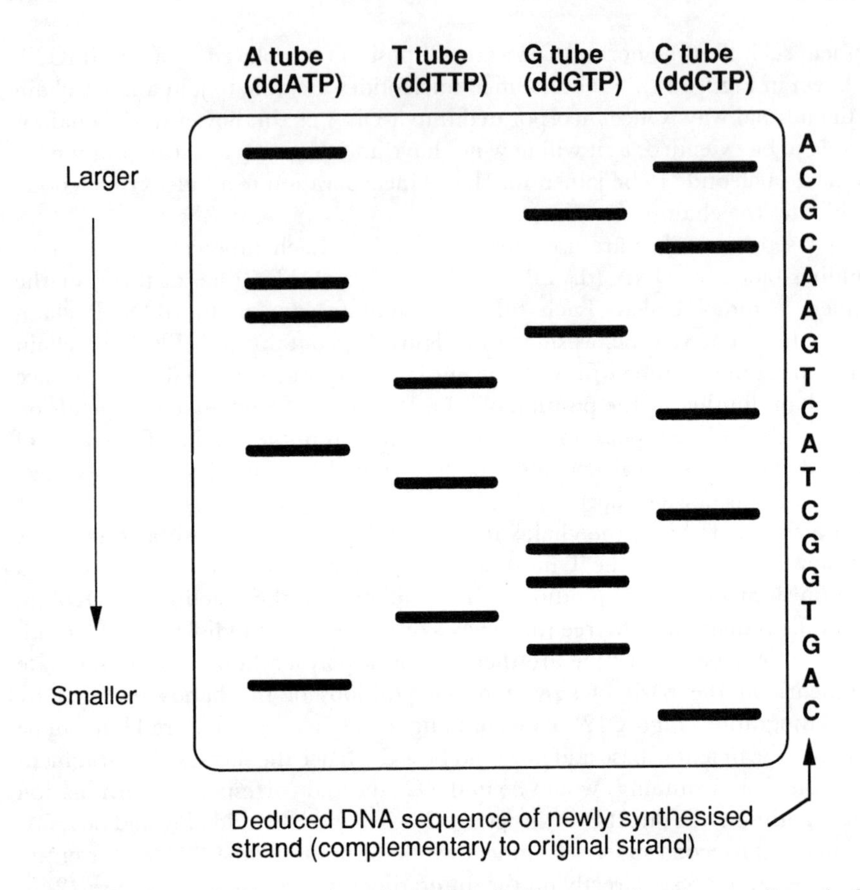

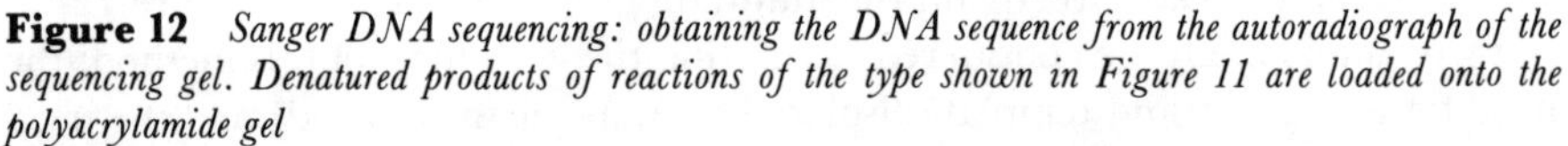

Figure 12 *Sanger DNA sequencing: obtaining the DNA sequence from the autoradiograph of the sequencing gel. Denatured products of reactions of the type shown in Figure 11 are loaded onto the polyacrylamide gel*

tool for generating the required single-stranded templates required for Sanger sequencing. The primer used in the sequencing reaction has been designed to be complementary to a region of the vector close to the point of insertion. Hence one primer can be used whatever the insert to be sequenced and no prior knowledge of the insert sequence is necessary.

From a single gel the sequence of a few hundred bases can normally be read. To obtain the sequence of a larger region, it is necessary to put together information from a number of different clones. In general, the most efficient way of doing this is to clone random, overlapping, fragments of the region into M13 vectors and to put together the final sequence on the basis of overlap. Computer technology plays a major role in this task. Indeed, computers play an indispensable role in the storage and analysis of vast quantity of data generated by sequencing programmes.[51]

[51] 'The Application of Computers to Research on Nucleic Acids' ed. D. Soll and R. J. Roberts, IRL Press, London, 1982, Vol. 1; 1984, Vol. 2; 1986, Vol. 3.

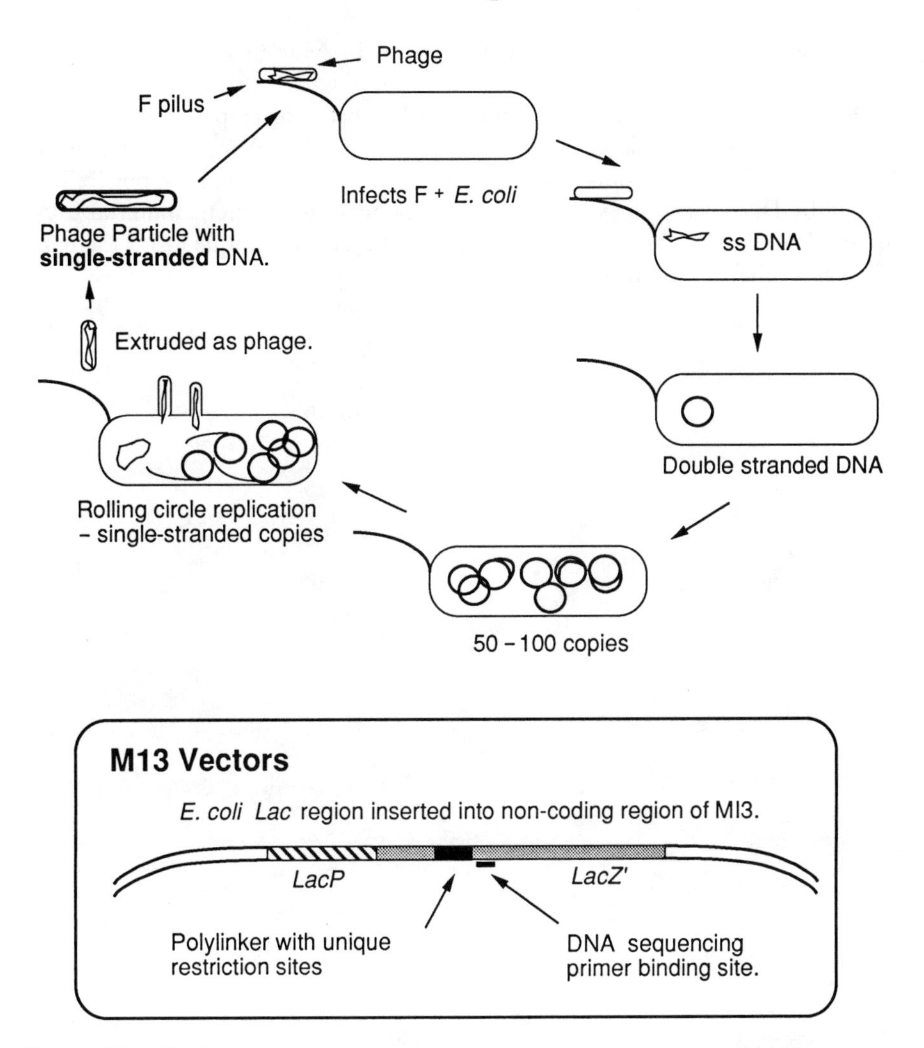

Figure 13 *The life cycle of the phage M13 and the modifications of this phage in M13 vectors*

The most ambitious sequencing programme announced to date is the plan to sequence the entire human genome.[52] To even make a start at achieving this it will be necessary to greatly increase the degree of automation in the sequencing procedure.[53]

Once DNA sequences for genes and surrounding regions are available, it is possible to start to answer a wide range of questions. Indeed, it is now easier to obtain the sequence of a protein from the sequence of its gene than directly from the protein. And it is also possible to start to intervene directly in the process of protein production. Techniques of *in vitro* mutagenesis allow changes to be made in the sequence of a DNA message before it is reintroduced into a cell. This topic

[52] J. D. Watson, *Science*, 1990, **248**, 44.
[53] G. L. Trainor, *Anal. Chem.*, 1990, **62**, 418.

is discussed later in this book in the chapter on Enzyme Engineering (Chapter 14).

Over the last few years a new approach has been developed which greatly assists in many of the tasks covered in this chapter. This technique, known as the Polymerase Chain Reaction (PCR) has even further expanded the possibilities offered by DNA technology by allowing amplification of genetic material to be undertaken without the need for cloning in cells. This technique forms the basis of the next chapter of this book.

CHAPTER 3

The Polymerase Chain Reaction

C. BUFFERY

1 INTRODUCTION

In order to study genetic structure and function, sufficient quantities of the DNA of interest must be available for analysis. Until recently this was accomplished by using micro-organisms to generate multiple copies of genes – a lengthy and laborious process. Now a cell free technique is available, known as the polymerase chain reaction (PCR).* This is the name given to the *in vitro* amplification of specific nucleic acid sequences by repeated cycles of primer directed synthesis. The mechanism is basically that which the cell uses during replication; denaturation enables a polymerase enzyme to generate a complementary copy of the parent DNA strand from deoxyribonucleotide triphosphates (dNTPs) in solution.

PCR finds a genetic needle in a haystack. A single gene sequence can be targeted amongst an abundance of other DNA, even in crude lysates, and will be amplified in an exponential fashion into an analysable quantity. The principle is astonishingly simple, yet nothing since the isolation of restriction enzymes has had more impact on the study and application of molecular biology.

The idea of repeatedly amplifying DNA was first published in 1971[1] but was only exploited some years later by scientists at the Cetus Corporation after Kary Mullis independently devised an amplification method[2,3] which he felt had potential. It is an exquisitely sensitive and fast technique, often replacing tedious and fickle cloning procedures, and is tolerant of poor quality DNA. 'Designer genes' can be created, at will, and one only has to consider the proliferation of literature describing its uses to realise that PCR is now a basic tool for the molecular biologist.

2 PRINCIPLES OF TECHNIQUE

An outline of the PCR is shown diagrammatically in Figure 1. Specific oligonucleotides are chemically synthesized, such that these primers are complementary

* PCR – Licence held by Hoffmann – La Roche

[1] K. Kleppe, E. Ohtsuka, R. Kleppe, I. Molineux, and H. G. Khorana, *J. Mol. Biol.*, 1971, **56**, 341.

[2] R. K. Saiki, S. Scharf, F. Faloona, K. B. Mullis, G. T. Horn, H. A. Erlich, and N. Arnheim, *Science*, 1985, **230**, 1350.

[3] K. B. Mullis, *Sci. Am.*, 1990, **262**, 36.

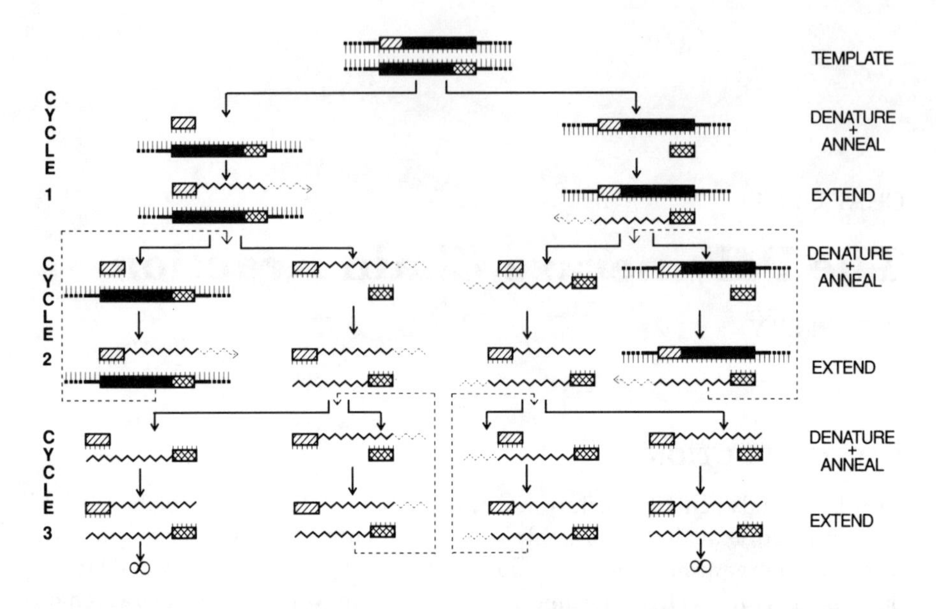

Figure 1 *Diagram representing the first three cycles of PCR.* ***Denature and Anneal:*** *This depicts the separation of the helix into two strands and the subsequent annealing of an oligonucleotide primer to each strand.* ***Extend:*** *This depicts the enzymic synthesis of a complementary strand commencing at the 3′ end of the bound primer. Dotted arrows indicate repetition of the processes involving 'long products' in subsequent cycles, whilst 'short product' accumulates exponentially*

to DNA flanking the sequence of interest. One primer is designed to anneal to the sense strand, the other to the antisense strand, with their 3′ ends pointing towards each other. Primers are mixed with a buffered solution of template DNA, dNTPs, magnesium, and a thermostable polymerase enzyme. The mixture is covered with a layer of mineral oil to prevent evaporation (or the vessels covered with a heated top-plate) and placed in a programmable heating block. The double-stranded DNA template is then denatured by heating to a temperature above its melting point. The temperature is lowered sufficiently for hybridization between primers and template to occur, yet kept high enough to prevent mismatch hybridization of the primers to similar sequences elsewhere in the genome. High concentrations of primer ensure that this reaction is favoured over re-annealing of template strands. The temperature is sometimes raised again towards the optimum for the polymerase, which has attached itself to the end of the primer–template duplex. Synthesis proceeds from the 3′ end of each primer until the reaction is stopped by heating to melting point for the second time. The product of this reaction is of indefinite length and is known as 'long product'. This completes the first cycle of PCR. The second cycle commences with this melting step, followed by primer annealing. This time, however, the primers not only anneal to the original DNA but also to the newly synthesized strands from the first reaction. These strands will possess the homologous primer sequence provided that synthesis in the previous cycle extended beyond the other primer binding site. The second cycle repeats the first with respect to the original DNA, but synthesis on the new strands can proceed only as far as the end of the molecule, which corresponds to the 5′ end of the opposite primer. After the third cycle of PCR, it is easy to see that synthesis directed by the products of the first two cycles will be bounded at both ends by the primer sequences and this short product will accumulate exponentially with subsequent cycles. Long product will continue to amass linearly. The major product of a PCR will therefore be of defined length, consisting of the sequence between the primers and the primer sequences themselves.

2.1 Limitations and Efficiency

Various parameters will affect the efficiency of the PCR. The length of the target DNA between primers is one limiting factor; generally the longer the sequence the less efficient the PCR, although amplifications of fragments > 10 kb have been reported.[4,5] The reaction can be hindered by the existence of complementarity between the primers themselves, causing an artefact known as 'primer dimer'. The partially hybridized primers become extended by the polymerase and amplified, sometimes monopolizing the reaction.

The sequence of the target DNA itself can also affect efficiency. Too much secondary structure in GC rich areas of the denatured strands can prevent the enzyme from reading the template – using a proportion of 7-deaza GTP (7-deaza guanosine triphosphate) has been shown to help in these situations.[6] This base analogue facilitates synthesis by reducing hairpin loops without compromising Watson–Crick base pairing. Substances such as glycerol and dimethyl sulfoxide (DMSO) may have a similar effect on secondary structure (but contrarily can inhibit polymerases) and additional proteins such as bovine serum albumin (BSA) can enhance PCR by protecting the polymerase and chelating inhibitors. Examples of inhibitors include humic acids, which are often found in archaeological specimens, and haem breakdown products, which may complex with the Mg^{2+} required for enzyme function.

The exponential phase of production does not continue indefinitely and after about 20 cycles most PCRs enter linear and then plateau phases. Reagent concentrations become limiting, the enzyme tires of repeated heating, and high concentrations of product tend to result in re-annealing rather than new primer binding.

In the early days of PCR the only polymerases widely available, such as the Klenow fragment of DNA polymerase 1, were easily denatured by the high temperatures required for melting. This meant that a fresh injection of enzyme had to be made at each cycle, which loaded the reaction mixture with increasing amounts of protein and was exceedingly tedious. A thermophilic bacterium inhabiting hot springs, *Thermus aquaticus*, was found to possess an excellent

[4] P. Kainz, A. Schmiedlechner, and H. Bernd Strack, *Anal. Biochem.*, 1992, **202**, 46.
[5] A. J. Jeffreys, R. Neumann, and V. Wilson, *Cell*, 1990, **60**, 473.
[6] L. McConlogue, M. A. D. Brow, and M. A. Innis, *Nucleic Acids Res.*, 1988, **16**, 9869.

enzyme – *Taq* polymerase. As well as enabling it to withstand repeated incubations at melting point, its heat tolerance allowed primer annealing and extension to be set at much higher temperatures than before; this dramatically increased the specificity of most PCRs by decreasing mispriming. This factor was particularly important when amplifying a sequence from multi gene families which show a lot of sequence similarity. High temperatures also prevented much of the inhibitory secondary structuring experienced and thereby increased the overall yield. Soon the enzyme was genetically engineered in *E. coli* and became commonplace. *Taq* polymerase, however, does not have the 3′–5′ exonuclease or 'proof-reading' activity of Klenow and hence incorporates more mismatched bases. The mutations are random and are not usually a problem unless they occur in very early cycles of PCRs with little parent DNA. In this case these unfaithful copies could become the major species of template. If the PCR product itself is to be cloned, the presence of mutations could affect expression. In these situations other thermostable enzymes with proof-reading ability (such as Vent™,* from *Thermococcus litoralis*) may be desirable.[7]

Apart from misincorporated bases and primer dimer, other artefacts can be produced during PCRs. If a primer is not fully extended before denaturation (which may occur if the polymerase has a low processivity or meets a damaged base)[8] then the partially synthesized chains could compete with the primers in subsequent cycles. This would be of no consequence if the DNA is homozygous at the locus being amplified, but in some circumstances two different alleles may be present in the template. In this case, the synthesis could jump from allele to allele if these unfinished products act as primers themselves, resulting in hybrid sequences.[9] However, this is unlikely to be a common occurrence if primer concentrations are carefully balanced.

2.1.1 Contamination. The very sensitivity which makes PCR such a useful technique is also its most serious limitation. Because exponential amplification of targets can be achieved even from single cells, any air-borne cellular debris or contaminated reagents present a serious risk of false results. By far the most important reservoirs of spurious template are the products of previous PCR reactions. These represent a highly enriched source of the target sequence, and can soon become ubiquitous in the laboratory after a period of time using the same primers. For this reason it is prudent to prepare and analyse PCR reactions in separate areas of the laboratory, using dedicated equipment (particularly pipettes), autoclaved and aliquoted reagents, and multiple controls with each reaction.

Some laboratories, using protocols which push the PCR to its limits of sensitivity, employ enzymatic or photochemical 'sterilization' to prevent amplification of product carryover. A simple method is to UV irradiate reaction vessels and reagents prior to preparing an amplification. This should cause pyrimidine

* Vent™ – Trade mark of New England Biolabs Inc.
[7] P. Mattila, J. Korpela, T. Tenkanen, and K. Pitkanen, *Nucleic Acids Res.*, 1991, **19**, 4967.
[8] S. Paabo, *Proc. Natl. Acad. Sci. USA*, 1989, **86**, 1939.
[9] R. K. Saiki, D. H. Gelfand, S. Stoffel, S. Scharf, R. Higuchi, G. T. Horn, K. B. Mullis, and H. A. Erlich, *Science*, 1988, **239**, 487.

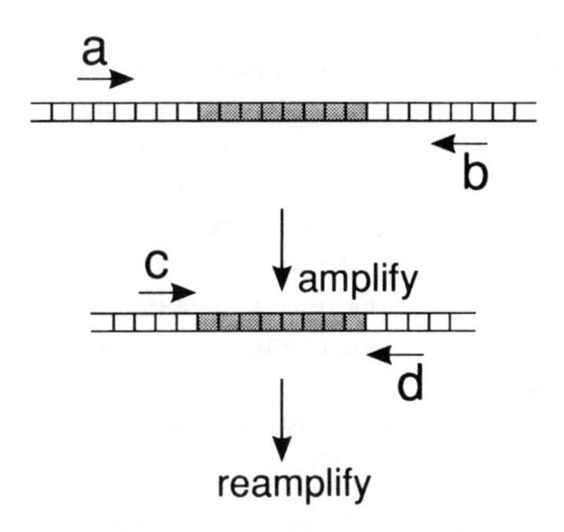

Figure 2 *Nested PCR: First amplification is performed using primer set* ***a*** *and* ***b****. A second amplification is performed on a sample of the products, using primer set* ***c*** *and* ***d****. This increases specificity*

dimers to form and make alien DNA unsuitable as template for the polymerase. However the effectiveness of this will depend on the ability of the radiation to penetrate where needed. Alternatively, photoactive isopsoralens can be added to the reaction which heighten the damaging effect of UV on DNA.[10] After cycling, the tubes are irradiated before opening so that any product which escapes will be disabled template.

With the enzymatic method, deoxyuridine triphosphate (dUTP) replaces dTTP in the polymerization so that all reaction products will be distinguishable from native T containing DNA. Prior to amplification the reaction mixtures are treated with uracil *N*-glycosylase which degrades any *U*-containing DNA from previous reactions but not normal DNA or RNA.[11] The enzyme is killed by the heat of the first cycle and therefore will not damage the newly accumulating product. These 'sterilization' methods are appealing but, as their effectiveness is not 100%, they must not be allowed to substitute for good laboratory technique and adequate controls.

2.2 Modifications of Technique

Variations of the basic PCR method have been developed to suit specific applications or to overcome problems. Tricks to improve the reaction specificity include 'nested', 'hot start', and 'touchdown' PCRs.

2.2.1 Nested PCR. If mispriming occurs in a PCR due to sequence similarities between target and related DNA, an aliquot of the products can be re-amplified with a second set of primers which have sequences 3′ to the original set, *i.e.* nested[12] (see Figure 2). Products of mispriming are unlikely to have sufficient

[10] Y. Jinno, K. Yoshiura, and N. Niikawa, *Nucleic Acids Res.*, 1990, **18**, 6739.
[11] M. C. Longo, M. S. Berninger, and J. L. Hartley, *Gene*, 1990, **93**, 125.
[12] T. M. Haqqi, G. Sarkar, C. S. David, and S. S. Sommer, *Nucleic Acids Res.*, 1988, **16**, 11844.

similarity to the correct locus to bind the second set of primers. The result then is a shorter product but without the contaminating sequences. A further modification of this method has been called 'drop in drop out' PCR in which the two sets of primers are added simultaneously.[13] The primers are designed such that the outer set produce a product with a much higher melting temperature than the inner set, and the inner set have a low annealing temperature. Cycling begins at high temperatures allowing only the outer primers to initiate synthesis. After several rounds the annealing temperature is dropped allowing the inner set to prime also. Further cycles later, when some shorter product has accumulated, the denaturation temperature is also dropped so that products of the outer primers can no longer melt to form templates, and eventually the major product becomes the sequence bounded by the inner primers.

2.2.2 Touchdown PCR. Touchdown PCR[14] is a method used to increase specificity without compromising yield. The principle is to initiate synthesis at very high annealing temperatures which permit only perfectly matched primer–template hybrids to form. The annealing temperature is dropped in a stepwise fashion with each cycle. Once copies of the target sequence have begun to accumulate over the first few cycles, high temperature annealing becomes much less critical for specificity, as it is previous products which form the major template. These products are unlikely to have any sites for mispriming. The benefit of decreasing the annealing temperature is to increase the probability of stable primer–target interaction. The result is improved yield of specific product compared to single high temperature annealing protocols.

2.2.3 Hotstart PCR. Most samples ready for amplification will have some single-stranded DNA present even before deliberate denaturation; primers can bind promiscuously to these areas at low stringencies. *Taq* polymerase shows a degree of activity at room temperature, despite being well below its optimum, which can lead to these imperfectly matched duplexes being extended before the reaction proper has begun. Hotstart PCR decreases artefacts by excluding a critical reagent (such as the enzyme or magnesium) from the reaction until it has reached denaturation point. This can be achieved by opening the tubes in the thermal cycler after the first melt and then adding the critical reagent, however this increases the risk of cross tube contamination. Another method is to cover the reaction mixture with a plug of inert wax and place the critical reagent on top.[15] Once the temperature reaches denaturation point the wax melts, thermal convention mixes the reagents, and synthesis begins.

2.2.4 Anchored and Inverse PCR. One constraint on the use of PCR is the need to know the sequence surrounding the DNA of interest. If the area to be amplified is beyond a known sequence, a way around the need for a flanking primer at the other end is to use an adaptation known as anchored PCR.[16] For example, a

[13] H. A. Erlich, D. Gelfand, and J. J. Sninsky, *Science*, 1991, **252**, 1643.
[14] R. H. Don, P. T. Cox, B. J. Wainwright, K. Baker, and J. S. Mattick, *Nucleic Acids Res.*, 1991, **19**, 4008.
[15] Q. Chou, M. Russell, D. E. Birch, J. Raymond, and W. Bloch, *Nucleic Acids Res.*, 1992, **20**, 1717.
[16] E. Y. Loh, J. F. Elliott, S. Cwirla, L. L. Lanier, and M. M. Davis, *Science*, 1989, **243**, 217.

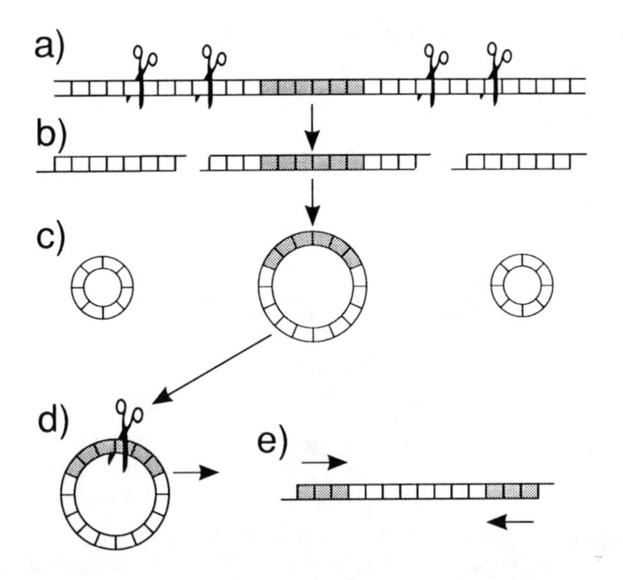

Figure 3 *Inverse PCR: (*a*) DNA containing a known sequence (shaded) is restricted with an enzyme cutting outside of that area. (*b*)–(*c*) Restriction fragments circularize and are ligated. (*d*) Closed circles are restricted within the known sequence using a rare cutting enzyme. (*e*) Primers complementary to either end of the known sequence are used to amplify the unknown area*

restriction fragment may only have a known sequence to which a specific primer can be made at one end of the target area. Using terminal deoxynucleotidyl transferase, the other end of the fragment can be given a poly G tail – the anchor. A poly C primer can now act as the partner to the specific primer. Alternatively, double-stranded target DNA can be modified at the end of an unknown sequence by the ligation of adaptors. As the sequence of the adaptor will be known, a primer can be made which will initiate synthesis at this restriction site within the unknown sequence.

Another method of amplifying into unknown regions is known as inverse PCR (Figure 3).[17] This can be used when there is known sequence in one area but amplification of the DNA either side of this area is desired. Initially the target DNA is cut with a restriction enzyme which does not cut within the known sequence, but at unknown points either side. The resulting DNA fragments can then circularize as the single strand overhangs, left by the restriction enzyme, anneal to one another. If the DNA concentration is low, this circularization is a more likely occurrence than rejoining in the original format or concatamerization. The circle is joined by the addition of a ligase and then re-opened using another rarer cutting restriction enzyme which cleaves only within the known sequence. The known area is now at both ends of the template and amplification can proceed as usual, however primers are generated in the opposite orientation to normal in relation to the original sequence because the area has been inverted by the circularization process.

[17] T. Triglia, M. G. Peterson, and D. J. Kemp, *Nucleic Acids Res.*, 1988, **16**, 8186.

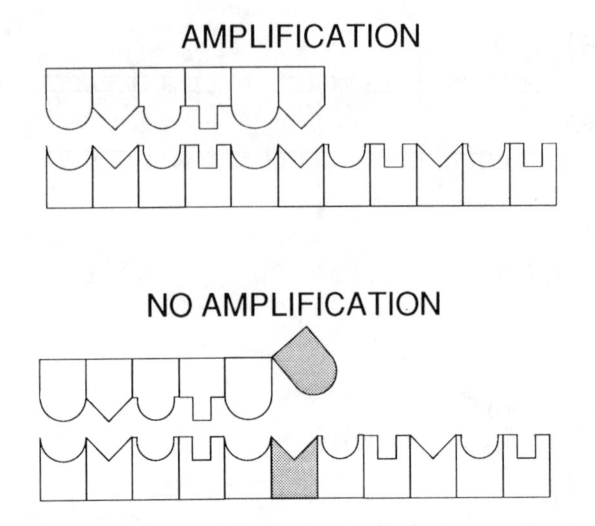

Figure 4 *Amplification Refractory Mutation System: Perfect base match at the 3′ end between the oligonucleotide primer and the template results in synthesis. Base mismatch at the 3′ end prevents the primer from initiating synthesis, therefore no amplification occurs*

2.2.5 ARMS PCR. Other adaptations of the technique have been devised which combine amplification with analysis. Amplification refractive mutation system (ARMS) PCR[18] is a means of sequence specific amplification. The rationale is that for *Taq* to initiate synthesis there must be perfect base pairing at the 3′ end of the primer–template hybrid (Figure 4). So if different alleles of the same locus exist it is possible to amplify them differentially by designing primers specific for each allele. Their 3′ end will be complementary to a position within the target which has an allele specific base difference. Enzymes with 3′–5′ exonuclease activity are likely to detect this mismatch and repair it, and therefore are not suitable for this purpose. One particularly neat application of this technique is minisatellite variant repeat analysis PCR.[19] It exploits multiple differences between alleles using amplification refractive PCR, producing a profile of hypervariable DNA which is virtually individual specific. For identity testing it may be superior to conventional 'fingerprinting' methods in its sensitivity and speed, ease of interpretation, and tolerance of poor quality template (see Chapter 10).

3 USES AND APPLICATIONS

The applications of PCR are too numerous to list here, but a synopsis of the major uses will be presented to give the reader a feeling for the scope of the technology.

[18] C. R. Newton, A. Graham, L. E. Heptinstall, S. J. Powell, C. Summers, N. Kalsheker, J. C. Smith, and A. F. Markham, *Nucleic Acids Res.*, 1989, **17**, 2503.
[19] A. J. Jeffreys, A. MacLeod, K. Tamaki, D. L. Neil, and D. G. Monckton, *Nature (London)*, 1991, **354**, 204.

3.1 Sequencing

One of the most common reasons for using PCR is to generate sufficient template for sequencing. It is much simpler and quicker than cloning, and amplification and sequencing may be done in one continuous operation if desired. The linear products of PCR will reanneal very quickly after denaturation which inhibits the binding of sequencing primers. For this reason most direct methods for sequencing PCR products involve separating the two template strands.

3.1.1 Asymmetric PCR. This method involves preferentially amplifying one template strand by controlling primer concentrations.[20] The primer for the strand to be sequenced is present in limiting concentrations whilst the opposite primer is present in abundance during the PCR. Initially amplification occurs as normal until the lesser primer is exhausted. This enriches the amount of template before subjecting it to further cycles during which the other primer continues to initiate synthesis of one strand only. The result is a moderate amount of double-stranded DNA and a vast amount of single-stranded template. The limiting primer can now be used as a sequencing primer with the Sanger method.

3.1.2 Strand Removal. Another way to prepare single-stranded template is to digest away one of the strands. This can be done by using one 5′ phosphorylated primer in the PCR – on adding λ exonuclease to the product the 5′ phosphorylated strand is digested leaving a single-stranded sequencing template.[21] Similarly, one primer can be linked to a ferrous bead.[22] The products are denatured in alkali and the beaded strand is removed with a magnet. It is also possible to recover the other strand using this technique – it is sometimes advisable to sequence both strands to confirm results.

3.1.3 Cycle Sequencing. Performing the sequencing reaction in a thermal cycler as a modified type of PCR is becoming increasingly popular. The methods described above all include a step during which one strand is removed; this is a time consuming process which is not always totally reliable. Cycle sequencing[23] can be performed directly on double-stranded DNA, in some cases even on crude lysates of plaques or colonies. In essence, dideoxy terminators are added to a normal type PCR mix; the extension and termination reactions are done together, one tube for each species of terminating nucleotide. The sequencing primer is 5′ end labelled with ^{32}P. Sequencing then proceeds in a cyclical fashion like PCR, melting followed by primer annealing followed by extension and termination. Re-melting allows the same template strand to be used many times, enhancing the sequencing signal in a linear amplification. The amount of starting template can be exponentially amplified if required, in a pre-sequencing PCR reaction. Fluorescent dyes have been used in conjunction with the dideoxy

[20] U. B. Gyllensten and H. A. Erlich, *Proc. Natl. Acad. Sci. USA*, 1988, **85**, 7652.
[21] R. Higuchi and H. Ochman, *Nucleic Acids Res.*, 1989, **17**, 5865.
[22] T. Hultman, S. Stahl, E. Hornes, and M. Uhlen, *Nucleic Acids Res.*, 1989, **17**, 4937.
[23] S. M. Adams and R. W. Blakesley, *Focus*, 1991, **14**, 31.

terminators such that each stop base carries a different colour.[24] This means that all four reactions can be carried out in the same tube and run in the same lane on a gel if an automated laser apparatus is available for analysis.

3.2 Gene Manipulation and Expression Studies

Oligonucleotide primers need not be complementary to the target at their 5′ end for synthesis to proceed during a PCR. This property provides the molecular biologist with a convenient way of altering a piece of DNA[25] and studying the effects. This is made possible because the 5′ primer tail becomes incorporated into the PCR product. Using this site directed mutagenesis the effects of elements (such as promoters or initiators) on gene expression can be investigated, and post PCR manipulations can be facilitated by tagging on restriction sites, adaptors, or sequences complementary to vectors. It is also possible to add a sequence coding for a monoclonal epitope[26] so that the protein of interest can easily be harvested once expressed.

PCR, although a technique for DNA amplification, has an important role to play in the study of mRNA's generated by gene expression. Firstly, the RNA is reverse transcribed into cDNA which is then amplified as normal. This process is sometimes referred to as RT PCR and in fact a thermostable polymerase from *Thermus thermophilus* has been isolated which has reverse transcriptase activity[27] in the presence of Mn^{2+}, and DNA polymerase activity in Mg^{2+}; thus both these steps can be linked. These expression studies can show, for example, when tumour or viral genes are being expressed and consequently whether any therapy is working. Sometimes predicting primer sequences for cDNA can be rather complicated and information is taken from the protein sequence itself.[28] Primers can be created with degeneracy (variable bases) where the third base in the codon is optional, or inosine is used at these positions as it will usually allow base pairing to occur whatever. Primer degeneracy is also used to search the genome for families of genes with similar sequences.

3.3 Evolutionary Biology and Mapping Studies

PCR has released a whole reservoir of information in evolutionary biology. With its capacity to deal with small amounts of poor quality DNA, the technique has enabled workers to analyse the preserved tissues of extinct species such as the woolly mammoth, quagga, moa, and marsupial wolf, as well as ancient human remains. By studying sequence similarlity in conserved genes, phylogenic trees can be mapped out. Such studies have shown that homeobox genes, which are

[24] J. M. Prober, G. L. Trainor, R. J. Dam, F. W. Hobbs, C. W. Robertson, R. J. Zagursky, A. J. Cocuzza, M. A. Jensen, and K. Baumeister, *Science*, 1987, **238**, 336.
[25] S. Scharf, G. T. Horn, and H. A. Erlich, *Science*, 1986, **233**, 1076.
[26] G. A. Martin, *Cell*, 1990, **63**, 843.
[27] T. W. Myers and D. H. Gelfand, *Biochemistry*, 1991, **30**, 7661.
[28] C. C. Lee, X. Wu, R. A. Gibbs, R. G. Cook, D. M. Muzny, and C. T. Caskey, *Science*, 1988, **239**, 1288.

central to the processes of cell differentiation and migration in early development, have been well conserved in many species throughout evolution.[29]

Mitochondrial DNA, with its high copy number, has been the target of choice for most evolutionary work and has been used to indicate the origin of human life in Africa and subsequent migration patterns.[30] These types of studies involve the PCR of DNA from hair, mummified tissue, and even bone,[31] and compare the information with that from contemporary samples. Ecologists and zoologists have also benefited from PCR – wild animals have been monitored by skin biopsy and urine analysis.

PCR has also assisted in the mapping of the human genome. Primers with homology to interspersed repetitive sequences (IRS) have helped in ordering the multiple fragments of human DNA being mapped. PCR has quickened the pace of study into human meiotic events and the mapping of closely linked markers because it has made the analysis of single sperm possible.[32] As these products of single meioses are present in abundance from one individual, they provide instant pedigree analysis.

3.4 Diagnostics

In human terms, diagnostics is probably the field in which PCR has had the greatest visible impact. Its speed and sensitivity make it the ideal technique for prenatal diagnosis of inherited disease – chorionic villus samples or cells from amniocentesis can be analysed within a day. Indeed, one of PCR's earliest applications was the detection of the sickle cell mutation in these samples.[33] Where the disease state is a single base mutation, results can be obtained by dot blots, often available in kit form. PCR product of the pertinent region is dotted onto a membrane and hybridized with an allele specific oligonucleotide (ASO) probe, giving a yes/no answer. In some cases it has been possible to diagnose disease at the pre-implantation stage by single cell biopsy or by drawing inference from the egg's polar body.[34]

PCR from cDNA provides information on the activity of tumour cells and also viruses. The PCR approach has been particularly useful in monitoring retroviral infections, where viral sequence may not be eliciting antibody production or actively producing new particles. It is often important to recognise both proliferative and latent phases, and PCR can distinguish them. In newborns, PCR has enabled the early detection of HIV infection. Babies born of infected mothers have sufficient maternal antibody to affect conventional detection methods; with PCR, proviral sequence itself can be searched out in the infant's blood.[35]

[29] S. S. Roberts, *J. Nat. Inst. Health Res.*, 1989, **1**, 129.
[30] L. Vigilant, R. Pennington, H. Harpending, T. D. Kocher, and A. C. Wilson, *Proc. Natl. Acad. Sci. USA*, 1989, **86**, 9350.
[31] E. Hagelberg and B. Sykes, *Nature (London)*, 1990, **342**, 485.
[32] H. H. Li, U. B. Gyllensten, X. Cui, R. Saiki, H. Erlich, and N. Arnheim, *Nature (London)*, 1988, **335**, 414.
[33] R. K. Saiki, C. A. Chang, C. H. Levenson, T. C. Warren, C. D. Boehm, H. H. Kazazian, and H. A. Erlich, *N. Eng. J. Med.*, 1988, **319**, 537.
[34] M. Monk and C. Holding, *Lancet*, 1990, **335**, 985.
[35] M. F. Rogers *et al.*, *N. Eng. J. Med.*, 1989, **320**, 1649.

Some pathogens, such as the syphilis organism, are very difficult to detect because they cannot be grown successfully in culture; amplifying diagnostic sequences is therefore an efficient method of identification. As PCR can be performed on formalin fixed tissue, collecting specimens is made easier when the organism in question is virulent and difficult to handle or where storage facilities are limited. A full discussion of the use of PCR in diagnostic medicine can be found in Chapter 11.

This has been a very brief overview of the types of uses to which PCR can be put and innovative applications are appearing with increasing frequency in the scientific literature. The technology will undoubtedly progress during the lifetime of this publication.

CHAPTER 4

The Expression of Foreign DNA in Bacteria

R. J. SLATER

1 INTRODUCTION

As can be seen from the previous chapters it is now a relatively easy matter to clone large quantities of specific DNA sequences regardless of origin. This greatly facilitates studies on gene expression. Restriction endonuclease cleavage sites can be mapped as reference points on the cloned DNA, the nucleotide sequence can be determined and coding sequences (open reading frames, or ORFs) identified. However, unless expression of cloned DNA can be obtained in a suitable host or system, many fundamental studies on the expression or function of genes cannot be carried out, and most commercial applications of genetic engineering could not be achieved.

The product of expression of foreign genes in host organisms is frequently referred to as 'recombinant protein' and its production is of enormous value to both fundamental studies and commercial biotechnology. Coding sequences are ligated into 'expression vectors' designed to allow replication and expression of inserted DNA in host cells. Currently the most popular hosts are *Escherichia coli*, *Bacillus subtilis*, yeast, and cultured cells of higher eukaryotes such as insect or mammalian cells. *E. coli* has, to date, been the host of choice for much work because of the sound base in appropriate genetic knowledge and ease of growth. Therefore, much of this article will concentrate on expression in *E. coli*; however, for many purposes the bacterium is not ideal. Proteins that are large (*i.e.* greater than 500 amino acids) or highly hydrophobic or require post-translational modification are best made in other cells. Also the reducing environment with *E. coli* is not conducive to the formation of disulfide bonds, proteins with many cysteines need to be made in a secretion system or alternative host.

To obtain expression of foreign genes in bacteria such as *E. coli*, it is first necessary to construct suitable vectors. Two principle types of vector are used: derivatives of viruses (*e.g.* bacteriophage λ) and plasmids (see Chapter 2). Specially designed bacteriophage constructs, such as λ gt11 or λZAP are expression vectors for the production of polypeptides specified by DNA inserts. These cloning tools produce recombinant protein as β-galactosidase fusion products (see Section 3.4). They are used, for example, to clone genes via the identification of gene products.[1] Such vectors are not suitable, however, for large-scale production because cell lysis occurs as a result of virus infection. They are used in small-scale production, for example in the production of large numbers of variant recombinant proteins for functional screening. Plasmids are the vectors of choice for the large-scale production of foreign proteins in bacteria. Consequently their use will be described in this chapter after some necessary background has been discussed.

It has been known since 1944 that it is possible, using classical techniques of bacterial genetics, to transform bacteria with DNA from the same or closely related species.[2] Recombinant DNA techniques, however, have given molecular biologists the opportunity to obtain expression of foreign DNA from totally unrelated species, even mammals and higher plants, in bacteria such as *E. coli*. The expectation that DNA should be expressed in an unrelated organism was based on the generally held belief that the genetic code is universal. That is, a DNA sequence coding for a protein in one organism, should function in any organism to produce a protein with the same amino acid sequence. This belief was given support when it was found that not only DNA from other bacteria but also DNA from lower eukaryotes, *Saccharomyces cerevisiae*[3] and *Neurospora crassa*,[4] could be expressed in *E. coli*. Unfortunately, however, when similar experiments were carried out with DNA from higher eukaryotes, expression of the foreign gene was not obtained; more sophisticated techniques, discussed later, were required.

The benefits that accrue from obtaining expression of foreign DNA in bacteria are considerable. Fundamental studies are made possible concerning the relationship between the protein's primary structure (the amino acid sequence) and its function. Large quantities of recombinant protein can be produced instead of purifying the protein from the original source, where supplies may be limited or purification difficult. In addition the DNA sequence of cloned DNA can be altered by *in vitro* mutagenesis and the effect of this mutation on subsequent protein properties studied. Amino acid sequences responsible for catalysis or membrane binding for example can be established. This brings us into the era of protein engineering, important not only for research purposes but also in the production of polypeptides with improved properties. An example is subtilisin, a serine protease used in laundry products. Site-directed mutagenesis was used to substitute the methionine at position 222, a site on the protein that was susceptible to oxidation by bleach.[5] The mutant genes were cloned in an expression vector and the products tested. An alanine substitution was effective in making the enzyme far more stable and active in bleach. Protein engineering is

[1] T. V. Huynh, R. A. Young, and R. W. Davis, in 'DNA Cloning, Vol 1; A Practical Approach', ed. D. M. Glover, IRL Press, Oxford and Washington, DC, 1985, p. 45.
[2] O. T. Avery, C. M. Macleod, and M. McCarty, *J. Exp. Med.*, 1944, **79**, 137.
[3] K. Struhl, J. R. Cameron, and R. W. Davis, *Proc. Nat. Acad. Sci. USA*, 1967, **73**, 1471.
[4] D. Vapnek, J. A. Hautala, J. W. Jacobson, N. H. Giles, and S. R. Kushner, *Proc. Natl. Acad. Sci. USA*, 1977, **74**, 3508.
[5] L. J. Abraham, J. Tom, J. Burnier, K. A. Butcher, A. Kossiakoff, and J. A. Wells, *Biochemistry*, 1991, **30**, 4151.

Table 1 *Some human proteins with therapeutic value already approved or in late clinical trials*

Protein	*Application*
growth hormone	pituitary dwarfism
insulin	diabetes
interferon-α 2b	hairy cell leukaemia and genital warts
erythropoietin	anaemia
tissue plasminogen activator	myocardial infarction
interleukin-2	cancer therapy

of great significance to research and development within the pharmaceutical industry, for example, the production of soluble receptors or ligand-binding sites for easier screening or rational drug design.[6,7]

Antibodies are attractive targets for protein engineering.[8] They have obvious potential as therapeutics and their structure is already well understood. Strategies include replacement of parts of the molecule with a toxin. Such fusions would be highly specific in delivering the toxin to target cells. An alternative strategy is to produce bi-specific antibodies that might, for example, bind a tumour cell with one domain and a T killer cell with the other.[9] This subject is discussed more fully in Chapter 14.

The prospect of being able to produce any protein, regardless of origin, in a fermentative organism in culture is a very attractive one, and has obvious commercial implications. Not surprisingly, proteins used in clinical diagnosis or medical treatment have received the most attention. Insulin produced in bacteria is already possible,[10] but this is only the beginning. Vaccine production is an obvious example, but a list of other candidates for early clinical use is given in Table 1. Apart from medically important proteins, enzymes such as proteases used by manufacturing and food industries, could be produced on a large scale by genetically engineered bacteria.[11] Chymosin (or renin), for example, can already be produced in this way for use by the cheese industry.[12,13]

Before the full potential of genetic engineering can be realized, the problems concerned with the expression of DNA from higher eukaryotes in bacteria have to be overcome. As was suggested earlier, DNA from an animal or higher plant cannot be used directly to transform *E. coli.* To overcome the problems and

[6] S. Marullo, C. Delavier-Klutchko, J. G. Guillet, A. Charbit, A. D. Strosberg, and L. J. Emorine, *Bio/Technology*, 1989, **7**, 923.

[7] B. C. Cunningham and J. A. Wells, *Proc. Natl. Acad. Sci. USA*, 1991, **88**, 3407.

[8] A. Pluckthun, *Nature (London)*, 1990, **347**, 497.

[9] J. Berg, E. Lotscher, K. S. Steiner, D. J. Capon, J. Baenziger, H. M. Jack, and M. Wabl, *Proc. Natl. Acad. Sci. USA*, 1991, **88**, 4723.

[10] D. V. Goeddel, D. G. Kleid, F. Bolivar, H. C. Heynecker, D. G. Yansura, R. Crea, T. Hirose, A. Kraszeuski, K. Itakura, and A. D. Riggs, *Proc. Natl. Acad. Sci. USA*, 1979, **76**, 106.

[11] W. Gilbert and L. Villa-Komaroff, *Sci. Am.*, 1980, **242**, 74.

[12] T. Beppu, *Trends Biochem.*, 1983, **1**, 85.

[13] M. L. Green, S. Angal, P. A. Lowe, and F. A. O. Marston, *J. Dairy Res.*, 1985, **52**, 281.

obtain expression of eukaryotic DNA in bacteria, it is necessary to understand as much as possible about the control of gene expression in bacteria and higher organisms. This is discussed in the next section.

2 CONTROL OF GENE EXPRESSION

The difficulties encountered in obtaining expression of genes from higher eukaryotes in bacteria can be explained if the control systems operating in the different cell types are compared. Although the basic machinery of gene action is the same, that is, protein synthesis directed by an RNA copy of a DNA template, there are a significant number of differences between the various types of organism to cause problems for genetic engineers. The mechanisms of gene expression that operate in prokaryotes and eukaryotes are described below and are summarized in Figure 1.

2.1 Prokaryotes

Probably the best understood system of gene control operating in bacteria is the operon model of gene expression originally proposed by Jacob and Monod in 1961.[14] The model, summarized in Figure 1a, is based on the control of genes responsible for the metabolism of lactose, and is an example of a mechanism of gene expression referred to as 'negative control'. The lactose, or *lac*, operon contains three structural genes (*i.e.* DNA sequences coding for structural proteins or enzymes) referred to as 'z', 'y', and 'a', which code for three enzymes involved in lactose metabolism: β-galactosidase to catalyse the cleavage of lactose to galatose plus glucose, lactose permease that facilitates the entry of lactose into the bacterial cell, and thiogalactoside transacetylase that catalyses the transfer of acetyl groups to galactosides.

The three structural genes are transcribed as a single, so-called polycistronic, mRNA which codes for all three proteins. This is a basic feature of the operon model, giving co-ordinated expression of a number of structural genes, in this case the z, y, and a genes. Transcription of the genes is catalysed by DNA-dependent RNA polymerase which binds to DNA at the promoter site. Promoters are specific DNA sequences that do not code for proteins but are essential for transcription of structural genes. Promoters are an important factor in the expression of foreign genes in bacteria and are discussed again later. Transcription begins just 'upstream' of the structural genes at a specific initiation point. The RNA transcript contains a nucleotide sequence at the 5′ end called the Shine Dalgarno (or S–D sequence) that is complementary to, and can therefore base-pair with, RNA in the ribosomes and acts as a ribosome binding site. Translation (protein synthesis) can then proceed. In prokaryotic organisms, protein synthesis begins before RNA synthesis is terminated, that is, transcription and translation are 'coupled'. Downstream of the final structural gene, in this

[14] F. Jacob and J. Monod, *J. Mol. Biol.*, 1961, **3**, 318.

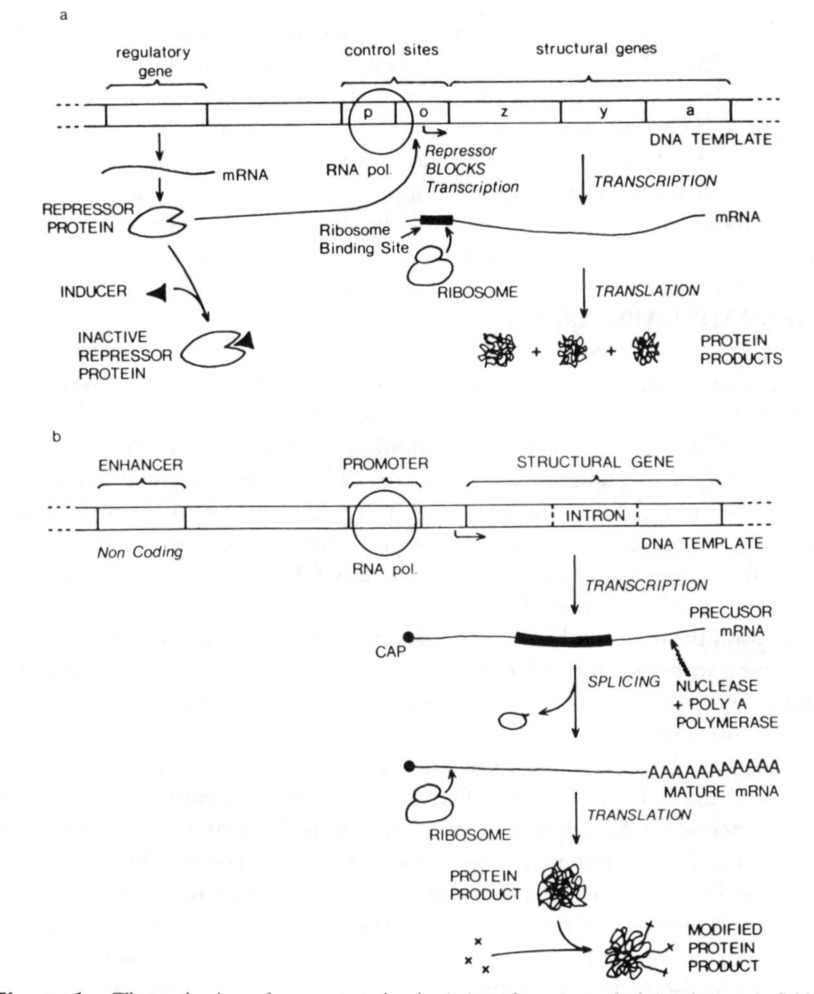

Figure 1 *The mechanism of gene expression in (*a*) prokaryotes and (*b*) eukaryotes: RNA pol, DNA dependent RNA polymerase; P, promoter; O, operator; →, transcription start site; see text for details*

case the 'a' gene, there is a termination signal for transcription, rich in thymidine nucleotides, where the DNA–RNA polymerase complex is unstable.

Transcription of the structural genes is controlled by a regulatory gene, called the 'i' gene in the *lac* operon, which codes for a protein, referred to as a repressor, that binds to a region of DNA adjacent to the promoter, called the operator. Repressor binding blocks the progress of the RNA polymerase and therefore inhibits transcription, preventing synthesis of enzymes encoded by the structural genes. An inducer acts by binding to the repressor. This alters the repressor's three-dimensional structure and prevents it from binding to the operator. In this case the RNA polymerase can continue to transcribe the structural genes coding for lactose metabolism. This system is referred to as a negative control because the protein (repressor) that interacts with the operator inhibits transcription.

Negative control is not the only mechanism of gene control acting in bacteria. There are additional control systems such as positive control and a process called attenuation. In positive control, an inducer binds to a protein that stimulates transcription. The best understood example is the system that operates when glucose is not available as an energy source. In this situation, the intracellular concentration of cAMP rises. The cAMP binds to a DNA-binding protein called CAP (catabolite activator protein) and the resulting complex binds to the promoters associated with operons involved with the breakdown of alternative sugars. The *lac* operon is therefore under both negative (*lac* repressor) and positive (cAMP–CAP) control.

Attenuation is a control system dependent on the coupled transcription/translation mechanism that occurs in bacteria. It is a system of control that operates on genes responsible for amino acid synthesis. When a particular amino acid, such as tryptophan, is in short supply, the ribosomes stall at tryptophan codons on the mRNA. This allows particular secondary structures in the mRNA to form that permit transcription to proceed. If the amino acid concerned is in plentiful supply the ribosome does not stall, a different secondary structure forms in the mRNA and transcription of that gene is prematurely terminated.

It is beyond the scope of this article to describe these methods of gene control in any greater depth. Several molecular biology texts[15–17] are available that give a general overview of the subject and more specific information can be obtained from specialist texts.[18–20]

It can be seen from the model shown in Figure 1a that many separate elements are required to obtain expression of structural genes: a promoter sequence for RNA polymerase binding, a transcription initiation site, a ribosome binding site, and an inducer. The latter can be the naturally occurring inducer, such as allolactose (6-*O*-β-D-galactopyranosyl-D-glucose, an isomer of lactose in which the galactosyl residue is present on the carbon 6 rather than the carbon 4 of glucose and is produced by basal levels of β-galactosidase) or an artificial inducer such as isopropyl-β-D-thiogalactopyranoside (IPTG). Both of these molecules are active in derepressing the *lac* operon. The advantage of artificial inducers in stimulating gene expression is that they may show greater activity. In this case, IPTG, active in its native form, is taken up by cells more readily than lactose and is not degraded by β-galactosidase. The *lac* operon has received so much attention since the operon model for gene expression was first proposed that it has been the obvious choice in the formation of expression vectors, as described below.

[15] D. Freifelder, 'Molecular Biology', 2nd Edn., Jones and Bartlett, Boston, 1987.
[16] B. Lewin, 'Genes IV', Oxford University Press, Oxford, 1990.
[17] J. D. Watson, M. Gilman, J. Witowski, and M. Zoller, 'Recombinant DNA', 2nd Edn., Scientific American Books, Freeman, New York, 1992.
[18] R. E. Glass, 'Gene Function: *E. coli* and its Heritable Elements', Croom Helm, London and Canberra, 1982.
[19] C. F. Higgins, in 'Genetic Engineering, Vol. 5', ed. P. W. J. Rigby, Academic Press, London, 1986, Chapter 1, p. 1.
[20] M. Ptashne, 'A Genetic Switch', 2nd Edn., Cell Press and Blackwell, Palo Alto, 1992.

2.2 Eukaryotes

The organization of genes and the mechanism of gene expression operating in eukaryotes is different in many respects from that in bacteria and is represented in diagrammatic form in Figure 1b. Although our knowledge is expanding at a remarkable rate, less detail is known about the control of gene expression in animals and plants in comparison with bacteria, and the mechanism of selective gene expression during cell differentiation in multicellular organisms, in which all cells contain the same genetic material, is only just beginning to be understood. DNA in higher organisms is maintained in the cell as a complex with histones and other proteins to form a structure referred to as chromatin. Significant alterations in chromatin structure occur during transcription and these changes are associated with gene control mechanisms. Many eukaryotic transcription factors have been isolated and cloned. They interact with specific DNA sequences and each other to initiate transcription in a highly organized and controlled manner. Details are beyond the scope of this chapter but more information can be found in specialist texts.[20,21]

No direct equivalent of the operon model has been found in animals and plants. Polycistronic mRNAs do not appear to exist: each structural gene is transcribed separately with its own promoter and transcription initiation and termination sites. The RNA polymerases of eukaryotes are more complex than in bacteria and exist in three forms, specific for pre-rRNA, tRNA, and 5S rRNA and mRNA synthesis. The promoter elements, while serving the same function as in bacteria, are organized in a different way and do not have the same DNA sequences. There does not appear to be a direct equivalent of the operator and once RNA polymerase has bound, RNA synthesis can begin. RNA polymerase binding is influenced by 'enhancer' elements which are non-coding DNA sequences which are *cis*-acting (*i.e.* influence the expression of genes on the same DNA molecule) and attract RNA polymerase to coding regions by binding transcription factors. The latter are referred to as *trans*-acting factors because the encoding genes can be on a different DNA molecule.[22,23] Following the onset of transcription, the RNA molecule is capped. That is, a 7-methyl guanosine residue is attached by a 5′–5′ phosphate linkage to the end of the RNA. Transcription is terminated and then significant RNA processing occurs prior to translation; it is generally accepted that coupled transcription–translation cannot occur in the nucleus of eukaryotes due to compartmentalization.

RNA processing involves several steps. Capping has already been mentioned, but there is also trimming of the RNA with ribonuclease and addition of between 20–250 adenine nucleotides (the 'poly(A)tail') to the 3′ end. Both of these processes enhance mRNA stability. Perhaps most significantly from the point of view of this article, processing involves the removal of introns. It has been known since the mid-seventies that many eukaryotic genes contain regions of DNA called intervening sequences, or introns, that do not code for an amino acid

[21] D. Latchman, 'Eukaryotic Transcription Factors', Academic Press, San Diego, 1991.
[22] T. Maniatis, S. Godbourn, and J. A. Fischer, *Science*, 1987, **236**, 1237.
[23] 'Transcription and Splicing', ed. B. D. Hames and D. M. Glover, IRL Press, Oxford, 1988.

sequence. The introns considerably increase the length of DNA required to code for a protein; there are many cases known where the total length of introns greatly exceeds the length of coding regions within a gene. The number of introns is highly variable: human globin genes, for example, have two introns, whereas the chicken ovalbumin gene has seven. All introns are transcribed and thus appear in the nascent mRNA. They must be removed from the mRNA molecule before protein synthesis can occur. This is carried out by splicing enzymes, present in the nucleus, that remove the intervening sequences and precisely ligate the coding sequences, or 'exons', back together again.[15–17,23,24] This has to be carried out in a very precise manner to maintain the correct reading frame of triplet codons for protein synthesis (for an explanation of reading frames see Figure 5). Splicing enzymes recognize precise sequences at the exon–intron boundaries. The following consensus sequence has emerged:

```
    C          A
5′  AG | GU   AGU---Y_N NAG | G3′
    A  |       G             |
  exon |          intron     |  exon
```

where Y_N represents a string of about nine pyrimidines and N denotes any base. The splicing enzymes act within large RNA–protein complexes that incorporate small nuclear RNAs that assist in sequence recognition. These complexes are called small nuclear ribonucleoproteins (snRNPs or 'snurps').

The problem from the point of view of genetically engineering bacteria is that prokaryotes do not have introns and therefore possess none of the necessary machinery for their removal. It is not surprising, therefore, that a eukaryotic gene, possessing an intron, cannot be expressed in *E. coli*.

Additional steps involved in the production of a mature protein in eukaryotes, are post-translational modifications such as peptide cleavage, addition of prosthetic group, glycosylation, or formation of multisubunit structures. These are specific to a cell type and are unlikely to be carried out by bacterial cells. This can cause significant problems in the production of complex eukaryotic proteins by genetically engineered organisms and is likely to be a topic of considerable research interest in the future.[25]

This section has attempted to give some background information which is necessary in order to understand the difficulties involved in obtaining expression of foreign genes in bacteria. The remainder of this article will describe how the problems have been overcome and give some examples of the success to date.

3 THE EXPRESSION OF EUKARYOTIC GENES IN BACTERIA

To obtain expression of eukaryotic genes in bacteria such as *E. coli*, the difference in mechanism of gene expression between the original organism from which the gene was obtained and the host bacterium must be overcome. The differences in gene expression mechanisms of particular importance to this discussion are: the

[24] T. Maniatis and R. Reed, *Nature (London)*, 1987, **325**, 673.
[25] R. B. Parekh and T. P. Patel, *Tibtechnology*, 1992, **10**, 276.

presence of introns in eukaryotic DNA, the difference in promoter sequences present in bacteria, animals, and plants, the absence of a ribosome binding site (Shine–Dalgarno sequence) on eukaryotic mRNA, preferential use of specific triplet codons in coding sequences, and, in many cases, the requirement for post-translation modification before the polypeptide is fully functional. The methods used to obviate these difficulties are discussed below.

3.1 Introns

It is apparent from the earlier discussion that a native eukaryotic gene cannot be expressed in bacteria when introns are present. There are two ways in which this problem can be overcome. Firstly, double-stranded DNA copies of mRNA molecules, referred to as complementary DNA or cDNA, can be generated by the use of an mRNA template and reverse transcriptase. This is a viral enzyme that produces a single strand of DNA, complementary in nucleotide sequence to the mRNA. The steps required to produce a double-stranded DNA molecule ready for cloning are described in Chapter 2.

The double-stranded cDNA molecule will not contain introns and can act as the coding sequence in expression vectors. There are, however, problems with the cDNA approach: if the mRNA is only present as a small constituent of a eukaryotic cell's RNA population, purification of the mRNA can be difficult; and the cDNA sequence synthesized by reverse transcriptase does not always include the 5′ end of the gene, random termination of reverse transcription, prior to completion of complementary strand synthesis can occur. The latter point is a serious problem but has been improved with refined cDNA synthesis approaches. Following synthesis the cDNA is tailored to the expression vector by the addition of restriction enzyme linkers and can thus be cloned.

A second approach which solves the intron problem is to synthesize the gene by chemical means without a template. If the amino acid sequence of the desired protein is known, it is possible to chemically synthesize a DNA molecule with the necessary sequence.[26–29] The advantages of this technique over the cDNA approach are considerable: the complete sequence is obtained, the DNA can be tailored to the vector as desired, and particular codons, preferred by the organism chosen as a host for the expression vector, can be incorporated into the gene. There is no theoretical limit to the size of DNA that can be synthesized, but in practice large genes must be synthesized as fragments which are subsequently ligated.

[26] M. H. Caruthers, S. L. Beaucage, C. Becker, W. Efcavitch, E. F. Fisher, G. Galluppi, R. Coldman, P. Dettaseth, F. Martin, M. Matteucci, and Y. Stabinsley, in 'Genetic Engineering', ed J. K. Setlow and A. Hollaender, Plenum Press, New York and London, 1982, p. 119.
[27] S. Narang, *J. Biosci.*, 1984, **6**, 739.
[28] M. D. Edge, A. R. Green, G.R. Heathcliffe, P. A. Meacock, W. Shuch, D. B. Scanlon, T. C. Atkinson, C. R. Newton, and A. F. Markham, *Nature (London)*, 1981, **191**, 756.
[29] D. G. Yansura and D. J. Henner, in 'Gene Expression Technology', *Methods Enzymol.*, ed. D. V. Goeddel, Academic Press, San Diego, 1990, **185**, pp. 54–60.

	"-35 sequence"		"Pribnow box"
p*trp*	TTGACA	--17bp --	TTAACTA - - transcription
p*lac* uv5	TTTACA	--18bp --	TATAATG - - transcription
p*tac*	TTGACA	--16bp --	TATAATG - - transcription
prokaryotic			
consensus	TTGACA		TATAAT
human β globin	CCAAT	--39bp --	CATAAA - - transcription

Figure 2 *DNA sequence of some characterized promoters*

3.2 Promoters

Promoters are sequences of DNA that are necessary for transcription. In *E. coli* the RNA polymerase recognizes the promoter as the first step in RNA synthesis. A similar system operates for the transcription of mRNA in eukaryotes but analysis of promoter sequences in bacteria and eukaryotes shows that there are important differences. The nucleotide sequence of some characterized promoters is shown in Figure 2. There is a marked similarity between the various promoter sequences found in *E. coli* represented by the consensus sequence given in Figure 2. The promoters for mRNA synthesis in eukaryotes, however, although similar in principle are more complex, show less overall similarity, and are not recognized by bacterial RNA polymerase. The important sequences for *E. coli* RNA polymerase are the TTGACA ('-35 sequence') and TATAAT ('Pribnow box') sequences found 35 and 10 base pairs upstream of the transcription initiation point respectively.

In order to obtain expression of a eukaryotic DNA sequence in *E. coli* it is necessary to place the coding sequence downstream from a bacterial promoter. The most commonly used promoters are those from the *lac* or *trp* operons (or a combination of the two called a '*tac*' promoter), the *rec*A promoter of *E. coli*, and the late promoter (λ P_L) from phage λ. Many vectors constructed using the *lac* promoter are based on the principle of using the entire lac control region and the first few nucleotides of the 'z' gene coding for β-galactosidase. The foreign DNA to be expressed is then inserted downstream of the β-galactosidase N terminal coding sequence giving rise to a fusion gene, discussed in more detail later. One of the effects of this is to maintain the control mechanism that normally operates on the *lac* operon, thus expression from the promoter is regulated by the *lac* repressor. Transcription, or derepression, of the gene is then brought about by the addition of an inducer such as IPTG. Similarly, expression vectors based on the *trp* promoter incorporate control regions normally in operation for the *trp* operon. The *rec*A promoter is induced by nalidixic acid. The λ P_L promoter is used in specially designed host cells that produce a temperature-sensitive repressor. Expression from this promoter is activated by raising the temperature of the culture.

Figure 3 *Base pairing between the Shine–Dalgarno sequence (in bold) on the mRNA and a complementary region on the 3′ end of 16S rRNA*

3.3 Ribosome Binding Site

The initiation of translation can be a significant limiting factor in expression of cloned genes.[30] An initiation codon, AUG is required, but other precise nucleotides particularly in the 5′ untranslated leader of the mRNA, are needed to facilitate suitable secondary and tertiary structures in mRNA and interaction between mRNA and the ribosome. Perhaps the best known of these sequences is the Shine–Dalgarno sequence (or S–D sequence or RBS: ribosome binding site). This is a stretch of 3–9 bases lying between 3 and 12 bases upstream from the AUG codon. It allows a complex to form between the mRNA and the 30S subunit of the ribosome via hydrogen bonding to the 16S rRNA. Not all *E. coli* mRNAs have an identical S–D sequence but a consensus can be identified (Figure 3). An S–D sequence is essential for translation, other sequences that help boost translation level include translation 'enhancers' (named after transcriptional enhancers because their sequence appears more important than their precise location) such as the Epsilon sequence in the g10-L ribosome binding site of bacteriophage T7.[31,32]

To obtain expression of foreign genes in *E. coli* it is necessary to incorporate ribosome binding motifs into the recombinant DNA molecule. Furthermore some sequences (such as the S–D sequence) must be located at an optimal distance from the translation start codon. This is most readily achieved by construction of fusion genes where an entire untranslated leader and 5′ coding sequence from a naturally-occurring gene is present. Nonetheless, all expression cassettes need to be tested thoroughly and sequences reorganized if necessary to optimize translation initiation.

3.4 Expression of Foreign DNA as Fusion Proteins

The problems associated with procuring a prokaryotic promoter and S–D sequence to obtain expression of eukaryotic DNA in *E. coli* can be obviated by constructing a fusion gene. The control region and *N*-terminal coding sequence of an *E. coli* gene is ligated to the coding sequence of interest. When introduced

[30] H. A. de Boer and A. S. Hui, in 'Gene Expression Technology', *Methods Enzymol.*, ed. D. V. Goeddel, Academic Press, San Diego, 1990, **185**, p. 103.

[31] P. O. Olins and S. H. Rangwala, in 'Gene Expression Technology', *Methods Enzymol.*, ed. D. V. Goeddel, Academic Press, San Diego, 1990, **185**, p. 115.

[32] P. O. Olins, C. S. Devine, and S. H. Rangwala, in 'Expression Systems and Processes for rDNA Products', ed. R. T. Hatch, C. Goochee, A. Moreira, and Y. Alroy, ACS, Washington, DC, 1991, p. 17.

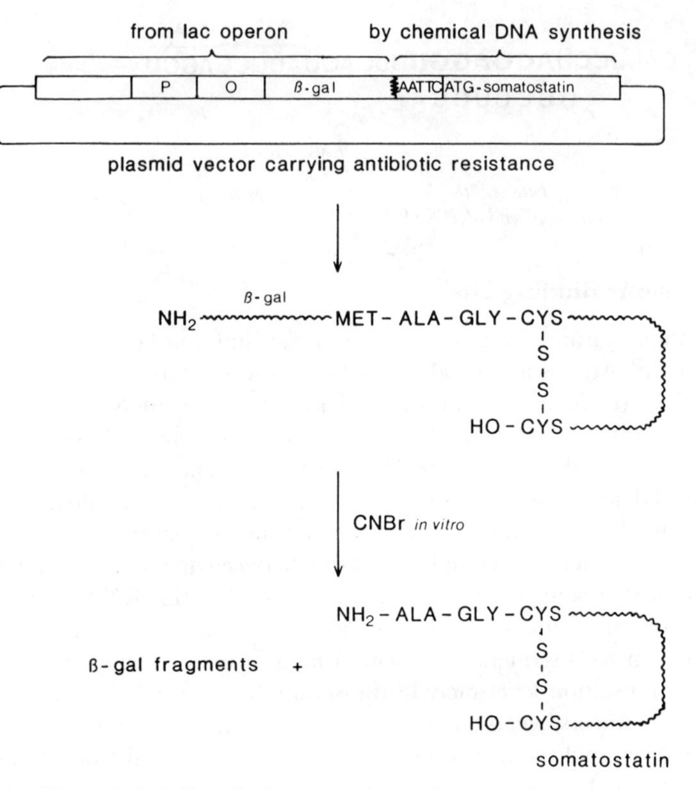

Figure 4 *The synthesis in* E. coli *of somatostatin as a fusion protein*

and cloned in *E. coli* RNA polymerase will recognize the promoter as native and will transcribe the gene. The 5′ end of the mRNA is also native and will consequently interact normally with a ribosome to commence protein synthesis. The protein that results will be chimaeric, the *N*- and *C*-terminals being derived from the prokaryotic and eukaryotic genes respectively. There are a number of advantages in taking this approach. Firstly, expression of the foreign DNA should be efficient. Secondly, the foreign gene can be placed under the control of the induction/repression system of the *E. coli* promoter/operator used. Thirdly, with some constructs, the fusion peptide may be exported to the periplasmic space via the signal sequence (discussed later) and fourthly, the protein should be relatively stable. This last point is important and worthy of some discussion. For reasons not fully understood, foreign proteins, particularly short peptides, in *E. coli* are recognized as such and are broken down by endogenous proteases. Foreign proteins expressed as fusion proteins, however, appear to be more stable. The *N*-terminal part of the chimera is recognized as 'self' by the cell. In this respect, the length of the *N*-terminal sequence coded for by *E. coli* DNA is important, the longer it is the more likely the fusion product is to be stable. Peptides shorter than about 80 amino acids are best synthesized as fusion proteins.

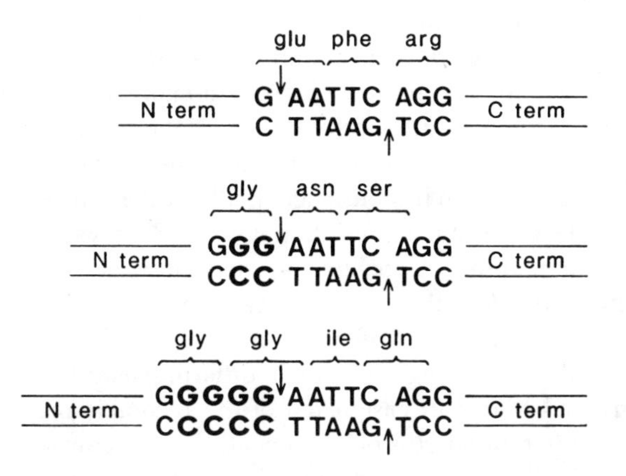

Figure 5 *The reading frame, based on triplet codons, of a fusion gene, constructed on an EcoR1 site (arrows), can be altered by the insertion of additional G–C pairs (bold)*

The majority of fusion genes that have been created for expression of foreign DNA in *E. coli* are based on the *lac* operon, using the *N*-terminal sequence of β-galactosidase to form the fusion peptide. The first example of this approach was to produce somatostatin in *E. coli*[33] by the approach illustrated in Figure 4. Somatostatin is a peptide hormone of 14 amino acids with the physiological role of inhibiting secretion of growth hormone, glucagon, and insulin. The coding sequence for somatostatin was obtained by chemical synthesis, based on knowledge of the somatostatin amino acid sequence. The DNA was constructed in such a way as to incorporate codons which are preferentially used in *E. coli*, and to leave single-stranded projections, corresponding to the cohesive ends produced by *Eco*R1 and *Bam*H1 digestion, at each end. The coding sequence was preceded by a codon for methionine and terminated by a pair of nonsense codons to stop translation.

The initial hybrid genes that were constructed should have produced a hybrid protein containing the first seven amino acids of β-galactosidase, but no somatostatin-like proteins could be detected. When an alternative hybrid gene was created by inserting the synthetic somatostatin gene at an *Eco*R1 site near the *C*-terminus of β-galactosidase, a stable fusion of this large β-galactosidase fragment and the somatostatin was synthesized. This illustrates very well the importance of fusion genes in maintaining protein stability.

The presence of the methionine residue at the *N*-terminus of the somatostatin amino acid sequence allows for the purification of somatostatin from the fusion protein. Cyanogen bromide treatment *in vitro* cleaves peptides at the carboxyl side of methionine residues. This process, of course, is only applicable to the production of eukaryotic proteins not containing internal methionines.

A foreign gene will only be expressed as a fusion protein if it is placed in the correct translational reading frame (Figure 5) since codons are based on triplets

[33] K. Itakura, T. Hirose, R. Crea, A. D. Riggs, H. L. Heyneker, F. Bolivar, and W. H. Boyer, *Science*, 1977, **198**, 1056.

of nucleotides there is only a one in three chance that two randomly selected coding sequences will be ligated in phase and this assumes that they are correctly oriented (a further one in two chance). To obtain expression, the foreign DNA can be ligated into a position, known by the DNA sequence to be in the exact reading frame, as in the somatostatin experiment described above, or several recombinant molecules need to be constructed, all in different reading frames, to enable selection of the successful clone. There are different ways of achieving this second approach; for example, successive additions of two G–C base pairs to the *Eco* R1 fragment of the *lac* 'z' gene has created vectors with three different reading frames.[34] Alternatively, infection of foreign DNA can employ the method of homopolymer tailing, in which differing lengths of a linker are randomly constructed.[35] In this case, each of the recombinant molecules constructed will have different lengths of the repeating G–C pair and at least some (one in six) should be in the correct reading frame and orientation.

The fusion peptide approach has been used in a considerable number of cases following the somatostatin experiment, for example in the synthesis of thymosin,[36] neo-endorphin,[37] and human insulin.[10] In the latter case the initial approach was to synthesize the A and B chains as separate fusion products. This was later refined to a strategy where a single β-galactosidase–insulin fusion was cleaved in one step.

In all these cases the desired peptide was prepared from the fusion peptide by CNBr cleavage which depends on the absence of methionine within the required polypeptides. The procedure, however, is not the only method that has been used for the cleavage of fusion peptides. β-Endorphin was synthesized as a β-galactosidase fusion protein in *E. coli*.[38] In this case, because of the presence of internal methionine residues in β-endorphin, the native hormone was prepared by citraconylation and trypsin treatment. For most larger proteins, however, it is not possible to remove the β-galactosidase residues, and the fusion protein is the final product.

Not all fusion genes are based on the *lac* operon. For example, fusion genes have also been constructed using the β-lactamase gene carried in plasmid pBR322. Although expression from this promoter is not particularly efficient, its use is of interest because β-lactamase is a secretory protein, responsible for conferring ampicillin resistance, and carries a signal sequence. Signal peptides are sequences of amino acids at the *N*-terminus of proteins; they have an affinity for the cell membrane and are responsible for the export of proteins that carry them; they are cleaved by enzymes present in target organelles of eukaryotes or the periplasmic space in bacteria. Fusion genes, such as β-lactamase–proinsulin hybrid, have been constructed at the unique *Pst* 1 restriction site located between

[34] P. Charney, M. Perricaudet, F. Galibert, and P. Tiollais, *Nucleic Acid Res.*, 1978, **5**, 4479.
[35] L. Villa-Komaroff, A. Efstratiadas, S. Broome, P. Lomedico, R. Tizard, S. P. Naber, W. L. Chick, and W. Gilbert, *Proc. Natl. Acad. Sci. USA*, 1978, **75**, 3727.
[36] R. Wetzel, H. L. Heyneker, D. V. Goeddel, G. B. Thurman, and A. L. Goldstein, *Biochemistry*, 1980, **19**, 6096.
[37] S. Tanaka, T. Oshima, K. Ohsue, T. Ono, S. Oikawa, I. Takano, T. Noguchi, K. Kangawa, N. Minamino, and H. Matsuo, *Nucleic Acid Res.*, 1982, **10**, 1741.
[38] J. Shine, I. Fettes, N. C. Y. Lan, J. L. Roberts, and J. D. Baxter, *Nature (London)*, 1980, **285**, 456.

codons 183 and 184 of the β-lactamase coding sequence.[35] The resulting chimaeric proteins contain a substantial proportion of the β-lactamase amino acid sequence and they are, not surprisingly therefore, often found in the periplasmic space.

An alternative approach is to produce expression vectors incorporating a sequence from the gene for haemolysin.[39] This is a protein product of some pathogenic *E. coli*: it is secreted by the cells via a signal at the *C*-terminus of the protein. Ligation of this signal onto mammalian prochymosin and immunoglobulin genes has resulted in successful secretion of fusion products.

Clearly, secretion of recombinant protein has considerable advantages for production on a commercial scale: extraction and purification are simpler, unwanted *N*-terminal methionines are removed, and proteins fold better. Secretion is one of the reasons for the interest in *Bacillus subtilis* as a host organism (see Section 6).

3.5 Expression of Native Proteins

It is possible to obtain expression of native proteins using a nucleotide sequence that only codes for the peptide required and contains no codons from a prokaryotic gene. The advantage of this approach is that the amino acid sequence produced should be identical to the naturally occurring eukaryotic protein and therefore should exhibit full biological activity without the need to remove a fusion peptide; a difficult if not impossible task in most cases. Direct expression is best suited to proteins between 100 and 300 amino acids, without a great number of cysteines.

To obtain direct expression of a native protein, it is necessary to place a coding sequence, with an ATG translation initiation codon, downstream from a bacterial promoter and ribosome binding site. There will be difficulties if a cDNA is being used for the coding sequence. Depending on the nature of the mRNA template used, there may be a long leader sequence upstream of the ATG codon. As it is most unlikely that there will be a restriction site in exactly the right position to allow this leader to be removed and the coding sequences joined to the vector perceiving the correct SD–AUG spacing, a variety of methods have had to be developed. The human growth hormone gene has been successfully expressed in bacteria using the procedure[40] as outlined in Figure 6. The methods used are a good example of where many different techniques or approaches can be combined together in one experiment. The human growth hormone (HGH) gene was rather too long for complete synthesis by chemical means. The bulk of the coding sequence was therefore obtained by cDNA synthesis and cloning. The cDNA was then cut with *Hae* III to give a defined length lacking the mammalian *N*-terminal signal sequence which would not operate in *E. coli*. The fragment was tailored to the expression vector using a chemically synthesized fragment that

[39] L. Gray, K. Baker, B. Kenny, N. Mackman, R. Haugh, and I. B. Holland, in John Innes Symposium 'Protein Targeting', ed. K. Charter, *J. Cell Science*, Supplement, 1989.

[40] D. V. Goeddel, H. L. Heyneker, T. Hozumi, R. Arentzen, K. Itakura, D. G. Yansura, M. J. Ross, G. Miozzari, R. Crea, and P. H. Seeburg, *Nature (London)*, 1979, **281**, 544.

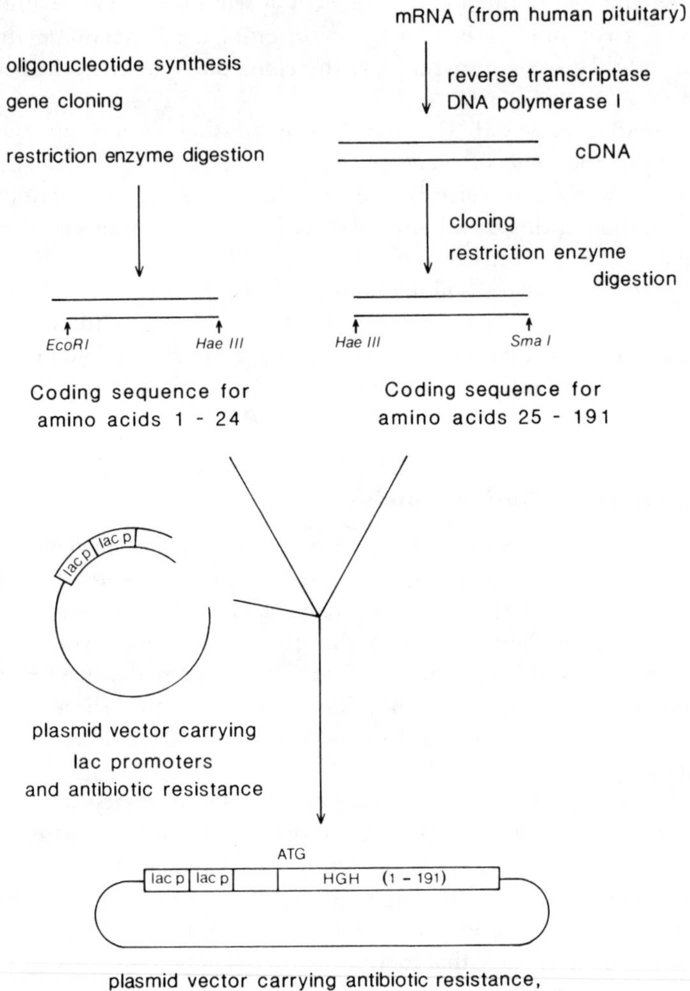

Figure 6 *Construction of an expression vector that directs the synthesis of human growth hormone (HGH) in* E. coli

coded for the first 24 amino acids of HGH and combined an *Eco* R1 site for attachment to the plasmid vector. The cDNA and synthetic fragments were ligated together and inserted via *Eco* R1 and *Sma* 1 sites into an expression vector containing two copies of the *lac* promoter. The resulting plasmid contained a ribosome binding sequence, AGGA, eleven base pairs upstream from the ATG initiation codon for the HGH gene. In the *lac* operon, the Shine–Dalgarno sequence lies seven base pairs upstream of the β-galactosidase initiation codon. A derivative of the original expression vector was therefore constructed in which four base pairs between the AGGA and ATG sequences were removed. This was achieved by opening the plasmid with *Eco* R1 and digesting the single-stranded tails with S1 nuclease and religating the blunt ends. Surprisingly, this new plasmid produced less HGH when introduced into *E. coli* than the original construct, containing the full 11 base pair sequence that had been deliberately shortened. This illustrates the subtleties involved in the relationship between the leader sequence and the initiator codon in protein synthesis and is discussed further in Section 5.

The HGH produced as above in *E. coli* is a soluble protein and can be readily purified. It has the same biological activity as the HGH from human pituitary and apparently differs in only one respect: the presence of an extra methionine residue at the *N*-terminus.[41] This amino acid would normally be removed by enzymes in the pituitary gland. The presence of this additional methionine does not interfere with the protein's biological activity but, ideally, bacterial products that are likely to be used for clinical purposes need to be as close as possible in structure to their natural counterparts, to avoid complications such as reaction by a patient's immune system.

The hormone can be produced in bacteria without the extra methionine by cloning the coding sequence next to a bacterial signal sequence. This specifies secretion of the protein to the periplasmic space between the inner and outer membranes of the bacterial cell. A periplasmic protease cleaves the signal sequence including the methionine leaving native HGH which can be extracted by a hypotonic periplasmic shock.

Human growth hormone was the first of many genes to be expressed as a native protein. Those that quickly followed include human leucocyte interferon (LeIF-A), murine, and human prolactin and human interleukin, cloned downstream from a *trp* promoter,[29,40] human fibroblast interferon, using both the *trp* and *lac* promoter,[42] human perproinsulin downstream from a *lac* promoter,[27] and mouse dihydrofolate reductase, using β-lactamase promoter.[43] The approach is now the one of choice for the synthesis of medium-sized proteins.

4 DETECTING EXPRESSION OF FOREIGN GENES

Using the methods of classical microbial genetics, it is relatively simple to detect expression of foreign prokaryotic genes in host organisms. A host is chosen for a transformation experiment that carries a particular mutant, such as a deficiency in the synthesis of histidine. The mutant would therefore be a *his* auxotroph which would normally require a histidine supplement in its growth medium. If the foreign gene to be introduced into the cells codes for histidine synthesis (his^+) successful transformants can be detected by their ability to grow without the histidine supplement, *i.e.* they can be 'selected'.

This relatively simple technique, however, cannot generally be employed

[41] D. V. Goeddel, E. Yelverton, A. Ullrich, H. L. Heyneker, G. Miozzari, W. Holmes, P. H. Seeburg, T. Dull, L. May, N. Stebbing, R. Crea, S. Maeda, N. McCandliss, A. Sloma, J. M. Tabar, M. Cross, P. C. Familletti, and S. Pestka, *Nature (London)*, 1980, **287**, 411.

[42] T. Taniguchi, L. Guarente, T. M. Roberts, D. Kimelman, J. Douhan, and M. Ptashne, *Proc. Natl. Acad. Sci. USA*, 1980, **77**, 5230.

[43] A. C. Y. Chang, H. A. Erlich, R. P. Gunsalas, J. H. Nunberg, R. J. Kaufman, R. T. Schimke, and S. N. Cohen, *Proc. Natl. Acad. Sci. USA*, 1980, **77**, 1442.

when expression of foreign eukaryotic DNA is desired. The reason is simple: it is unlikely that expression of a eukaryotic gene will confer an advantage to its bacterial host. Insulin synthesis, for example, is hardly likely to enhance the growth advantage of an *E. coli* cell. Alternative methods are required to detect expression of eukaryotic genes. If the function of the desired gene is known, a suitable test can be developed. This is relatively straightforward if the protein required is an enzyme, an assay can be applied to a host-cell extract although renaturation may be required. If, however, the desired protein is not an enzyme a more complex test is necessary. These tests usually employ immunodetection techniques.[1,44] For example, bacterial colonies, suspected of containing the desired protein, are grown on cellulose nitrate paper as a replica of a master agar plate. The colonies are lysed, their contents bound to the cellulose nitrate and the filter incubated with a solution of radioactively-labelled antibody, specific to the protein required. Following thorough washing of the filter and autoradiography, colonies containing the required protein can be identified. The relevant colony or colonies on the master plate can then be selected and cultured. Alternatively, proteins can be extracted and separated on a polyacrylamide gel. Western blotting is then used to transfer the protein to a membrane for immunodetection.

Detecting expression of foreign eukaryotic genes where there is no enzyme assay, or when an antibody is not available, is more difficult. In this case, novel proteins over and above the natural background of host proteins need to be detected by protein separation techniques such as polyacrylamide gel electrophoresis.[29] If expression levels are low, as was the case in early experiments a system is required to lower the background. Such systems include the use of mini-cells, maxi-cells, or a coupled transcription–translation system, discussed elsewhere.[45–47]

5 MAXIMIZING EXPRESSION OF FOREIGN DNA

Until now this article has been concerned with describing the principles and techniques involved in obtaining detectable levels of expression of foreign genes in bacteria. Commercial applications of genetic engineering, however, depend on obtaining high levels of expression such that production of, for example, hormones or vaccines is a realistic economic proposition. The number of cases where genuinely high levels of expression have been obtained is relatively small. Not surprisingly, the achievement of high production levels is a result of considerable research effort and investment. Some of the best examples are the production of insulin, growth hormone, and interferon with levels of expression approaching

[44] S. Broome and W. Gilbert, *Proc. Natl. Acad. Sci. USA*, 1978, **75**, 2746.
[45] B. Oudega and F. R. Mooi, in 'Techniques in Molecular Biology', ed. J. M. Walker and W. Gaastra, Croom Helm, London and Canberra, 1983, Chapter 13, p. 239.
[46] J. M. Pratt, in 'Transcription and Translation, A Practical Approach', ed. B. D. Hames and S. J. Higgins, IRL Press, Oxford and Washington, DC, 1984, Chapter 7, p. 179.
[47] R.J. Slater, in 'Techniques in Molecular Biology, Vol 2', ed. J. M. Walker and W. Gaastra, Croom Helm, London and Canberra, 1987, Chapter 12, p. 203.

10^5–10^6 molecules per cell.[48] In the case of insulin this is equivalent to nearly 40 mg of product per 100 g wet weight of cells.

Several factors can be identified as being involved in influencing the level of expression.[49] These include: promoter strength, codon usage, secondary structure of mRNA in relation to position of a ribosome binding site, efficiency of transcription termination, plasmid copy number and stability, and the host cell physiology. These factors are discussed below.

Optimal promoters need not necessarily be the naturally occurring ones. Sequences nearest to the consensus sequence are the most efficient, as illustrated by experiments employing the *tac* promoter[50] a hybrid between the *lac* and *trp* promoters (Figure 2). Mutations can be made in promoters to alter their characteristics and thereby influence expression. For example, L8 and uv5 mutations in the *lac* promoter render the promoter insensitive to catabolite repression and improve RNA polymerase binding respectively.[51]

Codon usage influences levels of expression and can be accommodated if genes are synthesized chemically. The genetic code is degenerate, so for many amino acids there is more than one codon. Efficiency is related to the abundance of tRNA in the cell and the codon–anticodon interaction energy.[52,53]

The distances between the promoter, ribosome binding site, and ATG initiation codon can have a profound effect on levels of expression; greater than 2000-fold differences have been recorded. This is probably due to different secondary structures forming in the mRNA following transcription. It appears that to obtain high levels of expression the initiation codon (AUG) and the ribosome binding regions need to be present as single-stranded structures. Secondary structures can be predicted with computer analysis based on the maximum possible changes in free energy associated with folding of an RNA chain, but this is not easy to apply. The most pragmatic approach to the problem, therefore, is to construct a series of vectors with different distances, for example, between the S–D and ATG sequences. A series of clones can then be screened for levels of expression and the optimum selected. 'Cassette' vectors have now been available for several years, all the relevant signals are optimally situated with a unique restriction site for the inclusion of a coding sequence.[54]

Screening can be difficult if there is no convenient assay for the desired product. A novel solution is to produce a recombinant DNA molecule containing the desired sequence upstream from the carboxy terminal sequence of β-galactosidase placed in the same reading frame. In this case production of a fusion protein can be detected by its β-galactosidase activity. The most suitable of several clones, incorporating varying distances between the S–D and ATG

[48] B. R. Click and G. K. Whitney, *J. Ind. Microbiol.*, 1987, **1**, 277.
[49] S. J. Coppella, G. F. Payne, and N. Dela Cruz, in 'Expression Systems and Processes for rDNA Products', ed. R. T. Hatch, C. Goochee, A. Moreira, and Y. Alroy, ACS, Washington, DC, 1991, Chapter 1, p. 1.
[50] H. A. de Boer, L. J. Comstock, and M. Vasser, *Proc. Natl. Acad. Sci. USA*, 1983, **80**, 21.
[51] F. Fuller, *Gene*, 1982, **19**, 43.
[52] M. Gouy and C. Gautier, *Nucleic Acid Res.*, 1982, **10**, 7055.
[53] J. Grosjean and W. Fiers, *Gene*, 1982, **18**, 199.
[54] N. Panayotatos, in 'Plasmids: A Practical Approach', ed. K. G. Hardy, IRL Press, Oxford and Washington, DC, 1987, p. 163.

sequences, can then be selected and the carboxy terminal of β-galactosidase removed to leave the required gene sequence as desired.[30,55]

Levels of expression are also influenced by transcription terminator (A–T rich) sequences which must be included at the end of coding regions. Read-through of the RNA polymerase may interfere with other genes downstream or may produce unnecessarily long mRNA molecules that could have reduced translation efficiency and be an undue strain on the cell's energy resources.[56]

Plasmid copy number and stability are important factors in successfully exploiting genetic engineering.[49,56] There is little to be gained from constructing expression vectors that are lost during large scale fermentation. Losses of plasmids occur by several means: a slow rate of plasmid replication compared with cell division, insertion or deletion events in regions of a plasmid necessary for replication, inefficient partitioning at cell division, and plasmid multimerization. The stability and copy number of plasmids in bacteria is dependent on the plasmid construction, growth conditions, such as temperature, and growth rate. Plasmid stability can be maintained by antibiotic selection ('active selection') but this may be undesirable during mass production because of costs and waste disposal problems. The plasmid pBR322, one of the original cloning vectors, segregates randomly during division but naturally occurring plasmids contain a partitioning function, *par*, which ensures segregation at cell division. Incorporation of features such as *par* regions into expression vectors is advantageous to large-scale production.[49,57] Alternatively, genes could be cloned in expression vectors that confer an advantage to cells carrying the plasmid; this is called passive selection. Such a gene, for example *valS*, might code for an essential function (in this case a tRNA synthetase) deficient in the host chromosome.[58]

Once expression of foreign DNA has been successfully achieved, the host organism will need to be maintained in culture for as long as required. The growth conditions that produce optimum levels of expression at maximum stability for the least expenditure are required. Effects of batch or continuous culture, choice of growth medium, *etc.* need to be considered. The choice of media can have significant effects on the production of recombinant protein. For example, conditions that are required to select for maintenance of the plasmid may be poor for synthesis of high levels of product. Alternatively, conditions optimal for production may be limiting cell growth. In short, media can affect the production rate, yield, and secretion efficiency.[59]

Further problems are apparent even after the protein is synthesized. Frequently it has been found that foreign proteins expressed at high levels are deposited into insoluble inclusion bodies. Extraction then involves denaturing agents such as thiourea (hardly conducive to the extraction of biologically active products) followed by protein purification and refolding. Redox couples such as

[55] L. Guarente, G. Lauer, T. M. Roberts, and M. Ptashne, *Cell*, 1980, **20**, 543.
[56] P. Balbas and F. Bolivar, in 'Gene Expression Technology', *Methods Enzymol.*, ed. D. V. Goeddel, Academic Press, San Diego, 1990, **185**, p. 14.
[57] G. Skogman, J. Nilsson, and P. Gustafsson, *Gene*, 1983, **23**, 105.
[58] J. Nilsson and S. G. Skogman, *Bio/Technology*, 1986, **4**, 901.
[59] T. Kohno, D. F. Carmichael, A. Sommer, and R. C. Thompson, in 'Gene Expression Technology', *Methods Enzymol.*, ed. D. V. Goeddel, Academic Press, San Diego, 1990, **185**, p. 187.

a mixture of oxidized and reduced glutathione can be used to form the disulfide bridges within the protein. Furthermore, foreign proteins can be partially degraded by protease.[60] Such degradation products are potentially immunogenic and present a problem in the production of therapeutics.

6 ALTERNATIVE HOST ORGANISMS

The commercial exploitation of genetically engineered bacteria will involve, in most cases, the purification of the foreign product. This is best achieved using a host that is capable of secreting protein products into the growth medium. *Bacillus subtilis* is such an organism: it is a non-pathogenic, gram-positive bacterium, and is amenable to large-scale fermentation technology.[61] It has a well developed natural transformation system (*i.e.* for taking up linear fragments of homologous DNA) which was invaluable in early genetic analysis. Importantly, if *B. subtilis* were to be a host cell for the synthesis of products relevant to the food and beverage industries, it has the official status of being 'generally regarded as safe' (GRAS) accorded by the US Food and Drug Administration.

Despite the attractions the applications of *B. subtilis* for the production of recombinant protein has been limited. There are some difficulties which must be overcome. Our fundamental knowledge and application of molecular biology is behind that of *E. coli*. Nothing equivalent to the λ vectors exists for example. An improvement, however, has been the production of shuttle vectors that can be used both in *E. coli* and *B. subtilis*. Promoters and ribosome binding sites show subtle differences to *E. coli*, in other words it is not possible to simply use expression vectors already developed for this organism in *B. subtilis*. However, constitutive and controllable promoters have been developed for *B. subtilis*, for example, from growth-phase dependent genes for *Bacillus* α-amylase and protease. An alternative approach has been to create hybrid systems such as a *Bacillus* penicillinase promoter linked to the *E. coli lac* operator. This system requires a host cell containing the *lac* repressor gene under the control of a *Bacillus* promoter. Expression of foreign proteins is controlled by IPTG as with the *E. coli* systems.

A disadvantage of *B. subtilis* is that cells produce several different extracellular proteases which inevitably cause degradation of secreted foreign proteins.[60,61] Much work has gone into producing strains with reduced levels of this protease activity. Another limitation on *B. subtilis* is our poor understanding of the secretion pathway. The basic mechanism involves a signal sequence present on pre-proteins but it has been observed that some foreign proteins are exported far more efficiently than others even if they contain the same pre sequence. It is not clear at which stage transport is blocked.

Despite these difficulties, *B. subtilis* has great potential and already there are cases of high levels of expression of foreign proteins. These include α-interferon, growth hormone, epidermal growth factor, and pepsinogen secreted to levels of 15, 200, 240, and 500 mg dm^{-3} of medium respectively.[61]

[60] S-O. Enfors, *Tibtechnology*, 1992, **10**, 310.

[61] C. R. Harwood, *Tibtechnology*, 1992, **10**, 247.

7 FUTURE PROSPECTS

Clearly there are considerable commercial and scientific opportunities in this area of molecular biology. Many of the basic techniques required to obtain expression of any DNA sequence in any host organism are now available and levels of expression are commercially viable. Future work is likely to concentrate on maximizing expression, characterizing the best and most cost-effective growth conditions for mass production, the synthesis of novel, genetically engineered proteins, such as designer antibodies or receptor fragments for drug development, and the further development of alternative host organisms, particularly *B. subtilis*. As discussed earlier, however, bacteria are not capable of carrying out many of the functions required to produce protein altered by post-translational modifications. In this respect eukaryotic hosts have the advantage. Their use is discussed in the following chapters.

8 FURTHER READING

R. W. Old and S. B. Primrose, 'Principles of Gene Manipulation: An Introduction to Genetic Engineering', 4th Edn., Blackwell, Oxford, 1989.

J. D. Watson, M. Gilman, J. Witkowski, and M. Zoller, 'Recombinant DNA', 2nd Edn., Freeman, New York, 1992.

'Systems for Heterologous Gene Expression', in *Methods Enzymol.*, ed. D. V. Geoddel, Academic Press, New York, 1990, **185**.

CHAPTER 5

Yeast Cloning and Biotechnology

B. P. G. CURRAN AND V. C. BUGEJA

The yeast *Saccharomyces cerevisiae* has been increasingly employed over the past ten years for the production of heterologous proteins. *S. cerevisiae* was developed as the first eukaryotic expression system because as a unicellular organism it could be genetically manipulated using many of the techniques commonly used for *Escherichia coli* and as a eukaryote it was a more suitable host for the production of authentically processed eukaryotic proteins. *S. cerevisiae* was also acceptable as safe for use in the production of recombinant pharmaceutical products because unlike *E. coli* it did not produce pyrogens or endotoxins and it had a long history of safe use in large scale fermentation processes in the brewing industry.

S. cerevisiae's amenability to recombinant DNA manipulation coupled with its well characterized physiology, biochemistry, and genetics has made this organism a cornerstone of contemporary biotechnology.

1 TRANSFORMATION IN *S. CEREVISIAE*

1.1 Introducing DNA into Yeast

A pre-requisite for gene cloning in any organism is that DNA can be introduced into cells. There are several procedures that are available to achieve efficient transformation of *S. cerevisiae*.

It is possible to treat exponentially growing cells enzymatically to remove their cell walls. The resulting spheroplasts or protoplasts are exposed to DNA in the presence of calcium ions and polyethylene glycol which results in protoplast aggregation, localized membrane fusion, and DNA uptake. After transformation, the protoplasts are embedded in hypertonic selective agar which allows transformants to regenerate their cell walls and to form colonies. Frequencies of up to 10^6 transformants μg^{-1} DNA can be obtained using this technique.[1,2]

Alternative procedures are also available that do not require the removal of the cell wall. Intact yeast cells can be transformed by techniques which rely on

[1] A. Hinnen, J. B. Hicks, and G. R. Fink, *Proc. Natl. Acad. Sci. USA*, 1978, **75**, 1929.
[2] J. D. Beggs, *Nature (London)*, 1978, **275**, 104.

Table 1 Saccharomyces cerevisiae *selectable markers for yeast transformation*

Auxotrophic markers

Gene	Chromosomal mutation	Reference
HIS3	*his3-Δ1*	a
LEU2	*leu2-3*	b
	leu2-112	
TRP1	*trp1-289*	c
URA3	*ura3-52*	a

Dominant markers

Gene	Selection	Reference
CUP1	copper resistance	d
G418[R]	G418 resistance (kanamycin phosphotransferase)	e
TUN[R]	tunicamycin resistance	f

[a] K. Struhl, D. T. Stinchcomb, S. Scherer, and R. W.Davis, *Proc. Natl. Acad. Sci. USA*, 1979, **76**, 1035. [b] J. D. Beggs, *Nature (London)*, 1978, **275**, 104. [c] G. Tschumper and J. Carbon, *Gene*, 1980, **10**, 157. [d] S. Fogel and J. Welch, *Proc. Natl. Acad. Sci. USA*, 1982, **79**, 5342. [e] T. D. Webster and R. C. Dickson, *Gene*, 1983, **26**, 243. [f] J. Rine, W. Hansen, E. Hardeman, and R. W. Davies, *Proc. Natl. Acad. Sci. USA*, 1983, **80**, 6750.

alkali cations (usually lithium) in a procedure analogous to *E. coli* transformation.[3] The efficiency is approximately 10-fold lower than with the protoplasting method but allows cells to be spread directly onto selective plates (rather than embedded in agar) facilitating the use of colony screens.

Recently, high efficiency transformation has been achieved through electroporation of both intact yeast cells and protoplasts. This simple and rapid technique involves a brief voltage pulse which reversibly permeabilizes cell membranes facilitating entry of DNA molecules into the cell. Transformation efficiencies of 10^7 transformants μg^{-1} of plasmid DNA have been reported for intact yeast cells.[4]

1.2 Yeast Selectable Markers

Many yeast selectable markers are genes which complement a specific auxotrophy. This requires the host cell to contain a recessive, non-reverting mutation. The most widely used selectable markers and their chromosomal counterparts are listed in Table 1. There are also a limited number of dominant selectable

[3] H. Ito, Y. Fukada, K. Murata, and A. Kimura, *J. Bacteriol.*, 1983, **153**, 163.
[4] E. Meilhoc, J-M. Masson, and J. Teissié, *Bio/Technology*, 1990, **8**, 223.

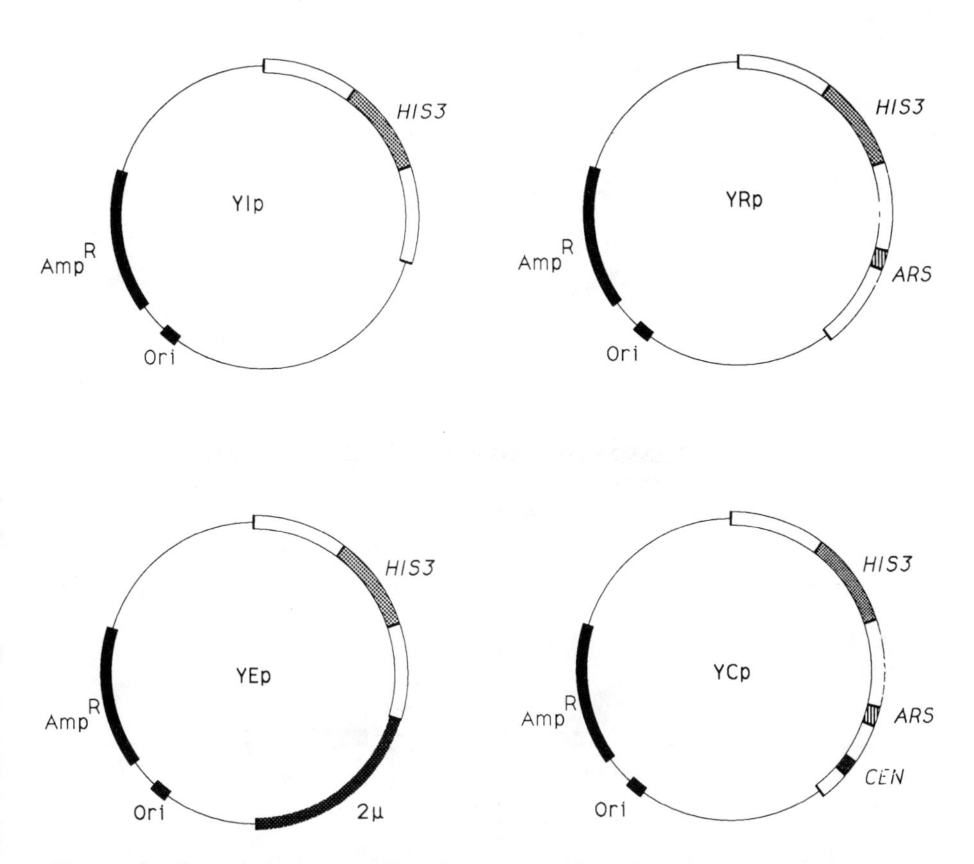

Figure 1 *Yeast cloning vectors: Yeast Integrating (YIp), Yeast Replicating (YRp), Yeast Centromeric (YCp), and Yeast Episomal (YEp) vectors indicating prokaryotic gene for resistance to ampicillin (Amp^R), prokaryotic replication origin (ori), yeast auxotrophic marker (*HIS3*), autonomous replication sequence (*ARS*), yeast centromere (*CEN*), and yeast 2μ DNA sequences. Additional bacterial (———) and yeast (▭) DNA sequences*

marker systems available but they are troublesome and not widely used. They are, however, useful for transforming polyploid brewing yeast. A number of these are also included in Table 1.

2 YEAST VECTOR SYSTEMS

All *S. cerevisiae* cloning plasmids are shuttle vectors which contain a segment of bacterial plasmid DNA to allow manipulation and large-scale DNA preparations to be performed in *E. coli* and a yeast selectable marker to allow selection of yeast transformants. The vectors can be divided into a number of different basic types depending on their structural elements. Schematic diagrams of each type are shown in Figure 1.

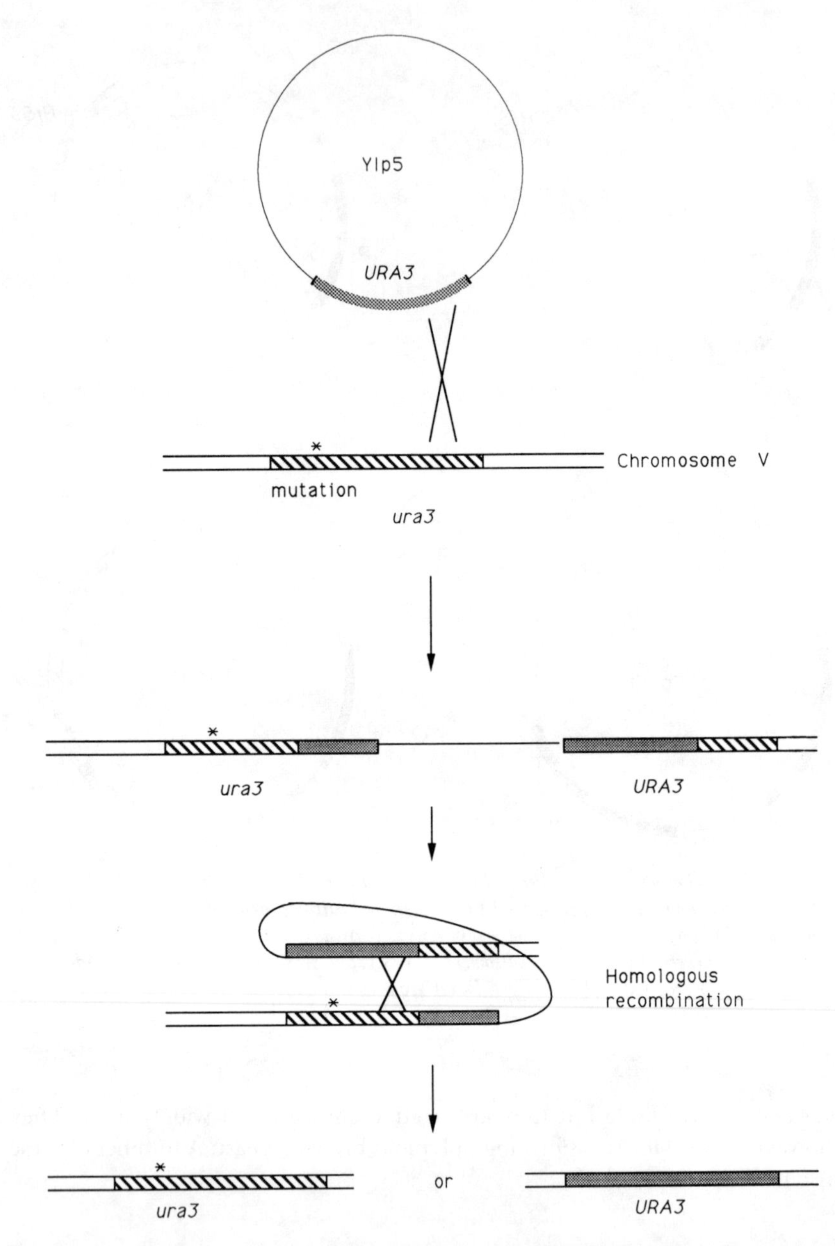

Figure 2 *Integration of a YIp vector. Schematic diagram showing the integration following transformation of plasmid YIp5 carrying a functional* URA3 *gene into a homologous region on chromosome V. The asterisk indicates a mutation at the* URA3 *chromosomal locus (*ura3 *gene). The duplicated structure which results from the integration event is unstable and undergoes homologous recombination returning the original allele (*ura3*) or replacing it with the fully functional copy of the gene (*URA3*). In both cases the excised plasmid cannot replicate in yeast and is therefore lost*

2.1 Integrating Plasmids

YIp (Yeast Integrative plasmid) vectors lack a yeast replication origin and must be integrated into the yeast genome in order to be maintained during cell division. They are normally present at one copy per cell and are very stable. A typical YIp can be seen in Figure 1. It consists of a segment of the *E. coli* plasmid pBR322 containing a selectable antibiotic resistance gene (Ampicillin) and a bacterial replication origin which facilitate selection and amplification in *E. coli*. The plasmid also contains unique restriction sites and the *S. cerevisiae HIS3* gene which provides a selectable phenotype in yeast.

YIps integrate by recombination between homologous sequences on the YIp plasmid and the host genome. Figure 2 shows integration of YIp5 (which carries a functional *URA3* gene) at a homologous mutated *ura3* region on chromosome V of *S. cerevisiae*. Consequences at the locus of integration can be quite complex: the entire plasmid may be incorporated into the chromosome resulting in duplication of the homologous sequences; recombination can occur after integration resulting in the loss of one or other of the single copy integrants; or multiple copies of plasmids can be integrated in tandem arrays at the target site. These events can be distinguished from one another by Southern hybridization of appropriate restriction digests.

Linear DNA fragments can also undergo homologous integration in the genome because their ends are highly recombinogenic.[5] An example of such an integration event is shown in Figure 3. This high frequency homologous integration has facilitated the development of powerful genetic tools to test the functions of sequences (Gene Disruption), to test *in vitro* modified sequences (Gene Replacement), and to rescue mutant alleles by genetic recombination.[6]

YIps are mainly used to introduce vectors at low copy number into chromosomal loci where they are stably maintained and inherited (frequency of loss is $< 0.1\%$ per generation).

2.2 Independently Replicating Plasmids

2.2.1 Yeast Replicating Plasmids. YRp (Yeast Replicative plasmid) vectors contain an Autonomous Replicating Sequence (*ARS*) of chromosomal origin in addition to the elements found in YIps (Figure 1). YRps are capable of autonomous replication, are present at 3–30 copies per cell, exhibit extreme meiotic and mitotic instability, and require constant selection.[7] As a result of this instability, YRp vectors are now rarely used.

2.2.2 Yeast Episomal Plasmids. YEp (Yeast Episomal plasmid) vectors are based on the *ARS* sequence from the endogenous yeast 2μ plasmid which contains genetic information for its own replication and segregation (for a review, see reference 8). They are capable of autonomous replication, are present at 20–200

[5] T. L. Orr-Weaver, J W. Szostak, and R. J. Rothstein, *Proc. Natl. Acad. Sci USA*, 1981, **78**, 6354.
[6] R. Rothstein, *Methods Enzymol.*, 1991, **194**, 281.
[7] A. W. Murray and J. W. Szostak, *Cell*, 1983, **34**, 961.
[8] A. B. Futcher, *Yeast*, 1988, **4**, 27.

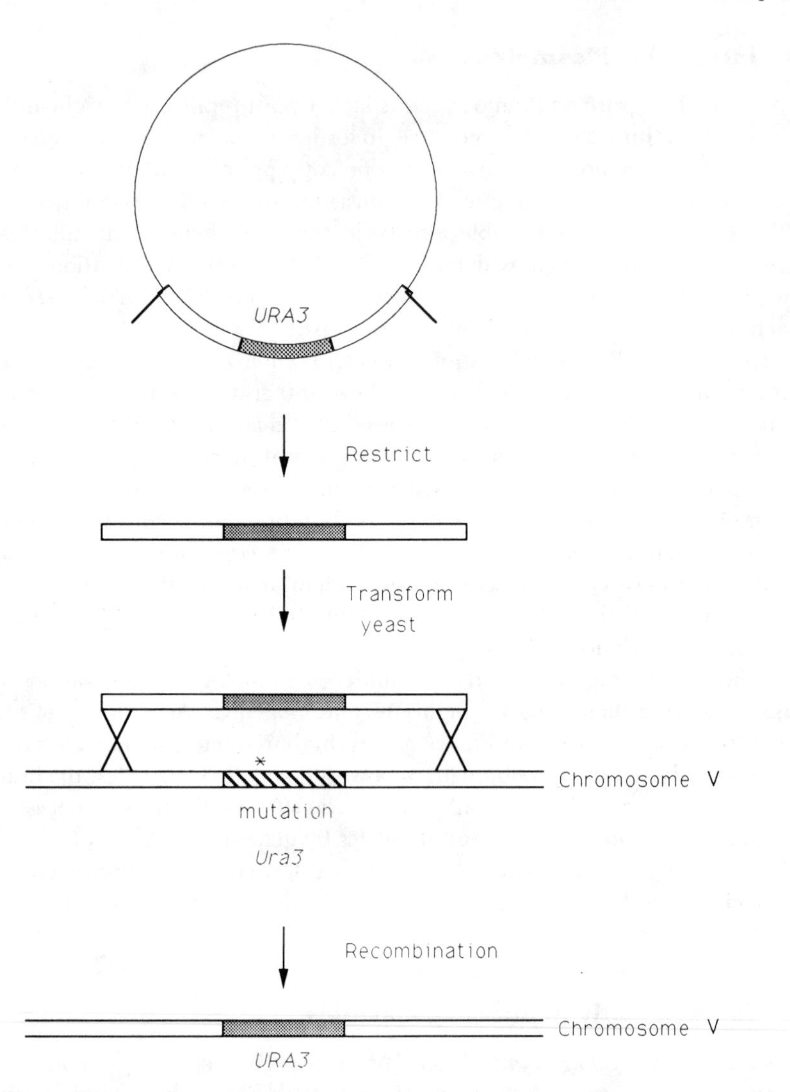

Figure 3 *Schematic diagram showing the integration of linear yeast DNA into a homologous region on the chromosome. The linear fragment isolated from plasmid DNA carries a functional* URA3 *gene flanked by chromosomal DNA sequences. Following transformation the fragment is targeted to the chromosomal* URA 3 *locus (the asterisk indicates a mutation at the* URA3 *chromosomal locus giving a* ura3 *gene). Recombination results in replacement of the mutant gene with the wild type gene (*URA3*)*

copies per cell, are ten times more stable than YRps, and under selective conditions are found in 60% to 95% of the cell population (Figure 1). These plasmids are the most commonly used yeast vectors and are extensively exploited in yeast expression systems (see Section 3.2.1).

2.2.3 Yeast Centromeric Plasmids. YCp (Yeast Centromeric plasmid) vectors contain both an *ARS* and a yeast centromere (Figure 1). The *ARS* sequence can be either chromosomal or 2μ in origin. YCps are normally present at one copy

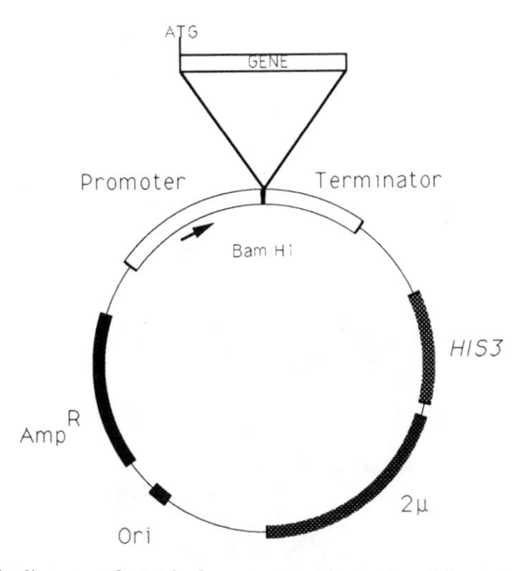

Figure 4 *Schematic diagram of a typical yeast expression vector. The vector is constructed on a YEp (see Figure 1) with yeast promoter and terminator regions. Heterologous genes are expressed by cloning coding sequences into the unique BamH1 cloning site. Transcription initiation is indicated by the arrow*

per cell, can replicate without integration into a chromosome and are stably maintained during cell division even in the absence of selection. Apart from their use as general yeast cloning vectors this type of plasmid forms the basis of a very specialized type of yeast cloning vector, the Yeast Artificial Chromosome (see Section 2.3.2).

2.3 Specialized Plasmids

2.3.1 Yeast Expression Vectors. The plasmids detailed in Sections 2.1 and 2.2 can be modified for use as expression vectors by the addition of suitable regulatory sequences. A typical expression vector is shown in Figure 4. It consists of a YEp vector carrying yeast promoter and terminator sequences on either side of a unique restriction site. Promoterless heterologous genes can be expressed by inserting them into this site in the appropriate orientation (see Section 3.1).

2.3.2 Yeast Artificial Chromosomes. YACs (Yeast Artificial Chromosomes) are specialized vectors capable of accommodating extremely large fragments of DNA (100 Kb–1000 Kb).[9] Schematic diagrams of a YAC and its use as a cloning system are shown in Figure 5. YACs contain a centromere, an autonomous replicating sequence, two telomeres, and two yeast selectable markers separated by a unique restriction site. They also contain sequences for replication and selection in *E. coli*. YACs are linear molecules when propagated in yeast but must be circularized by a short DNA sequence between the tips of the telomeres for propagation in bacteria. When used as a cloning vehicle, the YAC is cleaved with restriction enzymes to generate two telomeric arms carrying different yeast

[9] D. T. Burke, G. F. Carle, and M .V. Olsen, *Science*, 1987, **236**, 806.

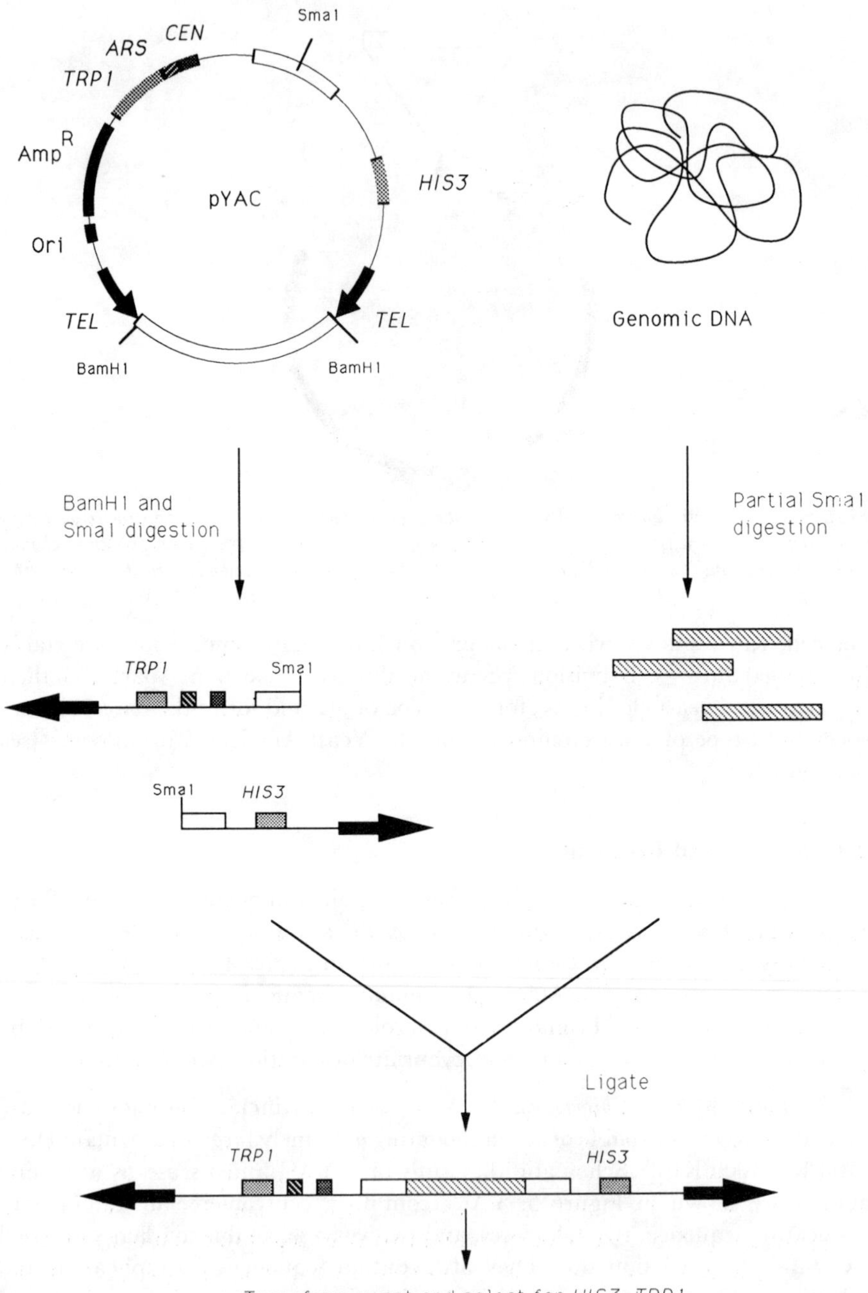

Figure 5 *Schematic diagram of a YAC cloning vector, indicating prokaryotic gene for resistance to ampicillin* (AMP^R)*, prokaryotic replication origin* (*ori*)*, yeast auxotrophic markers* (HIS3, TRP1)*, autonomous replicating sequences* (ARS)*, yeast centromeres* (CEN)*, and telomeres* (TEL)*.*

selectable markers. These arms are then ligated to suitably digested DNA fragments, transformed into a yeast host, and maintained as a mini chromosome. YACs have become indispensable tools for mapping complex genomes because they accommodate much larger fragments of DNA than bacteriophage or cosmid cloning systems.

3 PROTEIN PRODUCTION FROM *S. CEREVISIAE*

S. cerevisiae's most valuable contribution to modern biotechnology has been in its role as a host for the production of heterologous proteins (for reviews, see 10–12). Nevertheless, despite the versatility of yeast expression systems, the successful recovery of high levels of authentic, biologically active, heterologous proteins is still largely a matter of trial and error. Satisfactory recovery of heterologous proteins depends on a variety of factors including: the source of the DNA, the overall level of mRNA present in the cell, the overall amount of protein present in the cell, and the nature of the required product.

3.1 The Source of Heterologous DNA

As a general rule *S. cerevisiae* neither recognizes regulatory sequences of genes from other species nor efficiently excises heterologous introns. It is therefore necessary to obtain the gene of interest as a cDNA clone and splice it into an expression vector to obtain acceptable levels of protein production.

The level of expression may also be affected by the codon bias in the original organism. Impressive production of many heterologous genes in yeast has been achieved without optimizing codon usage patterns, but in other cases a correlation between the presence of rare codons and a drop in heterologous protein production has been noted.[13,14]

3.2 The Level of Heterologous mRNA Present in the Cell

The overall level of heterologous mRNA in the cell is a balance between the production of mRNA (which depends on the copy number of the expression vector and the strength of the promoter) and its stability in the cytoplasm (which depends on the RNA sequence and whether or not the expression vector has a suitable terminator sequence).

3.2.1 mRNA Production. Large scale production of foreign proteins by yeast requires the stable maintenance of the heterologous gene in the host cell. Expression vectors based on YEp technology are reasonably stable under selective conditions, but the copy number and overall stability of YEp expression

[10] J. E. Ogden, in 'Applied Molecular Genetics of Fungi', ed. J. F. Peberdy, C. E. Caten, J. E. Ogden, and J. W. Bennett, Cambridge University Press, 1991, p. 66.
[11] M. F. Tuite, in 'Biotechnology Handbooks, 4. *Saccharomyces*', ed. M. F. Tuite and S. G. Oliver, Plenum Press, New York, 1991, p. 169.
[12] D. J. King, E. F. Walton, and G. T. Yarranton, in 'Molecular and Cell Biology of Yeasts', ed. E. F. Walton and G. T. Yarranton, Blackie, London, 1989, p. 107.
[13] A. Hoekema, R. A. Kastelein, M. Vasser, and H. A. diBoer, *Mol. Cell Biol.*, 1987, **7**, 2914.
[14] L. Kotula and P. J. Curtis, *Bio/Technology*, 1991, **9**, 1386.

vectors depends on the size of the plasmid (small plasmids are more stable than larger ones), the gene used as a selectable marker in yeast (*e.g.* plasmid copy number is increased 8- to 10-fold by replacing a complete *LEU2* gene with a *LEU2* gene carrying a deleted promoter region (*LEU2-d*)[15]) and the nature of the expressed protein (see Section 3.4).

Plasmid instability associated with 2μ-based vectors can be circumvented by introducing a centromere into the vector but at the cost of reducing the plasmid copy number to 1–2 copies per cell. Mitotic stability can also be achieved by integrating the expression unit into the yeast genome but the number of expression units per site is limited. The number of integrated heterologous expression units can be increased by targeting them to multicopy genes. This strategy was successfully used to integrate 20 copies of a human nerve growth factor expression unit into *S. cerevisiae* using the multicopy Ty transposable element as the target site.[16]

The number of transcripts produced per plasmid is the second major factor in determining the overall level of mRNA production in the cell. This is controlled by the promoter being used to express the heterologous gene. Yeast promoters have been extensively studied over the last few years but an analysis of their structure and function is beyond the scope of this chapter. The reader is referred to recent reviews.[17]

The major prerequisite for an expression vector promoter is that it is a strong one. The vast majority of expression vectors therefore use promoters from glycolytic enzyme genes because they encode some of the most abundant mRNAs and proteins in the cell. The most frequently encountered ones are phosphoglycerate kinase (*PGK*), alcohol dehydrogenase 1 (*ADH1*), and glyceraldehyde-3-phosphate dehydrogenose (*GAPDH*). All of these promoters are constitutively expressed and so the heterologous protein is made throughout the growth of the cells.

However, constitutive expression can be a drawback to production of high levels of heterologous product when the foreign protein has a toxic effect on its host. Many expression vector promoters are therefore regulable, allowing the cells to be grown to maximum biomass before heterologous gene expression is induced. These include vectors based on the promoter of the acid phosphatase gene (*PHO5*) which is induced by the removal of phosphate from the culture medium, the protease B (*PRB1*) promoter which is induced when cells are released from glucose repression at the end of exponential growth in a glucose medium (when the cell switches to respiratory metabolism), and promoters from galactose inducible genes, *e.g.* galactokinase (*GAL1*) which is induced by removing glucose from the medium and replacing it with galactose. A number of regulable promoters have been successfully used for regulated heterologous gene expression and these are listed in Table 2.

3.2.2 Heterologous mRNA Stability. One of the major rate determining steps in the expression of a foreign protein in yeast is the stability of the heterologous mRNA.

[15] S. D. Emr, *Methods Enzymol.*, 1990, **185**, 231.

[16] A. Sakai, F. Ozawa, T. Higashizaki, Y. Shimizu, and F. Hishinuma, *Bio/Technology*, 1991, **9**, 1382.

[17] J. Mellor, in 'Molecular and Cell Biology of Yeasts', ed. E. F. Walton and G. T. Yarranton, Blackie, London, 1989, p. 1.

Table 2 *Promoters used to direct heterologous gene expression in yeast*

Promoter		Examples of heterologous genes expressed using promoter	Reference
PGK	3-Phosphoglycerate kinase	Human-IFN-α	a
ADH1	Alcohol dehydrogenase 1	Human-IFN-α	b
GAPDH	Glyceraldehyde-3-phosphate dehydrogenase	Human EGF	c
PHO5	Acid phosphatase	HBsAg	d
TRP1	*N*(5′-Phosphoribosyl)-Anthianilate isomerase	Human-IFN-α	e
GAL1, 10	Galactokinase	Calf chymosin	f
MFα1	α-Mating pheromone	Human EGF	g
CUP1	Copperthionein	Mouse IG Kappa chain	h

[a] M. F. Tuite, M. J. Dobson, N. A. Roberts, R. M. King, D. C. Burke, S. M. Kingsman, and A. J. Kingsman, *EMBO J.*, 1982, **1**, 603. [b] R. A. Hitzeman, F. E. Hagie, H. L. Levine, D. V. Goeddel, G. Ammerer, and B. D. Hall, *Nature (London)*, 1981, **293**, 717. [c] M. S. Urdea, J. P. Merryweather, G. T. Mullenback, D. Coit, U. Hebertein, P. Valenzuela, and P. J. Barr, *Proc. Natl. Acad. Sci. USA*, 1983, **80**, 7461. [d] A. Mijanohara, A. Toh-E, C. Nozaki, F. Hamada, N. Ohtomo, and K. Matsubara, *Proc. Natl. Acad. Sci. USA*, 1983, **80**, 1. [e] M. J. Dobson, M. F. Tuite, J. Mellor, N. A. Roberts, R. M. King, D. C. Burke, A. J. Kingsman, and S. M. Kingsman, *Nucleic Acids Res.*, 1983, **11**, 2287. [f] C. G. Goff, D. T. Moir, T. Kohno, T. C. Gravius, R. A. Smith, E. Yamasaki, and A. Taunton-Rigby, *Gene*, 1984, **27**, 35. [g] A. J. Brake, J. P. Merryweather, D. G. Coit, U. A. Heberlein, F. R. Masriarz, G. T. Mullenbach, M. S. Urdea, P. Valenzuela, and P. J. Barr, *Proc. Natl. Acad. Sci. USA*, 1984, **81**, 8642. [h] L. Kotula and P. J. Curtis, *Bio/Technology*, 1991, **9**, 1386.

mRNA half-lives cannot be accurately predicted from primary structural information (although some 'instability elements' have been identified[18]) or from the role of the encoded protein, so the stability of heterologous mRNAs must be empirically determined. Unstable transcripts curtail high level expression.

The absence of correct termination information in the 3′ coding region of a gene can also adversely affect heterologous mRNA stability. *S. cerevisiae* often fails to recognize heterologous transcriptional terminator signals because its own genes lack typical eukaryotic terminator elements. In the absence of an appropriate signal at the 3′ end of a heterologous gene, the RNA polymerase continues to transcribe the template and produces abnormally long mRNA molecules which are often unstable.[19] The lack of appropriate termination information can cause as much as a 10-fold drop in heterologous protein yield.[20] Expression vectors therefore frequently contain the 3′ terminator region from a yeast gene (*e.g.* *PGK* or *ADH1*) to ensure efficient mRNA termination (Figure 4).

[18] R. Parker and A. Jacobson, *Proc. Natl. Acad. Sci. USA*, 1990, **87**, 2780.
[19] K. S. Zaret and F. Sherman, *J. Mol. Biol.*, 1984, **177**, 107.
[20] J. Mellor, M. J. Dobson, N. A. Roberts, A. J. Kingsman, and S. M. Kingsman, *Gene*, 1985, **33**, 215.

3.3 The Amount of Protein Produced

A high level of stable heterologous mRNA does not necessarily guarantee a high level of protein production. The protein level depends on the efficiency with which the mRNA is translated and the stability of the protein after it has been produced.

3.3.1 Translation of Heterologous mRNA. The site of translation initiation in 95% of yeast mRNA molecules corresponds to the first AUG codon at the 5′ end of the message. Unlike prokaryotes (which require a Shine–Dalgarno sequence), but in common with mammalian systems, yeast does not require a sequence element to mediate translation. However, in contrast to mammalian systems, regions of dyad symmetry or the presence of an upstream AUG in the mRNA leader sequence effectively inhibits the initiation of translation. It is therefore advisable to eliminate these features from heterologous cDNAs for expression in yeast. The overall context of the sequence on either side of the AUG codon (with the exception of an A nucleotide at the -3 position) and the leader length do not appear to affect the level of translation (for a review, see reference 21).

3.3.2 Stability of Heterologous Proteins. Achieving high level transcription and translation of a heterologous gene in any expression system does not necessarily guarantee the recovery of large amounts of heterologous gene products: some proteins are degraded during cell breakage and subsequent purification; others are rapidly turned over in the cell. The powerful tools provided by a detailed knowledge of yeast genetics and biochemistry can be used to minimize these problems in *S. cerevisiae*: genes for proteolytic enzymes can be inactivated to prevent degradation during cell lysis, the ubiquitin-dependent degradative pathway for intracellular proteins manipulated to increase the half-life of proteins, and the yeast secretory pathway exploited to smuggle heterologous proteins out of the cell into the culture medium where protease levels are low.

The yeast vacuole contains several *endo-* and *exo-*proteases which gain access to the heterologous protein when the cells are harvested and lysed. The use of protease inhibitors can alleviate the problem to some extent but a genetic approach in combination with protease inhibitors is even more effective. Many of the genes which code for these proteins have been cloned and mutant host strains isolated. One of these mutant genes (*PEP4-3*) is widely used because it prevents the activation of a range of vacuolar zymogen proteases. The use of protease deficient host strains can improve both the yield and the quality of heterologous proteins.[22,23]

The intracellular turnover of cytoplasmic proteins can also affect the levels of protein produced during heterologous gene expression. A variety of factors combine to modulate protein half-life thus conferring stabilities on different

[21] T. F. Donahue and A. M. Cigan, *Methods Enzymol.*, 1990, **185**, 366.
[22] T. Cabezon, M. De Wilde, P. Herion, R. Loriau, and A. Bollen, *Proc. Natl. Acad. Sci. USA*, 1984, **81**, 6594.
[23] S. Rosenberg, P. J. Barr, R. C. Najarian, and R. A. Hallewell, *Nature (London)*, 1984, **312**, 77.

proteins that can vary by orders of magnitude. In eukaryotes a ubiquitin-mediated proteolytic pathway appears to play a major role in regulating the rate of turnover of cytoplasmic proteins. When ubiquitin (a highly conserved eukaryotic protein of 76 amino acids) is conjugated to amino groups of other proteins, the ubiquitin–protein conjugates are subjected to one of two fates: the removal of the ubiquitin molecule by a ubiquitin specific carboxyl terminal hydrolase with release of ubiquitin and the intact protein or the addition of further ubiquitin molecules and the complete proteolysis of the protein with the liberation of ubiquitin and free amino acids. It has been proposed that the half-life of heterologous proteins could be modified by manipulating the ubiquitin-mediated proteolytic pathway[24] and it has been demonstrated (at least in some cases) that the sensitivity of proteins to proteolysis by the ubiquitin degradative pathway can be dramatically altered by changing the *N*-terminal amino acid of the protein: the half-life of a β-galactosidase protein was changed from two minutes to 20 hours by replacing an *N*-terminal arginine with a methionine, serine, or alanine residue.[25] It should be noted however that alterations in amino acid sequences may not be desirable when heterologous proteins are produced for use as therapeutic products because they may cause unacceptable immunogenicity problems.

Vacuolar proteases and the ubiquitin degradative pathway can be circumvented in many instances by exploiting the yeast secretory pathway. If a heterologous protein is efficiently secreted from the cell, its exposure to intracellular protease activity is minimized and its recovery and purification are greatly facilitated because only low levels of native yeast proteins are secreted into the culture medium.

Entry into the secretory pathway is determined by the presence of a short hydrophobic 'signal' sequence on the *N*-terminal end of secreted proteins. This 'signal' sequence directs the protein to sites on the endoplasmic reticulum (ER) membrane where translocation into the lumen of the ER takes place. The signal sequence is cleaved by a luminal signal peptidase[26] and the protein passes through the ER and Golgi apparatus, before being packaged into secretory vesicles and exported beyond the cell membrane. If an appropriate 'signal' sequence is attached (by gene manipulation) to the *N*-terminus of a heterologous protein it will also enter the secretory pathway.

The 'signal' sequences from *S. cerevisiae*'s four major secretion products have been used with varying degrees of success to direct secretion of heterologous proteins. The different systems available are detailed in Table 3: invertase and acid phosphatase signal sequences target proteins to the periplasmic space, whereas α-factor and killer toxin signals target the proteins to the culture medium. The secretion of heterologous proteins is empirically determined, but when successful it minimizes the protease problem. A typical secretion vector is shown in Figure 6.

[24] D. Wilkinson, *Methods Enzymol.*, 1990, **185**, 387.
[25] A. Bachmair, D. Finley, and A. Varshavsky, *Science*, 1986, **234**, 179.
[26] M. G. Waters, E. A. Evans, and G. Blobel, *J. Biol. Chem.*, 1988, **263**, 6209.

Table 3 *Signal sequences used to direct secretion of heterologous proteins from yeast*

Signal sequence	*Location of gene product*	*Examples of heterologous proteins secreted from yeast*	*Reference*
Invertase	Periplasmic	α-1-Antitrypsin	a
Acid phosphatase	Periplasmic	α-Interferon	b
α-Factor mating pheromone	Culture medium	Epidermal growth factor	c
Killer toxin	Culture medium	Cellulase	d

[a] D. T. Moir and D. R. Dumais, *Gene*, 1987, **56**, 209. [b] A. Hinnen B. Meyhack, and R. Tsapis, in 'Gene Expression in Yeast', Kauppakirjapaino, Helsinki, 1983, p. 157. [c] A. J. Brake, J. P. Merryweather, D. G. Coit, U. A. Heberlein, F. R. Masiarz, G. T. Mullenbach, M. S. Urdea, P. Valenzuela, and P. J. Barr, *Proc. Natl. Acad. Sci. USA*, 1984, **81**, 4642. [d] N. Skiper, M. Sutherland, R. W. Davies, D. Kilburn, R. C. Miller, A. Warren, and R. Wong, *Science*, 1985, **230**, 958.

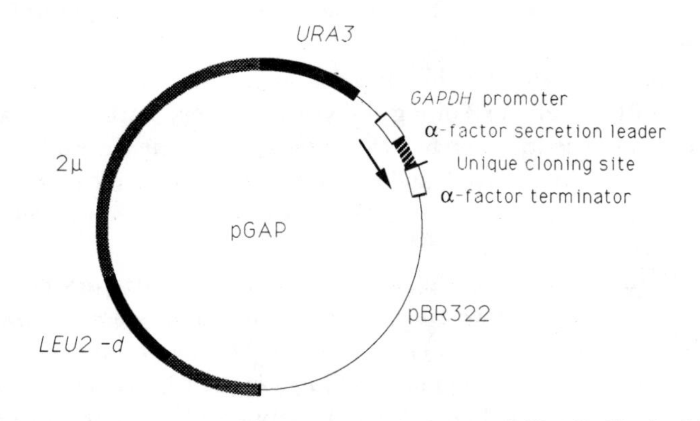

Figure 6 *Schematic representation of the yeast secretion vector pGAP (J. Travis, M. Owen, P. George, R. Carrell, S. Rosenberg, R. A. Hallewell, and P. J. Barr,* J. Biol Chem., *1985,* ***260****, 4384.). The vector contains* LEU2-d *and* URA3 *yeast selectable marker genes, pBR322 sequences for amplification in* E. coli *and 2μ sequences for autonomous replication in yeast. The expression 'cassette' contains a unique cloning site flanked by* GAPDH *promoter, α-factor secretion leader, and α-factor terminator sequences. Transcription initiation is indicated by the arrow*

3.4 The Nature of the Required Product

The object of heterologous gene expression is the high level production of biologically active, authentic protein molecules. It is therefore important to consider the nature of the final product when choosing the expression system. The protein size, hydrophobicity, normal cellular location, need for post-translational modification(s), and ultimate use must all be assessed before an appropriate expression system is chosen. The expression vector can be a high or low copy number plasmid, contain a constitutive or regulated promoter, and it may or may not carry a secretory signal sequence. A high copy number vector with a constitutive promoter can frequently be used but a regulated promoter is preferable if high level of the heterologous protein are toxic to the cell. Many proteins can be secreted, but an accurate prediction of suitable candidates is difficult. Secretion can be a rate-limiting step in heterologous protein production and low copy number (or integrated) expression vectors can enhance the yield of some secreted proteins.[27] The secretory pathway is often chosen for heterologous protein production because it enhances protein stability (see Section 3.3.2), facilitates the accurate folding of large proteins, and contains the machinery for post-translational modification.

The biological activity of a protein is critically dependent on its 3-dimensional structure. Disulfide bond formation readily occurs in the non-reducing environment of the ER and the passage of a heterologous protein containing disulfide bonds through the secretory pathway facilitates accurate folding. A direct comparison between the intracellular production and extracellular secretion of prochymosin and human serum albumin resulted in the recovery of small quantities of mostly insoluble, inactive protein when they were produced intracellularly but the recovery of soluble, correctly folded, fully active protein when they were secreted.[28,29]

The biological activity and/or stability of heterologous proteins can also be affected by the post-translational addition of carbohydrate molecules to specific amino acid residues. Glycosylation in yeast is of both the *N*-linked (via an asparagine amide) and *O*-linked (via a serine or threonine hydroxyl) types, occurring at the sequences Asn-X-Ser/Thr and Thr/Ser respectively. Inner core *N*-linked glycosylation occurs in the ER and outer core glycosylation in the Golgi apparatus. However, it is important to note that the number and type of outer core carbohydrates attached to glycosylated proteins in yeast are different to those found on mammalian proteins. In many cases these differences can be tolerated but if the protein is being produced for therapeutic purposes they may cause unacceptable immunogenicity problems. One approach to overcoming this problem is to remove the glycosylation recognition site by site directed mutagenesis. This strategy was successfully used to produce urokinase type plasminogen activator.[30]

Secretion can also be used to produce heterologous proteins that have an amino acid other than methionine at their *N*-terminus. If a secretory signal is spliced onto the heterologous gene at the appropriate amino acid (normally the penultimate one), then the *N*-terminal methionine which is obligatory for translation initiation will be on the secretory signal. Proteolytic cleavage of this signal from the heterologous protein in the ER will generate an authentic *N*-terminal amino acid (Figure 7).

Different secretion signals work with varied efficiencies when fused to different proteins, and in some instances heterologous signals can be more effective than

[27] J. Ernst, *DNA*, 1986, **5**, 483.
[28] R. A. Smith, M. J. Duncan, and D. T. Moir, *Science*, 1985, **229**, 1219.
[29] T. Etcheverry, W. Forrester, and R. Hitzeman, *Bio/Technology*, 1986, **4**, 726.
[30] L. M. Melnick, B. G. Turner, P. Puma, B. Price-Tillotson, K. A. Salvato, D. R. Dumais, D. T. Moir, R. J. Broeze, and G. C. Avgerinos, *J. Biol. Chem.*, 1990, **265**, 801.

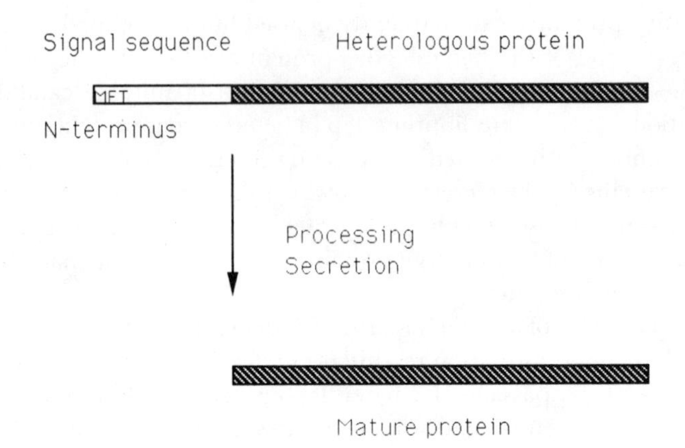

Figure 7 *Schematic diagram to show the secretion of a heterologous protein using a signal sequence*

yeast ones.[31] An empirical approach is needed with each protein. As a general rule, the α-factor system (see Section 3.3.2) is efficient for peptides or small proteins but the situation with larger proteins (especially hydrophobic ones) is very complex and unpredictable.

Despite the advantages secretion offers for the production of heterologous proteins in yeast, higher overall levels of protein production are often possible using intracellular expression. Some proteins form insoluble complexes when expressed intracellularly in *S. cerevisiae* but many others do not. Human superoxide dismutase was recovered as a soluble active protein after expression in yeast. It was also efficiently acetylated at the amino terminus to produce a protein identical to that found in human tissue.[32] Other proteins can be produced as denatured, intracellular complexes which can be disaggregated and renatured after harvesting. The first recombinant DNA product to reach the market was a hepatitis B vaccine produced in this way.[33]

When large quantities of heterologous proteins are produced intracellularly, important secondary modifications may be absent. In eukaryotes, the *N*-terminal methionine is not always removed and when it is retained it may or may not be modified by acylation.[34] These modifications are often essential for biological activity and their absence can cause immunogenicity problems in therapeutic proteins. Removal of an unwanted *N*-terminal methionine from an intracellularly expressed heterologous protein may occur naturally in yeast, or the residue can be removed after harvesting either by cyanogen bromide cleavage or by using cloned *E. coli* methionine amino peptidase (*MAP*). A more elegant method is to engineer a fusion protein consisting of ubiquitin joined to the penultimate amino acid of the heterologous protein. The removal of the *N*-terminal ubiquitin

[31] D. Sleep, G. P. Belfield, and A. R. Goodey, *Bio/Technology*, 1990, **8**, 42.
[32] R. A. Hallewell, R. Mills, P. Tekamp-Olsen, R. Blacker, S. Rosenberg, F. Otting, F. R. Masiarz, and C. J. Scandella, *Bio/Technology*, 1987, **5**, 363.
[33] D. E. Wampler, E. D. Lehman, J. Boger, W. J. McAleer, and E. M. Scolnick, *Proc. Natl. Acad. Sci. USA*, 1985, **82**, 6830.
[34] R. L. Kendall, R. Yamada, and R. A. Bradshaw, *Methods Enzymol.*, 1990, **185**, 398.

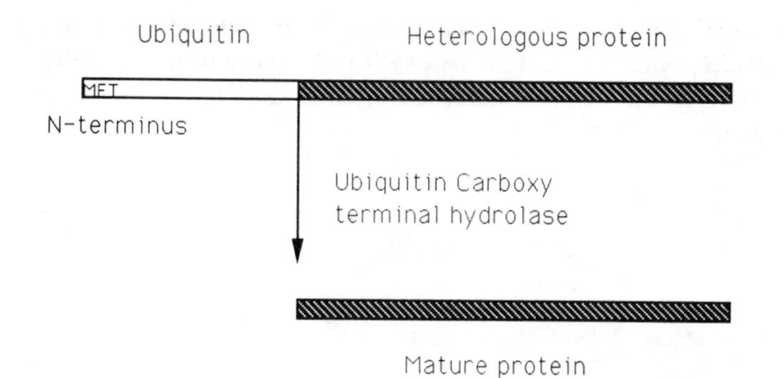

Figure 8 *Schematic diagram to show a ubiquitin fusion protein in yeast which is cleaved* in vivo *by an endogenous ubiquitin-specific proteinase to yield mature protein*

Table 4 *Commercially exploited heterologous proteins produced in* S. cerevisiae

Company	*Product*
Chiron	Tumour necrosis factor
Immunex	Granulocyte macrophage/colony stimulating factor
Merck & Co./Chiron	Hepatitis B vaccine
Novo – Nordisk	Human insulin

extension by ubiquitin carboxy terminal hydrolyase generates a heterologous protein with an authentic *N*-terminal amino acid (Figure 8). High level expression of human *IFN* (interferon) and *aPL* (proteinase inhibitor) have been achieved in yeast using this system.[35]

4 FUTURE PROSPECTS

A wide range of heterologous proteins have been expressed in *S. cerevisiae*. Some have been produced commercially (Table 4)[36] and of these the hepatitis B vaccine has already carved itself a lucrative niche in the marketplace. *S. cerevisiae* expression systems are powerful and versatile but they are not without limitations. The choice of expression system to ensure successful production of high levels of heterologous product is still largely a matter of trial and error. It is often necessary to test a number of different systems before optimal expression is achieved and even then the production of an authentically glycosylated protein may not be possible. Alternative yeast hosts (*e.g. Pichia pastoris*) are currently being developed for heterologous gene expression[37] because they can achieve higher expression levels than *S. cerevisiae* and are also less likely to hyperglycosy-

[35] E. A. Sabin, C. T. Lee-Ng, J. R. Shuster, and P. J. Barr, *Bio/Technology*, 1989, **7**, 705.
[36] N. Rau, *Genetic Eng. News*, July 1990, **10**, 14.
[37] M. A. Romanos, C. A. Scorer, and J. J. Clare, *Yeast*, 1992, **8**, 423.

late the expressed proteins. These may eventually replace the traditional host in some industrial applications but *S. cerevisiae*'s well characterized genetic systems will ensure its central role in eukaryotic cloning and biotechnology into the foreseeable future.

CHAPTER 6

Cloning Genes in Mammalian Cell-lines

E. J. MURRAY

1 INTRODUCTION

DNA cloning in mammalian cells originated from observations that naked, uncoated protein-free viral DNA, when presented to cells, resulted in the initiation of the viral life-cycle within a small number of cells in the population.[1,2] These cells are said to be transfected. The transfection efficiency of such experiments was increased by the use of facilitators. Thus, effective transfection only became apparent when presenting the viral DNA with DEAE-dextran,[3] or as a calcium phosphate co-precipitate.[4] Of course, such techniques may be used to transfect cells with non-viral DNA, and since those early times other transfection techniques have been used, such as micro-injection,[5] liposome or protoplast fusion,[6] electroporation,[7] polycation assisted transfection,[8,9] and scrapefection[10] (see later sections for details). To molecular biologists, DNA transfections in mammalian cells have been invaluable to the current understanding of the mechanisms underlying gene expression. The biotechnologist will want to use this knowledge to generate altered cell-lines which produce high yield, and maximum activity, of a commercially viable product.

E. coli offers many attractive features when used as a host for the production of recombinant proteins, including the ability to generate a high biomass in a short time and the relative low cost of maintaining a microbial fermentation.[11] In

[1] F. Graham and A. Van der Eb, *Virology*, 1973, **52**, 456.
[2] J. M. McCutchan and J. S. Pagano, *J. Natl. Cancer Inst.*, 1968, **41**, 351–6.
[3] L. M. Sompayrac and K. J. Danna, *Proc. Natl. Acad. Sci. USA*, 1981, **78**, 7575.
[4] M. Wigler, S. Silverstein, L. Lih-Syng, A. Pellicer, C. Yung-Chi, and R. Axel, *Cell*, 1977, **11**, 223.
[5] M. Capecci, *Cell*, 1980, **22**, 479.
[6] M. Rassoulzadegan, B. Binetruy, and F. Cuzin, *Nature (London)*, 1982, **295**, 257.
[7] E. Neuman, M. Schafer-Ridder, Y. Wong, and P. Hofschneider, *EMBO J.*, 1982, **1**, 841.
[8] P. Felgner, T. Gadek, M. Holm, R. Roman, H-W. Chan, C. Wenz, J. Northrop, G. Ringold, and M. Danielson, *Proc. Natl. Acad. Sci. USA*, 1987, **84**, 7413.
[9] R. Aubin, M. Weinfeld, and M. Paterson, *Somatic Cell Molec. Genet.*, 1988, **14**, 155.
[10] M. Fechheimer, J. F. Boylan, S. Parker, J. E. Sisken, G. L. Patel, and S. G. Zimmer, *Proc. Natl. Acad. Sci. USA*, 1987, **84**, 8463.
[11] R. E. Spier, in 'Molecular Biology and Biotechnology', 2nd Edn., ed. J. M. Walker and E. B. Gingold, Royal Society of Chemistry, London, 1985, p. 119.

addition, a wide variety of prokaryotic expression vectors are commercially available which enable high inducible levels of recombinant protein fusions which bear either maltose binding protein, glutathione *S*-transferase, or hexa-histidine peptides. The specific binding properties of the fusions can be exploited by using specific affinity chromatography to facilitate purification procedures. Other prokaryotic expression vectors can exploit the secretion pathway to the periplasmic space to enable recombinant proteins to re-fold in a relatively non-reducing environment. Nevertheless, it is clear that the activity of some eukaryotic proteins are defective when expressed in *E. coli* and although some experimental modifications may be employed to increase the activity,[12] the problems may be insurmountable. Thus, the use of mammalian cell-lines offer the only option as host for the production of recombinant proteins with high specific activities.

There are two key aims in this chapter. Firstly, the variety of transfection techniques will be described and compared, and secondly, the parameters governing DNA transfection will be evaluated to permit an understanding of how the expression of a transfected gene can be optimized and modulated.

Before discussing the above aims in depth, let us now consider the fate of DNA during a transfection experiment. During calcium phosphate co-precipitation transfection, up to 100% of the cells adsorb the DNA upon their membrane and enter the cell probably by a phagocytic mechanism.[13] The function of most facilitators is to increase the ease of one or both of these processes. The DNA is able to migrate to the nucleus in a proportion of these cells and become complexed with a variety of histonal and non-histonal nuclear proteins.

The physical fate of the DNA from this point is unclear, but in general circular DNA templates become nicked and subsequently linearized within 48 h.[14] Some laboratories note that the use of the calcium phosphate co-precipitation procedure leads to concatemerization,[15] resulting in the formation of high molecular weight ligates called pekeliosomes.[16]

Within the 48 h time span, many copies of the DNA are present and exist in an episomal (extra-chromosomal) state. At this stage, expression of some DNA templates is apparent. This is called *transient* expression, and can be modulated by a variety of experimental factors and DNA sequences/topology which will be discussed in later sections. After a few days, the observed expression decreases presumably due to the loss of extra-chromosomal copies which are unable to replicate and segregate in a dividing population of cells. However, a small proportion of these copies are able to integrate into the host cell genome, upon which they become a stably inherited genetic unit. Usually, the integration occurs at random sites (*i.e.* non-homologous) in the genome and multimeric copies of the DNA are present.[16] However, transfection techniques like micro-injection and electroporation tend to favour single copy integrations at a fre-

[12] C. S. Schein, *Curr. Opinion Biotechnol.*, 1991, **2**, 746.
[13] A. Loyter, G. E. Scangos, and F. Ruddle, *Proc. Natl. Acad. Sci. USA*, 1982, **79**, 422.
[14] H. Weintraub, P-F. Cheng, and K. Conrad, *Cell*, 1986, **46**, 115.
[15] F. Weber and W. Schaffner, *Nature (London)*, 1985, **315**, 75.
[16] M. Perucho, D. Hanahan, and M. Wigler, *Cell*, 1980, **22**, 309.

quency which enables integration into homologous sites to be detected.[17,18] This may be important to potential clinical uses of DNA transfections in gene therapy. The process of stable integration is inefficient in that only 10^{-4} cells (or much fewer) of the initial transfected population eventually produce stable clones containing the foreign donor DNA. Thus, the success of such experiments depends upon the design of a stringent selection for the integration event. A variety of selection techniques are available which are either dominant or recessive (see Section 6). Some selections cause the integrated foreign DNA to become amplified by several orders of magnitude, an observation which may suit the biotechnologists' purpose to achieve high yields of gene products.

2 METHODS OF DNA TRANSFECTION

A variety of methods are currently available for introducing DNA into mammalian cells, namely calcium phosphate coprecipitation,[4] DEAE-dextran,[3] microinjection,[5] protoplast fusion,[6] electroporation,[7] lipofection,[8] polybrene–DMSO treatment,[9] and scrapefection.[10] Most cell-lines respond better to one or other particular technique, which has to be determined empirically. Figure 1 shows the 4 stages involved in a generalized transfection, and which are utilized by the different techniques.

2.1 Calcium Phosphate Co-precipitation

For mainly historic reasons, this technique is probably the most popular. The technique has been studied by Wigler *et al.* in mouse L-cells, and has been optimized to the extent that 10^{-1}–10^{-4} cells of the population may be transfected depending upon cell-types and laboratory.[19–22] As a general rule, there is an upper limit of approximately 20 μg DNA to be transfected. This amount is near the optimum for co-precipitate formation,[13] thus, if necessary carrier DNA (salmon sperm DNA) is added to the test gene to achieve this total. In essence, a calcium phosphate precipitate is slowly induced to form in the presence of DNA. It is also possible to incorporate bacteriophage λ particles in the resulting precipitate.[23] This is then added directly to the cells and incubated for 4–16 h. Alternatively, the co-precipitate may be formed directly in the tissue culture dish.[24] The cells are then rinsed free of the precipitate, and harvested 30–48 h later for transient expression, or passaged in selective media for 10–14 days to

[17] O. Smithies, R. Gregg, S. Boggs, M. Koralewski, and R. Kucherlapati, *Nature (London)*, 1985, **317**, 230.
[18] K. Thomas, K. Folger, and M. Capecchi, *Cell*, 1986, **44**, 419.
[19] M. Wigler, S. Silverstein, L-S. Lee, A. Pellicer, Y-C. Cheng, and R. Axel, *Cell*, 1977, **11**, 223.
[20] M. Wigler, A. Pellicer, S. Silverstein, and R. Axel, *Cell*, 1978, **14**, 725.
[21] M. Wigler, R. Sweet, G-K. Sim, B. Wold, A. Pellicer, E. Lacy, T. Maniatis, S. Silverstein, and R. Axel, *Cell*, 1979, **16**, 777.
[22] C. Gorman, in 'DNA Cloning–A Practical Approach', Vol. 2, ed. D. M. Glover, IRL Press, Oxford and Washington, DC, p. 143.
[23] M. Ishiura, S. Hirose, T. Uchida, Y. Hamada, Y. Suzuki, and Y. Okada, *Mol. Cell. Biol.*, 1982, **2**, 607.
[24] C. Chen and H. Okayama, *Mol. Cell. Biol.*, 1987, **7**, 2745.

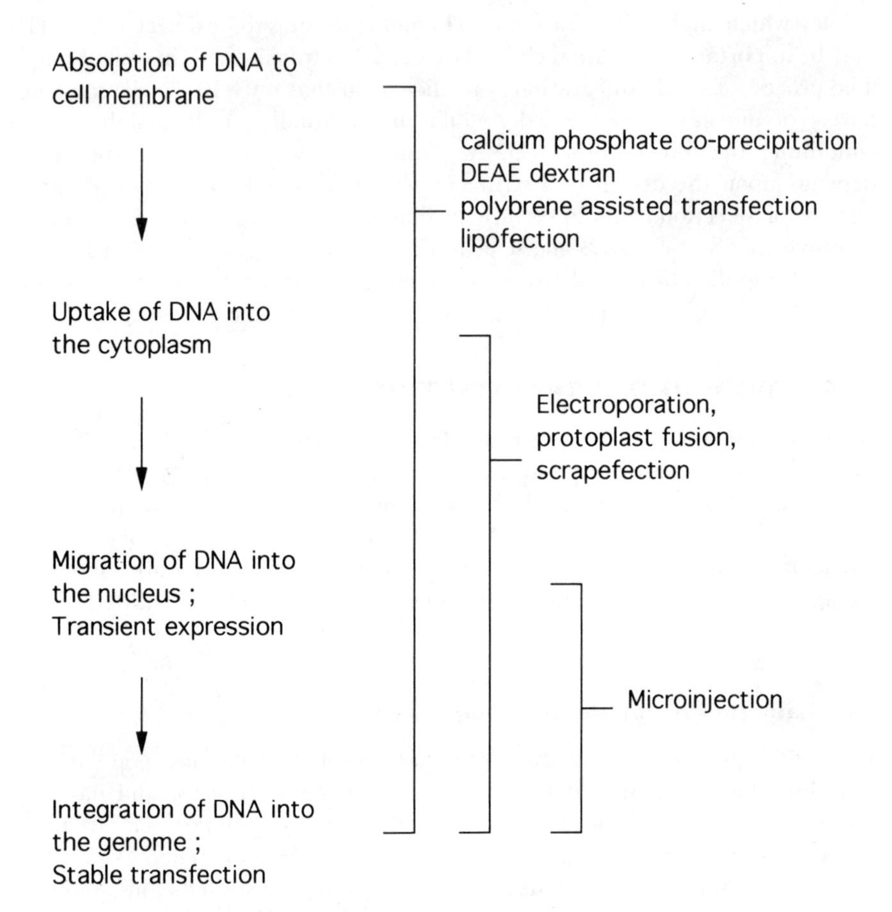

Figure 1 *The stages of DNA transfection which are involved in a variety of techniques are shown*

isolate stable clones. The function of the calcium component is not known, but it possibly concentrates the DNA on the cell membrane to further facilitate uptake by phagocytosis, and also to protect the DNA from hydrolytic nucleases.[13] The transfection efficiency, or level of gene expression depends upon both the efficiency of uptake by the cells, and the transcriptional capability of each DNA template introduced into the successfully transfected cells. The efficiency of DNA uptake is critical upon the pH of the calcium phosphate DNA co-precipitate formation, and the amount of DNA in that complex.[13,20] A higher proportion of the cells can take up DNA after the application of a glycerol shock.[22] The inherent transcriptional capability of DNA templates can also be modulated by a number of factors. Sodium butyrate treatment increases expression via hyperacetylation of associated histonal proteins,[25] and enhancer sequences may be used as part of the vector (see Section 3). The purity and isomeric form of the DNA preparation is also important. Transient expression is favoured by Form I DNA

[25] C. Gorman, B. Howard, and R. Reeves, *Nucleic Acid Res.*, 1983, **11**, 7631.

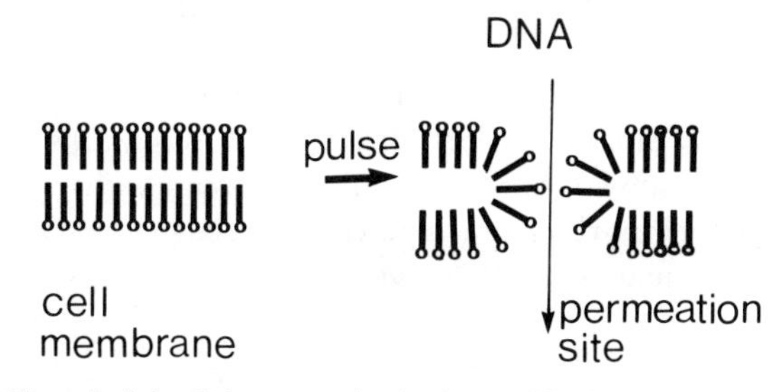

Figure 2 *The principle of electroporation is shown. The lipid bi-layer (cell membrane) interacts with an electrical pulse to generate a permeation site*

(covalenty closed, circular), whereas Form III DNA (linear) is successfully used for stable expression.

2.2 DEAE-Dextran

This technique has previously been used to increase the efficiency of viral infection of cell-lines. It is as simple and cheap as the calcium phosphate method, and can be used to transfect cells which will not survive even short exposures to calcium phosphate.[3]

Because the presence of serum may inhibit this method of transfection, the cells have to be extensively washed in phosphate-buffered saline prior to the addition of the DEAE-dextran–DNA solutions. The optimum DEAE-dextran concentration and length of incubation have to be experimentally determined to obviate the toxic effects of this treatment. Incubation times are in the range 1–8 h, after which excess DEAE-dextran–DNA is rinsed off the cell monolayer. As shown for calcium phosphate transfections, the transfection efficiency can be enhanced by glycerol-shock. Also, chloroquinine treatment has been used to increase the transient expression of the DNA template.[26] However, stable expression has been difficult to demonstrate using this technique.

2.3 Electroporation

This method is now commonly used and requires the purchase of specialist equipment. It is based upon the induction and stabilization of permeation sites within the cell membrane via an interaction of lipid dipoles in an electric field (see Figure 2). The nature and type of electrical pulse required to optimize such structures has been evaluated by commercial companies.[27] The DNA and cell concentration needs to be fairly high (*i.e.* 50 μg DNA/5 $\times$ 10^7 cells), but of course, this ratio has to be optimized for different cell-lines. After mixing, the DNA and cell suspension is subjected to 3 electric pulses at 8 kV cm^{-1} with a pulse decay of 5 μs. However, these spike potentials may be now superseded in

[26] H. Luthman and G. Magusson, *Nucleic Acid Res.*, 1983, **11**, 1295.
[27] G. Chu, H. Hayakawa, and P. Berg, *Nucleic Acid Res.*, 1987, **15**, 1311.

favour of a lower potential square-wave pulse which may increase the viability of the cells[27]

Electroporation does not require carrier DNA and is very quick compared to the above techniques (*i.e.* one may select for stable transformants 10 min after pulsing). Also, it has been shown that this technique gives a higher proportion of stable transformants which harbour a single copy of exogenous DNA at single integration sites. The frequency of these events allows the detection of homologous integration sites.[17]

2.4 Protoplast fusion

This is a general method for introducing macromolecules into cells and is therefore not restricted to DNA. The method is outlined in Figure 3. To obtain

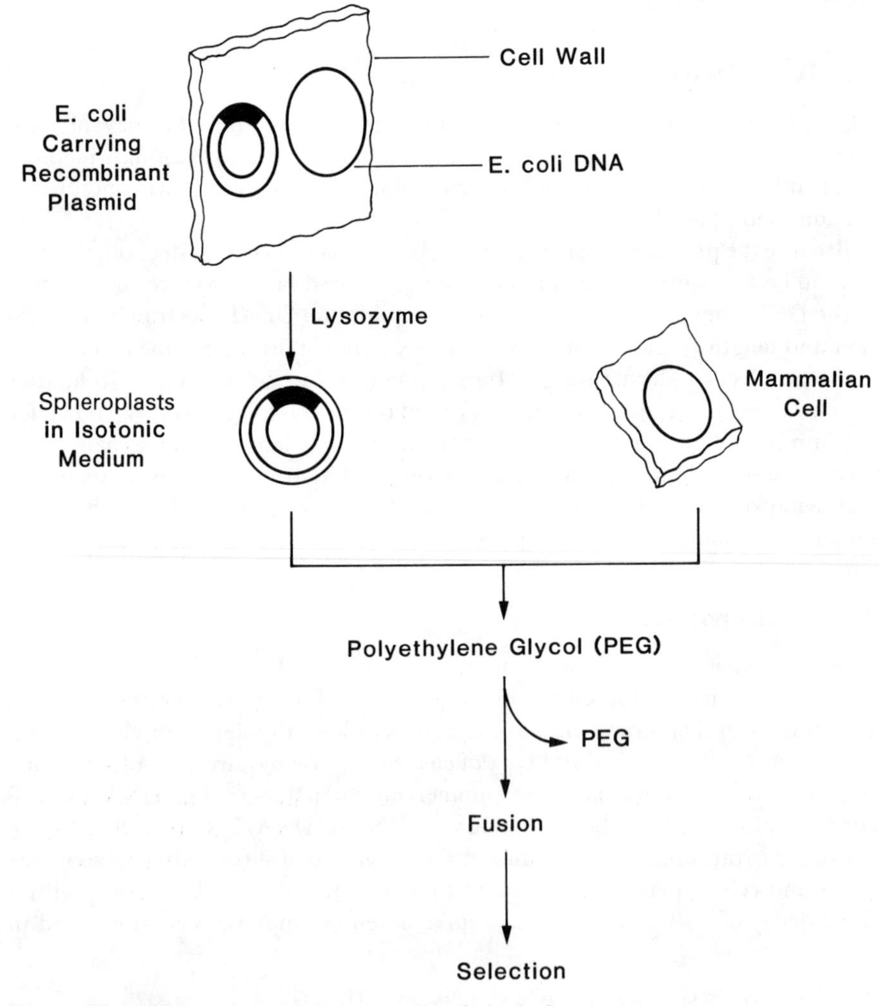

Figure 3 *A scheme illustrating cell or protoplast fusion is shown*

efficient transfection, care must be taken to ensure no lysis of spheroplasts, and that the optimal protoplast cell ratio is achieved (usually 10^4:1). One obvious advantage to this technique is that the cloned DNA to be transfected, need not be purified from *E. coli*.

2.5 Lipofection

Synthetic cationic lipids can be used to generate liposomes in the presence of DNA.[8] The resulting unilamellar structures entrap 100% of the DNA present. These vesicles may be fused to cell membranes and thus deliver the DNA directly to the cell interior. Serum appears to inhibit this process. The rationale behind this procedure is that a single DNA plasmid is surrounded by sufficient cationic lipid to completely neutralize the negative charge of DNA and provide a net positive charge which facilitates an association with the negatively charged surface of the cell. Depending upon the cell-line, lipofection is reported to be 5–100 fold more effective than either the calcium phosphate or DEAE-dextran mediated transfection techniques.[8] The synthetic cationic lipids necessary for lipofection are available from commercial biotechnological companies.

2.6 Polybrene–DMSO Treatment

Polybrene is a synthetic polycation which can be used to facilitate DNA transfection by serving as an electrostatic bridge between the negatively charged DNA and positively charged components on the host cell membrane.[9] As such, this transfection technique exploits the same strategy as described above for DEAE-dextran and lipofection. Thus, similar to DEAE-dextran, the uptake of the adsorbed DNA on the cell surface is enhanced by treatment with DMSO which, in the manner of the glycerol shock, serves to permeabilize the DNA coated cells. In contrast to DEAE-dextran, the use of polybrene as a facilitator enables the investigator to generate stably transfected clones.

2.7 Micro-injection

The term transfection is a misnomer when applied to this technique. The DNA is introduced directly into the nucleus with a glass micro-pipette with the aid of a micro-manipulator. This is obviously very time-consuming with respect to the techniques previously described. All the viable micro-infected cells transiently expressed the DNA and 1–30% will become stable clones. No carrier DNA is required and the form and number of DNA molecules present in the host cell may be strictly controlled. Injection in the cytoplasm gives 10^{-3} fewer stable transformants compared to nuclear injections.[5]

2.8 Scrapefection

This technique was initially described as a general method for introducing functional macromolecules into adherent cell-lines.[28] It has since been demon-

[28] P. L.McNeil, R. F. Murphy, F. Lanni, and D. L. Taylor, *J. Cell Biol.*, 1984, **98**, 1556.

strated to function well in a DNA transfection experiment.[10] The protocol is very simple and involves incubating a washed monolayer of cells with a buffered DNA solution (1–50 $\mu g\ cm^{-3}$) prior to scraping with a rubber policeman (a type of rubber spatula). The scraped cells are then distributed to fresh culture plates and analysed for transient expression of the transfected DNA 1 to 5 days later.[10] Up to 80% of scrapeloaded cells were shown to express the DNA and cell viability of 70% was obtained. Stable expression of co-transfected DNA is also possible. The method is rapid, efficient and has a low cost compared to other transfection procedures. It does not require the formation of a DNA complex or co-precipitate and therefore is a relatively facile operation. As such, this is probably the first choice in determining which transfection technique is adequate for a cell-line of choice.

3 REQUIREMENTS FOR GENE EXPRESSION

Although the nature and variety of DNA sequences which modulate eukaryotic transcription is beyond the scope of this chapter, brief mention should be made of some salient points.

The DNA sequences required for efficient initiation of gene expression can be broadly classified into three categories (*a*) simple promoters, (*b*) enhancers, and (*c*) regulatory regions. Figure 4 illustrates some well-characterized sequences required for expression for a number of genes.

In the simplest terms promoters may be regarded as being a patchwork of common consensus sequence motifs, which are found across a wide variety of different genes, with some sequences being more prevalent than others.[29] The TATA motif is amongst the most common and is usually found at a fixed characteristic position at ~30 base-pairs (bp) upstream (*i.e.* 5′) of transcriptional start-sites (cap-sites).

Other promoter sequence motifs include CAAT and CCGCCC, which are usually found upstream from the TATA box and are less fixed in their relative position or orientation to the cap-site. These three sequence motifs are operated upon by fairly ubiquitous transcription factors – namely TF11D, CTF (also called NF-1), and Sp1 respectively.[30]

Enhancer sequences were first identified in viral genomes, and may also be regarded as a patchwork of sequence motifs, of which some are also found in promoters.[30,31] They are operationally defined as DNA sequences which augment transient gene expression in *cis*, and in an orientation-independent manner.[32] Other characteristics of enhancers include the ability to exert the effect over large stretches of DNA, and the ability to activate different (heterologous) promoters than those by which they are naturally found. Host cellular genes have been found to contain enhancer sequences, which appear to

[29] P. Bucher and E. N. Trifonov, *Nucleic Acid Res.*, 1986, **14**, 10009.
[30] N. Jones, E. Ziff, and P. Rigby, *Genes and Dev.*, 1988, **2**, p. 000.
[31] H. Singh, R. Sen, D. Baltimore, and P. A. Sharp, *Nature (London)*, 1986, **319**, 154.
[32] J. Banerji, S. Rusconi, and W. Schaffner, *Cell*, 1981, **27**, 299.

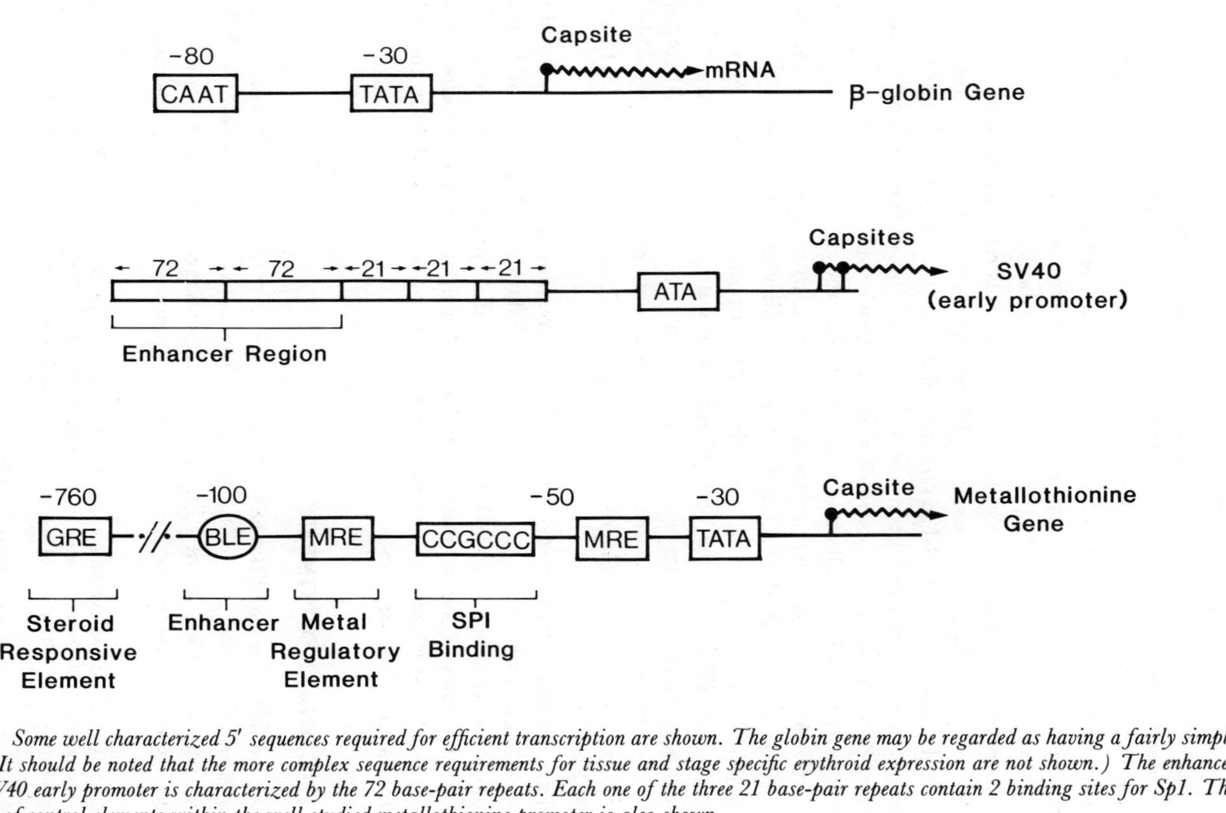

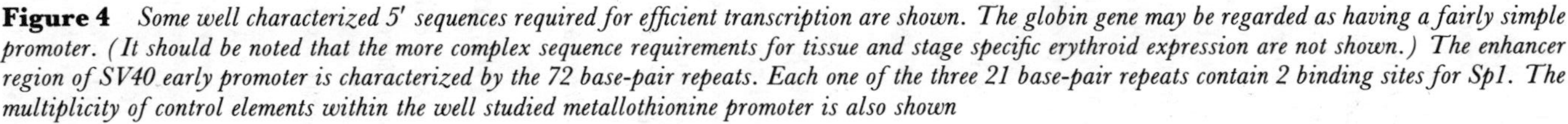

Figure 4 *Some well characterized 5′ sequences required for efficient transcription are shown. The globin gene may be regarded as having a fairly simple promoter. (It should be noted that the more complex sequence requirements for tissue and stage specific erythroid expression are not shown.) The enhancer region of SV40 early promoter is characterized by the 72 base-pair repeats. Each one of the three 21 base-pair repeats contain 2 binding sites for Sp1. The multiplicity of control elements within the well studied metallothionine promoter is also shown*

function in a tissue specific manner.[33,34] Viral enhancers have a wider, more varied host range, although some retroviral enhancers do have a preference for a certain cell-type under certain conditions.[25,35]

Regulatory sequences may be defined loosely as DNA sequences which interact with tissue-specific *trans*-acting proteins to ensure correct tissue-specific or temporal-specific expression. Included in this category are sequences which are able to induce gene expression under an appropriate stimulus, *i.e.* heat-shock,[36] presence of heavy metals,[37] or steroid hormones.[3338] This ability to induce, and therefore control, gene expression is obviously very useful to the biotechnologist especially if the protein products are toxic to the host cell. Thus, populations of stable clones harbouring such genes need only be induced prior to harvest.

A novel type of enhancer/regulatory sequence has recently been identified in the human β-globin gene cluster.[34,39] These sequences, called dominant control regions (DCRs), are required to commit adjacent genes to specific expression in erythroid tissue, and have been identified by studies of the sequence requirements of human globin gene expression in transgenic mice.[34] Phenotypically they appear to function in a similar fashion to tissue specific enhancers, but exhibit the novel property of isolating adjacent chromatin from the effect of the host integration site. Thus, they have the unusual ability of manifesting a degree of exogenous gene expression which is dependent upon the exogenous gene copy number. The DCR has now been identified to within a few kilobases[39] which make it possible to construct retroviral vectors (see Sections 4.3 and 4.5) containing these important regulatory regions. Such tools may prove invaluable to correct gene defects by replacing defective globin genes in, for example, thalassaemic host bone marrow.

There are many more stages which affect expression of the final gene product such as post-transcriptional processing, translation efficiency, and post-translational modifications, (which this chapter shall not deal with) and which will be of greater or lesser importance depending upon the gene product under study. Note that *E. coli* will not necessarily perform these functions in the same manner as in eukaryotic hosts.

4 THE DNA COMPONENT

So far, we have discussed the DNA component in a transfection without defining its nature. Basically, it is possible to transfect cells with chromosomal DNA, naked genomic DNA, cloned DNA, or bacteriophage λ. Many elegant experiments have been performed by transfecting genomic DNA into cell-lines. For

[33] M. Walker, T. Edland, M. Boulet, and W. Rutter, *Nature (London)*, 1983, **306**, 557.
[34] F. Grosveld, G. Blom van Assendeltf, D. Greaves, and G. Kollias, *Cell*, 1987, **51**, 975.
[35] C. Gorman, G. T. Merlinoi, M. C. Willingham, I. Pastan, and B. Howard, *Proc. Natl. Acad. Sci. USA*, 1982, **79**, 6777.
[36] H. Pelham and M. Bienz, *EMBO J.*, 1982, **1**, 1473.
[37] F. Lee, R. Mulligan, P. Berg, and G. Ringold, *Nature (London)*, 1982, **294**, 228.
[38] D. H. Hamer and M. J. Walling, *J. Mol. Appl. Genet.*, 1982, **1**, 273.
[39] D. Talbot, P. Collis, M. Antoniou, M. Vidal, F. Grosveld, and D. Greaves, *Nature (London)*, 1989, **336**, 352.

instance, if one isolates total genomic DNA from a tissue expressing gene X to transfect an X-negative cell-line, then it is possible to characterize gene X by selecting for a stable tranfectant now exhibiting the phenotype associated with X expression. Secondary transfections will permit the eventual isolation and cloning of the gene. This type of approach has been used to isolate activated cellular oncogenes[40] and is amenable to isolate genes which express their products on the transfected cell surface, providing antibodies are available as probes for their presence. Similar strategies have been employed in the cloning of T-cell specific genes in fibroblasts.[41]

4.1 Use of Vectors

The use of vectors allows considerable flexibility in regulating the expression of cloned genes. Although, in theory, any simple pBR322 plasmid vector will permit a degree of transient and subsequent stable expression of the passenger gene, many different types of vector have been constructed and imbued with characteristics which render them particularly amenable for gene cloning. Vectors can be classified into either viral-based or plasmid-based, and each class may be sub-divided into general purpose cloning vehicles or expression vectors (for cDNA inserts). Expression vectors are those which contain an efficient promoter upstream of a cloning site such that transcription of the inserted cloned cDNA sequences can be driven by the promoter in the vector. A comprehensive listing of cloning vectors can be found in reference 42.

4.2 Plasmid-based Vectors

A common feature of all plasmid-based vectors is the presence of a prokaryotic replication origin (replicon) and a selectable marker gene to permit the recombinant DNA molecule (*i.e.* vector and insert) to be amplified in *E. coli*.

Most vectors have, in addition, a eukaryotic replicon and/or a eukaryotic selectable marker. This latter feature is required for stable expression and will be discussed in depth in Section 6. The selectable marker need not be present on the vector, but may be provided, in *trans*, on a separate vector, to be co-transfected at an appropriate ratio.[21]

The eukaryotic replicon is usually derived from viruses such as bovine papilloma virus (BPV), Simian virus 40 (SV40), or Epstein Barr virus (EBV). It should be noted that viral replicons require additional viral gene products to initiate DNA replications of the recombinant. Such proteins may be supplied in *trans* either by co-transfecting with the viral gene or, more uniformly, by choosing a cell-line which constitutively expresses the required viral functions. For instance, a monkey cell-line CV-1 has been stably transformed with the gene coding for the appropriate SV40 replicative functions to generate COS cell-lines,

[40] C. Tabin, S. Bradley, C. Burgmann, and R. Weinburg *et al.*, *Nature (London)*, 1982, **300**, 143.
[41] D. R. Littman, Y. Thomas, P. Maddon, L. Chess, and R. Axel, *Cell*, 1985, **40**, 237.
[42] P. H. Pouwels, B. E. Enger-Valk, and W. J. Brammar, 'Cloning Vectors–A Laboratory Manual', Elsevier Science Publishers B.V., Amsterdam, 1985.

which are able to replicate plasmid vectors containing the SV40 replicon up to 4×10^5 copies per transfected cell. However, the viability of these transfected cells appears to decrease after a few days probably due to this excessive DNA synthesis.[43] It has been observed that certain pBR322 sequences act in *cis* to inhibit extra chromosomal replication (so called 'poison' sequences).[44] Derivatives of pBR322 that lack these sequences are available and should therefore provide the basis for any replicative vector.

An interesting feature of chimeric plasmid-BPV vectors is their ability to induce stable transformation of recipient mouse cells, thereby reducing the serum requirement for cell-propagation and also providing an intrinsic selection for the transfection event without applying any external selection, although one is restricted in the choice of a semi-permissive cell-line.

The recombinant BPV vector is stably maintained *without* integration and exists as a stable extrachromosomal element at 10–100 copies per cell.[45]

Chimeric vectors have been constructed which contain both SV40 and BPV origins, and also BPV genes required in *trans* for the regulated replication from the BPV replicon.[43] Upon transfection of COS cells with the chimera, it appears that the regulated controlled replication from the BPV origin is dominant over the runaway SV40 replication. Thus, whereas plasmid vectors containing the SV40 replicon alone undergo excessive replication leading to cell death, the dual replicon chimera is maintained extrachromosomally at 500–1000 copies per cell leading to the generation of stable cell-lines. Deletion of the BPV transacting function results in runaway replication. It is evident that if conditional mutations in these transacting proteins are found, this system offers considerable control over recombinant copy number and cell-line viability.

Some plasmid based vectors also contain other eukaryotic regulatory sequences such as a splicing site to promote mRNA stability and a poly-adenylation sequence to ensure appropriate termination of transcription.[46]

Vectors which contain secretory signals for the production and generation extracellular secreted proteins would have the advantage of enabling easy harvest of the cloned gene products and these are now available from commercial biotechnological companies.

4.3 Virus Vectors

Animal virus vectors offer an alternative to DNA transfection as a means of enabling a foreign gene to be expressed in different cell types.[47] Recombinant viral genomes carrying the cloned gene may be packaged into infectious virions and used to infect cell-lines or tissues. Factors influencing choice of viral vector include, types of cells or animals to be infected, whether cell transformation or

[43] J. M. Roberts and H. Weintraub, *Cell*, 1986, **46**, 741.
[44] M. Lusky and M. Botchan, *Nature (London)*, 1981, **293**, 79.
[45] P. D. Mathias, H. U. Bernard, A. Scott, G. Brady, T. Hashimto-Gotoh, and G. Schutz, *EMBO J.*, 1983, **2**, 1487.
[46] P. Gruss and G. Khoury, *Nature (London)*, 1980, **286**, 634.
[47] P. W. J. Rigby, *J. Gen. Virol.*, 1983, **64**, 255.

cell lysis is required, the size of DNA to be cloned, and whether virus infectivity is to be retained.

Small DNA viruses such as SV40 have a restricted host range and limited capacity for cloning due to the constraints of genome size able to be packaged. In most cases, viral genes are deleted to permit incorporation of foreign DNA, and as a consequence these viral replacement vectors are replication defective requiring helper virus or special cell-lines for their propagation and replication.

Large viruses like human adenovirus and poxvirus have a greater capacity for foreign DNA without necessarily reducing their infectivity. Murine retroviruses have also been successfully employed as replacement vectors.

4.4 Adenoviral Vectors

Infection by wildtype adenovirus induces cytopathic changes in human cell-lines and tissues (in particular adenoidal and other lymphoid organs). A variety of viral strains have been isolated and it has been shown that some strains are more oncogenic than others in their ability to induce tumours in newborn rodents and rat primary cells. However, no strain which exhibits oncogenic potential in any human tissue has been reported. The virus may be propagated in Hela cells to a high titre and evince characteristic cytopathology (*i.e.* nuclear lesions). Adenovirus has a 35 kb linear genome which makes the manipulation as potential vectors difficult, although variant strains which contain single cloning sites have been developed to overcome these difficulties.[48] Their upper DNA packaging limit is 5% greater than the wild type genome and therefore can be used as insertion vectors for small inserts and, as such, can be propagated without a helper virus. Extra sequences can be cloned by deleting some early viral genes and replacing these with larger inserts. These replacement recombinant vectors have to be propagated in helper cell-lines (such as 293) in which the deleted viral functions are provided by chromosomally integrated sequences. Thus, naked recombinant viral DNA is transfected into these cells and a productive infection ensues releasing infectious recombinant virions that may be harvested for future infections.

4.5 Retrovirus Vectors

To understand the potential advantages of using retrovirus as cloning vectors, brief mention must be made of their life-cycle. Infectious wildtype replication competent retroviruses contain a single-stranded 7–8 kb RNA genome which is packaged by viral coded envelope protein. The type of envelope protein determines the host range infectivity of the virus. After infection, the RNA genome is translated and the resulting viral proteins are able to convert the RNA genome into double-stranded DNA (reverse transcription). During this process, the ends of the viral genome are dupliated to generate long terminal repeats (LTR). LTRs represent powerful promoter/enhancer elements. The DNA intermediate

[48] N. Jones and T. Shenk, *Proc. Natl. Acad. Sci. USA*, 1979, **76**, 3665.

is able to integrate into the host genome by a specific mechanism not available to DNA viruses. The genome is now described as a provirus. The integration is precise in that the resulting proviral genes are co-linear and unrearranged, being flanked by 5′ and 3′ LTRs. The integration site in the host genome is random in most cell-lines. The proviral state is stable, and the cell remains viable.

If the LTRs remain active, a genomic RNA molecule is produced which contains all the coding information represented between the two LTRs. This polycistronic message undergoes a variety of different fates. It may be packaged into an infectious particle, or translated to produce proteins, or spliced into sub-genomic mRNAs and subsequently translated. The packaging of the RNA is dependent upon the presence of a signal sequence at the 5′ end. This is called the ψ-sequence and mRNAs which lack this signal cannot be packaged. This observation has lead to the production of special helper cells called ψ-2 and ψ-AM.[49] These cell-lines have been infected with a retrovirus which lacks the ψ-sequence. Thus, the resulting provirus produces all the necessary functions for packaging but is unable to package its own genomic RNA. Therefore, transfection of these cell-lines with *in vitro* constructed recombinant proviral DNA, which contains the gene to be cloned between two LTRs, results in the sole production of infectious recombinant viral particles.[49] However, in some rare instances, recombination does occur during the propagation cycle and results in wildtype infectious particles. The ψ-AM cell-line permits the recombinant virus to exhibit a very broad host range.

From the above discussion, it should be evident that a recombinant proviral DNA, into which a gene which contains introns has been inserted, will be generated as an infectious particle which now contains the cloned gene as a cDNA copy.[50]

4.6 Poxviral Vectors

Poxviruses, such as vaccinia, have a very large DNA genome (187 kb) and considerable flexibility is observed in the amount of genomic DNA to be packaged; therefore it has the capacity to accommodate very large inserts (up to 35 kb in some strains). No helper virus is required for propagation and vaccinia strain exhibit a wide host range for infectivity. The genetic make-up of vaccinia differs considerably from the viruses previously mentioned, such that a vaccinia gene cannot be recognized by host RNA polymerases. Therefore naked vaccinia DNA is non-infectious. Additionally, vaccinia coded RNA polymerases do not transcribe promoters recognized by host RNA polymerases. Thus vaccinia promoters must be used to obtain efficient expression of foreign cDNAs. Indeed only cDNAs may be cloned in vaccinia vectors because the virus replicates in the cytoplasm of infected cells, whereas splicing is a nuclear event.

In order to construct an infectious vaccinia recombinant, a two-step strategy is employed (Figure 5). First, an intermediate plasmid is constructed *in vitro* which contains the foreign cDNA linked to a vaccinia promoter. This chimeric tran-

[49] R. D. Cone and R. C. Mulligan, *Proc. Natl. Acad. Sci. USA*, 1984, **81**, 6349.
[50] K. Shimotohno and H. M. Temin, *Nature (London)*, 1982, **299**, 265.

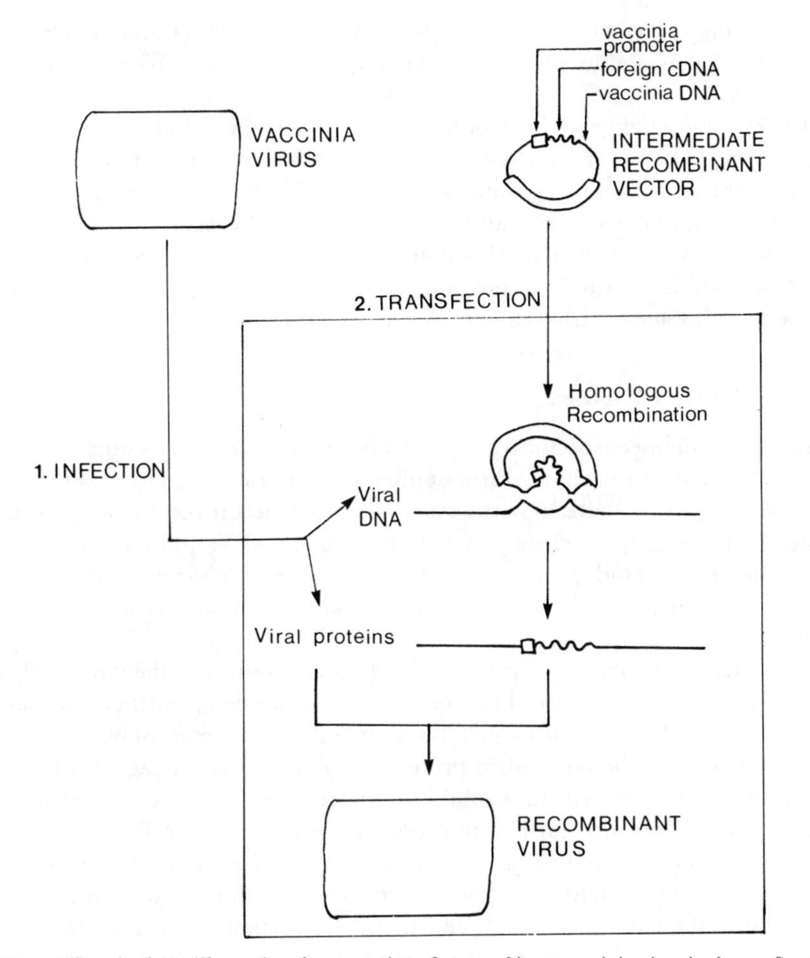

Figure 5 *A scheme illustrating the generation of a recombinant vaccinia virus is shown. See text for details*

scription unit is also flanked by vaccinia DNA from a non-essential region of the viral genome. General intermediate vectors of this sort are available which contain either one or two vaccinia promoters.[51] Additionally, these intermediate vectors may contain a selectable marker driven by a viral promoter. Secondly, the recombinant intermediate vector is transfected into eukaryotic cells which have been previously infected with vaccinia. In a cell which contains an infecting virus, the flanking homologous sequences in the intermediate vector and virus genome enable recombination to occur with the resulting transfer of the chimeric foreign cDNA from the plasmid to the virus genome. The recombinant viral genomes are then packaged into infectious particles. Because the flanking DNA in the plasmid was from non-essential regions, the recombinant virus produced does not require any helper functions from this point on.

[51] M. Mackett and G. L. Smith, 1986, **67**, 2067.

At best, only 0.1% of total viral progeny are recombinant, but selective techniques are available to ensure relatively easy isolation of the required virus.[51]

A variety of foreign genes have been cloned in vaccinia including rabies virus glycoprotein, HTLV III envelope protein, Hepatitus B virus surface antigen, and influenza virus haemagglutinin (see reference 51 and references therein).

The WHO have been investigating the possible use of vaccinia as a basis for a live recombinant vaccine. Experimental animals which have been inoculated with the recombinant vaccinia mentioned above have been protectively immunized against influenza virus, Hepatitus B virus, and rabies.

4.7 Baculovirus Vectors

Baculovirus is an insect specific agent which is unable to infect any other species. Viral vectors have been derived from either *Bombyx mori*[52] or, more commonly *Autographa californica*.[53] Baculoviruses are classified according to their visible characteristics namely: nuclear polyhedrosis viruses (NPVs), granulosis viruses (GVs), and non-occluded viruses (NOVs). *Autographa californica* (AcNPV) is composed of a nucleocapsid core which encompasses a circular DNA genome of ~ 130 kb.

During part of the natural viral particle maturation process, the viral nucleocapsid is enveloped by host cellular membrane before being further packaged into large polyhedral structures (occlusion bodies) via the expression of a single viral gene encoding the polyhedrin protein (~ 30 kD). To propagate the virus *in vitro*, the polyhedrin protein is solubilized to release the infectious enveloped nucleocapsids. Therefore further infection of neighbouring cells is achieved independently of the polyhedral protective structure, although the polyhedrin gene continues to be extensively expressed in concert with other late viral genes, most notably the p10 gene, which encode other components of the protective polyhedral structure. It is clear that these two genes are rendered non-essential for viral propagation and therefore may be replaced by other exogenous coding sequences. Due to the complexity of the baculoviral genome, the preferred method of gene replacement is similar to the strategy described above for pox viruses. A transfer vector is initially generated by ligating a deleted version of the polyhedrin gene into a bacterial plasmid. The viral features essential for polyhedrin gene expression are retained, as in the vector pAcYM1 (see reference 53 for a comprehensive listing of available vectors). The gene sequences coding for the desired protein are appropriately ligated into the transfer vector, and subsequently co-transfected into a host insect cell-line (*Spodotera frugiperda*) with infectious AcNPV DNA. Ensuing recombination within the transfected cell generates a helper virus independent polyhedrin-negative infectious virus which is identified by the presence of a clear polyhedrin-negative plaque upon the host

[52] S. Maeda, T. Kawai, M. Obinata, H. Fujiwara, T. Horiuchi, Y. Saeki, Y. Sato, and M. Furusawa, *Nature (London)*, 1985, **315**, 592.

[53] V. Luckow and M. Summers, *Bio/Technology*, 1988, **6**, 47.

cells. The ability of insect cells to correctly process exogenous gene products is reviewed in reference 54.[54]

It is clear that other non-essential genes of the recombinant virus obtained may be similarly manipulated. Additionally, if essential viral gene functions are provided in *trans* by the host *Spodoptera frugiperda* cell-line, then these may also be replaced giving rise to a multiply recombined infectious expression system which should be able to subvert the biochemistry of the host cell to the investigators purpose. Since baculovirus specifically infects insects and is non-pathogenic to animals and plants, such vector systems may be used in field studies to combat agricultural pests.

Finally, the option of growing the baculoviral recombinant vectors in insect larvae offers the biotechnologist a relatively cheap but effective host to provide large and continuing amounts of exogenous protein.[52]

5 SOME CONSIDERATIONS IN CHOICE OF CELL-LINE

Established cell-lines vary considerably in the ability to take up DNA via transfection. Amongst the more susceptible are human Hela cells and mouse L-cells, whereas human K562 and mouse MEL cell-lines are relatively refractory. It has been noted that transfection efficiency will decline upon prolonged passage of some cell lines *in vitro* (say > 15 passages).

It is obviously desirable to have a quick and easy assay for transfection efficiency, especially with respect to determining the experimental conditions which permit optimal expression with a specific promoter. The CAT assay is a simple transient expression system which can be used to evaluate these parameters. CAT vectors contain a prokaryotic gene coding for chloramphenicol acetyl transferase (CAT), which is absent in all mammalian cell-lines. The variety of CAT vectors available allows easy cloning of either enhancer or promoter sequences.[55] After transfection the cells are lysed by successive freeze–thaw cycles and aliquots of the resulting cellular extract can be rapidly assayed for CAT activity. Alternatively, staining the transfected cell population with anti-CAT antibodies enables the fraction of successfully transfected cells to be calculated and therefore maximized.

A number of other reporter gene systems may be used including tissue plasminogen activator (tPA) or β-galactosidase or the luciferase gene.[56]

Obviously the chosen cell-line has to be compatible with the gene system being used by the investigator. Thus if a regulatable promoter (such as metal-ion or steroid hormone inducible) is being employed to direct gene expression, then the host cell must have the appropriate receptors to allow entry of the inducing agent. Also if the gene product has to undergo specific modifications or proteolytic cleavage to achieve full activity, then it is necessary to choose a cell-line able to perform these processes.

[54] L. K. Miller, *Ann. Rev. Microbiol.*, 1988, **42**, 177.
[55] C. Gorman, L. Moffat, and B. H. Howard, *Mol. Cell Biol.*, 1982, **2**, 1044.
[56] N. Wrighton, G. McGregor, and V. Giguere, in 'Methods in Molecular Biology, Volume 7', ed. E.J. Murray, Humana Press Inc., Clifton, New Jersey, 1991.

Finally if the objective of the DNA transfection is to generate stable clones, one has to pick a cell-line appropriate for the selection system to be used (see Section 6.1).

6 TRANSIENT *VERSUS* STABLE EXPRESSION

It is apparent from the preceding sections that the advantages of transient expression are rapidity and the obviation of any selection. However much these features are appreciated by the molecular biologist, the disadvantages to the biotechnologist include the requirement of very clean, pure Form I DNA and the high background of untransfected cells. Selection for stable clones overcome both these problems.

6.1 Selection by Host Cell Defect Complementation

A number of selective techniques are based upon alleviation of inhibited synthesis of host cell nucleotides. For instance, aminopterin is a potent antagonizer of dihydrofolate reductase (dhfr) which is required for both purine and pyrimidine *de novo* biosynthesis, and is therefore extremely toxic to all cells. The lethal effect of aminopterin can be overcome by supplementing the growth medium with the appropriate nucleotides precursors which are utilized via the cell by the salvage pathway enzymes namely: thymidine kinase (tk) for thymidine, adenine phosphoribosyl transferase (aprt) for adenine, and hypoxanthineguanine phosphoribosyl transferase (hgprt) for guanine (Figure 6). Thus tk^- cell-lines will only survive in HAT medium (containing hypoxanthine–aminopterin–thymidine) if they have been transfected with the tk gene. Similar selective criteria can be defined for the aprt and hgprt genes for the corresponding negative cell-lines (for a full discussion, see reference 56). These selectable marker genes have been introduced in a variety of cloning vectors.[42]

6.2 Dominant Selective Techniques

A major disadvantage of HAT selection is the restriction of being limited to the relatively few tk^-, $aprt^-$, or $hgprt^-$ cell-lines. Thus, a number of dominant selection systems have been designed which can be more universally applied.

The *E. coli* xgprt gene is functionally analogous to the mammalian hgprt gene, but differs from its mammalian counterpart in the ability to use xanthine as a GTP precursor via a XMP intermediate (Figure 6). Mycophenolic acid (MA) inhibits the *de novo* biosynthesis of the XMP intermediate. This inhibition is more pronounced by the addition of aminopterin. Therefore only cells transfected with the *E. coli* xgprt gene will survive in medium containing MA and aminopterin supplemented with xanthine and adenine, in the absence of guanine.[57] (Thymidine must also be included in the selective medium to overcome the concomitant inhibition of pyrimidine biosynthesis by aminopterin.)

[57] R. Mulligan and P. Berg, *Proc. Natl. Acad. Sci. USA*, 1981, **78**, 2072.

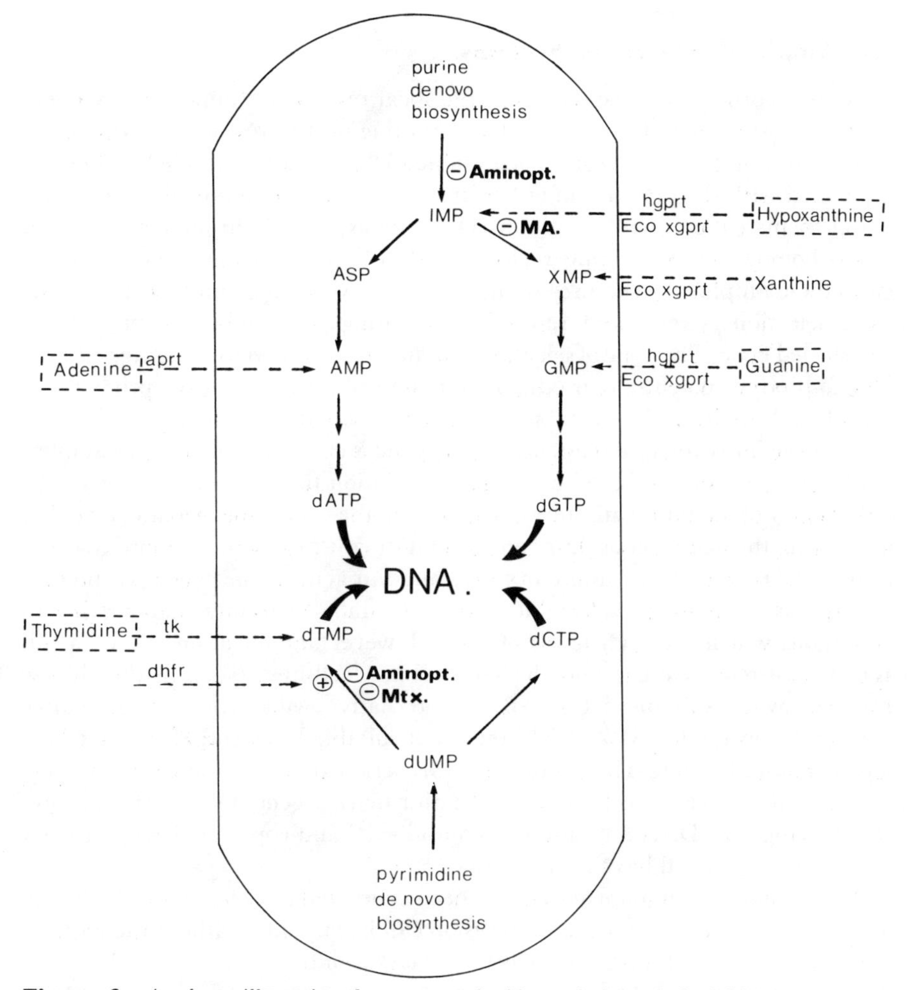

Figure 6 *A scheme illustrating* de novo *nucleotide synthesis and the inhibition of various intermediates by aminopterin (aminopt.), methotrexate (mtx), and mycophenolic acid (MA) is shown. The inhibitory effects can be alleviated by the presence of nucleotide precursors and the appropriate salvage pathway enzyme genes as indicated by dotted lines. The salvage pathway selections in which the precursor is enclosed by a dotted line are dependent upon the transfected cell being negative for the corresponding selectable marker gene. The other two selections are dominant. Eco =* E. Coli. *See text for details*

Other dominant selections are based upon conferring resistance to aminoglycosides such as G418 and hygromycin. These potent inhibitors of eukaryotic translation can be inactivated by phosphotransferases, and the genes coding for these enzymes have been isolated and cloned into a number of different vectors.[58] However, these antibiotics are relatively expensive and some cell-lines are relatively resistant to their uptake.

[58] F. Grosveld, T. Lund, E. J. Murray, A. Mellor, and R. A. Flavell, *Nucleic Acid Res.*, 1982, **11**, 6715.

6.3 Amplifiable Selection Systems

Some transformed cell-lines can undergo local reiterative rounds of DNA replication at particular chromosomal loci, resulting in the presence of many copies of genes residing at those regions of amplified DNA. These amplified regions are associated with the presence of either unstable extra-chromosomal self replicating elements called double minutes (DMs) or expanded chromosomal regions called homogenously staining regions (HSRs). The mechanism which propagates these amplified structures is unclear, but nevertheless can be exploited by some selection systems and result in generating thousands of copies of the transfected gene. This type of selection can therefore be used to generate cell-lines able to produce potentially prodigious amounts of transfected gene products.[59]

Although many such selectable markers are available (see reference 55 for a review) the dihydrofolate reductase (dhfr) gene is commonly used. This amplifiable selection system is based upon the observation that cells can overcome the lethal effect of methotrexate (mtx), which is another analogue of folate, by either amplifying the endogenous dhfr gene or mutating it to generate a more resistant form of the protein.[60] A mutant mtx-resistant dhfr gene has been cloned and may be used as a dominant selectable marker in $dhfr^+$ cell-lines under stringent conditions which use high levels of mtx. However amplification is low in this system (see reference 61) and the use of $dhfr^-$ cell-lines (such as the chinese hamster ovary sub-line DUKX-B11) and native wild type dhfr selectable marker, for example pSV2.dhfr,[62] represent suitable hosts and vectors for high levels of amplification. The transfected cells are treated with increasing doses of mtx, and the surviving cells amplify the dhfr marker gene. During this process, the flanking host DNA had also been amplified[55] and copy numbers of greater than 2000 copies/cell have been obtained.[63]

An alternative dominant system has been generated at Celltech in which high levels of amplification and gene expression can be obtained without the requirement of a mutant cell-line. The selectable marker in this system is the glutamine synthetase (gs) gene which is amplified in response to methionine sulfoximine (msx) even in cell-lines containing an active gs gene.[61] Up to 180 mg dm^{-3} of recombinant protein have been achieved with this system.

[59] R. J. Kaufman, in 'Genetic Engineering Vol 9', ed. J. Setlow, Plenum Press, New York.
[60] M. Wigler, M. Perucho, D. Kurtz, S. Dana, A. Pellicer, R. Axel, and S. Silverstein, *Proc. Natl. Acad. Sci. USA* 1980, **77**, 3567.
[61] M. I. Cockett, C. R. Bebbington, and G. T. Yarranton, *Biol Technology*, 1990, **8**, 662.
[62] S. Subramani, R. Mulligan, and P. Berg, *Mol. Cell Biol.*, 1981, **1**, 854.
[63] G. F. Crouse, R. N. McKewan, and M. L. Pearson, *Mol. Cell. Biol.*, 1983, **3**, 257.

CHAPTER 7

Plant Biotechnology

M. G. K. JONES AND K. LINDSEY

1 INTRODUCTION

A wide range of plants has been selected over a long period for use in agriculture and horticulture. For each crop there are different aims for improvement, and within a crop species there may be a range of more specific requirements. To this may be added more biotechnological applications of plant cells, such as the production of secondary metabolites by cells cultured in bioreactors. Thus the scope for applying new technologies of genetic manipulation is great. Two main interrelated disciplines are required for genetic manipulation of plants – tissue culture and cell biology, and molecular biology. Many plant species can be regenerated from single cells or explants via tissue culture. This process can itself produce genetic changes. Individual wall-less cells (protoplasts) can be fused together to produce novel hybrids and genetically modified transgenic plants can also be produced by the introduction of foreign DNA into protoplasts or cells, from which whole transgenic plants can be regenerated. In this chapter, both cell and molecular biological approaches to genetic manipulation of plants will be considered. However, to set the scene, it is first necessary to consider the aims, current practices, and limitations of conventional plant breeding.

1.1 Plant Breeding, Practices and Prospects

Conventional plant breeding usually involves the production of variability by making sexual crosses between selected genotypes possessing characters to be combined, to produce a population of plants that include superior genotypes. This is followed by extensive selection from the progeny to identify those genotypes which will eventually lead to the development of new varieties. Thus input of new characters or generation of new genetic combinations is limited to the early stages, and most of the effort is taken up by field selection over many years (eight or more, depending on the crop). Although the genetic basis of some useful characters is understood, in most instances this is not the case and the process has a considerable empirical element. Nevertheless, over the past 40 years or so, significant yield increases have been achieved, about half being contributed by genetic improvement and the other half by improved agronomic practices.

The main objectives of plant breeding are to obtain increased yield, improved quality (both nutritional and technological), disease and pest resistance, stress tolerance (*e.g.* heat, cold, drought, salinity, waterlogging), and herbicide resistance.[1]

A problem of conventional breeding is that agronomically desirable characters are frequently genetically ill-defined, and many are polygenic. The best current varieties usually have specific defects that breeders wish to improve. Thus there may be a sacrifice of quality to obtain a high-yielding variety, or the variety may be susceptible to particular pathogens that limit its range of use. In order to correct the defect, further sexual crosses are required, with a consequence that extensive selection is necessary to produce another variety. Thus by conventional approaches it is not possible to upgrade a specific variety directly. Such an approach is now made possible by genetic engineering.

A further problem of conventional breeding is that the range of genes that are accessible is limited to related species that can be crossed sexually. Useful genes present in sexually incompatible wild species, for example, can now be made available either by protoplast fusion (somatic hybridization) or by cloning and transformation. In particular, genetic engineering allows access to and introduction of genes from virtually any other organisms, *i.e.* viruses, bacteria, yeasts, algae, fungi, animals, and unrelated plants.

In addition to the modification of plants by various types of genetic manipulation, tissue culture and molecular biology can be used to facilitate aspects such as handling and multiplication of plants, and to speed up the screening and selection processes. In particular, the production of genetic maps using molecular techniques should aid identification and selection of useful genes. Entirely new applications of crop plants, such as the production of pharmaceuticals by using plants as bioreactors, can now be achieved.

1.2 Tissue Culture and Plant Manipulation

Whole plants, parts of plants, and also single cells (Figure 1) can be grown in sterile culture, supported by media that contain salts, vitamins, a carbon source, and plant growth regulators. There are two types of applications of tissue culture that reflect a fundamental consequence of the growth of plant cells in culture. The first involves culture of organized tissues or meristems, from which whole plants can be regrown, with the aim of maintaining strict genetic fidelity. The second involves applications where variability occurs in regenerated plants, either as a result of growing disorganized callus cultures, or from manipulative procedures.[2]

[1] M. G. K. Jones, in 'Plant Genetic Engineering', ed. J. H. Dodds, Cambridge University Press, Cambridge, 1985, p. 269.

[2] M. G. K. Jones and A. Karp, in 'Advances in Biotechnological Processes', ed. A. Mizrahi and A. L. van Wezel, vol. 5, A. R. Liss, New York, 1985, p. 91.

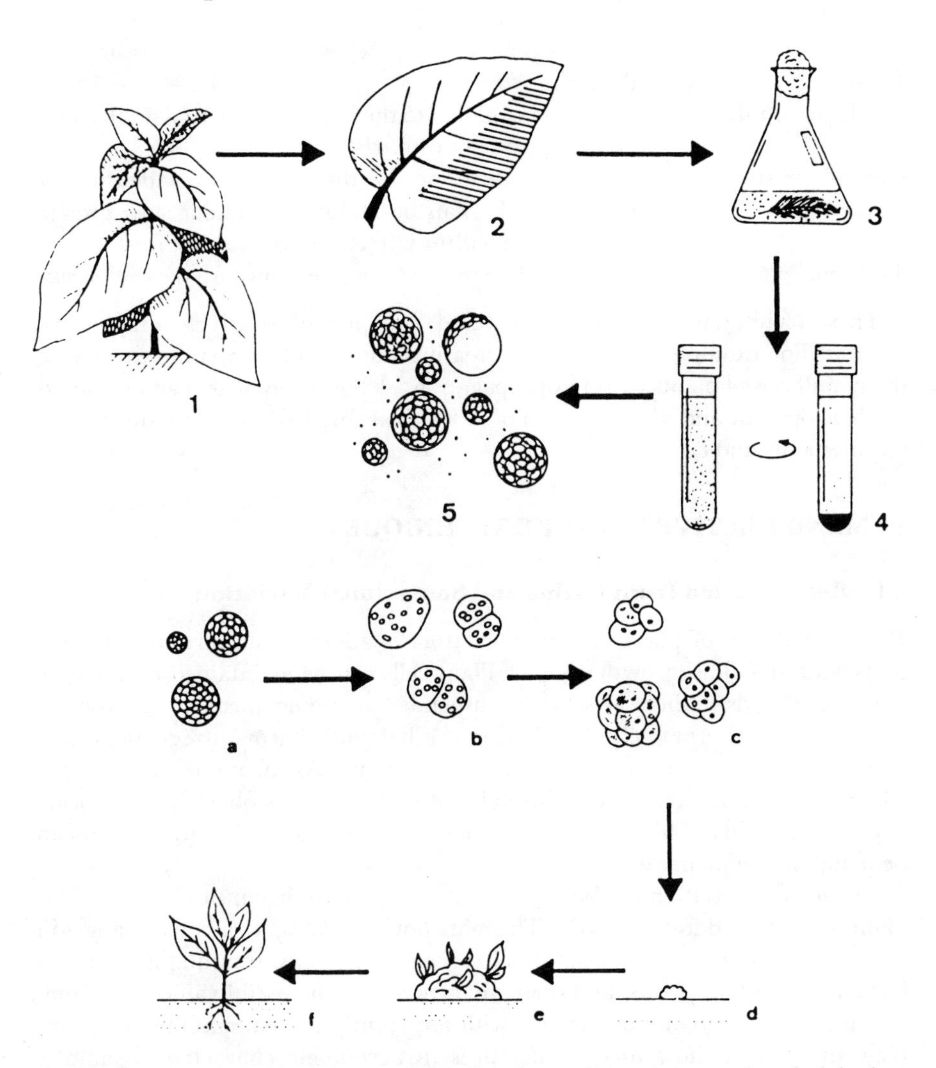

Figure 1 *(top) Isolation of plant protoplasts from leaves; (bottom) regeneration of whole plants from protoplasts*
(Courtesy of M. de Both)

1.3 Applied Techniques for Maintenance of Genetic Fidelity

The range of culture techniques available for plant breeders in which genetic fidelity is maintained includes:[3]

Micropropagation	– clonal multiplication *in vitro*
Virus elimination	– the removal of latent virus infections by meristem culture and/or heat or chemotherapy
Embryo rescue	– the recovery of hybrid plants in culture, usually from

[3] M. G. K. Jones, in 'The World Biotech Report: Agricultural Economics & Technology', Online Publications, London, 1987, p. 45.

	wide sexual crosses, when endosperm tissues are deficient
Ploidy manipulation	– reduction of ploidy to the haploid level, and doubling to give homozygous diploids
Germplasm storage	– mother plants for outbreeding crops, or a wide range of genetic stocks, can be maintained under slow growth conditions in culture, free from pathogens, frost, *etc.*
Transport	– pathogen-free material can be transported easily *in vitro*

These techniques are now widely used and involve relatively simple procedures. For example, there is an increasing number of companies producing many millions of plants by micropropagation.[4,5] The approaches can also speed up development and dissemination of new or existing varieties, or reduce space requirements and costs.

2 MANIPULATIVE CULTURE TECHNIQUES

2.1 Regeneration from Callus and Somaclonal Variation

If an imbalance of plant growth regulators is added to culture media, then disorganized 'callus' growth occurs.[6] Plant cells can be maintained in this state for long periods, either on solid or suspended in liquid media, with routine subculturing. Alternatively, by altering the balance of growth regulators, it is possible in many cases to regenerate intact plants. As mentioned above, one consequence of passage of cells through even a brief callus phase is the appearance of variability. This variability is known as 'somaclonal variation', and can be manifested in many ways.

Somaclonal variation has been widely described both in cultured cells and for plants regenerated from culture.[7] The more obvious examples involve changes in chromosome number and structure of regenerated plants, resulting in morphologically abnormal plants, and these are generally of no useful value. Nevertheless, it is possible to obtain variants with apparently normal chromosome complements, which show useful differences in agronomic characters. Examples include crops such as potato, tomato, cereals, carrot, and celery, and there are examples of commercial products on the market for the latter two crops. This approach is of particular interest for crops like potato, which are vegetatively propagated, tetraploid, and extremely heterozygous. This means that a variety is lost if a sexual cross is made, and since the use-span of a potato variety may be 75 years (*e.g.* varieties like Russet Burbank and King Edward), it is important to develop methods to upgrade existing varieties. The usefulness of somaclonal variation[8] is summarized in Table 1.

[4] M. G. K. Jones, in 'Plant Products and the New Technology', ed. K. W. Fuller and J. R. Gallon, Phytochemistry Society, Oxford University Press, Oxford, Vol. 26, p. 215.
[5] K. L. Giles and W. W. Morgan, *Trends Biotechnol.*, 1987, **5**, 35.
[6] K. Lindsey and M. M. Yeoman, in 'Cell Culture and Somatic Cell Genetics of Plants', Vol. 2, ed. I. K. Vasil, Academic Press, Orlando, Florida, 1985, p. 61.
[7] 'Somaclonal Variation and Crop Improvement', ed. J. Semal, Nijhoff, Dordrecht, 1986.
[8] A. Karp, M. G. K. Jones, D. Foulger, N. Fish, and S. W. J. Bright, *Am. Potato J.*, 1987, **66**, 669.

Table 1 *Problems and advantages of somaclonal variation*

Problems	*Advantages*
1. Not all characters change 2. Many characters change in a negative direction 3. Not all variation is novel 4. Not all changes are stable 5. Field evaluation is still required	1. Changes can occur to produce useful agronomic characters 2. Changes occur at high frequency 3. Some changes are novel, and cannot be achieved by conventional technology

For potato, variants with altered yield, disease resistance, flower colour, tuber skin colour, sugar content, and storage characteristics have been obtained[2,7] with up to 3% of regenerants exhibiting potentially useful variation.[8] This figure is much higher than for selections following a comparable programme involving sexual crosses.

Somaclonal variation can occur after any manipulation that involves a callus phase, and thus also presents a complicating factor that must be borne in mind when other manipulative techniques such as somatic hybridization or transformation are carried out.

2.2 Mutant Selection from Culture

It is possible to include selective agents in culture media to try to select desired mutants that may be induced by treatment with physical or chemical mutagens, or by exploiting somaclonal variation.[9] The potential advantage of selection *in vitro* is that millions of individual cells can be treated and selected under controlled conditions in a relatively small space. The main problem is, however, that a good understanding of the biochemical basis of the desired mutant is required in order to design the best selection system, but at present this is not available for most agronomically desirable characters. Successful selection can be achieved for specific traits such as resistance to herbicides, toxins, heavy metal ions, and amino acid analogues. Thus maize and alfalfa plants resistant to herbicides have been selected in this way, but if a less specific selection pressure (*e.g.* salt) is applied, physiological changes resulting in resistance in culture may not be reflected in regenerated plants.

2.3 Protoplast Fusion

Once the walls of plant cells have been removed to produce protoplasts, it is possible, by fusion techniques, to obtain hybrid cells of widely different origins, and then to culture the heterokaryons to produce somatic hybrid plants. Chemical fusogens (*e.g.* polyethylene glycol; Ca^{2+} ions at high pH) have been used to induce fusion, but now it is possible to use electrical techniques to fuse protoplasts

[9] P. J. Dix, in 'Plant Cell Culture Technology', ed. M. M. Yeoman, Blackwell, Oxford, 1986, p. 143.

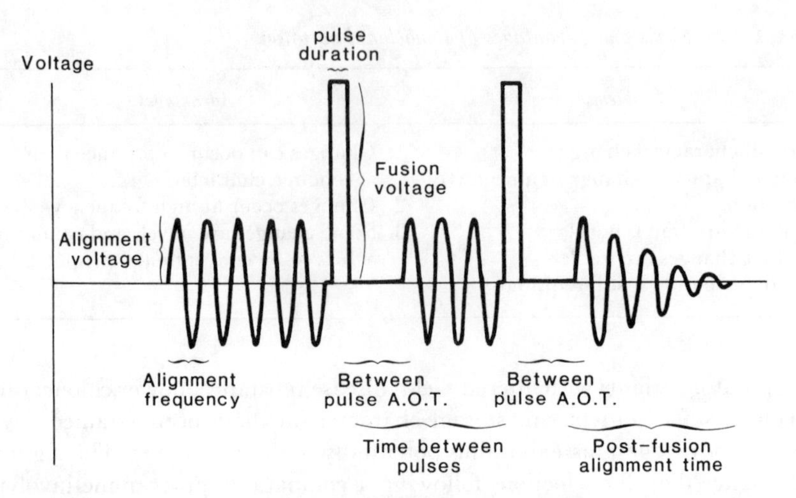

Figure 2 *Electrical profile required for electrofusion. The alternating (alignment) field causes protoplasts to align in chains, and fusion is induced by one or more direct current pulses. The alignment field is then briefly switched on to promote fusion, but then rapidly reduced to zero (AOT = alignment of time)*

at relatively high frequencies.[10,11] The procedure of 'electrofusion' (Figure 2) involves placing protoplasts between electrodes in a non-conductive medium, and applying first a high-frequency alternating field (about 1 MHz) which causes protoplasts to align in chains by 'dielectrophoresis', and then applying direct current pulse(s) (10–100 μs, 1–3 kV cm^{-1}) to cause reversible membrane breakdown ('pore' formation). Contacting protoplasts can then fuse together. Useful applications of protoplast fusion technology include:

(i) the combination of complete genomes (*e.g.* of sexually incompatible species).
(ii) partial genome transfer from a donor (genome fragmented) to a recipient.
(iii) manipulation of organelles (*e.g.* chloroplasts, mitochondria): of particular interest in relation to cytoplasmic male sterility and herbicide resistance.

Taking potato as an example, the transfer of resistance to the plant viruses Potato Leaf Roll Virus (PLRV), Potato Virus X (PVX), and Potato Virus Y (PVY) has been achieved by fusing together protoplasts of the virus-resistant wild species (*Solanum brevidens*) with those of an agronomically selected potato (*Solanum tuberosum*) line to produce somatic *S. tuberosum* hybrid lines that exhibit PLRV, PVX, and PVY resistance (Figure 3).[12,13] *S. brevidens* is sexually incompatible with potato, but the hybrids are female fertile. Thus useful genes for resistance to the viruses can be introduced into potato germplasm without the need for detailed knowledge of the underlying molecular basis of the resistance.

[10] M. J. Tempelaar and M. G. K. Jones, *Planta*, 1985, 205.
[11] M. J. Tempelaar and M. G. K. Jones, *Plant Cell Rep.*, 1985, 92.
[12] A. Austin, M. A. Baier, and J. P. Helgeson, *Plant Sci.*, 1985, **39**, 75.
[13] M. G. K. Jones *et al.*, 'Progress in Plant Cellular and Molecular Biology', ed N. J. J. Nijkamp *et al.*, Kluwer, Dordrecht, 1990, p. 286.

Figure 3 *Somatic hybrid plant (centre), obtained by fusing protoplasts of* S. tuberosum *(left) and* S. brevidens *(right)*

This approach is being used for other crop species, to access useful characters in sexually incompatible wild species, or to combine selected characters from different genotypes within the same crop but avoiding recombination.

2.4 Transformation – Production of Transgenic Plants

Various approaches are available for the introduction of foreign genes into plant cells or protoplasts. The tissue culture aspects of this process are considered in this section, and molecular aspects are discussed later. The aim is to introduce foreign genes such that they are stably integrated within the genomes of plants and the gene products are expressed in a heritable manner.

2.4.1 Transformation Using Agrobacterium *as a Gene Vector*. The most widely used approach is to make use of the natural genetic engineering ability of the soil bacteria *Agrobacterium tumefaciens* and *A. rhizogenes* to transfer genes into plant cells. Wild type *A. tumefaciens* induces tumours at sites of wounds of dicotyledonous plants. This process depends on the presence in the bacterium of a tumour-inducing (Ti) plasmid (Figure 4). The presence of phenolic compounds from the wounded cells induces transcription of genes in the virulence (VIR) region of the plasmid that are involved in the excision of part (*ca.* 10%) of the plasmid known as the T (transferred) region. The T-DNA is then transferred to the plant cell and stably integrated into its genome. Wild type T-DNA contains genes that code for the production of unusual amino acids, opines, and also for plant hormones (auxin and cytokinin). The production of excess hormones causes a gall to form at the site of infection, and the bacteria can utilize the opines produced as a carbon and nitrogen source. Similarly, *A. rhizogenes* transfers Ri T-DNA that induces the formation of transformed 'hairy roots' at infection sites,

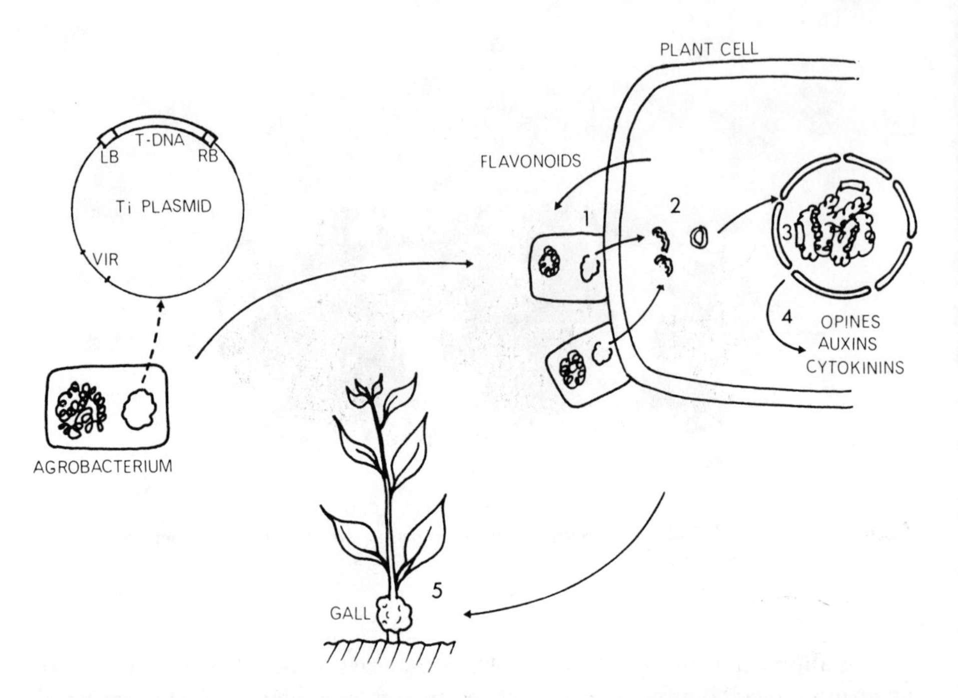

Figure 4 *Diagrammatic representation of* Agrobacterium tumefaciens, *containing chromosomal DNA and the tumour-inducing (Ti) plasmid, and of the processes involved in the genetic colonization of plants. (1) Attachment of bacteria to the plant cell, on wounding. (2) T-DNA is excised at its left (LB) and right (RB) borders, by gene products of the virulence (VIR) region, and transferred to the plant cell. (3) Integration of T-DNA into the plant cell genome. (4) Transcription and translation of the T-DNA. (5) Induction of cell division in the wounded region, resulting in tumour (gall) formation*

from which transformed plants can be regenerated (Figure 5) (see Bevan and Chilton[14] for a review).

A series of vectors has been developed based on the Ti plasmid transfer system. In these, the wild type genes have been excised (to give 'disarmed' vectors) and they can be replaced by genes of interest to be introduced into plants. Removal of the genes that code for phytohormone overproduction also means that phenotypically normal transformed plants can be regenerated. Any inserted gene must have appropriate promoter and polyadenylation regions to be successfully expressed in the host plant, and using these systems many dicotyledonous plants, transformed with a range of gene constructs, have been produced.

To obtain whole plants, however, it is necessary to be able to regenerate plants from tissue explants in culture, or from isolated protoplasts. Thus basic studies in tissue culture of regeneration systems for genotypes of interest are required.

A major problem with using *Agrobacterium* as a gene vector is that it does not transform all dicotyledonous species with equal efficiency and it does not nor-

[14] M. W. Bevan and M. D. Chilton, *Ann. Rev. Genet.*, 1982, **16**, 357.

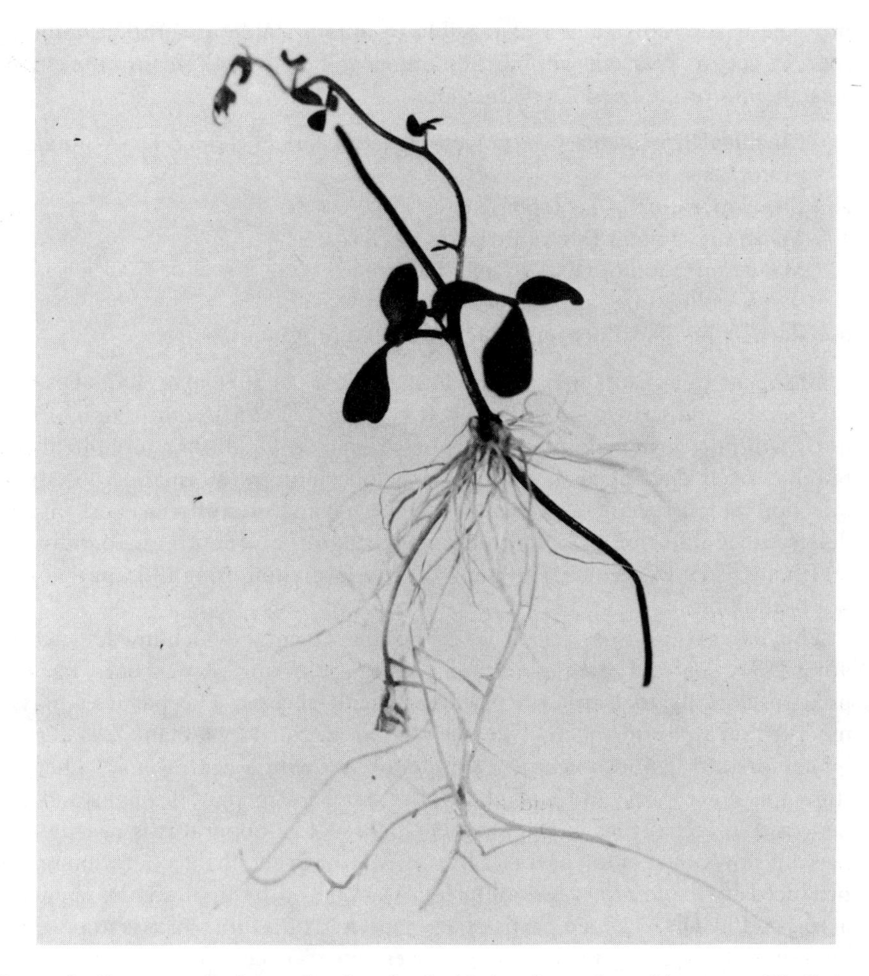

Figure 5 Lotus corniculatus *plant inoculated with* Agrobacterium rhizogenes. *The 'hairy roots' that grow from the site of inoculation contain transferred DNA from the bacterial Ri plasmid. Transformed plants can be regenerated from these roots*

mally infect monocotyledonous plants, such as cereals and grasses, which are the most important food crops. However, some monocot exceptions are now known.[15–18] This fact has led to studies of alternative methods of gene insertion, which are discussed next.

2.4.2 Direct Gene Transfer. From the need to develop transformation systems for monocotyledonous plants, and the knowledge that viral nucleic acids could be introduced into protoplasts, study of direct introduction of gene constructs into

[15] J. P. Hernalsteens, L. Thia Toong, J. Schell, and M. Van Montagu, *EMBO J.*, 1984, **3**, 3039.
[16] G. M. S. Hoykaas-Van Slogteren, P. J. J. Hoykaas, and R. A. Schilperoort, *Nature (London)*, 1984, **311**, 763.
[17] A. C. F. Graves and S. L. Goldman, *Plant Mol. Biol.*, 1986, **7**, 43.
[18] W. Schäfer, A. Görz, and G. Kahl, *Nature (London)*, 1987, **327**, 5239.

protoplasts showed that it was possible to obtain stable integration without a specific vector. Various approaches are now being used to introduce DNA directly into protoplasts. These include:

1. Chemical treatments (*e.g.* polyethylene glycol) to induce DNA uptake by protoplasts.
2. Electroporation of protoplasts.
3. Micro-injection of DNA into plant nuclei.
4. Macro-injection of DNA to soak meristems.
5. 'DNA-pollination' approaches.
6. Particle bombardment using DNA coated tungsten particles.

Chemical treatments involve the formation of calcium phosphate–DNA co-precipitates and treatment with polyethylene glycol.[19] Micro-injection of DNA into plant nuclei (in protoplasts or tissues) is more difficult than for animal cells, but has been developed as a procedure for plant transformation.[20] 'Macro-injection' is less specific and involves soaking immature inflorescences in DNA delivered by infection of DNA solutions through the structure surrounding the meristems.[21] However, early promise of transformation using this approach has not been realized.

The 'biolistic' approach, *i.e.* particle bombardment,[22] which involves accelerating DNA coated 1 μm tungsten or gold particles into plant tissues, has been used successfully to transform various difficult species. The particles may be mounted on a plastic 'macroprojectile' and accelerated by a blank .22 charge or by compressed helium towards a stopping plate with a central hole. The plate stops the macroprojectile and allows the particles to pass through, and they penetrate target tissues in the chamber below. The apparatus is evacuated to prevent slow down of the particles by air. Alternatively, the acceleration may be produced by explosive evaporation of a water drop by discharge of capacitors across it. The DNA-coated particles are deposited on a film which is stopped by a mesh in this case,[23] and again the particles penetrate cells of the target tissues.

The biolistic approach has been used successfully to produce transgenic plants of maize and wheat. The constructs used have included strong constitutive promoters (*e.g.* the Cauliflower Mosaic Virus 35S promoter or rice actin promoter) driving selectable marker genes such as neomycin phosphotransferase (nptII), which confers resistance to the antibiotic kanamycin, or phosphinothricin acetyl transferase (bar), which confers resistance to the herbicide basta (phosphinothricin). The appropriate antibiotic or herbicide is used to select transgenic tissues. The target tissue is usually embryogenic cell suspension culture cells layered over agar. Transgenic plants of the grain legume soybean, have also been produced by this approach.[23]

The other most widely applied method of gene transfer into plant protoplasts is

[19] K. Linsdey, M. G. K.Jones, and N. Fish, *Methods Mol. Biol.*, 1988, **4**, 519.
[20] T. J. Reich, V. J. Tyer, and B. L. Miki, *Bio/Technology*, 1986, **4**, 1001.
[21] A. de la Peña, H. Lörz, and J. Schell, *Nature (London)*, 1987, **327**, 70.
[22] T. M. Klein, E. D. Wolf, R. Wu, and J. C. Sanford, *Nature (London)*, 1987, **327**, 70.
[23] D. E. McCambe, W. F. Swain, B. J. Martinell, and P. Christou, *Bio/Technology*, 1988, **83**, 923.

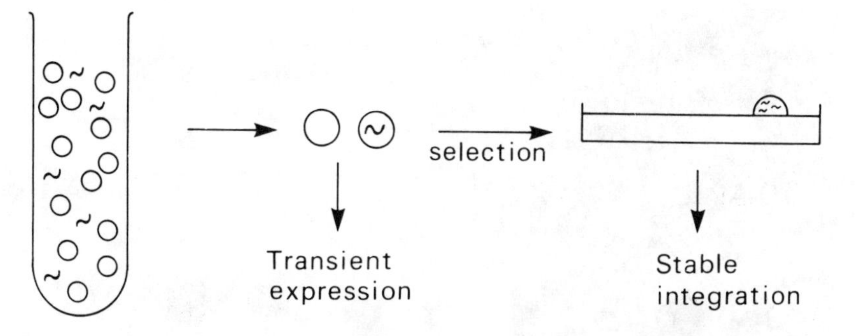

Figure 6 *Direct gene transfer into protoplasts. DNA is introduced directly into protoplasts, and the gene products can be assayed by transient expression or for integration after selection*

direct gene transfer (Figure 6) using other chemical (polyethylene glycol) or electrical approaches to introduce DNA into protoplasts.

Electroporation involves the introduction of DNA through pores formed by the application of DC electrical pulses, using the same principle as in electrofusion.[24]

Using direct gene transfer to introduce genes encoding selectable markers, a range of transformed plant cells, including monocots and dicots, have been selected.[25,26] The most notable of these is rice, where many transgenic plants have been regenerated because it is relatively easier to regrow whole plants from single cells than for other cereals. Another use of direct gene transfer is to study 'transient expression' of molecular constructs (*e.g.* reference 27). To obtain enough stably transformed tissues for analysis of gene expression may take several months, but the behaviour of introduced, but non-integrated gene constructs can be studied after 1–2 days culture, following chemical introduction or electroporation into protoplasts (Figure 7) or after particle bombardment. This approach allows a much more rapid analysis of molecular constructs.

The main drawback of direct gene transfer methods that involve protoplasts is the requirement for the regeneration of protoplasts back to plants if transgenic plants are required. This is more time consuming than using *Agrobacterium* as a vector.

However, with the range of techniques available, and with rapidly accumulating knowledge of the regeneration of plants from various sources in culture, it should be possible to transform most crop plants. Some applications of these techniques are described in the following sections.

2.5 Plant Cell Cultures for the Production of Useful Chemicals

A large number of commercially valuable chemicals in use today are derived directly from plant material, including about 25% of all prescription pharma-

[24] H. Jones, M. J. Tempelaar, and M. G. K. Jones, in 'Oxford Surveys of Molecular and Cell Biology', ed. B. J. Miflin, Oxford University Press, Oxford, 1987, **4**, p. 347–357.

[25] M. E. Fromm, L. P. Taylor, and V. Walbot, *Nature (London)*, 1986, **319**, 791.

[26] R. S. Boston, M. R. Becwar, R. D. Ryan, P. B. Goldsborough, B. A. Larkins, and T. K. Hodges, *Plant Physiol.*, 1987, **83**, 742.

[27] H. Jones, G. Ooms, and M. G. K. Jones, *Plant Molec. Biol.*, 1989, **13**, 503.

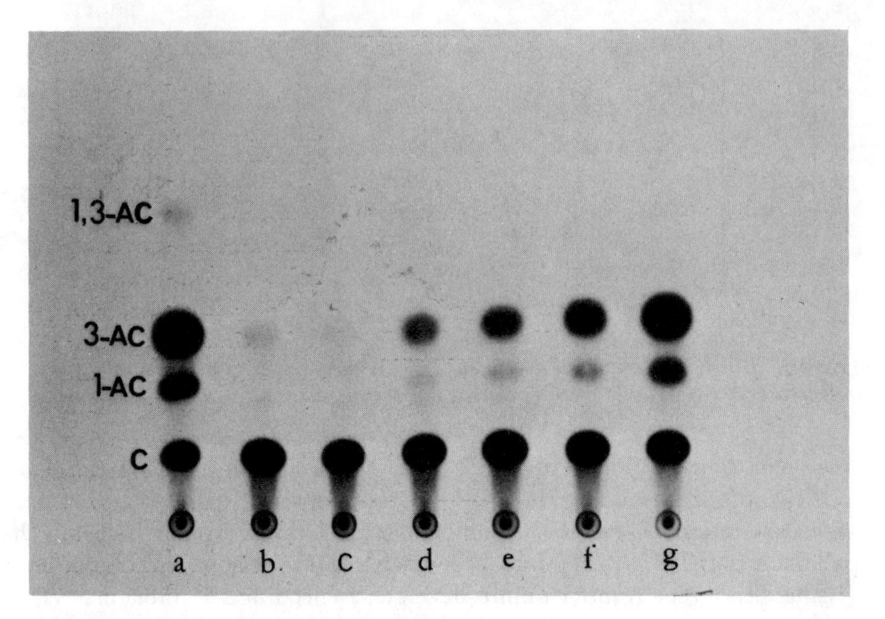

Figure 7 *Transient expression of chloramphenicol acetyltransferase (CAT) activity in electroporated protoplasts of sugar beet with a CaMV35S promoter–CAT construct:*
lane a: activity from 0.1 unit of CAT from E. coli
lane b: negative control, protoplasts
lane c: activity in 10^4 protoplasts
lane d: activity in 5×10^4 protoplasts
lane e: activity in 10^5 protoplasts
lane f: activity in 5×10^5 protoplasts
lane g: activity in 10^6 protoplasts
C = unconverted chloramphenicol; 1-AC = 1-acetylchloramphenicol; 3-AC = 3-acetylchloramphenicol; 1,3-AC = 1,3-diacetylchloramphenicol

ceuticals. The types of compounds obtained in this way fall into five broad categories of application: drugs, flavours, perfumes, pigments, and agriculturally useful chemicals (Table 2). Biochemically, the great majority of the compounds are secondary metabolites. Specific secondary products accumulate only in a restricted range of species, and sometimes only in a single species or genus (*e.g.* capsaicin, the pungent principle in the chilli pepper, *Capsicum frutescens*), and often within a specific organ or tissue, or at a specific stage of development in that species. In many cases the function of these compounds is uncertain, but their retention through evolution indicates some selective advantage to the plant.

Over the past 30 years, much effort has been put into by-passing the intact plant by exploiting cell culture techniques to produce specific secondary metabolites. The potential advantages of *in vitro* technology include production under defined environmental conditions, free from disease and pests, and on a continuous basis.[28] However, a number of technical and biological problems

[28] M. M. Yeoman, M. B. Miedzybrodzka, K. Lindsey, and W. R. McLauchlan, in 'Plant Cell Cultures: Results and Perspectives', ed. F. Sala, B. Parisi, R. Cella, and O. Cifferi, Elsevier-North Holland, Amsterdam, 1980, p. 327.

Table 2 *Some useful plant products*

Product	*Species*	*Application*
Atropine	*Atropa belladonna*	Ophthalmic
Scopolamine	*Datura* spp.	Anti-sea sickness
Vincristine	*Catharanthus roseus*	Anti-cancer
Digoxin	*Digitalis* spp.	Cardiatonic
Diosgenin	*Dioscorea deltoidea*	Contraceptive
Morphine alkaloids	*Papaver* spp.	Analgesic
Quinine	*Cinchona officinale*	Anti-malarial
Anthocyanins	Various spp.	Pigments
Saffron	*Crocus sativus*	Pigment/flavour
Shikonin	*Lithospermum*	Pigment
Rose oil	*Rosa* spp.	Perfume
Geranium oil	*Geranium* spp.	Perfume
Jasmine oil	*Jasminum* spp.	Perfume
Capsaicin	*Capsicum* spp.	Flavour
Glycyrrhizin (Liquorice)	*Glycyrrhiza glabra*	Flavour
Quinine	*Cinchona officinale*	Flavour
Ginger	*Zingiber officinale*	Flavour
Pyrethrins	*Chrysanthemum cinerarifolium*	Insecticidal

have been encountered in developing such processes, not least of which is the fact that, in the majority of cases studied so far, cultured cells have been found to accumulate very low levels of specific secondary metabolites, and usually lower levels than those normally found within the plant. Two factors appear to be important in determining the levels of product within cultured cells. The first is related to the variability (somaclonal variation) and instability of gene expression and metabolic activity within cultured plant cells. These factors affect the level of activity of secondary metabolic enzymes, and appear to play a role in determining the enzymological 'hardware'. A second major influence on production in cell cultures relates to the commonly observed inverse relationship between cell division rate and secondary metabolite accumulation *in vitro*. Accumulation in many examples is confined to the post-division phases of the sigmoidal growth cycle of cultured plant cells and evidence suggests that, in some examples at least, this is the result of changes in the direction of flux of precursor molecules common to both primary (cell-division related) and secondary (cell-expansion or cell-maturation related) pathways. It is also observed that multicellular organization in cell cultures favours the accumulation of secondary metabolites, perhaps partly as a result of the limitations imposed on cell divisions by the multicellular condition. This second set of factors further defines the biosynthetic behaviour of cultured plant cells which can also be modified by altering the nutritional and hormonal composition of the medium. Taken together, these genetical, physiological, and biochemical parameters can be used to construct large-scale production systems for specific chemicals.

The current strategies are based either on (1) fermenter systems, in which cells are grown up to a stationary phase in a one- or two-stage process and then harvested and the product extracted, or (2) fixed-bed systems, in which cells are immobilized as multicellular aggregates in biologically inert matrices such as gels, foams, or hollow fibres, and the product (which must be naturally released, or its release induced from the cells) is harvested in a continuous or semi-continuous process.

The mass culture system has been used successfully for the commercial production of shikonin, a red pigment, by the Mitzui Corporation of Japan. The success of the process is based first on the exploitation of *in vitro* biosynthetic variability, by the selection of cell-lines which produce high levels of the compound. The cells are subsequently cultured in a two-stage fermentation: in the first stage, the cells are grown in bulk, and in the second stage they are subjected to a change in medium composition, *i.e.* to one which favours shikonin production rather than cell division.

Immobilized cell culture systems have yet to be used in a fully commercial process, but there are potential advantages of both a physiological and chemical engineering nature over the mass culture system.[29] For example, the aggregated nature of immobilized plant cells in itself promotes enhanced yields (compared with freely suspended cells) of at least some secondary compounds and, once established, cultures can remain in a continuous or semi-continuous production phase for many months. Cultures which naturally export the product into the surrounding nutrient medium (*e.g.* capsaicin production by *Capsicum frutescens*) are ideal candidates for immobilization, but permeabilization techniques are being developed to induce the release of intracellular metabolites without damaging cell viability. The potential for genetically engineering plant cells for enhanced secondary metabolite production is discussed briefly below.

3 CROP MODIFICATION USING MOLECULAR TECHNIQUES

The rapid development of molecular biological techniques provides for major advances in the modification of agronomically useful traits in crop plants. It is now technically possible to identify, isolate, modify, transfer, and obtain expression of a range of specific genes, and there are now many examples in which a crop plant has been produced which possesses a stably modified, economically (or potentially economically) useful phenotype. The main drawback in the science has been that although specific breeding aims (desirable novel phenotypes) can be identified, the basis for the expression of desired characters is in many cases unknown or, at best, only incompletely understood. Many characters are polygenic, so making the isolation and identification process more difficult. However, the need to identify useful genes at a molecular level has stimulated molecular mapping and research aimed at identifying such genes.

[29] K. Lindsey, in 'Secondary Metabolism of Plant Cell Cultures', ed. P. Morris, A. H. Scragg, A. Stafford, and M. W. Fowler, Cambridge University Press, 1986, p. 143.

Molecular biological techniques can be employed in a variety of ways to facilitate crop improvements, both to speed up conventional breeding procedures and to upgrade existing varieties. Let us first consider the use of such techniques for the isolation and identification of specific genes.

3.1 Gene Identification and Isolation

The methods available can be categorized broadly into four groups: (1) the use of cDNA and genomic libraries, (2) the use of restriction fragment length polymorphisms, (3) transposon and insertional mutagenesis, and (4) use of the Polymerase Chain Reaction (PCR).

The construction and screening of cDNA and of genomic libraries has proved to be a powerful technique for isolating and characterizing the structure of genes, in a range of major crop species. The construction of gene libraries comprises a series of generalized steps: (1) the generation of fragments of DNA, which are then (2) joined to a cloning vector, such as a plasmid or bacteriophage, and (3) the introduction of the vector into a host cell, such as *E. coli*, in which it can replicate. Such a procedure results in a collection of cloned DNA fragments which can subsequently be screened for a particular gene. Genomic libraries of random DNA fragments can be produced by cutting total DNA with restriction enzymes, and it is possible to calculate the probability of including any specific DNA sequence in such a library.[30] Thus, in a library of 20 kb fragment size, 4.2×10^5 recombinants must be screened if a particular sequence is to be obtained with a certainty of 95%. In contrast to this so-called shotgun cloning technique, it is possible to improve the likelihood of obtaining a particular gene. Chromatographic or electrophoretic techniques have been investigated as a means of enriching fractions with DNA fragments,[31] and if the expression of a particular gene is enhanced in a particular tissue or at a particular developmental stage, such that the steady-state level of the transcripts is relatively abundant, cDNA cloning is a useful technique. Here, the enzyme reverse transcriptase is used to generate cDNA ('complementary' DNA) copies of extracted mRNAs. It may be possible to generate cDNA from purified mRNA, but for low-abundance transcripts it is possible to make a cDNA library which can be screened. Since cDNAs are produced from transcripts, *i.e.* from the products of active genes, the screening of 'redundant' (non-transcribed) DNA, of which the plant genome is largely composed, is avoided and the probability of recovering a sequence of interest from a population of recombinants is correspondingly increased.

The identification of particular genes in a library can be attempted using one or more of a number of techniques. If the amino acid sequence of the translation product is known, it may be possible to synthesize a radioactive oligonucleotide (usually 15–20 nucleotides long) from a unique region of the protein. This can be used as a selective probe by its ability to hybridize with the desired DNA

[30] L. Clarke and J. Carbon, *Cell*, 1976, **9**, 91.
[31] S. M. Tilghman, D. C. Tiemeir, F. Polsky, M. H. Edgell, J. G. Seidman, A. Leder, L. W. Enquist, B. Norman, and P. Leder, *Proc. Natl. Acad. Sci. USA*, 1977, **74**, 4406.

fragment. A potential difficulty is that most amino acids are encoded for by more than one codon; however, methionine and tryptophan are encoded by single triplets, and so an oligopeptide sequence containing these amino acids would be expected to increase the specificity of the synthetic probe.

It has been observed that the nucleotide sequences of genes encoding certain proteins important for basic cellular functions, such as respiration, have been conserved in evolution and are similar in species of diverse phylogenetic origins. It has therefore been possible to construct 'heterologous' probes, namely radioactive nucleotide sequences of particular genes, and use them to screen for the analogous gene in a library of a different species. This method has been employed, for example, to isolate the mitochondrion-encoded subunit II of cytochrome oxidase using a probe from yeast.[32]

Nucleic acid hybridization can be exploited further to recognize the genetic nature of a cDNA fragment by hybrid-select (hybrid released) translation. In this technique, a cloned cDNA is hybridized with total mRNA or even total cellular RNA, and after washing away unbound RNA the remaining mRNA can be eluted and translated in a cell-free system. The translation product can then be further characterized.[33] Forde *et al.*,[34] for example, have used this approach to identify B hordern genes in cDNA libraries of barley.

Since cDNAs are synthesized from mature mRNA, they do not contain introns or 5′ and 3′ sequences. The latter are of interest in the context of gene regulation, and cDNAs, identified for example by hybrid select translation, can be used to probe genomic libraries for fragments containing unspliced sequences of specific genes for further analysis. This technique has been used for the characterization of, for example, genes for storage proteins (*e.g.* reference 35) and for leghaemoglobin.[36]

3.2 Polymerase Chain Reaction

Development of the Polymerase Chain Reaction (PCR) has had a dramatic effect on approaches to gene isolation. The fact that microgram quantities of DNA may be produced of a given sequence, using appropriate primers, means that the more time consuming production of genomic libraries may not be required. For example, if sequences are available on a database for a particular gene, primers can be made, with redundancies if necessary, of consensus sequences, and the gene from a different organism amplified directly. To facilitate cloning, suitable restriction sites may be added to the primers. The possible approaches using PCR are too broad to detail here: they are well outlined in

[32] T. D. Fox and C. J. Leaver, *Cell*, 1981, **26**, 315.
[33] R. P. Ricciardi, J. S. Miller, and B. E. Roberts, *Proc. Natl. Acad. Sci. USA*, 1979, **75**, 4927.
[34] B. G. Forde, M. Kreis, M. B. Bahramian, J. A. Matthews, B. J. Miflin, R. D. Thompson, D. Bartels, and R. B. Flavell, *Nucleic Acids Res.*, 1985, **9**, 6689.
[35] M. Kreis, M. S. Williamson, J. Forde, D. Schmutz, J. Clark, B. Buxton, J. Pywell, C. Marris, J. Henderson, N. Harris, P. R. Shewry, B. G. Forde, and M. J. Miflin, *Philos. Trans. R. Soc. London, B*, 1986, **314**, 355.
[36] H. Baulcombe and D. P. S. Verma, *Nucleic Acids Res.*, 1978, **5**, 4141.

Chapter 3 and in a number of other publications (*e.g.* references 37 and 38). PCR approaches may also be used for sequencing, which has also advanced rapidly with the availability of automated DNA sequencers.

3.3 Restriction Fragment Length Polymorphisms (RFLPs) and Random Amplified Polymorphic DNA (RAPDs)

Although the techniques described above are invaluable for the isolation of single genes, it is more difficult to retrieve sequences encoding polygenic traits, *i.e.* for those phenotypes which are the result of interactions between two or more gene products. Such genes may be spatially separated on a single chromosome or on different chromosomes. One approach to tackle this problem is to use restriction fragment length polymorphisms (RFLPs) as genetic markers, and construct linkage maps. Digestion of genomic DNA from a single plant with a particular restriction endonuclease will produce a particular pattern of fragmentation, as observed when it is separated by electrophoresis and probed with a DNA clone of a gene of known or even unknown constitution. Should mutations be present in the same restriction site in different individual plants, the genetic heterogeneity (*i.e.* the RFLPs) can be used to produce molecular genetic maps.[39,40] Such maps can be constructed by analysing the segregation of RFLPs among the progeny of a sexual cross. Furthermore, it is possible to identify polymorphisms which are linked to quantitative (polygenic) traits, such as to genes controlling the soluble solids content of tomato fruit.[41]

RFLP approaches take time to carry out, and usually involve radioactive labelling of defined probes for hybridization to membranes following Southern blotting. A more rapid procedure, Random Amplified Polymorphic DNA (RAPD), uses PCR and avoids the need for lengthy extractions and radioactive labels.[42] It involves choosing random sequences of primers, of about 10 nucleotides, and using only one primer in the PCR reaction. If, by chance, the random sequence anneals and there is a similar sequence up to 2000 base pairs away, then a PCR fragment will be amplified. Thus there is no need for sequence data, primers can be screened rapidly for amplified bands, and these can be used and ordered as markers in mapping. They can also be used conveniently for strain identification or diagnostic purposes, *e.g.* of fungal or nematode pathogens.

[37] 'PCR Technology. Principles and Applications of DNA Technology', ed. H. A. Erlich, Stockton Press, 1989.

[38] M. J. McPherson, P. Quirke, and G. R. Taylor, 'PCR, A Practical Approach', IRL Press, Oxford, 1991, pp. 1–253.

[39] B. Burr, S. V. Evola, and F. A. Burr, in 'Genetic Engineering: Principles and Methods', Vol. 5, Plenum Press, New York, 1983, p. 45.

[40] T. Helentijaris, M. Slocum, S. Wright, A Schaefer, and J. Nienhuis, *Theor. Appl. Genet.*, 1986, **72**, 761.

[41] T. C. Osborn, D. C. Alexander, and J. F. Forbes, *Theor. Appl. Genet.*, 1987, **73**, 350.

[42] J. G. K. Williams, A. R. Kubeliak, K. J. Livak, J. A. Rafalski, and S. V. Tingey, *Nucleic Acids Res.*, 1990, **18**, 6531.

3.4 Transposon and Insertional Mutagenesis

Another method, which has potential for the identification of specific genes, is 'transposon and insertional mutagenesis', which relies on the inactivation or alteration of a gene by the insertion in it of a transposable nucleotide sequence (a 'jumping gene') or other sequence (*e.g.* T-DNA). Naturally occurring transposable elements have been well characterized in *Drosophila* and in maize[43] and their cloning has provided a tool for gene isolation. If transposon mutagenesis can be related to an altered phenotype, a radioactive cDNA to the transposable element can be used to isolate the sequence and its flanking regions. The transposable element may be inserted within or adjacent to the structural gene of interest. This technique is of particular value when the gene product is unknown, but is of less use for polygenic traits, when simultaneous mutagenesis of two or more loci might be required to alter the plant phenotype. Transposon mutagenesis has been used to identify, for example, the function of genes located on the Ti plasmid of *Agrobacterium* (*e.g.* reference 44) and on *Rhizobium* plasmids.[45]

One further strategy for gene identification which we will briefly consider is to incorporate the isolated sequence into an expression vector, transfect *E. coli* or perhaps a plant cell, and study either the gene product or its effect on the phenotype. This technique has been used to study proteins encoded by Ti plasmid genes, expressed in *E. coli* cells,[46,47] and this approach has been used to transform potato plants with Ti- and Ri-plasmid-encoded genes to study the effects on plant development.

It should be apparent that a number of strategies are now available to isolate and study specific genes. The next questions to be asked are, what genes are of interest to the plant breeder, or perhaps more pragmatically, what gene products have been characterized well enough that genes themselves can be isolated, modified, and reintroduced into the same or into a different species – and what modifications to these genes are desirable? The main aims of plant breeding were summarized previously. We will now consider approaches that are being used to achieve such aims by genetic engineering. In addition, these techniques enable advances in our understanding of plant differentiation and development, and how it can be manipulated. The breeding objectives will be discussed by reference to the best characterized examples.

3.5 Manipulating Yield

Although there are many factors which contribute to the overall yield of a crop plant, one of the most important and best characterized is the role of carbon fixation. Ribulose bisphosphate carboxylase–oxygenase (Rubisco) is the most

[43] H. P. Döring, E. Tillman, and P. Starlinger, *Nature (London)*, 1984, **307**, 127.
[44] C. T. Komro, V. J. Kirita, S. B. Gelvin, and J. B. Kemp, *Plant Mol. Biol.*, 1985, **4**, 253.
[45] P. R. Hirsch, M. Van Montagu, A. B. W. Johnston, N. J. Brewin, and J. Schell, *J. Gen. Microbiol.*, 1980, **120**, 403.
[46] G. Schröder, W. Klipp, A. Hillbrand, R. Ehring, C. Koncz, and J. Schröder, *EMBO J.*, 1983, **2**, 403.
[47] G. Schröder, S. Waffenschmidt, E. W. Weiler, and J. Schröder, *Eur. J. Biochem.*, 1984, **138**, 387.

abundant enzyme in the world. It is responsible for fixing CO_2 by combination with ribulose-5-phosphate, to give 2 molecules of 3-phosphoglyceric acid. However, in addition to this carboxylase reaction, it also catalyses a reaction between ribulose-5-phosphate and oxygen, to yield one molecule of 3-phosphoglyceric acid and one of the 2 carbon compound phosphoglycolic acid. The latter 'oxygenase' activity is responsible for the light-dependant loss of CO_2 known as 'photorespiration', and uses up energy as ATP. Photorespiration, which leads to loss of CO_2, therefore appears to reduce the efficiency of carbon fixation. An improvement in carbohydrate yield may be achieved by increasing the relative rate of the carboxylase to the oxygenase activity of Rubisco.

The Rubisco molecule is composed of eight identical large and eight identical small subunits, encoded in the chloroplast and nuclear genomes respectively. The genes, present as small multigene families, have been cloned and characterized, and are amenable to manipulation.

The large subunit of Rubisco contains the catalytic site of the enzyme, and it may be possible to manipulate its structure (or the structure of the small subunit, which may be involved in conformational changes in the large subunit) to raise carboxylase activity at a particular CO_2 concentration, *i.e.* to increase the affinity of the enzyme for CO_2. The interactions between the enzyme and metal cations or other cofactors may be critical in this context.[48] Since CO_2 and O_2 compete for the same catalytic site, photorespiration would be correspondingly reduced – it is known that by growing plants in an atmosphere enriched with CO_2 or depleted of oxygen, increased yields of reduced carbon can be obtained.[48] Alternatively, it may be possible to delete the oxygenase activity of the enzyme by protein engineering, or to select for mutant enzymes which possess no or reduced oxygenase activities, and transfer the encoding genes to crop plants. However, mutants lacking photorespiratory activity which cannot metabolize phosphoglycolate are not viable.

Of course, if the carboxylase activity is made more efficient, then other limitations in the flux of carbon may become apparent. Attention will be drawn increasingly to other enzymes, and eventually perhaps to the engineering of the more complex processes of partitioning carbon between the sites of production and the storage tissues.

3.6 Nitrogen Fixation

The biological machinery for nitrogen fixation is limited, in nature, to prokaryotic organisms, such as aerobic (*e.g. Azotobacter*) and anaerobic (*e.g. Klebsiella*) bacteria and to the cyanobacteria (*e.g. Anabaena*). However, higher plants such as legumes and *Acacia* can exploit these properties by entering a symbiotic relationship with *Rhizobium* and *Bradyrhizobium* species which are capable of forming root nodules. Root nodules involved in the establishment of nitrogen-fixing symbiosis between *Rhizobia* and legumes are specialized organs in which the *Rhizobia* reduce dinitrogen to ammonia, in exchange for carbon compounds derived from

[48] A. J. Keyes, *Pestic. Sci.*, 1983, **19**, 313.

photosynthesis. The symbiosis requires an exchange of appropriate molecular signals to co-ordinate recognition and developmental processes. Approaches to genetically manipulate nitrogen-fixing symbioses require an understanding of recognition, infection, and nodule development.[49] Genetic and molecular techniques have been applied extensively in studies of rhizobial genes involved in all steps of the interaction with legumes. The main genes involved are the nodulation genes (*nod*), polysaccharide biosynthesis genes (*exo* and *lps*), nodule development genes (*ndv*), and nitrogen fixing genes (*nif* and *fix*). On the plant side, plant genes specifically expressed during the symbiosis are termed 'nodulins' (*e.g.* leghaemoglobin, uricase II, glutamine synthase, choline kinase, sucrose synthase); these total more than 30 proteins.

Opportunities for the improvement of nitrogen yield in plants include (1) increasing the host range of symbiotic nitrogen-fixing bacteria; (2) increasing the efficiency of the fixation process in the symbiotic bacteria; (3) modifying N_2-fixing bacteria to maintain nitrogen fixation in the presence of exogenous nitrogen; and (4) transferring the nitrogen-fixing genes to non-symbiotic plants. The first of these objectives may be achieved by transferring the *nod* genes, which in some cases are located on plasmids,[50] from nodulating strains to non-nodulating strains of bacteria. Indeed, it has already been shown that, if a plasmid from the pea-nodulating species *Rhizobium leguminosarum* was transferred to *R. phaseoli* which normally nodulates only bean plants, it developed the ability to nodulate peas.[51,52] The role of the host plant in determining the symbiotic relationship is less clear, and has been investigated primarily by the characterization of nodule-specific plant gene products. The most abundant nodulin is leghaemoglobin, a protein related to myoglobin and the product of a multi-gene family, which represents 20–30% of total nodule cell protein. It has a high affinity for O_2, and binds oxygen, to maintain a stable low oxygen concentration within the nodule. The other nodulins can be categorized[53] as (*a*) Group 1 nodulins, representing structural proteins of the nodule; (*b*) Group 2 nodulins, enzymes such as glutamine synthase and uricase, involved in the metabolism by the host of fixed nitrogen; and (*c*) Group 3 nodulins, proteins involved in the support of bacteroid functions. It would be expected, therefore, that any improvements in the range of the symbiotic relationship would require the manipulation and transfer of these, as well as of the bacterial genes.

The efficiency of the nitrogen fixation process could conceivably be improved by limiting the loss of energy in the ATP-requiring process of hydrogen evolu-

[49] B. W. S. Sobral, R. J. Honeycutt, A. G. Atherly, and K. D. Noel, in 'Biology and Biochemistry of Nitrogen Fixation', ed. M. J. Dilworth and A. R. Glenn, Elsevier, Amsterdam, 1991, p. 229.

[50] M. P. Nuti, A. M. Ledeboer, A. A. Lepidi, and R. A. Schilperoort, *J. Gen. Microbiol.*, 1977, **100**, 241.

[51] A. B. W. Johnston, J. L. Beynon, A. V. Buchanan-Wollaston, S. M. Setchell, P. R. Hirsch, and J. E. Beringer, *Nature (London)*, 1978, **276**, 634.

[52] J. A. Downie, G. Hombrecher, Q.-S. Ma, C. D. Knight, B. Wells, and A. B. W. Johnson, *Mol. Gen. Genet.*, 1983, **190**, 359.

[53] F. Fuller, P. W. Kunster, T. Nguyen, and D. P. S. Verma, *Proc. Natl. Acad. Sci. USA*, 1983, **80**, 2594.

tion. Hydrogen also is a competitive inhibitor of nitrogenase. Evans *et al.*[54] have identified a gene whose product, an H_2-uptake hydrogenase, can recapture evolved hydrogen and so regain some of the otherwise lost energy. This gene has been transferred to various *Rhizobium* strains, with the apparent effect of increasing the efficiency of nitrogen fixation and improvement in legume growth.

Nodulation and fixation are usually induced only in conditions of low environmental nitrogen. It may be possible, when more is understood about the regulatory mechanisms, and about the interactions between bacteria and host, to modify the symbiotic processes so that they proceed under field conditions in which nitrogen is not a limiting factor.

With the advances that are being made in gene transfer technology, it may be possible to by-pass altogether the complications of the symbiosis, and introduce the genetic determinants for the fixation process directly into non-leguminous plants. However, given the complex nature of the interaction, before this can be achieved, much work must be done to isolate and characterize further the genes and their products, and their interactions with the plant nodulins, some or all of which presumably will also require to be transferred. There have been suggestions that some form of 'nodules' can be induced on roots of cereals such as wheat, but at present these structures certainly have no nitrogen-fixing activity, and their formation may have little relevance towards the goal of nitrogen-fixing wheat plants.

3.7 Improved Solute Uptake

Non-leguminous plants obtain their nitrogen as nitrate or ammonium ions from the soil. It is known that the growth and development of many crops can be improved by the application of fertilizers containing nitrogen, phosphate, and potassium,[55,56] and molecular biological techniques are now being applied to increase the efficiency of the uptake mechanisms in roots. The transport of nitrate ions, for example, across the plasma membrane is mediated by a membrane-bound protein but the encoding genes have yet to be identified. The use of uptake mutant plants should prove valuable in this respect. However, more is known about the enzymes in nitrate metabolism, including nitrate reductase, nitrite reductase, and glutamine synthase. These genes have now been cloned from plants, their promoters analysed, and the genes reinserted into transgenic plants.

Approaches to study other solute transporters include: (1) induction of transport systems, (2) extraction of the induced messages, and (3) micro-injection into cells for which transport measurements are easier (*e.g. Xenopus* eggs or giant algal cells). The latter provides a bioassay system for the message, which can be fractioned until specific mRNAs for induced transport proteins are identified.

[54] H. J. Evans, J. E. Emerich, J. E. Lepo, R. J. Maier, K. R. Carter, F. J. Hanus, and S. A. Russell, in 'Nitrogen Fixation', ed. W. D. P. Stewart and J. R. Gallon, Academic Press, London, 1980, p. 55.

[55] B. J. Miflin, J. M. Field, and P. R. Shewry, in 'Seed Storage Proteins', ed. J. Daussant, J. Mosse, and J. Vaughan, Academic Press, London, 1983, p. 215.

[56] J. Bingham, *Philos. Trans. R. Soc. London, B*, 1981, **293**, 441.

Such strategies indicate that molecular approaches may soon be available to manipulate solute transport processes, for example by over-expression of limiting transport proteins.

3.8 Quality

Since the biochemical and molecular basis of 'nutritional' or 'product' quality will be different for different types of crop (protein, carbohydrate, oil, secondary metabolites), both the molecular and the cell biological problems encountered in improving quality will be diverse. For example, whereas the basic biochemistry of cereal seed storage protein synthesis and accumulation is relatively well understood, the same cannot be said for the production of useful secondary metabolites, such as alkaloids, steroids, or terpenoids, making progress in the modification of quality, a slow and difficult task. For this reason, and also because of their economic importance, a great deal of effort has been concentrated on the study of the nutritional and technological qualities of the seed storage products of the cereals, legumes, and *Brassicas*. For the cereals and legumes, the interest in nutritional quality revolves around the fact that the storage proteins of a number of species are low in one or more essential amino acids. For example, wheat, barley, maize, and *Sorghum* are low in lysine,[57] while legume storage proteins are deficient in sulfur-containing amino acids. Barley and *Sorghum* are further lacking in threonine, and maize in tryptophan. The use of genetic engineering to improve cereal storage protein quality can be approached in several ways.[58] The first is by isolating and cloning the gene for the protein, inserting extra codons for the deficient amino acids, and then introducing the modified gene back into the plant. Many storage protein genes have been cloned, and have been transferred successfully across species barriers.[59] Improved nutritional quality of a cereal has yet to be demonstrated by this strategy. A second approach is to modify the expression of existing genes, so that proteins which are relatively rich in the deficient amino acids are preferentially synthesized. Completely artificial proteins, high in limiting amino acids, can also be synthesized. Regulatory sequences have now been identified for a number of storage proteins, and storage tissue specific expression of genes has been achieved for a number of species. The nutritional quality of cereals might also be improved by increasing the levels of specific soluble amino acids, and the generation and selection of mutant plants which overproduce these compounds has also been studied. Mutants have been generated in culture by the use of chemical agents or by irradiation, to generate enzymes which are insensitive to negative feedback control,[9] and cells resistant to toxic amino acid analogues have been selected which overproduce the natural amino acid.[60] The isolation, cloning, and modification of genes for key regulatory enzymes could lead directly to such changes, for both nutritional valuable

[57] S. W. J. Bright and P. R. Shewry, in 'Critical Reviews in Plant Science', Vol. 1, ed. B. V. Conger, CRC Press, Boca Raton, Florida, 1983, p. 49.

[58] P. R. Shrewry, B. J. Miflin, B. G. Forde, and S. W. J. Bright, *Sci. Prog. (Oxford)*, 1981, **67**, 575.

[59] P. B. Goldsbrough, S. B. Gelvin, and B. A. Larkins, *Mol. Gen. Genet.*, 1986, **202**, 374.

[60] R. A. Gonzales and J. M. Wildholm, in 'Primary and Secondary Metabolism of Plant Cell Cultures', ed. K.-H. Neumann, W. Barz, and E. Reinhard, Springer-Verlag, Berlin, 1985, p. 337.

amino acids and also for specific valuable secondary metabolites, such as alkaloids, phenolics, and flavonoids, derived from amino acids. Work has already started on the cloning and modification of tryptophan decarboxylase, involved in biosynthesis of serotonin and indole alkaloids, and of lysine decarboxylase, a key enzyme in lupin alkaloid metabolism.[61] Although still at an early stage, this approach opens up the possibility of engineering cultured plant cells and intact plants as 'green bioreactors' for the production of useful chemicals.

3.9 Technological Quality

The biochemical properties of crops can also, it is hoped, be modified to improve quality in relation to technological and industrial requirements. Two well studied examples, described in more detail by Jones,[1] are the breadmaking qualities of wheat, and the malting and brewing of beer (barley).

3.9.1 Breadmaking Quality of Wheat. Breadmaking quality depends largely on the visco-elastic properties of gluten, which consists of two major protein fractions, gliadin and glutenin. It is the gliadin factor which is responsible for the viscosity of dough, allowing it to rise, and the glutenin fraction provides the elasticity which prevents dough being overextended and collapsing during the rising and baking processes,[62] apparently as a result of interactions between high molecular weight (HMW) subunits of glutenin via disulfide linkages.[63] The manipulation of the proportion of HMW glutenins in wheat storage protein, particularly those known to be related to high-quality breadmaking, and of cross-linking by altering cysteine sites, have been proposed as strategies for improving the process.

3.9.2 Brewing. Beer production comprises (1) malting, in which the barley grain starch and protein reserves are enzymatically mobilized to sugars and flavour compounds, and (2) brewing, in which the sugars are fermented to alcohol. Malting quality appears to be adversely affected by the presence of hordein fraction proteins, particularly B hordeins and disulfide-linked compounds,[63] but more detailed knowledge of the functions of specific components in the process is required before genetic manipulation is possible. Improvements of the fermentation process may be more feasible. The wheat α-amylase gene, which breaks down starch, has been cloned[64] and transferred to yeast, cells of which were able to synthesize and export the active protein. This strategy could be adopted for other enzymes involved in starch and protein hydrolysis.

3.9.3 Genetic Engineering for Speciality Oil Production. By making mutants of *Arabidopsis thaliana*, as a model system to study oil production, it has been possible to dissect the pathway of oil production and modification in plants. All the properties found in oil crops have now been obtained in *Arabidopsis* mutants, in characters such as chain length (*e.g.* 16, 18, 20 carbons) and degree and position

[61] J. Berlin, H. Beier, L. Fecker, E. Forsche, W. Noé, F. Sasse, O. Schiel, and V. Wray, ref. 60, p. 272.

[62] P. I. Payne, ref. 55, p. 223.

[63] B. J. Miflin, J. M. Field, and P. R. Shewry, ref. 55, p. 215.

[64] S. J. Rothstein, C. M. Lazarus, W. E. Smith, D. C. Baulcombe, and A. A. Gatenby, *Nature (London)*, 1984, **308**, 662.

of double bonds. Many of the genes involved have been mapped, and some genes cloned and characterized.[65] Some of the aims of oil production include increased yield, better oils, and new oils for speciality uses. With the knowledge available it should now be possible to reach some of the aims using transgenic plants.

The application of molecular techniques to the improvement of technological quality in other crops is also progressing. For example, work is in progress on the mechanism of the sweetening of potato tubers on cold-storage, which causes discolouration of potato crisps during processing. The enzyme believed to be responsible has now been cloned and the aim is to substitute the cold-labile enzyme with one that is cold stable. Starch composition is also amenable to manipulation, for example, with the production of amylose-free transgenic potato plants.[66]

3.10 Resistance to Diseases and Pests

The major targets against which resistance is currently being engineered in plants are insects, viruses, fungi, bacteria, and nematodes. This subject has expanded rapidly: we will take selected examples of work from a range of crop plants, in particular potato, which is amenable to a range of manipulations.

3.10.1 Insect Resistance. Bacillus thuringiensis (B.t.) produces crystalline (*cry*) proteins (endotoxins) on sporulation. The proteins are harmless to humans, domestic animals, most insects, and bees. However, the *cry* protein product produced by different B.t. strains are toxic to specific classes of insects. For example *cry*I and *cry* II are toxic to Lepidopteran insects, *cry* III to Colorado potato beetle and other Coleopterans, and *cry* II and IV to Diptera.[67] Spore preparations of B.t. have been used as insecticides for many years, but these preparations had limited stability. Research carried out, particularly by the Monsanto Company, has modified expression of the *cry* IIIA genes to match potato codon usage. Expression of the modified B.t. protein genes in transgenic potato plants confers dramatic field resistance to Colorado potato beetles, and is likely to lead to a reduction in the need for application of chemical pesticide to control this pest. Further B.t. constructs are under development which make the possibility of acquired B.t. toxin resistance unlikely.[68] Similar approaches on the expression of B.t. *cry* proteins to protect a range of other crops, such as cotton, are in progress in many laboratories. Other approaches to insect resistance include expression of protease inhibitors in transgenic plants and compounds that deter insect feeding.

[65] B. Lemieux, B. Hange, and C. Sommerville, Proceedings of the International Society of Plant Molecular Biology Meeting, Tucson, Arizona, 1991, no. 727.

[66] R. G. F. Visser, E. Flipse, A. G. J. Kuipers, S. N. I. M. Salehuzzaman, and E. Jacobson, Proceedings of the International Society of Plant Molecular Biology Meeting, no. 56.

[67] M. Peferoen, S. Jansens, A. Raynaerts, and J. Leemans, in 'The Molecular and Cellular Biology of the Potato', ed. M. E. Vayda and W. D. Park, CAB, Wallingford, 1990, pp. 193–204.

[68] M. E. Vayda and W. R. Belknap, *Transgenic Res.*, 1992, **1**, 149.

3.10.2 Virus Resistance. The majority of plant viruses have single-stranded RNA genomes. A number of strategies have been employed to confer resistance to virus pathogens of crop plants. These include:

* protection by overexpression of viral coat protein genes;
* expression of antisense gene constructs;
* use of catalytic RNA (ribozymes) to cut and inactivate viral RNA;
* constitutive expression of viral replicase genes;
* constitutive expression of other viral genes.

Expression of viral coat protein genes in cells of transgenic plants is thought to interfere with uncoating of incoming virus particles on infection. Although the mechanism is not fully understood, there is ample evidence to show that constitutive expression of viral coat protein significantly reduces infection, with delayed development of symptoms and low virus accumulation in the few infected plants.[69,70] This has been shown for: Tobacco Mosaic Virus, Cucumber Mosaic Virus, Alfalfa Mosaic Virus, Potato Viruses X, Y, and Leaf Roll, Tobacco Streak Virus, Tobacco Rattle Virus, Soybean Mosaic Virus, Papaya Ringspot Virus, Watermelon Mosaic Virus, and Zucchini Yellow Mosaic Virus. Extensive field trials of 'coat protein' virus-resistant plants have been carried out, and resistance has been introduced to more than one virus using this approach (*e.g.* PVX and PVY resistance in potato).[70]

The expression of antisense constructs as a method leading to virus resistance has been less successful than the 'coat protein' approach, and it remains to be seen whether ribozymes ('gene shears') present in antisense sequences will improve this situation. The other approaches (*e.g.* expression of viral replicase or other genes) in host plants hold considerable promise. The aims in general are to find strategies that will make transgenic plants resistant to all strains of a particular virus, and preferably to a range of different virus pathogens.

3.10.3 Resistance to Fungal Pathogens. Because of the greater complexities of fungal-plant interactions, there has been less research on developing molecular approaches to confer host resistance to fungal than to other plant pathogens. However, this situation is changing. Approaches being used include expression of chitinase genes and β-1,3-glucanases in transgenic plants, which it is hoped will specifically degrade the hyphal walls of invading fungi.[68]

3.10.4 Resistance to Bacterial Pathogens. Taking potato as the example, transgenic potato plants expressing various anti-bacterial proteins have been generated.

The peptide fragment (*cecropin*) from a giant silkworm haemolymph protein, which is thought to act as an ionophore against a broad spectrum of bacterial species,[71] has been introduced and expressed constitutively in transgenic potato plants. Similarly, lysozyme, an enzyme (in this case from chicken) which

[69] A. Hockema, M. J. Huisman, L. Molendijk, P. J. M. van den Elzen, and B. J. C. Cornelissen, *Bio/Technology*, 1989, **7**, 273.

[70] C. Lawson, W. Kaniewski, L. Haley, R. Rozman, C. Newell, P. Sanders, and N. E. Turner, *Bio/Technology*, 1990, **8**, 127.

[71] L. Destafano-Beltram, P. G. Nagpala, P. G. Cetiner, J. H. Dodds, and J. M. Jaynes, in reference 67, p. 205.

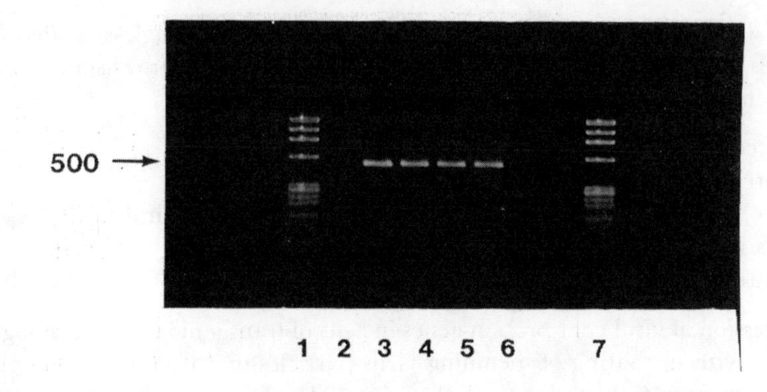

Figure 8 *PCR assay for Cucumber Mosaic Virus (CMV) in lupin seeds. Lanes 1 and 7: marker ϕX 174* Hae *III digest; lane 2: negative control; lane 3: one CMV-infected seed in 100; lane 4: one CMV-infected seed in 500; lane 5: one CMV-infected seed in 1000; lane 6: one CMV-infected seed in 2000*[73]

degrades the peptidoglycan bacterial cell wall has been expressed in transgenic plants. Field tests have indicated that expression of cecropin gave significant resistance to the bacterial diseases soft rot and blackleg (stem rot).[68]

3.10.5 Resistance to Nematode Pathogens. The rapid advances in molecular techniques, such as amplification of sequences by PCR, now means that economically important host-parasite relations of endoparasitic nematodes (*e.g.* cyst- and root-knot nematodes) can be studied in more detail. Approaches being followed include the disruption of nematode feeding cells to confer host resistance.[72]

3.11 Pathogen Identification and Testing

There have been rapid advances in using molecular techniques (1) for identification of strains of pathogens (*e.g. Rhizoctonia*) or (2) for pathogen testing (*e.g.* of viruses). Initially, probing dot blots with cDNA probes was used, but with the advent of PCR either using specific primers or random primers (RAPD), many new applications have emerged. An example is in routine testing for viruses, where PCR can be employed. Cucumber mosaic virus (CMV) is a serious seed-borne pathogen of lupins, the major grain legume grown in Australia. Routine testing of farmers seeds can be achieved with a sensitivity of detection better than 1 CMV-infected seed in a batch of 1000 lupin seeds (Figure 8).[73] This is considerably more sensitive than the previously used ELISA approach. The results of this test determine whether that seed can be planted the next season or not.

[72] P. R. Burrows and M. G. K. Jones, in 'Plant Parasitic Nematodes in Temperate Agriculture', ed. K. Evans, CAB, Wallingford, 1993, in press.

[73] S. Wylie, C. R. Wilson, R. A. C. Jones, and M. G. K. Jones, 'A Polymerase Chain Reaction Assay for Cucumber Mosaic Virus in Lupin Seeds', *Aust. J. Agric. Res.*, 1993, **44**, 41.

Table 3 *Herbicides and some strategies used to obtain resistance in transgenic plants*

Herbicide	*Target enzyme*	*Strategy for resistance*
Glyphosate [*N*-(phosphonomethyl)-glycine]	EPSP synthase	1. Overexpression of wild type EPSP synthase 2. Introduction of glyphosate-resistant EPSP synthase from *Salmonella*; expression in cytoplasm 3. As 2, but transit peptide used to direct resistant EPSP synthase expression into chloroplast
Sulfonyl urea	ALS (acetolactate synthase)	Introduction of resistant ALS gene
Imidazolinones	ALS (acetolactate synthase)	Introduction of resistant ALS gene
Basta L-Phosphinothricin (PPT)	GS (glutamine synthase)	1. Amplification of endogenous GS 2. Introduction of PPT acetyl transferase (PAT) to degrade PPT.
Atrazine	Blocks QB protein in Photosystem II electron transport	1. Introduce resistant *PSA* gene, targetted to chloroplast 2. Detoxification by introducing glutathione-*S*-transferase
Bromoxyinil		Detoxification by introducing nitrilase (*bxn*) gene

3.12 Resistance to Herbicides

The production of herbicide-resistant plants is relatively easy because the sites of action of major herbicides are known, and are usually a single enzyme. Three basic strategies have been employed to obtain resistance to herbicides in transgenic plants: (1) amplification of the target gene, *i.e.* the wild type enzyme; (2) introduction of a mutant, herbicide-insensitive gene into a plant; and (3) introduction of a gene for an enzyme that can inactivate or degrade the herbicide. All of these approaches have been used successfully to obtain transgenic, herbicide-resistant plants.[74] Resistance is now available to the major herbicides: glyphosate, sulfonyl ureas, imidazolinones, basta (phosphinothricin), and triazines: approaches used are summarized in Table 3.

3.13 Stress Tolerance

Specific responses can be elicited in plants subjected to a range of stressful conditions such as high and low temperature, anaerobiosis as induced by flood-

[74] J. Botterman and J. Leemans, *Trends Genet.*, 1988, 219.

ing, salinity, or UV light. The molecular basis of the responses are known to include the *de novo* synthesis of specific proteins (under temperature shock) and enzymes (*e.g.* alcohol dehydrogenase under anaerobiosis and phenylalanine ammonia lyase under UV irradiation). In some instances, the genes involved have been cloned and sequenced, and regulatory sequences identified. Examples include promoters of heat shock genes and genes whose expression is induced by anaerobic conditions.

3.14 Modification of Plant Development

Because of the complex and diverse nature of developmental processes, it is beyond the scope of this chapter to do more than indicate a few areas of progress. The use of *Arabidopsis thaliana*, the 'model' plant species with least DNA, for studies on control of developmental processes is rapidly providing information on genetic control of aspects such as organogenesis of flowers[75] (*e.g.* development of sepals, petals, anthers, and stigma), plant growth regulators and their receptors, light responses, *etc.* For practical biotechnological applications of modified plant development, three specific examples are discussed: control of male and female fertility/sterility, fruit ripening, and flower colour formation.

3.14.1 Control of Male and Female Sterility. It is well known that, for some crops, growth and yield of F1 hybrid plants may well be superior to that of inbred lines. In addition, since seeds from F1 hybrid plants do not breed true, growers need to purchase new stocks each year. These two aspects have attracted research on the control of male and female sterility. Control of male sterility has been most studied, because a male sterile line can only act as a female parent and needs to be cross pollinated to give F1 seed. Initial work on emasculation with chemical sprays ('gametocides') has been replaced by molecular approaches to achieve male sterility. The power of molecular techniques in this area has been demonstrated elegantly in collaborative work between laboratories in California and Belgium.[76] Pollen grains develop in anthers, nurtured by the tapetum layer. A tapetum-specific promoter, TA29, was isolated in California. When this was linked to a ribonuclease gene in Belgium, the latter was only expressed during tapetum development, leading to tapetum cell death and male sterility. Using a bacterial 110 amino acid ribonuclease called Barnase, TA29–Barnase transgenic plants are male sterile. Barstar is an 89 amino acid intracellular inhibitor of Barnase. It binds very tightly to Barnase and forms an inactive complex. Thus a cross between a TA29–Barstar male parent with a TA29–Barnase female parent (male sterile) yields an F1 hybrid which is fertile. This allows the production, for example, of F1 hybrid oilseed rape plants which are fertile and produce seed.

Applied to maize F1 hybrid seed production in the USA, this approach could replace hand de-tasselling which is now used to produce hybrid seed, and it could save the seed industry several hundred million dollars that is now spent on labour

[75] E. M. Meyeroritz, J. L. Bowman, L. L. Brockman, G. N. Drews, K. Goto, T. Jack, Z. Liu, M. Running, H. Sakai, L. E. Sieburth, and D. Wiegel, reference 65, no. 6.

[76] J. Leemans, A. Reynarts, M. DeBlock, M. De Beuckdeer, P. Rudelsheim, R. Goldberg, and C. Mariani, reference 65, no. 24.

for seed production. This system has been demonstrated to be effective for a range of crops: in field experiments on oilseed rape the transgenic plants were stable and exhibited no side effects.

3.14.2 Fruit Ripening. A considerable proportion of agricultural produce is lost after harvesting. Molecular techniques may be used to reduce this loss. For example, in fruit, such as tomato, which exhibit an autocatalytic climacteric rise in ethylene production on ripening, the pathway of ethylene production is simple. Ethylene is produced from *S*-adenosyl methionine via an intermediate ACC (aminocyclopropane-1-carboxylate). The two enzymes involved, ACC synthase and EFE (ethylene forming enzyme), have been cloned. Ethylene production in transgenic plants expressing antisense constructs to both ACC synthase and EFE is essentially switched off, and this can also be achieved using a bacterial enzyme that degrades ACC. Transgenic tomatoes from such plants can last without decaying for much longer than control plants, and they can be transported with less damage. A similar approach can be used to prolong transport and vase life of some flowers.

3.14.3 Modifying Flower Colours. The metabolic pathways that lead to flower colour development are now well characterized. The major anthocyanins which contribute to flower colours are the anthocyanidins delphinidin (mauve–blue), cyanidin (crimson–magenta), and pelargonidin (pink–scarlet–orange). The colour formed depends to a certain extent on the pH. It is relatively straightforward to switch off this colour pathway by introduction of an antisense chalcone synthase gene, and this may lead to white flowers, reduced colour, or developmentally-related switching off of colour, depending on the effectiveness of the antisense construct, its promoter, and the site of insertion.[77] The addition of enzymes missing from a particular flower can also be achieved: Calgene Pacific (Melbourne, Australia) is attempting to produce a blue rose by addition of the enzyme required to produce delphinidin driven by a petal specific promoter.

4 FIELD TESTING OF TRANSGENIC PLANTS

The first field trials of transgenic plants took place in 1986, and since that time there has been a dramatic increase in the number carried out. Field tests have been done both by public research organizations (Universities, Agricultural Research Institutes), and by commercial companies. As has been indicated there have been phenomenal advances in technology, and the emergence of regulatory frameworks to oversee and control field growth of transgenic plants. There are many crops that are potential candidates for commercial development, and companies involved have business strategies in place.

To date there have been field tests in at least 21 different countries, the majority (85%) being in North America (USA, Canada) and Europe (France, Belgium, UK, Holland). These tests involve 25 different crop plants, with over

[77] J. N. M. Mol, A. R. van der Krol, A. J. van Tunen, R. van Blockland, P. deLange, and A. R. Stuitje, *FEBS Lett.*, 1990, 427.

Table 4 *Transgenic crops which have been field tested (1986–1991, from Chasseray and Duesing, 1992[78])*

Alfalfa	Oilseed rape (summer)
Apple	Oilseed rape (winter)
Asparagus	Petunia
Birch	Poplar
Cantaloupe	Rice
Cauliflower	Soybean
Chrysanthemum	Squash
Chicory	Sugarbeet
Cotton	Sunflower
Cucumber	Tobacco
Flax	Tomato
Maize	Walnut
Melon	

400 field tests having been carried out. A list of plants that have been field tested (to 1991) is given in Table 4, and by numbers of tests in Table 5.

From Table 4, it can be seen that major crops (*e.g.* maize, rice), vegetable crops (*e.g.* potato, brassicas), horticultural crops (*e.g.* chrysanthemum, petunia), and tree species (*e.g.* birch, walnut) have all been field tested as transgenic plants. This list will no doubt be added to rapidly, for example, by other crops such as wheat.

The first experiments were of plants either with marker genes or genes for herbicide tolerance. As the science has progressed, the range of characters in transgenic plants has increased to include genes for resistance to viruses, insects, and fungi, and genes modifying product quality. A much wider range of potential modifications is in the pipeline. Many of the tests have been carried out at more than one site, as is of course necessary for commercial development of a transgenic crop. The largest scale release, so far, appears to have been in China, where transgenic virus-resistant tobacco has been grown over 500 hectares.[78]

Table 5 *Number of field tests of transgenic plants by crop species (modified from Chasseray and Duesing, 1992[78])*

Tobacco	75
Potato	71
Oilseed rape	71
Tomato	49
Maize	21
Sugarbeet	19
Alfalfa	18
Cotton	15
Others	54

[78] E. Chasseray and J. Duesing, *Agro. Food Ind. Hi-Tech*, 1992, **3**, 5.

Table 6 *Examples of potential benefits of transgenic plants*

Leading to environmentally friendly agriculture
* resistance to viruses * resistance to insects * resistance to nematodes * resistance to fungi * improved nutrient uptake/utilization * tolerance to salinity/waterlogging
Leading to higher yielding crops
* production of F1 hybrids * increased photosynthetic yield * diversion of photosynthates to harvestable organs
Leading to prolonged storage life
* delayed ripening of fruit – to enable exports * delayed senescence of flowers * reduced loss on storage * storage at room temperature leading to improved flavour
Leading to improved nutrition/processing
* improved nutritional quality * increased dry matter (*e.g.* starch in potatoes, more dense, absorbs less oil) * removal of unwanted or toxic compounds
Leading to increased understanding
* development processes * metabolic processes * cell communication
Leading to novel applications
* novel flower colours * production of antibodies in plants * use of plants as bioreactors * production of biodegradable plastics from cereals * expression of fish antifreeze protein for cold tolerance

4.1 Benefits and Risks

Potential commercial benefits of the generation and use of transgenic plants have been outlined earlier in this chapter and are summarized in Table 6. In addition to achieving the aims of plant breeders more rapidly and precisely, for resistance to pests and diseases, improved quality, and tolerance to environmental stresses, there could be significant benefit to the environment. This stems from the fact that if plants are inherently genetically resistant to pests and diseases then they will not need to be protected to the same extent by application of potentially toxic chemicals or sprays. Some of these are persistent, such as nematicides, and may enter the water table or their residues may be detected in produce.

As with any new technology, there are also some inherent risks, and these must be acknowledged and minimized. For example, introduction of insect resistance genes into tree species, such as eucalypts, which suffer badly from insect damage in Australia, could be transferred to trees in native forests. This would alter the insect population present, and thus have knock-on effects on the ecosystem. This problem can be solved by engineering the trees to be male and female infertile. The insect resistant transgenic trees could then be propagated *in vitro*, and grown in plantations without the transfer of introduced genes.

There are also groups that oppose new technologies, including genetic engineering, and the general public may also be wary of buying transgenic produce. There will therefore be a period when the public will have to adjust to the idea of transgenic produce, and it will be consumers and politicians, and not scientists, who will decide whether there will be limits to the exploitation of genetically engineered plants.

The first transgenic produce that consumers will find on supermarket shelves will probably be the 'Flav Savr' tomato.[79] This tomato has an antisense gene to polygalacturonase, an enzyme involved in wall softening during ripening. Switching off polygalacturonase synthesis has two consequences: (1) the tomatoes are less damaged on harvesting and last longer in transit and on the supermarket shelf; and (2) can therefore be picked red and ripe from the tomato plant rather than green as is the current practice, and this allows flavour compounds to develop before harvest. The response of consumers to this and other engineered produce is awaited with great interest.

[79] R. Hoyle, *Bio/Technology*, 1992, **10**, 629.

CHAPTER 8

Molecular Biology in the Pharmaceutical Industry

C. W. DYKES

1 INTRODUCTION

Commercial application of recombinant DNA (rDNA) technology began in the late 1970s, pioneered by small venture-capital biotechnology companies that set out to produce proteins such as tissue plasminogen activator, erythropoietin, and the myeloid colony stimulating factors, for direct use as therapeutic agents. Other fledging biotech companies sought to exploit the potential market for recombinant vaccines. The major pharmaceutical companies on the other hand were reluctant, in general, to depart from their traditional reliance on small molecule drugs and chose to utilize the new methodology for the provision of information and materials to facilitate the discovery of such compounds. Although this use of molecular biology as an enabling technology continues to predominate within the pharmaceutical sector, there is an increasing awareness of the potential of recombinant materials as products in their own right. In the next decade or so we may see the emergence of novel therapeutic agents based on engineered antibodies, antisense technology, and gene therapy products. However, it is the 'traditional' recombinant therapeutic proteins that are reaping the first financial rewards for the investments of the 1980s.

2 THERAPEUTIC PROTEINS

2.1 Native Proteins

Human growth hormone (hGH) and human insulin were two of the first recombinant protein products to be marketed.

The traditional source of hGH (brains of human cadavers) suffered from the double drawbacks of limited supply and the possibility of contamination by human pathogens. The successful production of recombinant hGH, in the bacterium *Escherichia coli*, was a major achievement that allowed the world demand for hGH (used to treat hypopituitary dwarfism) to be met easily, with a safer product.[1]

The case for production of recombinant human insulin was somewhat weaker, insofar as diabetes had been treated successfully for many years with animal insulins, which were in plentiful supply. However, it seemed logical that the human protein would be less likely to cause immunological complications and a convenient recombinant source appeared attractive. This product was first marketed in the United Kingdom in 1982 and by 1989 had become the most common form used by diabetic patients.

For many other proteins, such as the interferons, rDNA technology offered the only realistic means of producing the amounts required for therapeutic use.

The drive for efficient production of recombinant proteins resulted in the development of a range of expression vectors, for use in bacteria, filamentous fungi, yeasts, and cultured cells. Some of these, and the relatively new baculovirus,[2] drosophila,[3] and Semliki Forest Virus[4] expression systems, are capable of producing yields of over 1 g dm^{-3} of recombinant product, depending on the protein involved. Looking to the future, it has been predicted, probably optimistically, that the first therapeutic proteins produced in the milk of transgenic animals will appear on the market by the late 1990s![5]

The choice of expression system is governed to a large extent by the properties of the protein. Many human proteins are produced as insoluble aggregates in *E. coli* and higher yields of active material may be obtained using eukaryotic cells. It is particularly important to use mammalian cells if the activity of the protein is dependent on correct post-translational modification (*eg.* the addition of carbohydrate, or fatty acid moieties) since insect cells (*e.g.* Sf9 cells used in baculovirus expression systems) and yeasts (*e.g. Saccharomyces cerevisiae* or *Pichia pastoris*) exhibit different patterns of modifications. Thus, whereas *E. coli* is adequate for the production of hGH, it has been necessary to use mammalian cells for the production of active, authentic, human tissue plasminogen activator (t-PA). For other proteins, such as insulin, the choice is less critical and *E. coli* and *S. cerevisiae* have both been used.

Examples of recombinant proteins currently licensed for therapeutic use are listed in Table 1. In terms of sales, insulin and human growth hormone have been two of the most successful products, with obvious applications. However, newer products, such as the haemopoietic growth factors, appear to have greater potential. The latter proteins, such as erythropoietin (EPO) and the myeloid colony stimulating factors (CSFs) control the production of blood cells by stimulating the proliferation, differentiation, and activation of specific cell-types.[6] Erythropoietin stimulates the production of erythrocytes from immature

[1] D. V. Goeddel, H. L. Heyneker, T. Hozumi, R. Arentzen, K.Itakura, D. G. Yansura, M. J. Ross, G. Miozzari, R. Crea, and P. H. Seeburg, *Nature (London)*, 1979, **281**, 544.

[2] G. E. Smith, M. D. Summers, and M. J. Fraser, *Mol. Cell Biol.*, 1983, **3**, 2156.

[3] H. Johansen, A. van der Straten, R. Sweet, E. Otto, G. Maroni, and M. Rosenberg, *Genes Dev.*, 1989, **3**, 882.

[4] P. Liljestrom and H. Garoff, *Bio/Technology*, 1991, **9**, 1356.

[5] J. Hodgson, *Bio/Technology*, 1992, **10**, 86.

[6] G. C. Cowling and T. M. Dexter, *Trends Biotechnol.*, 1992, **10**, 349.

Table 1 *Recombinant proteins licensed for therapeutic use*

Protein	*Clinical use*
Insulin	diabetes
Growth hormone	hypopituitary dwarfism
Tissue plasminogen activator	clot lysis
Erythropoietin	anaemia
G-CSF	cancer chemotherapy
GM-CSF	bone marrow transplantation
Factor VIII	haemophilia
Interferon-α	cancers, hepatitis B, leukaemia
Interferon-β	cancers, ALS,[a] genital warts
Interferon-γ	cancers, ARC,[b] osteopetrosis
Hepatitis B surface antigen	hepatitis B vaccine

[a] Amyelotropic lateral sclerosis. [b] AIDS-related complex

erythroid progenitor cells. It has been used successfully to treat the anaemia resulting from renal failure and other acute or treatment-related anaemias. It may also have clinical utility in alleviating the platelet deficiency associated with cancer chemotherapy. Erythropoietin is expected to become the top-selling therapeutic protein by the mid-1990s with predicted annual sales of around $1200 million.

Granulocyte-colony stimulating factor (G-CSF), approved as an adjunct to cancer chemotherapy, generated sales of over $230 million in 1991 and is also in clinical trials for a further six indications, including infectious disease. Some estimates predict annual sales of over $1000 million by 1996. Similarly, granulocyte macrophage-colony stimulating factor (GM-CSF), already approved for use in bone-marrow transplants, is also in clinical trials for two additional indications.

The interferons are approved in a number of countries for treatment of viral disorders, and different types of cancer, often in combination with other agents. It is predicted that total sales of all three proteins will be about half that of EPO by the end of the decade.

Growth factors are expected to figure prominently in the next generation of recombinant therapeutic proteins. Healing wounds contain a mixture of growth factors and their receptors, often produced in a temporal pattern, such that sequential application of recombinant growth factors may represent the most effective treatment. Proteins such as EGF, NGF, FGF, IGF, PDGF, TGF-α, and TGF-β (epidermal, nerve, fibroblast, insulin-like, platelet-derived, and the transforming growth factors, respectively) all appear to be involved in wound healing. With the US market for ulcerating wound therapy expected to approach $1000 million per annum by the end of the decade the potential rewards in this area are considerable.

Potentially, recombinant vaccines have significant advantages over conventional products. However, to date, only one such vaccine, for treatment of hepatitis B, is in widespread use. Vaccines are covered in more detail in Chapter 12 and will not be discussed further here.

2.2 Modified proteins

Having cloned and expressed the cDNA for a therapeutic protein, it is relatively straightforward to prepare derivatives carrying altered amino acid sequences. Considerable effort has been invested in attempting to produce proteins with improved or novel biological activities (sometimes referred to as 'muteins'), by mutagenesis of conventional therapeutic proteins. Examples of muteins which have been evaluated in clinical trials, or in disease models, include derivatives of tumour necrosis factor (renal cell carcinoma[7]), G-CSF (METH-A fibrosarcoma[8]), interleukin-1β (haematapoiesis[9]), interferon-β (metastatic renal cell carcinoma[10]), basic fibroblast growth factor (acetic acid-induced gastric ulcers[11]), and interleukin-2 (stimulation of B and T lymphocytes[12]).

Attempts to generate 'humanized' antibodies represent a more challenging application of protein engineering technology. Potentially, monoclonal antibodies have numerous therapeutic applications, including septic shock and tumour therapy. However, currently available murine monoclonal antibodies, such as OKT3 (which is used to control allograft rejection in organ transplant patients) are of limited clinical value because patients frequently produce an immune reaction against the mouse proteins, sometimes after only one or two doses.[13] This could be avoided by using human monoclonal antibodies but production of such antibodies by conventional hybridoma technology, requiring human immunization, appears impractical at present.[14] An alternative approach, being explored by a number of research groups, is the production of genetically-engineered hybrid antibodies carrying the antigen-recognizing variable domains from rodent antibodies fused to constant domains from human antibodies.[15] Such hybrids should be less immunogenic than the original rodent molecules. This method is being further refined by transferring only the antigen-binding hypervariable complementarity determining regions (CDRs).[16]

Even more amibitious protein engineering projects aim to build totally new proteins using domain structures such as 'β-barrels' and 'α-helices' found in nature. However, the development of novel therapeutic entities using this approach is not imminent!

A recent report concluded that biotechnology drugs, especially rDNA products, have much higher clinical success rates and shorter development times

7 S. Conrad, U. Otto, H. Baisch, and H. Klosterhalfen, *J. Urol.*, 1992, **147**, (supplement 4) 282A.

8 Y. Nio, T. Shiraishi, M. Tsubono, H. Morimoto, C-C. Tseng, K. Kawabata, Y. Masai, M. Fukumoto, and T. Tobe, *Biotherapy*, 1992, **4**, 81.

9 J. R. Zucali, J. Moreb, R. C. Newton, and J. J. Huaog, *Exp. Haematol.*, 1990, **18**, 1078.

10 P. Kinney, P. Triozzi, D. Young, J. Drago, B. Behrens, H. Wise, and J. J. Rinehart, *J. Clin. Oncol.*, 1990, **8**, 881.

11 H. Satoh, A. Shino, N. Inatomi, H. Nagaya, F. Sato, S. Szabo, and J. Folkman, *Gastroenterology*, 1991, **100**, A155.

12 P. Ralph, I. Nakoinz, M. Doyle, M-T. Lee, K. Koths, R. Halenbeck, and D. F.Mark, *J. Cell. Biochem.*, 1986, Supplement 0 (10 part A), 71.

13 T. J. Shroeder, M. R. First, and M. E. Mansour, *Transplantation*, 1990, **45**, 48.

14 D. J. Chiswell and J. McCafferty, *Trends Biotechnol.*, 1992, **10**, 80.

15 J. D. Marks, A. D. Griffiths, M. Malmqvist, T. P. Clackson, J. M. Bye, and G. Winter, *Bio/Technology*, 1992, **10**, 779.

16 L. Reichmann, M. Clark, H. Waldman, and G. Winter, *Nature (London)*, 1988, **322**, 323.

than conventional drugs.[17] Taken together, rDNA products and monoclonal antibodies accounted for 1 in every 10 new drugs clinically tested during the 1980s. By 1989 this had risen to 1 in 3. Annual sales of biotechnology products may exceed $10 000 million by the end of the decade, which makes this application of genetic engineering extremely attractive to the biotechnology companies. However, this figure would be equivalent to less than 5% of the overall market for pharmaceutical products at that time, based on current predictions. Therefore, for the larger pharmaceutical companies, the use of molecular biology as a means of identifying new targets for disease therapy and improving the drug discovery process, is a more attractive proposition.

3 EXPLORING THE MOLECULAR BASIS OF DISEASE

Until fairly recently, immunologists trying to establish the roles of different mediators in haematopoiesis had little option but to use conditioned media as the source of their growth factors. Concentrations of the relevant factor were usually low and contamination with related activities was a constant problem. By the mid-1980s, their molecular biology colleagues were able to supply milligram quantities of pure proteins (often worth millions of pounds, at catalogue prices quoted then!). The quantity and purity of these materials revolutionized research in this field. This story was repeated across other therapeutic areas as previously rare proteins became available for study, in large amounts. This also provided a tremendous stimulus to the field of protein structure determination. *X*-Ray crystallographers suddenly found that they had access to unheard-of quantities of clinically important proteins and most pharmaceutical companies rapidly established in-house facilities to solve the structures of target proteins.

No sooner had the molecular biologists begun to satisfy the requests for pure, authentic proteins than they were being asked for non-authentic derivatives. The site-directed mutagenesis techniques being applied by the biotechnology companies to produce muteins were also extremely powerful tools for studying the modes of action and physiological roles of biological targets, particularly when activity data could be interpreted against an accurate three-dimensional model. However, this approach has been somewhat limited by the lack of available structures. Even at the time of writing, 3-D structures are available for only about 3% of the 17 000 or so proteins with known amino acid sequences.[18]

In the absence of an accurate structure, an alternative approach that has been used with some success is random mutagenesis, followed by screening or selection for mutants with the desired properties.[18] The isolation of an inactive derivative of the *E. coli* heat-labile enterotoxin A subunit, using this method was reported several years ago.[19] Having derived the inactive mutant by random mutagenesis, site-directed mutagenesis was used to demonstrate that loss of activity was due to

[17] B. Bienz-Tadmor, P. A. Dicerbo, G. Tadmor, and L. Lasagna, *Bio/Technology*, 1992, **10**, 521.
[18] D. Medynski, *Bio/Technology*, 1992, **10**, 1002.
[19] S. Harford, C. W. Dykes, A. N. Hobden, M. J. Read, and I. J. Halliday, *Eur. J. Biochem.*, 1989, **183**, 311.

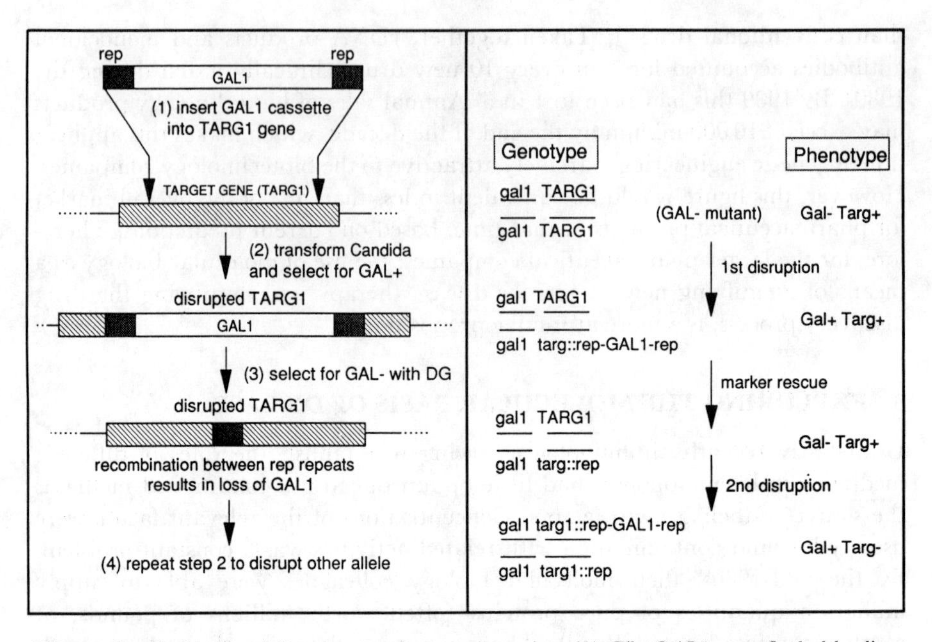

Figure 1 *Evaluation of fungal targets by gene disruption:* (1) *The GAL1 gene, flanked by direct repeats of another segment of DNA (shown here as REP), is inserted into the target gene, carried on a plasmid lacking fungal replication sequences.* (2) *The resulting construct is transformed into a gal- host, which is then incubated on galactose-deficient medium. This selects for clones in which the GAL1 cassette has integrated into the host chromosome, usually by homologous recombination into one of the two copies of the TARG1 gene. Disruption of the TARG gene in GAL1+ cells is verified by Southern blotting or by PCR.* (3) *Selected clones are transferred on to medium containing 2-deoxy galactose (DG), which is converted into a toxic metabolite by the GAL1 gene product.* (4) *Cells that have lost the GAL1 gene, usually by spontaneous recombination between the direct REP repeats, survive the DG selection. PCR or Southern blotting is used to verify that the GAL1 gene has been deleted in the expected manner and selected clones are retransformed with the plasmid carrying the disrupted TARG1 gene in a repeat of step 2. If GAL1+ cells are obtained, and are shown to carry two copies of the disrupted TARG1 gene, the gene product cannot be essential and is unlikely to be a useful target for antifungal therapy. If TARG1 is essential, no colonies will be obtained from the second transformation*

mutation of a critical serine residue, leading the authors to suggest that mutation of the corresponding serine residue in the closely-related cholera toxin A subunit could produce an inactive toxoid suitable for vaccine production. Subsequent determination of the crystal structure of the *E. coli* toxin revealed that this serine residue was in fact closely associated with the active site,[20] adding support to the conclusion from the earlier, random mutagenesis study.

Genetic manipulation techniques have also been extremely useful in identifying and validating novel targets, particularly in the infectious diseases area. Genes for potential viral targets, have been isolated from the viral genome, mutated to destroy activity, and then re-inserted, to determine if the active gene product was essential for pathogenicity. This approach was used to demonstrate that the HIV

[20] T. K. Sixma, S. E. Pronk, K. H. Kalk, E. S. Wartna, B. A. M. Van Zanten, B. Witholt, and W. G. J. Hol, *Nature (London)*, 1991, **351**, 371.

tat and *protease* genes were valid targets for AIDS therapy. Related approaches have also been used for the validation of novel antifungal targets (Figure 1).[21]

Target validation in mammalian systems has been more problematical. Antisense oligonucleotides can provide the only practicable alternative to transgenic animal technology for modulating the expression of a target gene to allow evaluation of any potential role in pathogenesis. These molecules, generally 18–30 nucleotides in length, and complementary to the 'sense' mRNA sequence, have been used to inhibit gene expression *in vitro* in cultured cells, and, recently, *in vivo*.[22] In the latter study, antisense oligonucleotides directed against the proto-oncogene *c-myb* were reported to inhibit intimal smooth muscle cell (SMC) proliferation in a rat arterial injury model. SMC proliferation is a pathological event responsible for restenosis after angioplasty and long-term failure of arterial grafts. In this example, the oligonucleotides were applied directly to the target tissue as a gel. The resulting inhibition of smooth muscle cell proliferation, clearly implicated *c-myb* in this mitogenic pathway and demonstrated the utility of this approach for exploring the molecular basis of disease.

Transgenic animals provide another extremely powerful means of analysing gene function and the regulation of gene expression. Using the developing technologies (described in more detail in Chapter 13) it is possible to delete or overexpress genes, or induce expression in inappropriate locations. Gene deletions can be used to mimic the effects of single gene defect diseases or to determine the physiological function of a particular gene product. Deletion of the *c-src* proto-oncogene caused osteopetrosis in transgenic mice[23] whereas deletion of the mouse homeobox gene *hox1-5* caused regionally-restricted abnormalities in organs such as thyroid, thymus, and throat musculature.[24] The role of the TGF-β-related growth factor, inhibin, was investigated by creating mice lacking a functional copy of the α-inhibin gene.[25] All of the homozygous mice developed gonadal stromal tumours, identifying inhibin as a critical negative regulator of gonadal stromal cell proliferation and the first known example of a secreted protein with tumour-suppressor activity.

The use of constructs carrying proto-oncogenes, growth factors, and cell-surface antigens, under the transcriptional control of heterologous promoters, has provided data about the pathological consequences of inappropriate expression and also provides clues as to the normal roles of these proteins in development and differentiation.[26,27] Preparation of a *bcr-abl* gene fusion mimicking the Philadelphia chromosome translocation was used to derive transgenic mice that developed leukaemia.[28] Such experiments serve two purposes: they validate the

[21] J. A. Gorman, W. Chan, and J. W. Gorman, *Genetics*, 1991, **129**, 19.
[22] M. Simons, E. R. Edelman, J-L. De Keyser, R. Langer, and R. D. Rosenberg, *Nature (London)*, 1992, **359**, 67.
[23] P. Soriano, C. Montgomery, R. Geske, and A. Bradley, *Cell*, 1991, **64**, 693.
[24] O. Chisaka and M. R. Capecchi, *Nature (London)*, 1991, **350**, 473.
[25] M. M. Matsuk, M. J. Finegold, J-G. J. Su, A. J. W. Hsueh, and A. Bradley, *Nature (London)*, 1992, **360**, 313.
[26] R. Jaenisch, *Science*, 1988, **240**, 1468.
[27] G. T. Merlino, *Transgenic Animals Res.*, 1991, **5**, 2996.
[28] N. Heisterkamp, G. Jenster, J. ten Hoeve, D. Zovich, P. K. Pattengale, and J. Groffen, *Nature (London)*, 1990, **344**, 251.

Table 2 *Disorders induced by inappropriate gene expression in transgenic animals*

Gene Deletions	
Phenotype	*Gene deleted*
loss of MHC class I antigen, loss of CD4-8$^+$ T-cell mediated toxicity	β2-microglobulin
osteopetrosis	*src* proto-oncogene
gonadal stromal tumours	α-inhibin
Dominant Negative Mutations	
Phenotype	*Gene expressed*
production of sickled erythrocytes	human sickle haemoglobin gene
osteogenesis imperfecta	mutant pro-α(1) collagen
Gene Additions	
Phenotype	*Gene expressed*
hypertension	mouse ren-2 gene, in rat adrenal
hypertension	rat renin and rat angiotensinogen in mouse
leukaemia	expression of recombinant *bcr-abl*
psoriasis	transforming growth factor-α
eosinophilia	interleukin-5
Kaposi's sarcoma	HIV-TAT (from HIV-LTR)

assumed causal relationship between the translocation and leukaemia and they also provide an animal model for testing therapeutic agents directed against the disease. Other examples of the use of transgenic animals in basic research are listed in Table 2.

These few examples illustrate the use of molecular biology to uncover the information necessary to identify novel targets for the treatment of disease. The other major application of the technology is to provide more powerful methods for discovering compounds active against the novel targets.

4 DRUG DISCOVERY

Traditionally, medicinal chemistry programmes aimed at deriving new therapeutic agents have been based on the structures of known ligands or active compounds identified by random screening (known as lead compounds). Both approaches have been used extremely successfully, in the absence of detailed structural information about the therapeutic target. However, the increasing availability of such information is beginning to make a major impact on the drug discovery process.

4.1 Protein Structure Determination

Advances in molecular biology now allow pharmaceutical companies to clone, express, purify, crystallize, and solve the structure of a target protein, by *X*-ray crystallography, in under a year, the actual time scale being dependent on the properties of the protein. Crystallography can also be used to obtain information about protein–ligand interactions either by soaking the ligand into a pre-formed crystal or by co-crystallizing the ligand with the protein, prior to *X*-ray analysis. Such information can be used to rationalize the effects of different modifications to lead compounds and also to predict which further structural alterations should improve the 'fit' of the compound to the target. Improvements are usually limited to increases in potency or selectivity since it is difficult to predict how pharmacokinetic and toxicological properties may be affected by structural modifications.

Clearly, structural information will play an increasingly important role in 'lead refinement'. The ultimate goal is to use such information to design drugs *ab initio*, from the structure of the target protein only, without having to start from a lead compound.

4.2 Rational Drug Design

There are few examples of receptor antagonists or protein inhibitors designed solely from the structure of the biological target. One of the successes and a notable failure both involve members of the same class of enzyme, the aspartyl proteases. HIV protease consists of a dimer of two identical subunits and this two-fold symmetry encouraged several groups to test the effects of symmetrical compounds. This resulted in the development of novel, nearly-symmetrical, inhibitors of the protease.[29,30]

The related protease, renin, which is involved in blood pressure regulation, has been the subject of intense study over the past decade, but there are still no effective inhibitors on the market. A recent report described the synthesis of a cyclic molecule designed to fit into the renin active site.[31] However, this compound was found to be ineffective because its conformation in solution was different from that of the renin-bound form, and more energetically favourable. This illustrates the point that design of novel inhibitors should take all aspects of the binding process into consideration, and not just the best fit between enzyme and inhibitor.

The synthesis of novel classes of inhibitors of thymidylate synthase has been another success for the structure-based approach.[32]

[29] J. Erickson, D. J. Neidhart, J. VanDrie, D. J. Kempf, X. C. Wang, D. W. Norbeck, J. J. Plattner, J. W. Rittenhouse, M. Turon, N. Wideburg, W. E. Kohlbrenner, R. Simmer, R. Helfrich, D. A. Paul, and M.Knigge, *Science*, 1990, **249**, 527.

[30] R. Bone, J. P. Vacca, P. S. Anderson, and M. K. Holloway, *J. Am. Chem. Soc.*, 1991, **113**, 9382.

[31] M. D. Reily, V. Thanabal, E. A. Lunney, J. T. Repine, C. C. Humblet, and G. Wagner, *FEBS Lett.*, 1992, **302**, 97.

[32] K. Appelt, R. J. Bacquet, C. A. Bartlett, C. L. J. Booth, S. T. Freer, M. A. M. Fuhry, M. R. Gehring, S. M. Herrmann, E. F. Howland, C. A. Janson, T. R. Jones, C-C. Kan, V. Katharde-

The number of structures available for clinically-important proteins is increasing rapidly, albeit from a very low baseline, and as the lessons learned from studies such as those on HIV protease and renin are applied to other proteins, it is inevitable that rational drug design, *ab initio*, will feature more prominently in the drug discovery process. However, until then, the pharmaceutical industry will continue to rely on its traditional tools.

4.3 Random Screening

Random screening remains the most productive approach to drug discovery. The input may consist of collections of small molecules amassed over decades of medicinal chemistry programmes, natural product samples collected from the four corners of the earth, or the more recent peptide and oligonucleotide (aptamer) libraries. The common denominator is that very large numbers of samples must be assayed quickly. The most usual assay format for the 'high-throughput screen' is the 96-well microtitre plate, which is compatible with automated assays. The application of rDNA technology has transformed the random screening method of lead generation in a number of ways.

4.3.1 Peptides as Lead Compounds. Low-molecular weight non-peptidic drugs are generally superior to peptides in terms of bioavailability, stability, and pharmacokinetics. However, despite the obvious failings of peptides as drugs they can be useful as a means of initiating medicinal chemistry programmes in the absence of small molecule leads.

The development of a bacteriophage expression system capable of displaying peptide epitopes[33] has lead to the preparation of epitope libraries containing hundreds of millions of different random peptide sequences to facilitate screening against target proteins. Mixtures of oligonucleotides are inserted into the pIII minor coat protein gene of a filamentous bacteriophage (*e.g.* M13) such that random peptide sequences of 6–15 amino acids are expressed at, or close to, the amino terminus of the coat protein. Recombinant phage are multivalent with respect to the peptide sequences since about 4–5 copies of the coat protein are found at one tip of each virion.

Phage are screened by 'panning' over immobilized preparations of specific antibodies, receptors, or other ligand-binding proteins. Interaction of the target protein with one of the peptide sequences results in retention of the appropriate phage which can be recovered and amplified by infection of *E. coli*. Sequencing the modified pIII genes reveals the sequence of the peptide involved. This technique can be used to delineate the interaction between antibodies and target proteins (epitope mapping) as well as for generation of peptide leads. The pIII protein can apparently accommodate quite large proteins allowing this approach to be used for screening a recombinant antibody library, to detect

[32] continued
kar, K. K. Lewis, G. P. Marzoni, D. A. Matthews, C. Mohr, E. W. Moomaw, C. A. Morse, S. J. Oatley, R. C. Ogden, M. Rami Reddy, S. H. Reich, W. S. Schoettlin, W. W. Smith, M. D. Vamey, J. E. Villafranca, and R. W. Ward, *J. Med. Chem.*, 1991, **34**, 1925.

[33] S. F. Parmley and G. P. Smith, *Gene*, 1988, **73**, 305.

derivatives with improved affinity for a hapten.[15] The phage pVIII protein has also been used in a similar manner.

Alternative screening methods for peptide leads, using random peptide synthesis have also been described.[34,35]

Having generated the peptide lead the major challenge still lies ahead. It is by no means straightforward to derive a non-peptidic molecule from even a small peptide and, using this technology, a single molecular biologist could keep several teams of medicinal chemists occupied into the next century!

4.3.2 Aptamers. Oligonucleotides may have potential as inhibitors of enzymes or non-enzymic proteins. Random oligonucleotide libraries containing of the order of 10^{13} different molecules have been screened against enzymes such as thrombin which is not known to contain a nucleotide binding site.[36] The library is incubated with an immobilized preparation of the enzyme, bound DNA is extracted, amplified by PCR (see Chapter 3), and re-incubated with immobilized protein. After a number of cycles of adsorption and amplification, the resulting population of oligonucleotides is cloned and sequenced. In the example quoted, individual oligonucleotides with sub-μMolar affinities for thrombin were obtained and a consensus sequence was identified by comparison of a number of cloned molecules. The affinities of the molecules described (termed 'aptamers') were too low to be of therapeutic utility. However, it may be possible to develop more potent compounds by treating these oligonucleotides as lead-compounds, as described in Section 4.1.

4.3.3 Recombinant Proteins for In Vitro *Assays.* Having identified an appropriate target, it is often difficult to obtain enough of the protein for a screening programme. This is especially true for many viral proteins such as the HIV protease which could not be isolated from natural sources in the necessary quantities. Without the recombinant material produced in *E. coli* it would not be feasible to screen against this protein in the high-throughput format favoured by pharmaceutical companies. Similarly, recombinant human proteins are now replacing the animal proteins used in screens where it had been impractical to isolate the native human homologue.

4.3.4 Whole-cell Screens. Whole-cell screens can be run as an alternative to *in vitro* assays if production of the target protein causes, or can be linked to, a phenotypic change in the host organism. One *E. coli* screen described for HIV protease involved co-expression of the HIV enzyme and a β-galactosidase derivative carrying a potential cleavage site for the protease.[37] Proteolytic inactivation of the β-galactosidase was prevented by inhibitors of the protease. The resulting increase in β-galactosidase activity was detected by growing the cells in the

[34] H. M. Geysen, S. J. Rodda, and T. J. Mason, *Mol. Immunol.*, 1986, **23**, 709.
[35] S. P. A. Fodor, J. L. Read, M. C. Pirrung, L. Stryer, A. Tsai Lu, and D. Solas, *Science*, 1991, **251**, 767.
[36] L. C. Bock, L. C. Griffin, J. A. Latham, E. H. Vermaas, and J. J. Toole, *Nature (London)*, 1992, **355**, 564.
[37] E. Z. Baum, G. A. Bebernitz, and Y. Gluzman, *Proc. Natl. Acad. Sci. USA*, 1990, **87**, 10023.

presence of the colorimetric substrate, X-Gal. An even simpler system exploited the cytotoxicity of the protease in *E. coli* by expressing the enzyme at levels which markedly inhibited growth. Inhibitors were detected by their ability to stimulate growth rate.

Once the screening organism has been constructed, a whole-cell microbial screen is cheap and simple to run. For some targets, *e.g.* those involving transcriptional or post-transcriptional events, there may be no realistic alternative. Although cultured human cells offer the most authentic cellular background for studying transcriptional phenomena, they are relatively susceptible to the toxic components often encountered in natural product samples. Yeasts such as *Saccharomyces cerevisiae* and *Schizosaccharomyces pombe* offer the advantages of eukaryotic cellular organization, but are markedly more robust in natural product screens than cultured cells. Soluble nuclear receptors such as the human oestrogen,[38] androgen,[39] and glucocorticoid[40] receptors have all been expressed functionally in *S. cerevisiae* and have retained their ability to transactivate expression from appropriate reporter constructs. These used hybrid yeast promoters, containing mammalian hormone response elements in place of the binding sites for native yeast transactivator proteins (Figure 2), controlling expression of easily-assayable enzymes such as β-galactosidase. Yeast-based hormone receptor screens offer more potential targets for interrupting hormone dependent gene expression than the *in vitro* receptor/ligand binding assays that were commonplace before this application of rDNA technology.

Since the first cloning of a human protein by functional complementation of a yeast mutant[41] a surprisingly large number of other human proteins have also been shown to be capable of complementation. Because these mutants depend on the activity of the homologue for growth or survival, specific inhibitors of the human protein will be lethal to the mutant but not to wild type yeast. This principle has been used to establish screens for a number of proteins including phosphodiesterases.[42] The method is particularly useful in screening against interactions for which no assay exists.

4.3.5 Membrane-bound Receptor Screens. Historically, standard assays for antagonists of membrane-bound receptors have involved radioligand binding assays using preparations of tissue enriched in the receptor of interest. Two major drawbacks have been that the tissues are usually animal in origin, not human, and that they almost invariably contain mixtures of receptors, some of which can be closely related to the target receptor. The molecular biologist's solution to these problems has been to clone hundreds of human receptors and express them in cultured cell-lines carrying few endogenous receptors. Chinese Hamster

[38] D. Metzer, J. H. White, and P. Chambon, *Nature (London)*, 1988, **334**, 31.
[39] I. J. Purvis, D. Chotai, C. W. Dykes, D. B. Lubahn, F. S. French, E. M. Wilson, and A. N. Hobden, *Gene*, 1991, **106**, 35.
[40] D. Picard, M. Schena, and K. R. Yamamoto, *Gene*, 1990, **86**, 257.
[41] M. G. Lee and P. Nurse, *Nature (London)*, 1987, **327**, 31.
[42] M. M. McHale, L. R. Cieslinkski, W. K. Eng, R. K. Johnson, T. J. Torphy, and G. P. Livi, *Mol. Pharmacol.*, 1991, **39**, 178.

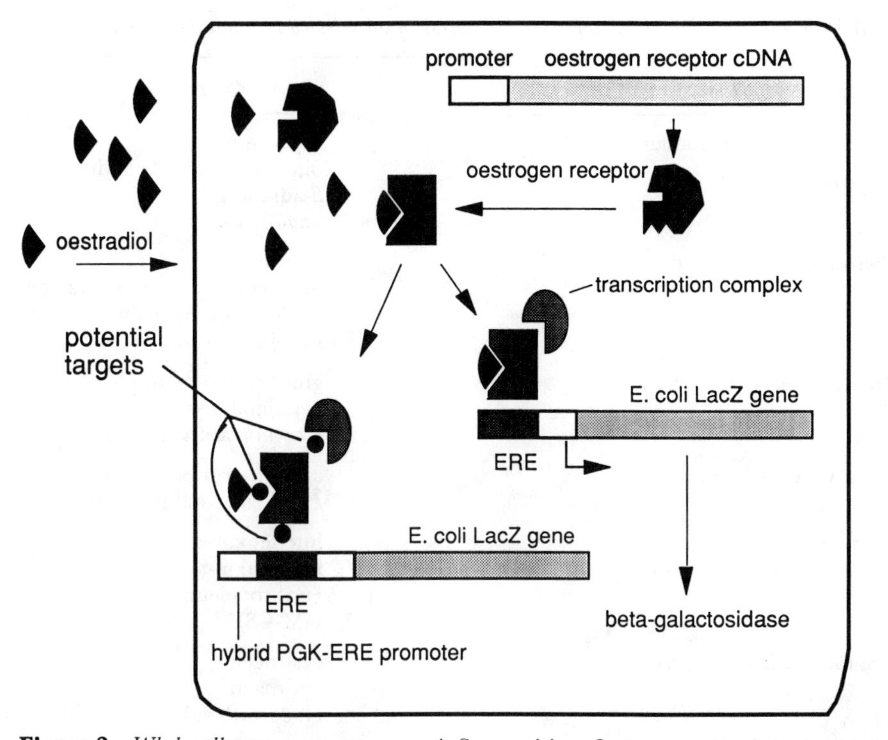

Figure 2 *Whole-cell oestrogen receptor screen in* S. cerevisiae. *Oestrogen receptor is produced in* S. cerevisiae *from a yeast promoter (*e.g. *CUP1). A second construct carries the* E. coli *LacZ gene, under the transcriptional control of a hybrid yeast promoter (*e.g. *phosphoglycerate kinase – PGK) containing one or more copies of the oestrogen response element (ERE). In the presence of oestradiol, hormone/receptor complexes can interact with the ERE, and the host transcriptional machinery, to drive expression of the LacZ gene. Compounds which interfere with the interaction between the receptor and* (a) *oestradiol,* (b) *the ERE, or* (c) *the transcription complex may all be detected as antagonists. This system may also be used to screen for novel agonists by omitting oestradiol*

Ovary (CHO) cells have been the first-choice host for the majority of these cloned receptors but *E. coli*[43] and *S. cerevisiae*[44] have also been used.

Examples of receptor classes of interest to the pharmaceutical industry are shown in Table 3.

The G-protein-linked membrane-bound receptors form one of the largest and most important classes of receptors with over 100 having been cloned. Receptors of this type are involved in a range of physiological processes (Table 4) including responses to taste, odorants, and light. Olefactory epithelium alone is believed to contain over 1000 distinct odorant receptors of this class.

One disadvantage of microbial expression systems for the G-protein-linked receptors is that the binding affinity of the receptor for agonists alters depending on whether the appropriate heterotrimeric G-protein is associated with the

[43] S. Marullo, C. DeLavier-Klutchko, J-G. Guillet, A. Charbit, A. D. Strosberg, and L. J. Emorine, *Bio/Technology*, 1989, **7**, 923.
[44] K. King, H. G. Dohlman, J. Thorner, M. G. Caron, and R. J. Lefkowitz, *Science*, 1990, **250**, 121.

Table 3 *Receptors types of therapeutic interest, with selected examples of each class*

Receptor Type	*Receptor*
7-transmembrane-domain (G-protein-linked)	rhodopsin muscarinic acetyl choline β-adrenergic angiotensin
Tyrosine kinase-linked	insulin platelet-derived growth factor epidermal growth factor fibroblast growth factor
Ion channel	glutamate (ionotropic) γ-amino butyric acid nicotinic acetyl choline
Guanylate cyclase-linked	atrial natriuretic factor *E. coli* heat-stable enterotoxin
Lymphokine	interleukin-2 interleukin-3 erythropoietin GM-CSF
Soluble nuclear receptor	oestrogen androgen glucocorticoid thyroid hormone

intracellular region of the receptor. Neither *E. coli* nor *S. cerevisiae* possess G-proteins which associate with mammalian receptors. However, *S. cerevisiae* has an endogenous G-protein-linked signalling pathway involved in mediating responses to mating pheromones[45] and academic and industrial groups around the world have invested considerable effort in attempting to couple expression of mammalian G-protein-linked receptors into this pathway. There has been one report of successful β-agonist-mediated activation of the pathway in cells expressing both the β-2-adrenergic receptor and the cognate mammalian G-protein α-subunit (Gsα).[44] However, this has proved difficult to repeat and it is generally believed that the system as described is too variable to have general utility.

4.3.6 Transcriptional Screens. Inhibition of the expression of a gene is an alternative to inhibiting the activity of the gene product. This potential approach to therapy is particularly relevant to cancers, immunoregulatory disorders, inflammation, and metabolic diseases where aberrant gene expression plays a major role in pathogenesis. In some of these disorders even a slight modulation in expression levels may have marked beneficial effects. Gene expression can be regulated by a number of mechanisms, such as RNA transport, splicing, or

[45] I. Herskowitz, *Microbiol. Rev.*, 1988, **52**, 536.

Table 4 *Members of the G-protein-linked, 7-transmembrane domain receptor class*

Receptor[a]	*Physiological Roles*	*Therapeutic Targets*
Bradykinin	pain, inflammation	analgesia
Endothelin	vascular tone	vascular diseases
Gastrin/CCKB	gastric acid secretion	gastric ulcer
Dopamine	various, CNS	*e.g.* schizophrenia
Angiotensin	blood pressure regulation	hypertension
Adrenergic	various	*e.g.* asthma, hypertension
Serotonin	various	*e.g.* migraine, depression, emesis
Thrombin	platelet activation, mitogenesis, vascular proliferation	thrombosis
Neurokinin	smooth muscle contraction	*e.g.* asthma, anxiety

[a] Families of subtypes exist for most of the receptors listed. Different subtypes may be involved in different physiological processes, only some of which are shown here. Furthermore, agonists and antagonists of the same receptor subtype may have applications in different diseases, *e.g.* β-2 adrenergic antagonists (β-blockers) are used in hypertension and migraine, whereas β-2 agonists are used for treating asthma

stability but, in general, the major control is exerted at the level of transcription, by proteins known as transcription factors.[46]

Many transcription factors bind to specific DNA sequences in the region of a gene immediately upstream from the transcription start site and by interacting with RNA polymerase II, can either stimulate, or inhibit, transcription, depending on the factors involved. The activities of these proteins can be regulated by a number of mechanisms including temperature, phosphorylation, ligand-binding, and interactions with other transcription factors, as well as by their own steady-state levels. Thus genes carrying the appropriate regulatory elements may be expressed in a constitutive, inducible, tissue-specific, or developmentally-regulated manner depending on whether sufficient levels of the relevant activated transcription factors are present (Table 5).

The complexity of these interactions, and the known examples of ligand-mediated regulation, has persuaded many pharmaceutical companies that it should be possible to find compounds capable of modulating the transcription of specific genes, or classes of genes, known to be involved in pathogenic disorders. The application of rDNA technology, over the past decade, has provided the information and the tools necessary to begin the search for such compounds.

Transcriptional regulation screens utilize cultured cells that have been transfected either transiently or stably with a transcriptional unit carrying the promoter of interest linked to a reporter gene. The activity of the promoter is

[46] D. S. Latchman, 'Eukaryotic Transcription Factors', Academic Press Ltd., London, 1991.

Table 5 *Examples of regulated gene expression*

Gene	*Control*	*Transcription Factor*
Metallothionein	inducible – heavy metals	unknown
Somatostatin	inducible – cyclic AMP	CREB/ATF
α-1-Antitrypsin	inducible – phorbol ester	AP1
Myosin heavy chain	inducible – retinoic acid	retinoic acid receptor
Interleukin-2	tissue-specific – activated T cells	NFAT-1
Growth hormone	tissue-specific – pituitary	Pit-1
Haemoglobin	tissue-specific – erythroid cells	DF1
Interleukin-2 α-receptor	tissue-specific – activated T cells	NFκB

monitored by measuring the levels of the reporter gene product in the presence of test samples (see Figure 3). Although such screens can only approximate expression of the target gene *in vivo*, they provide a relatively convenient means of testing large numbers of potential modulators rapidly and there is considerable optimism that this approach will lead to the development of an important new class of drug.

4.4 Drug Development

Molecular biology is also having an impact on the analysis of compounds selected as positives in primary high-throughput screens. A good example of this is the characterization of compounds active against receptor targets, where it is important to know if a compound is selective against receptor subtypes. Two subtypes of the endothelin receptor, ET-A and ET-B, have been reported in humans[47–49] and rats.[50,51] These have been cloned, and couple to endogenous signalling pathways when expressed in CHO cells. Therefore, this set of four cell-lines may be used to determine whether a compound has agonist or antagonist activity, shows any selectivity between the ET-A and ET-B subtypes, or exhibits any differences in activity towards the rat and human receptors. The latter point may be important when compounds are evaluated in rats before being tested in humans.

Other applications of rDNA technology in drug development include cloning and expression of cytochrome p450 genes. Since the liver is the primary site of drug metabolism, candidate drugs are normally incubated with rat liver extracts,

[47] Y. Ogawa, K.Nakao, H. Arai, O. Nakagawa, K. Hosoda, S-I. Suga, S. Nakanishi, and H. Imura, *Biochem. Biophys. Res. Commun.*, 1991, **178**, 248.

[48] A. Sakamoto, M. Yanigasawa, T. Sakurai, Y. Takuwa, H. Yanigasawa, and T. Masaki, *Biochem. Biophys. Res. Commun.*, 1991, **178**, 656.

[49] M. Adachi, Y-Y. Yang, Y. Furuchi, and C. Miyamoto, *Biochem. Biophys. Res. Commun.*, 1991, **180**, 1265.

[50] T. Sakurai, M. Yanigasawa, Y. Takuwa, H. Miyazaki, S. Kimura, K. Goto, and T. Masaki, *Nature (London)*, 1990, **348**, 732.

[51] Y. H. Lin, E. H. Kaji, G. K. Winkel, H. E. Ives, and H. F. Lodish, *Proc. Natl. Acad. Sci. USA*, 1991, **88**, 3185.

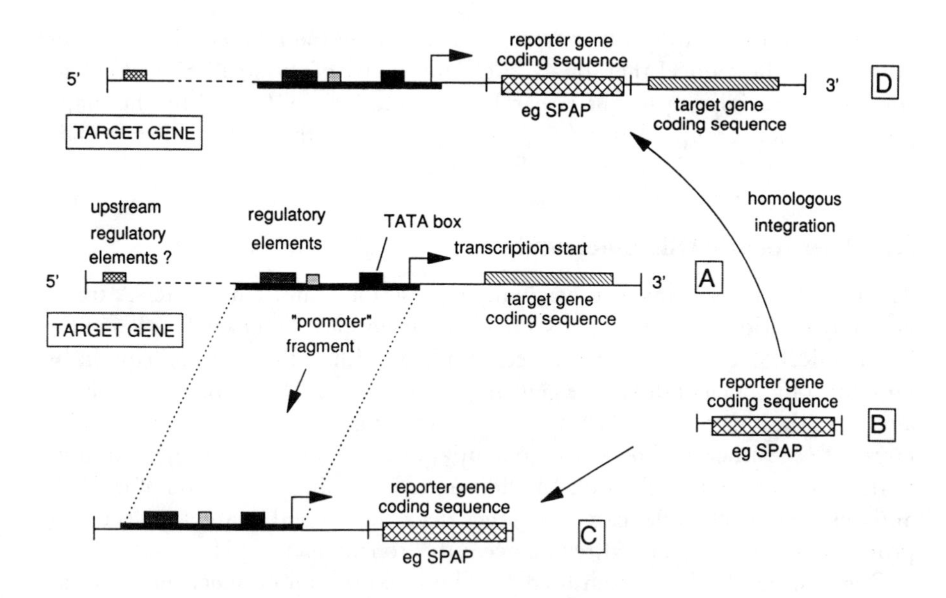

Figure 3 *Transcriptional Screens. A reporter gene construct (C) is prepared by placing the coding sequence of an easily assayable protein, such as secreted placental alkaline phosphatase (SPAP – construct B), under the transcriptional control of the promoter from the target gene (A). It is assumed that the 'promoter' fragment contains all of the relevant regulatory elements, although it is difficult to be certain that there are no intra-intronic, 3′ or 5′ elements outside the subcloned region. The reporter construct is then transfected into an appropriate host cell where SPAP production is used as a convenient measure of promoter activity.*

An alternative approach is to insert the SPAP coding region into one copy of the target gene in its chromosomal locus in the appropriate host cell, by homologous integration, to produce construct D. in order to achieve this, the SPAP sequence must first be inserted into a large (> 2 kilobase) subcloned fragment of the target gene, (not shown), immediately downstream from the promoter, to produce a clone carrying the target gene sequences required for homologous recombination into the chromosome. None of the 5′, 3′, and intragenic regulatory elements should be deleted in this procedure

and also tested in rats, to determine the range of metabolites produced. An increasing trend in drug metabolism laboratories is to utilize panels of recombinant cell-lines expressing individual, cloned p450s to demonstrate the range of metabolic conversions that could occur and also to ascertain if a drug is capable of inducing the expression of any of the p450s known to correlate with hepatocarcinogenesis. The cloning of a soluble nuclear receptor activated by peroxisome proliferators, *e.g.* clofibrate, was an elegant example of the increasing use of molecular biology to study mechanisms of hepatic toxicity and carcinogenicity.[52]

5 FUTURE PROSPECTS

At the time of writing, nucleic-acid-based therapeutics have yet to go on sale and there is doubt in some quarters about whether such products will ever rival the

[52] I. Issemann and S. Green, *Nature (London)*, 1990, **347**, 645.

conventional drugs and proteins that now dominate the market. However, the potential of these novel therapies is such that many of the small biotechnology companies specializing in this research are being heavily funded by the major pharmaceutical groups. Antisense technology and gene therapy are the two main categories of potential nucleic-acid-based therapies.

5.1 Therapeutic Oligonucleotides

The term 'antisense' was used originally to describe inhibition of mRNA translation by hybridization of an antisense oligonucleotide to a selected region of the 'sense' mRNA. The term has also been applied to inhibition of transcription by targeting oligonucleotides to the gene itself, such that a DNA triplex is formed as a result of the oligonucleotide binding in the major groove of the double-stranded target DNA. Either method has the potential to block gene expression by binding to the appropriate nucleic acid. In addition to inhibiting translation of mRNA, antisense oligonucleotides can also induce degradation of the target molecule by providing a double-stranded region accessible to ribonuclease H.

Oligonucleotides have been used to inhibit expression of oncogenes, normal cellular genes, and recombinant reporter genes, and to inhibit growth of viruses in culture.[53–55] In theory, any disease characterized by inappropriately high gene expression ought to be amenable to oligonucleotide therapy. Viral diseases are particularly attractive because many viral genes have no cellular homologue and there is a clear causal relationship between the virus and the disease. Table 6 lists a selection of diseases that have been proposed as candidates for oligonucleotide therapy. Clinical trials of the first antisense drugs have begun, with treatments for genital warts (directed against Human Papilloma Virus) and Cytomegalovirus being tested.

The power of these techniques lies in the exquisite specificity available in the interaction between the oligonucleotide and the target nucleic acid sequence. Antisense molecules are capable of discriminating between RNA sequences differing by only a single base, as was shown recently in a study on the inhibition of mutant *ras* in cultured cells.[53] Furthermore, once a gene has been clearly implicated in disease development, therapeutic agents can be designed solely on the basis of the nucleic acid sequence: detailed knowledge of the mode of action of the gene product is not essential.

The potential of these molecules to inhibit the expression of target genes specifically is beyond doubt. The major obstacle is delivery to the target tissue. Much of the current research into therapeutic oligonucleotides is aimed at solving the problems of instability in serum and poor uptake into cells. Because of these problems, very high oligonucleotide concentrations have to be used in order to achieve the desired effect and at today's prices therapy would be very expensive. The former problem is being addressed by using chemically modified backbone derivatives instead of the standard phosphodiester linkage to improve

[53] S. T. Crooke, *Bio/Technology*, 1992, **10**, 882.
[54] J. M. Chubb and M. E. Hogan, *Trends Biotechnol.*, 1992, **10**, 132.
[55] S. Agrawal, *Trends Biotechnol.*, 1992, **10**, 152.

Table 6 *Examples of diseases thought to be amenable to oligonucleotide therapy, with suggested target genes*

Therapeutic Area	*Disease*	*Possible Target Genes*
Cancer	Lymphoma	*Bcl-2*
	Leukaemia	*bcr/abl*
	Melanoma	basic fibroblast growth factor
	Carcinoma	*myc*, *myb*
Cardiovascular	Hypertension	renin
		angiotensinogen
		endothelin precursor
Immune disorders	arthritis	autoimmune immunoglobulins
	lupus erythromatosus	autoimmune immunoglobulins
Central nervous system	Alzheimer's disease	β-amyloid
	depression	monoamine oxidase
Metabolic diseases	acromegaly	growth hormone
	ulcer	pepsinogen
	psoriasis	transforming growth factor-α
Infectious diseases	AIDS	HIV-TAT and others
	Malaria	haem polymerase
	Human papilloma virus	various
	Influenza A and B	various
	cytomegalovirus	various
	Herpes Zoster	various
	HSV-1, HSV-2	ICP4 and others

resistance to nucleases.[55] Possible solutions to the uptake problem include exploiting the ubiquitous transferrin receptor for internalization of transferrin–oligonucleotide complexes,[56] or the use of liposomes.[57] Significant improvements in both of these areas will have to be made before antisense therapy becomes a realistic, affordable alternative to conventional drugs.

5.2 Gene Therapy

Since the first successful expression of a heterologous gene in a microbial host, the medical world has dreamed about the possibility of using gene transfer methods to treat human genetic defects. Those dreams are already beginning to be realized in current 'gene therapy' trials in the USA and in Europe. Furthermore, it is becoming clear that gene therapy will not be restricted to the recessive, single-gene defects originally envisaged as the initial targets but will also find applications in cancer and infectious diseases.

[56] M. Zenke, P. Steinlein, E. Wagner, M. Cotten, H. Beug, and M. L. Birnstiel, *Proc. Natl. Acad. Sci. USA*, 1990, **88**, 8850.
[57] E. Wickstrom, *Trends Biotechnol.*, 1992, **10**, 28.

The major obstacles to effective gene therapy are efficient transfer of 'foreign' DNA into the chosen target cells and stable expression of the transferred DNA. Retroviral vectors have been used successfully to achieve efficient transfer of the human adenosine deaminase (ADA) gene into lymphocytes. Lack of ADA leads to accumulation of 2-deoxyadenosine, which is toxic to B and T lymphocytes, resulting in severe immunodeficiency. In the initial clinical trials in the USA, blood cells removed from a patient suffering from ADA deficiency were treated with a retroviral vector carrying the ADA gene, in an *in vitro* procedure. Over a 10 month period during which the patient received eight transfusions of treated cells, expression of the inserted ADA gene resulted in a marked clinical improvement with significant numbers of the recombinant cells persisting for over 6 months after the last infusion.[58]

Although retroviral vectors allow highly-efficient and stable gene transfer into dividing cells, they appear to be unable to infect non-replicating cells.[59] However, this has been achieved using adenovirus vectors, which have the ability to infect a wide range of tissues. An adenovirus vector has been used to obtain expression of the human cystic fibrosis transmembrane conductance regulator gene in rat lungs.[60] In this experiment the vector preparation was delivered by intratracheal distillation and expression of the human gene was detectable for 6 weeks after infection. Since these vectors do not normally integrate into chromosomal DNA to a significant extent, it might be expected that long-term expression of the target gene would not be possible using this system. Nevertheless initial results are promising and the first trials of this system in human cystic fibrosis patients have recently been approved by the United States Recombinant DNA Advisory Committee.

Other non-viral DNA transfer methods are being explored. There are plans to inject liposome-encapsulated DNA, carrying the *HLA-B7* gene, directly into melanoma tumour nodules, in the first attempt at *in vivo* gene therapy. Expression of this class I MHC molecule by tumour cells is expected to render them more susceptible to attack by cytotoxic T-cells.

Exploitation of tissue-specific, cell-surface receptors may aid gene targeting *in vivo*. DNA complexed with asialoglycoprotein may be directed to the liver for uptake by the asialoglycoprotein receptor found only on hepatocytes. However, targeting to the liver is complicated by the apparent necessity to perform a partial hepatectomy, to stimulate cell division, in order to achieve stable expression.[61] The use of adenovirus vectors to achieve infection of a range of tissues, with constructs carrying cell type-specific promoters, provides another possible approach to obtaining gene expression in the desired tissue. For diseases where treatment requires the systemic action of a secreted protein product, targeting the corrective gene to the cell type where it would normally be expressed may not

[58] A. D. Miller, *Nature (London)*, 1992, **357**, 455.

[59] D. G. Miller, M. A. Adam, and A. D. Miller, *Mol. Cell Biol.*, 1990, **10**, 4239.

[60] M. A. Rosenfeld, K. Yoshimura, B. C. Trapnell, K. Yoneyama, E. R. Rosenthal, W. Dalemans, M. Fukayama, J. Bargon, L. E. Stier, L. Stratford-Perricaudet, M. Perricaudet, W. B. Guggino, A. Pavirani, J-P. Lecocq, and R. G. Crystal, *Cell*, 1992, **68**, 143.

[61] G. Y. Wu, J. M. Wilson, F. Shalaby, M. Grossman, D. A. Shafritz, and C. H. Wu, *J. Biol. Chem.*, 1991, **266**, 14338.

Table 7 *Potential disease targets for gene therapy*

Disease	*Gene to be Used in Treatment*
Immune deficiency	adenosine deaminase
Cystic fibrosis	cystic fibrosis transmembrane regulator
Hypercholesteraemia	low density lipoprotein receptor
Emphysema	α-1-antitrypsin
Sickle-cell anaemia	β-globin
Thalassaemia	β-globin
Haemophilia	factor XI, factor VIII
Cancer – melanoma	HLA-B7
AIDS	'TAR-loop'[a]
Phenylketonuria	phenylalanine hydroxylase
Diabetes	GLUT-2 glucose transporter, glucokinase

[a] 'TAR-Loop' refers to short nucleic acid sequences designed to mimic the 'TAR' region in HIV mRNA transcripts, intended to sequester the HIV tat transactivator protein

be necessary. Secretion from other somatic cells may provide adequate levels of the desired gene product.

Some of the diseases in which gene therapy could be applied are listed in Table 7. As noted above, preliminary trials have produced some very encouraging results and its seems likely that treatments for at least some of the diseases listed will become available in the next decade. Exactly how the pharmaceutical/biotechnology companies will handle development and commercialization of these new technologies is far from clear. The specialist biotechnology companies currently working on *in vivo* gene therapy expect to sell a DNA or viral product, analogous to a conventional drug. However, for *ex vivo* treatments, *i.e.* those in which cells will be removed from the patient prior to insertion of the appropriate gene, different approaches are possible. The end product could be a DNA or viral product carrying the DNA of interest, with the necessary procedures being performed in hospitals and clinics as for conventional medical treatments. Alternatively, the product could be a patient's genetically-engineered cells. Some biotech companies may seek to establish a service role in preparing such cells, probably in close association with major medical centres. However it is organized, if gene therapy can be applied successfully to cancer and infectious disease as well as the inherited single-gene defects originally targeted, this may prove to be one of the most significant applications of rDNA technology.

6 CONCLUSIONS

Recombinant DNA technology has already made a significant impact on most areas of biological research within the pharmaceutical industry and its influence will spread further in the next few years. Known DNA sequences can now be cloned fairly easily using polymerase chain reaction methodology and with the anticipated explosion in sequence data from the human genome project just around the corner, molecular biologists will be able to tap into a rich new source of regulatory elements, promoters and cDNAs. The current trend is for small, specialist biotech companies to adopt and develop the novel, high-risk technologies such as antisense and gene therapies, with financial support from the pharmaceutical giants. This appears to be a mutually-beneficial arrangement which leaves the pharmaceutical companies free to concentrate their molecular biology resources on other activities that are more certain to benefit their drug discovery programmes. The companies that most successfully integrate these resources into their research activities will have a significant advantage in the intensely competitive market of the 1990s and beyond.

CHAPTER 9

The Current Impact of Recombinant DNA Technology in the Food Industry

R. K. PAWSEY

1 INTRODUCTION

It is now over 15 years since research into biotechnological processes and their application began. A comprehensive review[1] in the previous edition of this book, published in 1988, surveyed the potential impact of biotechnology on the food industry. Biotechnology was seen to offer, and still does, the prospect of changes in the raw materials supplying the food industry – that is to say, through genetic manipulation, the production of new seed varieties of staple monocotyledenous food grains and of dicotyledenous plants such as tomatoes. Changes which were foreseen were in the resistance of plants to adverse growth conditions, in yields, in food storage qualities, and in food processing qualities. Biotechnology could, and can, offer changes in food animals – in their productivity, in the composition of their meat, and in their milk or egg yields. The research into the genetic modification of organisms to produce new starter strains, new enzymes, and new products has reached the point where the dairying and brewing industries in particular may be expected to soon be making regular and increasing use of these developments. There are prospects for the development of novel foods – single cell protein, and for both new and well established food ingredients from new, often microbial sources. The development of microbial products, particularly of enzymes, and the use of immobilized organisms including those produced by genetic modification offer prospects of processing or monitoring food quality in faster or cheaper ways.

Yet currently, in 1992, the impact of biotechnology on the food industry remains small – and the reasons for this are three-fold. One reason is that in the interest of protecting the health and safety of consumers and of protecting the environment the legal framework is still being developed. Secondly, the timescale for product launching lengthens as a consequence of the requirement for ade-

[1] R. K. Pawsey and D. J. Cox, in 'Molecular Biology and Biotechnology', 2nd ed., ed. J. M. Walker and E. B. Gingold, Royal Society of Chemistry, London, 1988, 157–193.

quate safety testing, and also commercial demands for adequate legal protection of inventions;[2,3] and thirdly because consumers are largely uninformed about biotechnology the food industry fears they are likely to resist foods in which biotechnological products and processes are used. However, some impact is being seen and as the obstacles are resolved the impact is likely to escalate.

2 LEGAL REQUIREMENTS IN THE PRODUCTION OF NOVEL FOODS AND PROCESSES IN THE UK

Mycoprotein ('Quorn'®), the novel food produced from the mould *Fusarium graminareum*, was launched in 1986 after approval by the Ministry of Agriculture, Fisheries, and Food (MAFF) that it was safe for human consumption. It was first released only in premanufactured foods such as stews and curries; in 1992 however the raw material became available through retail food outlets for home use. A big publicity campaign was launched and the product appears to be enjoying commercial success through consumer acceptance. In a sense this was a project which, while not involving genetically engineered organisms, laid the ground rules for approval procedures for all other novel products and processes.

Current law aims to cover three aspects of the development of genetically modified organisms – the experimentation and development in the laboratory; the experimental, accidental, and full-scale release into the environment; and the health and safety for consumers of food products in which they are involved.

Regulations to control the safety of genetic modification have been in place since 1978 and the current operative regulations are the genetic Manipulation Regulations, 1989, made under the Health and Safety at Work, *etc.* Act, 1974. Although these Regulations protect human health and safety they are not considered to be adequate and new Regulations 'The Genetically Modified Organisms (Health and Safety) Regulations' are currently undergoing the consultative process. The PROSAMO[4] program (Planned Release of Selected and Modified Organisms) is a collaborative government/industry programme, supported in part by the Agricultural and Food Research Council and is intended to provide the necessary scientific data and reassurance on key questions associated with the possible release of micro-organisms and higher plants into the general environment.

As far as marketing of products consisting of or containing GMOs is concerned Part VI of the Act includes a clearance system. The draft Genetically Modified Organisms (Environmental Protection) Regulations, currently under consultation, would implement in Great Britain the detailed environmental requirements of the Act and the relevant EC Directives and would replace the 1989 Regulations referred to above.

In the UK in October 1988 the Advisory Committee on Irradiated and Novel Foods was reconstitued as the Advisory Committee on Novel Foods and Processes (ACNFP) to reflect more accurately its interest in the rapidly developing area of

[2] R. S. Crespi, *Trends Biotechnol.*, 1991, **9**, 117. [3] R. S. Crespi, *Trends Biotechnol.*, 1991, **9**, 151.
[4] PROSAMO programme – details from the Co-ordinating agents, Rarfon Ltd., Elm Tree House, Southover High St., Lewes BN7 1JB or MAFF.

food biotechnology. The ACNFP defines novel foods as: '. . . foods or food ingredients which have not hitherto been used for human consumption to a significant degree in Western Europe and/or which have been produced by extensively modified or entirely new food production processes'.

The types of products in which the committee is interested include food produced by technology involving genetic manipulation, *e.g.* using recombinant DNA methods; food produced by technology involving mutations which are not site specific; synthetic food items and foods produced by significantly or entirely, new processes.[5] Thus any food which contains transgenic organisms, or material derived from transgenic organisms is embraced in the definition 'novel'. Where a food related novel process has been or is being developed it may be submitted to ACNFP which advises MAFF and the Department of Health (DoH) Ministers on approvals of novel processes.

To date the requirement on companies to submit their novel products to the approval process has been a voluntary process. This is because under the existing food law companies in the UK must themselves be satisfied of the safety of their products and must themselves evalute whether foods or processes should be referred to the Committee for consideration. The food legislation of the UK is primarily concerned to ensure that food consumed is safe to eat. The Food Acts (1990, 1984) make it an offence to 'add any substance to food, use any substance as an ingredient in the preparation of food, abstract any constituent from food, or subject food to any process or treatment so as (in any case) to render the food injurious to health'. Regulations made under these Acts cover what may be added to food and these have the scope of control biotechnologically produced materials such as enzymes, colours, texture modifiers, and so on.

However, in contrast to the previous situation the Food Safety Act, 1990 contains statutory powers to regulate novel foods, although no Regulations under this power have yet been made.

Additionally, the voluntary nature of applying for approval of foods in which genetic engineering techniques are involved is likely to change. Currently EC Regulations for such foods are being drafted, and are certain to include the requirement that such foods and processes are scrutinized and approved before their release onto the consumer market.

Since 1986, the time of final approval of 'Quorn', a number of novel foods and processes have been considered by ACNFP, which has subsequently made recommendation to the Ministers regarding their approval. These are summarized in Table 1. The ACNFP is an advisory body whose advice the Government can either accept or decline.

3 FOOD CROPS

Genetic engineering (GE) of food crops is a rapidly developing field and is discussed more fully in Chapter 7. Theoretically, strains of food plants specifically designed to possess almost any combination of desirable qualities are possible

[5] Advisory Committee on Novel Foods and Processes, Annual Report 1989, Department of Health and the Ministry of Agriculture, Fisheries, and Food, 1990.

Table 1 *Some matters considered by ACNFP and/or FAC since 1989*

	Matters considered	*Outcome*
1989	A genetically modified bakers yeast (*Saccharomyces cerevisiae*) for use in bread dough. Applicant Gist–Brocades. Involved a genetic transfer between two strains of *S. cerevisiae*.	Recommended for food use, 1989. Press release 1/3/90. Government approval 17/1/91.
1989	A chymosin enzyme from a genetically manipulated source organism *Kluveromyces lactis*.	Clearance given 17/1/91. Since the enzyme is identical to the one found in calf rennet no special food product labelling is deemed necessary.
1989	A code of practice on taste trials for beers produced from genetically modified yeasts and genetically modified tomatoes was drawn up.	Press release 20/8/90.
1989	Transgenic animals – consideration of the general issues involved.	
1989	A consideration of the issues relating to labelling of food produced using techniques of genetic manipulation.	
1990	A second chymosin enzyme – from transgenic *Aspergillus niger* var. *awamori*.	Considered by the Food Additives Committee (FAC) 6/12/90. Government approval 2/5/91. No special food product labelling required.
1990	Tasting of genetically modified tomatoes – an application for an 'in-house' taste trial.[a]	Approved by ACNFP.
1990	Consumer concerns workshop.	
1991	Transgenic animals – an application to allow food products derived from a transgenic animals breeding program onto the food market.	Application refused.
1991	Honey containing pollen from genetically modified plants (*e.g.* pest or pesticide resistant strains).	Pending (1992) – implications under consideration.
1991	Guildelines on assessment of novel foods and process (includes guidelines on labelling).	Published by MAFF 1991.[a]
1991	Consideration of the potential for dissemination of marker genes from genetically modified organisms.	
1991	Application received for a third chymosin – from transgenic *E. coli* K-12.	Government accepted advice to approve it on 2/3/92.

Table 1 *(continued)*

	Matters considered	*Outcome*
1991	Implications of the use of microbially derived enzymes in foods.	
1991	Draft guidelines drawn up on the conduct of taste trials for novel foods.	
1992	MAFF consumer panel considered the use of genetically modified micro-organisms in foods.	

[a] Guidelines on assessment of novel foods and processes. Published by the Ministry of Agriculture, Fisheries, and Food, 1991.

because the advantage that GE has over conventional plant breeding is that genes can be transferred between any pair of organisms, plant or not. The current possibilities are summarized in Table 2.

The first stage of the development of the transgenic strain relies on the insertion of specific genes derived from donor sources; and the second stage requires the subsequent growth of whole plants from these modified undifferentiated cells.

Gene insertion (see Chapter 7) has largely been achieved by using modified strains of the bacterium *Agrobacterium tumiefaciens* as the vector. Subsequently the successfully modified cells are cultured into plantlets, then into mature plants. Greatest success has been achieved with dicotyledenons. Monocotyledenous plants – barley, maize, *etc.* – are resistant to *A. tumiefaciens* so other techniques (so far less successful) have been devised. Additionally the regeneration of monocotyledenous plants which has been proving far more difficult than for dicotyledenons has not been fully solved.

Currently there are about 30–40 crops[6,7] which have been genetically modified and some of these are listed in Table 3. However, very few of these are at the commercially useable stage. For example, all the monocotyledenous plants listed in Table 3 have only been successfully modified by the transfer of model marker genes. The monocots are very economically important encompassing many of the world's staple food crops – rice, wheat, barley – and the development of mature genetically modified monocotyledenous plants still remains to be achieved. So it is highly probable that genetically engineered dicotyledenous plants – such as tomatoes – in which a number of stable genetically engineered changes have been achieved will be on the market sooner than barley. For example, the tendency of stored tomatoes to soften during storage has been modified. The activity of the gene encoding for the enzyme polygalacturonase has been sub-

[6] J. L. Jones, *Trends Food Science Technol.*, 1992, **3**, 55. [7] R. Fraley, *Bio/Technology*, 1992, **10**, 40.

Table 2 *Goals for biotechnological improvements of crop plants*

Goals	*Comments*
1 Resistance to disease (insect pests, microbial pathogens, competing weeds).	Several companies are already trading plants with these characters.
2 Resistance to drought and soil salinity.	Will offer improved prospects for Third World agriculture.
3 Nitrogen fixing ability.	Long term goal.
4 Cheaper to grow (requiring less herbicide, pesticide).	Less environmental damage.
5 Higher yielding.	Traditional methods of plant breeding have given higher yields of:
– more tons per acre (the traditional goal)	rice in India, wheat in Europe and India
– higher yields of essential components.	Cloned palm oil,[a,b,c] soy bean[d]
6 Easier handling – rapid and synchronous ripening.	Cloned sesame seeds.[e] Tomatoes.[f]
7 Better nutritional properties.	Higher protein content; increased amounts of essential amino acids, *e.g.* methionine, lysine.
	Different lipid composition of oil seeds.[g]
8 Better storage properties.	Freezing resistance, *e.g.* tomatoes.[h] Reduce post harvest losses through delaying ripening, *e.g.* tomatoes.[h]
9 Better processing qualities.	Retention of texture. Improvement of colour. Changes in viscosity. Changes in elasticity. Changes in emulsifying properties.
10 Absence of allergens.	Gluten free grains.
11 Novel products.	Unique tasty processed vegetable snacks, *e.g.* 'vegisnax'.[i]
12 Novel sources of proteins.	From seeds, plants, and single celled organisms.
13 Modified sensory attributes.	Increased sweetness due to genetic engineering, *e.g.* thaumatin expression in temperate climate plants,[i] *e.g.* monellin.[k]

[a] Unilever. [b] S.B. Altenbach, S. S. M. Sunn, and J. B. Mudd, in 'Genetic improvements of agriculturally important crops', ed. N. M. Fraley and J. Schell, Cold Spring Harbour Laboratories Press, 1988, p. 61. [c] V. C. Knauf, *Trends Biotechnol.*, 1987, **5**, 40. [d] P. Christou, D. E. McCabe, B. J. Martinell, and W. F. Swain, *Trends Biotechnol.*, 1990, **8**, 145. [e] In Israel. [f] Campbells Soups. [g] 'Biotechnology of plant fats and oils', ed. J. Rattay, American Oil Chemists Society, 1991. [h] J. L. Jones, *Trends Food Science Technol.*, 1992, **3**, 55. [i] Cross and Blackwell. [j] M. Witty, *Trends Biotechnol.*, 1990, **8**, 113. [k] L. Penarrubia, R. Kim, J. Giovanni, S. H. Kim, and R. L. Fischer, *Bio/Techology*, 1992, **10**, 561.

Table 3 *Some crops so far successfully genetically engineered*[a,b]

Alfalfa	Apple	Aspargus	Barley	Cabbage	Capsicum
Carrot	Celery	Clover	Cotton	Cowpea	Cucumber
Frenchbean	Grapevine	Lettuce	Liquorice	Maize	Oilseedrape
Pear	Potato	Rice	Rye	Soybean	Sprouts
Strawberry	Sugarbeet	Sunflower	Tomato	Turnip	Wheat

[a] J. L. Jones, *Trends Food Science Technol.*, 1992, **3**, 55. [b] R. Fraley, *Bio/Techology*, 1992, **10**, 40.

stantially reduced by the insertion of antisense RNA to polygalacturonase[8] resulting in red, but firm tomatoes.[9,10]

However, it is not known whether the tomatoes with genetically modified storage properties have retained the good flavour properties of the donor strains (such as Ailsa Craig) from which they derive. This is because in the UK the work on tomatoes has been carried out under strict control and large scale taste trials have not yet taken place. Taste trials could through the dissemination of seeds in human faeces lead to the uncontrolled release of genetically modified seeds and thus have not yet gained the approval of the Government's advisory bodies. Until trials are demonstrably satisfactory to both the producers and the legislators the tomatoes will not be available for sale.

4 FOOD ANIMALS

The large scale production of anabolic steroid hormones such as bovine somatotrophin (BST) in micro-organisms through recombinant DNA techniques remains possible. Yet in spite of the huge investment by the manufacturers in the development of BST, the commercial success of the product remains to be seen. In the UK milk from cows so treated during a trial has been permitted to mingle with the milk supply – and some sections of the public reacted to this with alarm. Surveys[11] show that the public do not accept such products – partly because they do not believe that the use of BST would result in cheaper milk, partly because they are legitimately concerned as to the long term effect on human health,[12] and partly because of the social and ethnical issues involved. Although leaner pork would probably be acceptable to the public even if produced through the use of PST (pork somatotrophin), generally consumers are extremely wary of genetic engineering of animals.[13]

The development of transgenic animals remains contentious. While much of the research work is related to the elucidation of the development of embryos, some is more directly concerned with the development of food species. For

[8] J. E. Blalock, *Trends Biotechnol.*, 1990, **8**, 140. [9] G. E. Hobson, *Acta Hortic.*, 1989, **258**, 593.
[10] A. Flaherty, *Grower*, 1988, **11**, 16.
[11] B. Senauer, E. Asp, and J. Kinsey, 'Food trends and the changing consumer', Egon Press, Baldock, 1991.
[12] J. H. Hulse, *IFST Proc.*, 1986, **19**, 11.
[13] A. H. Scholten, M. H. Feenstra, and A. M. Hamstra, *Food Biotechnol.*, 1991, **5**, 331.

example, transgenic catfish carrying growth hormone genes from rainbow trout have been developed in the USA. The idea is to farm the catfish whose growth rate would be at that of the rainbow trout. However, the environmental implications of the accidental release of such farmed, transgenic fish could be serious. The ABRAC (Agricultural Biotechnology Research Advisory Committee) of the US Department of Agriculture gave limited approval[14] to the continued research by only permitting study of the catfish in specially constructed containment ponds. In the UK the general issue associated with the consumption of food products derived from transgenic animals are considered by ACNFP. Again the acceptability to the public of such products on taste, ethical, and environmental grounds remain debated in some of the sicentific press but largely untested.

5 CURRENT TRENDS IN MANUFACTURED GOODS

It can be seen from Table 1 that currently the foods involving genetic manipulation and now reaching the consumer in the UK are only two: bread (GE yeast) and cheese (GE chymosins).

Yeasts have been used in food fermentations for centuries – in breadmaking and in the production of fermented beverages. Recent developments in understanding the genetics and biochemistry of yeasts are leading to the development of new yeast strains through both exploitation of classical yeast genetics and through development of recombinant DNA techniques.[15]

In yeast production new rapid growing, high cell yielding strains have potential value; in breadmaking rapid growth and good carbon dioxide yield in different dough systems, including those of high osmotic pressure are goals already achieved in the laboratory. In the UK one strain of genetically manipulated baker's yeast has progressed beyond development to achieve Government approval (Table 1). The genes responsible for maltase and maltose permease in the recipient yeast have been modified by sequences from the donor strain resulting in a recombinant strain which expresses the enzymes independent of the presence of maltose and not inhibited by glucose, thus permitting more efficient utilization of maltose.[16] Short synthetic sequences have been used to link the implant into the recipient genome but apart from this the manipulated yeast contains no genetic material other than that from strains of *Saccharomyces cerevisiae*. Because of this the Committee (ACNFP) considering the first genetically manipulated food organisms to be evaluated for sale was satisfied that the risk that the manipulated strain will produce toxic metabolites is no greater than that of unmodified strains used previously. The risk of genetic transfer from the modified yeast to human consumers or their gut flora was also considered no greater than might be anticipated from any other strain of baker's yeast.[5]

In the brewing industry a number of changed attributes in yeasts such as the production of strains for the production of light beers through more efficient amylase activity, control of the development of off-flavours produced by the

[14] J. Fox, *Bio/Technology*, 1992, **10**, 492.
[15] J. W. Chapman, *Trends Food Sci. Technol.*, 1991, **2**, 176.
[16] Gist–Brocades NV, European patent A2 0 306 107, 1989.

production of di-acetyl and phenolics, improvement of filtrability through reduction in residual β-glucans, and improvement in storage quality by the control of haze development have been foreseen as being valuable to the industry.

A number of enzymes (amylases, aceto-lactate dehydrogenase, and β-glucanases) which would contribute to the solution of these problems, derived from a range of both microbial and non-microbial sources, have successfully been cloned into yeasts.[17] So far in the UK, ACNFP has only received an application for the taste trials of beer made from genetically modified yeasts, and has drawn up a code of practice (Table 1). No application has yet been made for the sale of beer so made.

Equally in dairying a number of targets can be foreseen for dairy starter cultures; in bacteriophage resistance, in resistance to antibiotics and herbicides, in the ability to produce acid steadily and facilitate cheese making, in the production of enhanced flavours and cheese texture, and in the acceleration of ripening. A number of new strains produced through non-recombinant techniques such as conjugation and cell fusion are currently on commercial trial. One new patented strain, *Lactobacillus casei* spp. *rhamnosum* GG, has the following characteristics; ability to attach to human intestinal mucosal cells, stability to acid and bile, *in vitro* production of an antimicrobial substance, and hardly *in vitro* growth. It is to be used to produce a fermented whey drink, and a yoghourt-type product both of which are aimed at promoting health benefits in human beings (being both helpful in the control of diarrhoea and constipation[18]). This organism is illustrative of new opportunities perceived by the dairy industry. Being a human strain, and not produced by GE techniques its launch onto the market is, in spite of extensive clinical trials, simpler than for GE organisms of similar potential.

So far not one application for the approval for food use of new lactic organisms produced by recombinant DNA techniques has been received.

The most significant change in the food industry in the UK – has been the approval for food use of the enzyme chymosin – the principle component of calf rennet (Table 1) produced by three separate transgenic organisms – *Kluveromyces lactis*, *Aspergillus niger* var. *awamori*, and *E. coli* K-12. This enzyme from new sources performs satisfactorily in the production of cheeses from cows',[19] and ewes'[20] milk. The organisms have been achieved by the cloning of appropriate gene sequences from calf cells. The chymosin, which is responsible for coagulation of milk in the manufacture of cheese, has been shown to be identical with that produced from calves, and because of this, in the UK, the advisory committees did not recommend specific labelling of such products.[21] The labelling of such foods has not yet become a public issue – but the Consumers Association holds the view that consumers should be able to make informed choices about

[17] W. J. Donnelly, *J. Soc. Dairy Technol.*, 1991, **44**(3), 67.

[18] S. Salminen, S. Gorbach, and K. Salminen, *Food Technol.*, June 1991, 112.

[19] G. Van den Berg and P. J. De Koning, *Neth. Milk Dairy J.*, 1990, **44**, 189.

[20] M. Nunez, M. Medina, P. Gaya, A. M. Guillen, and M. A. Rodriguez-Marin, *J. Dairy Res.*, 1992, **59**, 81.

[21] Advisory Committee on Novel Foods and Processes, Annual Report 1990. Department of Health and the Ministry of Agriculture, Fisheries, and Food, 1991.

whether they wish to buy such foods or not.[22] Plant and fungal rennets (used for vegetarian cheese) have their own properties and confer their own characteristic flavours on cheeses in which they are used. But chymosin produced from transgenic micro-organisms in addition to being identical to the natural product has other advantages. It can be more readily harvested than rennet from the traditional source – the calves stomach, may also contain a higher concentration of active enzyme, yet produces cheese with texture and flavour properties comparable to traditionally produced cheese. So currently the incentive for the production of chymosin from these new sources is in the economics – the estimated chymosin market currently is 18 tonnes of chymosin of value £100 million,[17] so clearly consumer acceptance is vital to realize this market.

More generally there are many processes in the food industry where enzymes can be used. The sources of the enzymes are varied, but many are microbially derived. Clearly the potential exists for the genetic modification of microbial species to improve expression, stability, specificity, activity, and yield. Currently a number of food enzymes (β-glucanase, glucan α(1-4) glucosidase, glucoamylase, amylase, pullulanase, acetolactate decarboxylase, β-galactosidase, chymosin, α-galactosidase, phospholipase, lipoxygenase) have been cloned in yeasts.[15] In use the enzymes may be applied as additives, immobilized, or encapsulated. But where the sources are from genetically modified organisms approval will have to be sought and gained before commercial food use is possible. As far as the dairy industry is concerned the applications of enzyme technology are very varied. In his review of the applications of biotechnology in the dairy industry, Donnelly[17] discusses the use of enzymes in the hydrolysis of milk proteins tailored to products in health care and baby foods, a market related to that referred to earlier in respect of patented *Lactobacillus casei* spp. *rhamnosum* GG. There is potential too for the use of enzymes to accelerate cheese ripening. But, in both situations the production of bitter flavours – due to high contents of hydrophobic amino acids such as leucine, proline, and phenylalanine in casein, tend partly to counteract the advantages offered.

Greater knowledge of the composition of the bitter peptides and in the potential to tailor enzymes to degrade them is needed. Although attempts are being made to define the flavour changes in maturing cheese[23] and although some progress is being made[24] the need for more detailed knowledge of the enzymology of protein breakdown is a major disadvantage and until that at least is rectified it will not be possible to exploit the potential of GE in dairy products. Additionally, the use of added enzymes for flavour development in non-processed cheeses is not permitted under current UK legislation. Yet since addition is permitted in some EC countries, harmonization of Regulations[25] during 1992 may permit their use in the UK and would add impetus to the desire to use GE enzymes in cheeses and dairy products.

[22] 'Another GMO for cheese', *Food Process.*, April 1992.
[23] N. Y. Farkye and P. F. Fox, *Trends Food Sci. Technol.*, 1990, **1**, 37.
[24] J. Kok, *FEMS Microbiol. Rev.*, 1990, **87**, 15.
[25] 'Regulatory aspects of microbial food enzymes,' 3rd edn., The Association of Microbial Food Enzyme Producers, Bruxelles, 1991.

Looking again at Table 1, beer, honey, and tomatoes whose production has in some way involved GE organisms may be expected to be the next products on the UK market. The wide range of other products – food additives such as polysaccharides, low calorie sweeteners, flavour modifiers, vitamins, colourings, water binding agents, and nutritional supplements are still waiting their entry onto the market. Some are nearer than others. For example, a patented strain of *Acetobacter* producing a novel form of bacterial cellulose was expected in 1991 to soon receive approval for food use in the USA. The cellulose produced is claimed through its water binding properties to be an effective thickener.[26] There is much interest too in microbially produced flavouring materials – some from organisms associated with traditional fermentations, others through the genetic improvement of organisms. A strain of *Streptococcus diacetylactis*, for example, has been genetically modified to produce greater than normal quantities of diacetyl through elimination of the gene coding for the enzyme α-acetolactate decarboxylase.[27] This could then lead after approval to the use of such strains in ensuring a better flavour profile for some foods.

In food quality monitoring, recombinant DNA technology offers the prospect for more rapid tests for detection and sometimes enumeration or quantification of pathogens, for spoilage organisms, for chemical indications of spoilage, for contaminants, and for adulterants. Some systems which exploit these techniques are already on the market, which because they are exclusively used in the laboratory analysis of food samples, do not have to go through the same approvals procedures as GE affected foods for consumption. A number of DNA probes (some produced by cloning) are available for the detection of food borne pathogens such as Salmonella, Listeria, *Escherichia coli*, Yersinia, Campylobacter, Staphylococcus, Pseudomonas, Shigella, Vibrio, and viruses.[28] It is thought that the use of such probes being quicker than conventional techniques may help in the evaluation of the quality of raw materials, and of that of the finished products.[29] There are some disadvantages however – in that some information such as biotype or strain identity and serotype gained by traditional techniques may not be obtained through the use of gene probe methods. Woolcott[28] points out that while it can be argued that such information is not really required 'traditional attitudes still compel microbiologists to put a name to every isolate'.

Another technique offering a rapid and sensitive method for the detection of antimicrobial substances in food is through the production of GE bioluminescent sensor strains. These organisms, because they die in the presence of the inhibitor, fail to emit light. By monitoring the reduction in luminosity the presence of antimicrobial material can be determined. This clearly has value in the cheese industry, for example where traces of antibiotic inhibit the starter organisms. Equally, in defined conditions the same detector can be used to evaluate preservative systems in foods, or the efficacy of biocides.[30,31]

[26] R. A. Kent, R. S. Stephens, and J. A. Westland, *Food Technol.*, June 1991, 108.
[27] M. Y. Toonen, M. C. M. Zoetmulder, and A. M. Ledeboer, *FEMS Microbiol. Rev.*, 1987, **46**, 26.
[28] M. J. Woolcott, *J. Food Prot.*, 1991, **54**(5), 387.
[29] J. Leighton Jones, *Trends Food Sci. Technol.*, 1991, **2**, 28.
[30] G. S. A. B. Stewart, S. P. Denyer, and J. Lewington, *Trends Food Sci. Technol.*, 1991, **2**, 7.
[31] J. M. Baker, M. W. Griffiths, and D. L. Collins-Thompson, *J. Food Prot.*, 1992, **55**(1), 62.

The potential for biotechnological techniques in monitoring food quality is wide and offers many techniques for the detecting and quantifying analytes at trace levels not possible by other techniques. The wider use of these techniques should lead to a raising of standards of food quality.[32]

6 CONSUMER ACCEPTANCE

It is a more recent development in the history of biotechnology that the UK consumer has been able to exercise any influence on the direction in which biotechnological development has, or will occur.

The Consumer's interests are now represented through the Consumer Panel, set up in 1991 by MAFF, and which is an avenue through which consumer concerns can be raised at the highest level. This is a welcome development but is still essentially a reactive, rather than pro-active, Committee.

MAFF has recently (11/3/92) published a series of free booklets for the public and for consumer groups which explain issues raised by genetic modification. The commercial success of biotechnological innovation in the food industry is certainly dependent on the consumer acceptability of products, which in turn depends on consumer understanding. Surveys still show that consumer understanding of biotechnology is low. It has been shown, for example, that a very high proportion of people across Europe have 'never considered whether or how biotechnology will affect their lives'. But the fact that there are currently products on the market – such as cheese containing chymosin from GE organisms, of which the public is unaware, may be attributed to the reluctance of manufacturers to mention it on their labels.[33]

Not only should public awareness of biotechnology be raised, but assurance and confidence in the safety of the products needs to be achieved. The caution of Government bodies is justified and it is welcome that the UK Government is depositing the data supporting submissions for novel foods (including those involving GE organisms) in the British Library, available for public perusal.[5] The issues of safety continue to be debated in the technical press[34] but consumer understanding is still lacking.[35] That will not change until real effort is put into educating the public in the issues concerned. The Citizens audit of EC Policy on Biotechnology run by the Euro Citizen Action Service[36] of the EC (May, 1992) may have helped. But what is really needed is greater public debate through the press and the media to help both inform and influence public opinion particularly where there are ethical issues (such as the development of transgenic animals) and environmental issues (such as the release of transgenic organisms) to be considered.

Then, without undue secrecy on the part of commerce, and unjustified

[32] 'Biotechnology and Food Quality', ISBN 0 409 90222 5, ed. S. Kung, D. B. Bills, and R. Quatrano, Butterworths, 1989.
[33] 'Sex and the Single Market', *Bio/Technology*, 1991, Autumn Supplement.
[34] M. W. Pariza *et al.*, 'Safety issues: foods of new biotechnology *vs.* traditional products. A symposium', *Food Technol.*, 1992, **3**, 100.
[35] C. M. Bruhn, *Food Technol.*, 1992, **3**, 80.
[36] Euro Citizen Action Service (ECAS), Rue du Trone, 98, B-1050 Bruxelles.

consumer prejudice on the other side, the real benefits, and potential profitability[37] of biotechnology and recombinant DNA technology beginning to be shown will develop and flourish.

[37] B. J. Spalding and B. Shriver, Jn., *Bio/Technology*, 1992, **10**, 497.

CHAPTER 10

DNA Profiling in Forensic Science

M. J. GREENHALGH

1 INTRODUCTION

Many forensic examinations still follow the principle established by Edmund Locard in the early years of this century, that there may be a transfer of trace evidence between the criminal and scene of the crime and *vice versa*. Finding such trace evidence may point to the guilt of a suspect whilst its absence may indicate innocence. The forensic biologist is mainly concerned with trace evidence such as blood, semen, saliva, and hairs found in cases of murder, rape, and serious assault. DNA profiling has provided a means of identifying the source of such trace evidence with a degree of certainty previously thought impossible.

Before 1985 and the discovery of DNA profiling, forensic biologists used a series of blood grouping tests to decide whether or not blood or semen could have originated from a particular individual. Serological tests such as the ABO and Rhesus blood groups were followed by protein polymorphism testing in the 1960s and 1970s. To obtain strong evidence that a sample originated from an individual it was necessary to perform a large number of these tests always assuming that sufficient material was available. Even then, the chance of another person having the same set of groups could be high. The stability of many protein markers was also limited and therefore it was rarely possible to type body fluid stains more than a few months old.

2 DNA PROFILING

In 1986 Alec Jeffreys of Leicester University published a paper describing a technique that he called DNA fingerprinting.[1] He described areas of non-coding DNA called minisatellites which consisted of a short 'core' sequence laid end to end many times (Tandem repeats) (see Figure 1). The exact number of tandem repeats varies greatly between individuals and consequently the lengths of the minisatellites are different in different people. Others had described similar regions but Jeffreys was the first to exploit their immense variability.[2]

The human haploid genome contains approximately three thousand million

[1] A. J. Jeffreys, V. Wilson, and S. L. Thein, *Nature (London)*, 1985, **314**, 67.
[2] A. R. Wyman and R. Whyte, *Proc. Natl. Acad. Sci. USA*, 1980, **77**, 6754.

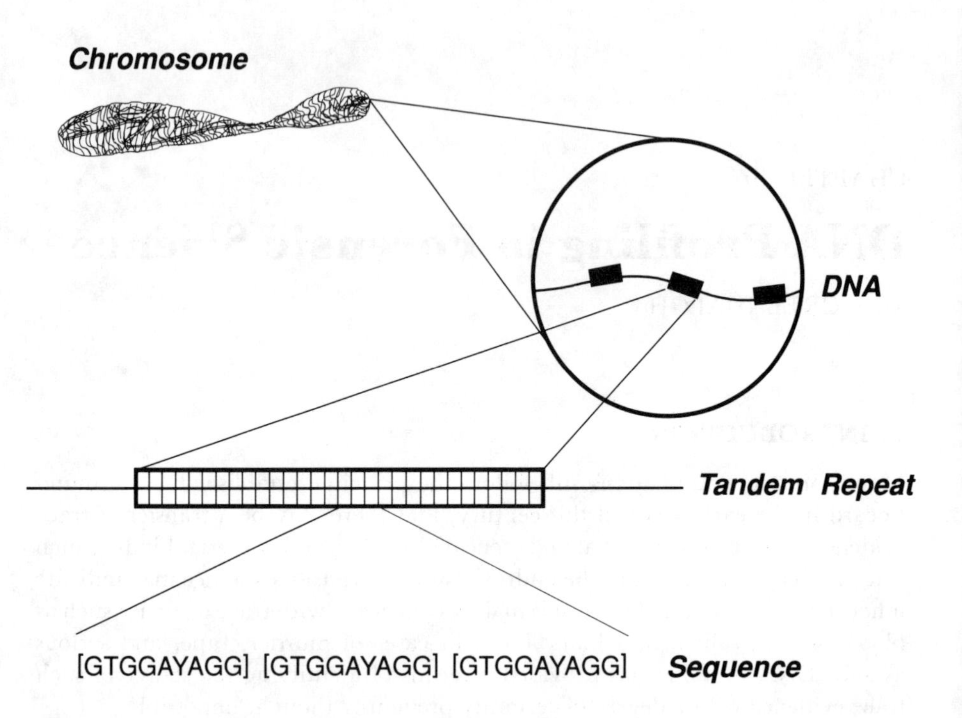

Figure 1 *A schematic diagram showing the basic structure of variable number tandem repeat (VNTR) polymorphisms. The supercoiled DNA of the chromosome is unravelled, producing a linear strand. Interspersed along this are regions of minisatellite DNA. These consist of a number of repeats laid end to end (Tandem Repeats). The nucleotide sequence of three repeats of the MS1 locus (D1S7) is shown. This is one of the most polymorphic regions yet found. The number of tandem repeats can vary from 200 to more than 2000 in different individuals. NB Y represents a pyrimidine base (T or C)*

base-pairs. Of these some 10% are sequences coding for proteins, enzymes, *etc.* Scattered throughout the remaining non-coding regions are areas of minisatellite repeats. The core sequences of many of these regions are similar and a possible role in recombination during meiosis has been suggested.[1] Jeffreys original method looked at many of these minisatellites simultaneously to obtain 'bar code' type patterns. This is known as the multi-locus probe method (MLP) (see Figure 2). More recently the single-locus probe (SLP) method has been used in preference in most crime laboratories because of its simplicity, increased sensitivity, and the ease with which the results can be stored and compared on computer[3] (see Figure 3). This method uses a series of sequential tests that each detect a single minisatellite. In general only two bands are seen at each stage of the test.

3 METHOD

Most of the DNA profiling methods now in use are a variation on the Southern blotting procedure. The method is schematically represented in Figure 4.

[3] Z. Wong, V. Wilson, I. Patel, S. Povey, and A. J. Jeffreys, *Ann. Hum. Genet.*, 1987, **51**, 269.

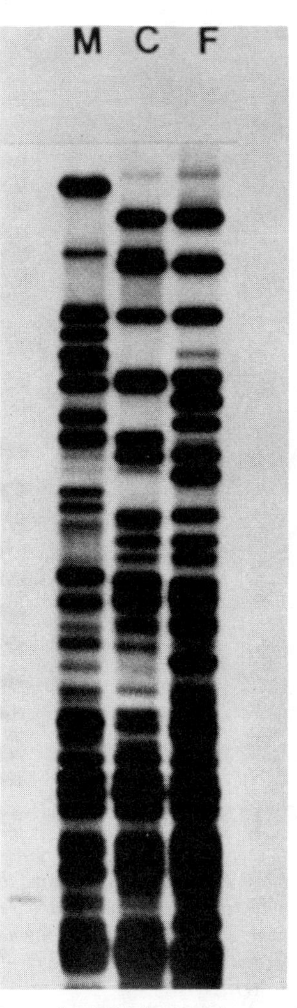

Figure 2 *An example of the use of a multi-locus probe to confirm the paternity of a child. Hinfl digested DNA from the mother (M), child (C), and putative father (F) has been separated by electrophoresis and then hybridized with probe 33.15 labelled with ^{32}P. The minisatellite regions detected show Mendelian inheritance. All the bands present in the child's profile can be seen either in that of the mother or the putative father confirming that he is indeed highly likely to be the true father of the child*
(Photograph reproduced by kind permission of Cellmark Diagnostics, Abingdon, UK)

3.1 Purification of the DNA

DNA is present in all nucleated cells and consequently a wide range of biological materials can be profiled, *e.g.* blood (DNA source is the white blood cells), semen, hair roots, and the buccal cells in saliva. The exact methods vary depending on the type of body fluid, however the essentials are similar. Cellular membranes are disrupted by a detergeant such as SDS (sodium dodecyl sulfate) and proteins are digested with a non-specific protease (proteinase K). Repeated

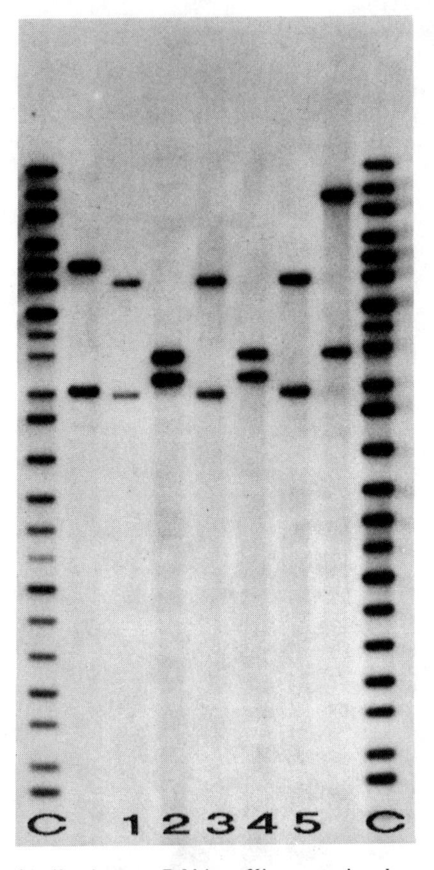

Figure 3 *Results of a chemiluminescent DNA profiling test using the probe MS43A (D12S11) in a sexual assault case. The profiles from the victim are in lanes 2 and 4 and those from the suspect are in lanes 1 and 5. The profile obtained from semen on the victim's vaginal swab is in lane 3. The profile from the semen matches that of the suspect. The molecular weight of the bands can be calculated by reference to the control marker ladders (C).*

Only the profile of the semen was obtained from the vaginal swab. Contaminating vaginal epithelial cells have been preferentially lysed using a detergent and the more resistant spermatozoa collected by centrifugation. Obtaining a single profile from a mixture of body fluids simplifies the interpretation of results[4]

extractions with phenol and chloroform precipitate cellular debris leaving the DNA in the aqueous phase. This can then be precipitated by the addition of ethanol. As the amounts of DNA are very small it is useful to add a carrier such as glycogen to the solution to ensure efficient precipitation. Typically a small bloodstain (0.5 cm diameter) would yield approximately 500 ng of DNA. The DNA is centifuged to the bottom of the micro tube, and the pellet which may be too small to see is dissolved in a suitable buffer for restriction.

[4] P. Gill, A. J. Jeffreys, and D. J. Werrett, *Nature (London)*, 1985, **318**, 557.

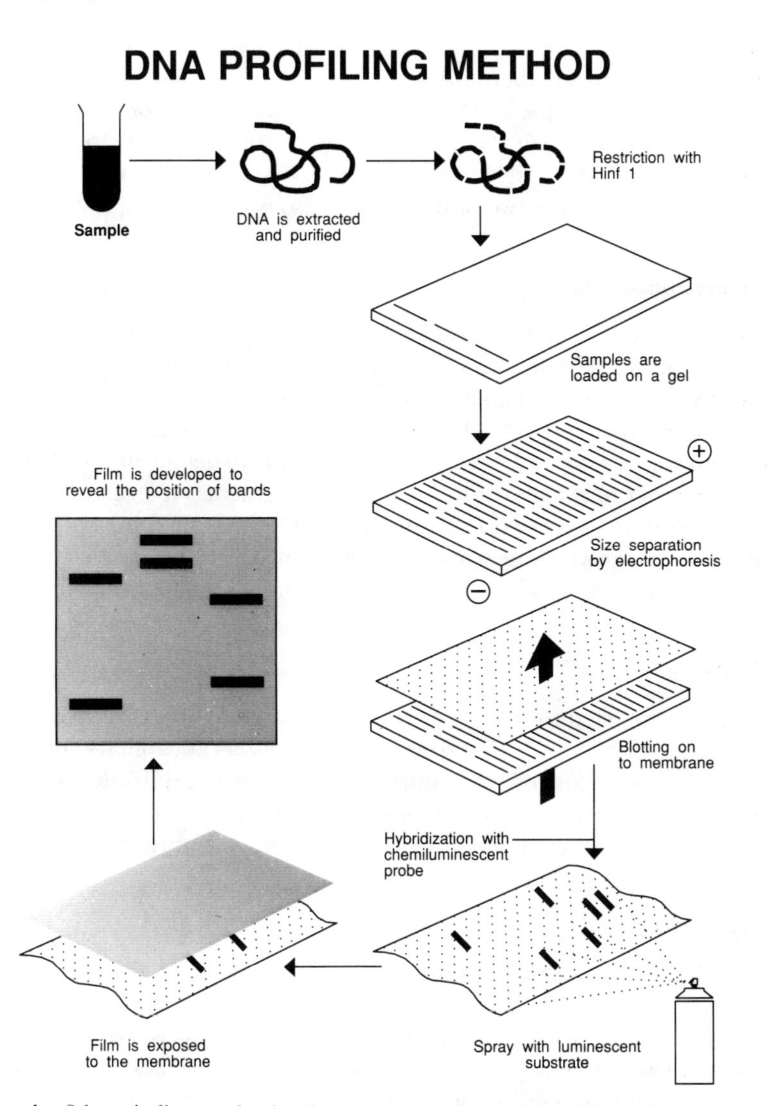

Figure 4 *Schematic diagram showing the major steps in the production of a single-locus probe DNA profile*

3.2 Restriction

In order to detect the differences in the length of the tandem repeats it is necessary to cut the DNA into fragments. A restriction enzyme is chosen that cuts in the flanking DNA either side of the tandem repeat. Ideally this enzyme should cut close to the ends of the tandem repeat but not within it. It should also be resistant to inhibition by contaminants that are often present in forensic samples. In Europe the standard enzyme is HinfI which cuts at the sequence GA*N*TC, where *N* can be any base. The use of a common enzyme allows the comparison of results obtained in different laboratories. In the United States HaeIII is the enzyme of choice. The results obtained with this enzyme are not compatible with those from the European laboratories.

The quality of the sample DNA and the progress of restriction are checked by examining small portions of the pre- and post-restriction sample on a test gel. Lack of high molecular weight, undegraded DNA in the original sample and unsuccessful restriction are two of the major reasons for profiling tests to fail.

3.3 Electrophoresis

The size separation of the DNA fragments is achieved using submarine agarose gel electrophoresis. The minisatellite regions that are commonly investigated produce DNA fragments ranging in size from approximately 1 kb to 20 kb. The resolution of the system is not sufficient to distinguish fragments that only differ by a few repeats and consequently it is very difficult to define all the alleles that can exist. In order to calculate the size of DNA fragments on a gel it is necessary to include a number of reference size markers. These usually consist of fragments of viral DNA (phage λ) whose sizes are known. The electrophoresis continues for approximately 20 hours at 70 volts in order to obtain maximum resolution.

3.4 Blotting

When this is completed the DNA is made single-stranded by alkali treatment of the gel and transferred to a nylon membrane by Southern (capillary) or vacuum blotting. To prevent loss of DNA during subsequent hybridization stages it is fixed by exposure to a measured amount of short wave ultra violet light which initiates a chemical reaction binding the DNA to the membrane.

3.5 Hybridization

The multi-locus (MLP) and single-locus (SLP) techniques are processed in the same way up to this point. The difference is in the type of hybridization. With the MLP method, the probe for the tandem repeat of interest is hybridized under conditions of low stringency. The higher salt concentration and lower temperature allow the probe to bind not only to the complimentary sequence of the target DNA but also to similar related sequences. Consequently many different fragments are detected and 'bar code' type pattern is produced.

Single-locus probes use higher stringency conditions which only allow the probe to bind to a single site on each of the homologous chromosomes producing the characteristic two band pattern. The discrimination power of the multi-locus probe is much greater than that of one single-locus probe and it is therefore necessary to use a series of probes sequentially. After the first results have been obtained, the probe is removed from the membrane by washing in alkali or by heating in a detergent solution. It is now possible to use a probe for a different tandem repeat loci and obtain a second set of results. This process is repeated until results from three or four loci are obtained, giving a level of evidential value similar to the multi-locus method. Despite the disadvantage of having to perform several hybridizations, the SLP method is preferred in almost all crime cases. As only two bands are obtained from each sample at any time, it is possible to analyse mixtures of body fluids such as might be found in cases of multiple rape or in assault cases where more than one person has bled. The complex nature of the MLP patterns make it difficult to interpret mixtures. The sensitivity of the single-locus probes are also greater and it is often possible to obtain a profile when the DNA is too degraded to produce a successful MLP pattern.

3.6 Detecting the Probe

Until recently the positions of the minisatellite fragments have been detected by using a probe containing radioactive phosphorus.[5] The ^{32}P decays emitting β-particles which fog an X-ray film adjacent to the site of the bound probe. The method is sensitive and robust but the exposure time for the autoradiographs can be as long as 10 days when very small DNA samples (50 ng) are being analysed. Recently a much more rapid chemiluminescent method of equivalent sensitivity has been introduced.[6] This relies on using chemically synthesized oligonucleotide probes which each have a molecule of alkaline phosphatase attached by a linker arm. Hybridization takes place at a lower temperature to avoid denaturing the enzyme and following a series of washes at controlled stringency, the luminescent substrate (AMPPD) is sprayed on to the membrane. A phosphate group on the substrate is removed by the enzyme and the unstable intermediate that is produced gradually decays with the emission of light. Again this is detected by placing the membrane in a cassette, in contact with a sheet of photographic film. An exposure of 1 to 2 days is equivalent to an autoradiographic exposure of 10 days. This is a considerable step forward as forensic scientists often have to provide results rapidly during a police enquiry.

4 INTERPRETATION OF PROFILING RESULTS

4.1 Multi-locus Probes

The interpretation of these results is relatively simple. It is usual to have both the stain from the scene of the crime and control sample from the suspect present on the same gel. It is then a case of performing a visual comparison between the two tracks and if they match, counting the number of bands that are clearly visible. Previous experiments have shown that the chance of two unrelated individuals sharing band is approximatley 0.25 and therefore the chances of two individuals sharing a complete profile is 0.25^{N} where N is the number of matching bands. The band matching statistic must be derived by experiment for each laboratory as it can vary depending on the resolution of the gels and the type of hybridization conditions used. Typically if a profile has 11 bands the chances of finding another unrelated person with the same profile is approximately 1 in 4 million.

[5] A. P. Feinberg and B. Vogelstein, *Anal. Biochem.*, 1983, **132**, 6.

[6] A. F. Giles, K. J. Booth, J. R. Parker, A. J. Garman, D. T. Carrick, H. Akhaven, and A. P. Schaap, *Adv. Forensic Haemogenet.*, 1990, **3**, 40.

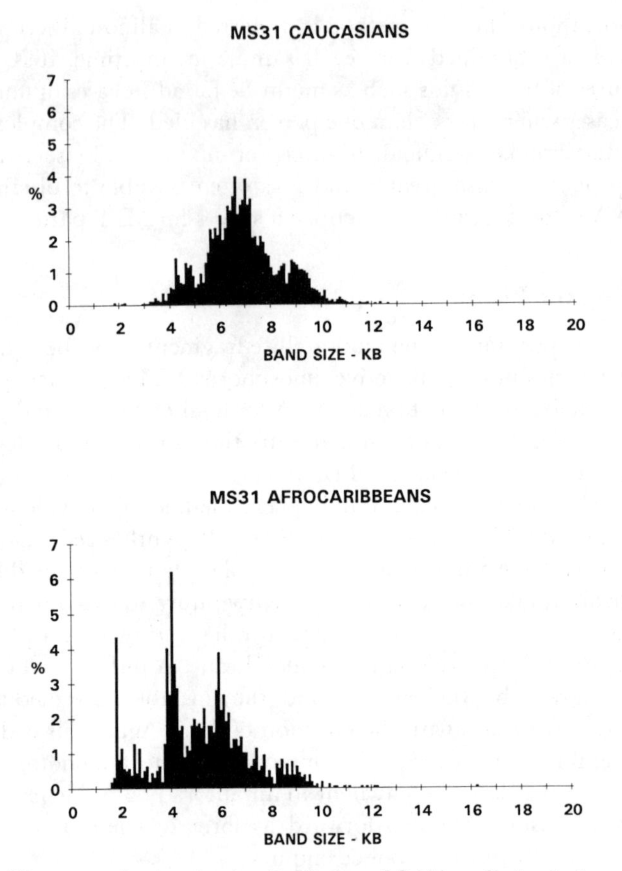

Figure 5 *Histograms showing the size distribution of DNA profile bands detected by the MS31 probe (D7S21) in two different racial groups in the London area*

4.2 Single-locus Probes

The initial stage in SLP analysis is again a visual comparison of the suspect and crime samples. If these do not match it will usually be readily apparent. However if they are similar it is necessary to measure the molecular weights of the bands. Most laboratories use a computerized video scanner to detect the centres of the bands and to compare them with the known control marker bands.[7] Band weights are measured in base-pairs or kilo base-pairs (kb).

Following the subjective decision that two bands match this is confirmed by checking that their molecular weight values do not differ by more than some predetermined error margin.

To estimate how common a particular SLP profile is, it is necessary to have a collection of several hundred profiles from different unrelated individuals. This is referred to as a database. After measuring the molecular weights of the bands the

[7] T. Catterick and J. R. Russell, *Lab. Microcomputer*, 1990, **9**, 105.

computer program calculates their frequency of occurrence by reference to the database. The rarity of the complete profile is calculated by multiplying together frequencies at each locus.[8] It is not possible to uses a simple band share statistic as with MLP as some band weights are significantly more common than others and there are also differences between major racial groups (see Figure 5).

5 DATA STATISTICS

There has been considerable debate in the scientific and legal world about the value of DNA evidence in court cases. Suggestions have been made that the figures quoted for the chances of obtaining a matching profile from a random member of the population have been exaggerated. These theories assume that sub-populations exist within the major racial groups within which particular DNA profiles may be considerably more common.[9] It must be said however that no clear evidence for the existence of sub-groups of this type exists and experiments where an artificially stratified population has been made show that the effects are limited.[10,11] The strength of DNA profiling is based on the fact that the regions of DNA analysed have extremely high mutation rates.[3] As the DNA is not transcribed into functional proteins there is no selection pressure and mutations can accumulate rapidly. This means that even if a population is very small and interbreeding, such as might be the situation in a remote community, the limited number of DNA alleles present can rapidly become more diverse because of mutations.

6 INTELLIGENCE DATABASES

A single-locus DNA profile can be represented by a series of band weight values and these can be readily stored on a computer. It is possible to search a computer file containing profiles from unsolved crimes against any new crime to see if the same criminal is responsible. Also the profile from a previously convicted criminal can be compared against outstanding cases. The main use for this approach is with cases such as rape where the offender often commits a series of crimes before being caught and frequently reoffends. Already the Metropolitan Police Laboratory has had considerable success in linking offences committed by the same person using this technique. Collating evidence from all the crimes in a series may greatly assist the investigating team.[12]

7 POLYMERASE CHAIN REACTION

The major problems with current DNA profiling techniques are that they are slow (several weeks even with the new chemiluminescent method) and often not sensitive enough to detect the minute amounts of trace evidence found in some

[8] C. Buffery, F. Burridge, M. Greenhalgh, S. Jones, and G. Willott, *Forensic Sci. Int.*, 1991, **52**, 53.
[9] R. C. Lewontin and D. Hartl, *Science*, 1991, **254**, 1745.
[10] R. Chakraborty and K. K. Kidd, *Science*, 1991, **254**, 1735.
[11] I. W. Evett and P. Gill, *Electrophoresis*, 1991, **12**, 226.
[12] J. E. Allard, *Adv. Forensic Haemogenet.*, 1991, **4**, 295.

cases. The polymerase chain reaction (PCR) is likely to be the answer to these problems (see Chapter 3). The ability to synthesize many copies of a particular DNA sequence makes it possible to analyse materials that are not suitable for conventional profiling, such as hairs and saliva staining on cigarette ends. Also, it may be possible to analyse very old items where much of the DNA has become degraded. Recently PCR methods have been used to analyse DNA from bones that have been buried for many years.[13] Although the DNA is fragmented into short lengths, enough intact copies of the region of interest may remain to enable amplification to take place.

The great sensitivity of PCR methods could be a problem in some instances as small amounts of contaminating DNA may lead to false results. It may be necessary to handle control blood samples from witnesses in a different laboratory from where the items recovered from the crime scene are being examined. Separation of items in this way will prevent any accidental contamination.

7.1 Short Tandem Repeats (STR's)

It is not possible to use the current tandem repeat regions for PCR as they are too large to be efficiently amplified. Instead a series of 'Short Tandem Repeats' (STR's) can be used.[14] These are similar in general structure but they consist of a core repeat only 3 or 4 base-pairs long. Their overall length is often only a few hundred base-pairs making them ideal for amplification. They are less variable than current systems and it is necessary to use a battery of 9 or 10 loci to obtain the same degree of evidential value. Using a high resolution sequencing gel to separate the amplified product enables one to distinguish between DNA fragments differing by as little as one repeat unit.

7.2 Mitochondrial Sequencing

Another possible approach is to sequence the DNA from a particularly variable region of DNA such as the 'D' loop of the mitochondrion.[15] Although mitochondrial DNA is present in many tissues the prime uses of this technique is likely to be with hair. Unlike genomic DNA which is only found in hair roots, mitochondrial DNA is found in the hair shaft. Hairs without root material are often found in forensic cases but at present very little information can be obtained from them.

7.3 Minisatellite Variant Repeat (MVR) Mapping

A new method published by Jeffreys 1991 exploits a further level of variation in minisatellite DNA.[16] Conventional single-locus profiling makes use of the variation in the number of tandem repeats present at a locus. MVR also uses

[13] E. Hagelberg, B. Sykes, and R. Hedges, *Nature (London)*, 1989, **342**, 485.
[14] A. Edwards, A. Civitello, H. Hammond, and T. Caskey, *Am. J. Hum. Genet.*, 1991, **49**, 746.
[15] R. Higuchi, C. H. Beroldingen, G. F. Sensabaugh, and H. A. Erlich, *Nature (London)*, 1988, **332**, 543.
[16] A. Jeffreys, A. MacLeod, K. Tamaki, D. Neil, and D. Monckton, *Nature (London)*, 1991, **354**, 204.

variation within the sequence of the repeat unit. The locus D1S8 (MS32 probe) has two different types of repeat unit, which differ by a single base.

Type 'A' [G**G**CCAGGGGTGACTCAGAATGGAGCAGGY]
Type 'T' [G**A**CCAGGGGTGACTCAGAATGGAGCAGGY]

The order in which these units are arranged on each of the homologous chromosomes is variable and it is possible to express this as a digital code. The different repeats are mapped using a complex PCR reaction which utilizes primers that will only initiate extension from one or other of the core sequences. Two amplification reactions are performed. In each case there is a common upstream primer which is located outside the repeat region. In one of the reactions the second primer is specific for 'A' type repeats and in the other it is specific for the 'T' type. The initial few cycles of PCR produce all the possible lengths terminating at the appropriate primer (A or T). Further amplification of the fragments is obtained using the common upstream primer and a primer which binds to a unique length of DNA which is attached as a tail on the A and T primers. This prevents shorter and shorter products being produced at each cycle. The products are run in adjacent wells on an electrophoresis gel and the results visualized by Southern blotting and chemiluminescent probe hybridization. A 'ladder' of bands is produced in each track and it is a simple matter to read down the gel looking at each position whether there is a band in the 'A' track only – code 1; 'T' track only – code 2; both tracks – code 3.

The method has the great advantage that no standard markers or measures of molecular weight are required. The digital code simplifies computer storage and inter-laboratory comparison of results. However there are likely to be problems in cases where there is a mixture of body fluids as this might not be readily apparent.

8 CONCLUSION

DNA analysis in forensic science has progressed very rapidly in recent years. The methods now available enable scientists to provide police investigators and the courts with very strong evidence in cases where body fluid traces exist. It is possible to state with a high degree of certainty the likelihood that a body fluid came from a particular person and also to definitely exclude other people who could not be the source. Similarly the technique can provide very powerful evidence in cases of disputed paternity.

The field is still rapidly advancing and in the next few years the sensitivity and speed of the method are likely to increase with the introduction of PCR technology.

CHAPTER 11

Molecular Diagnosis of Inherited Disease

C. G. MATHEW AND S. ABBS

1 INTRODUCTION

One of the most satisfying aspects of the dramatic advances in molecular biology over the past decade has been the speed with which they have been applied to the diagnosis of inherited disease. Less than ten years ago, families suffering from some of the most common genetic disorders such as cystic fibrosis and muscular dystrophy could be offered little or nothing in the way of predictive or prenatal testing. In sex-linked disorders, where the abnormal gene is located on the X chromosome and only boys are affected, most parents chose to terminate male pregnancies. In diseases like cystic fibrosis, where the risk of having a second affected child is one in four, many families decided against further pregnancies.

The application of molecular technology to the study of such diseases has changed all this. Disease genes have been hunted down and the causative mutations identified. Accurate prenatal diagnosis can be done early in pregnancy for an ever expanding number of disorders. Diagnosis has even been possible for disorders for which the causative gene has not been identified by using harmless variations of DNA sequence that are quite close to the disease gene to mark the chromosome which carries the mutation in the family. Screening of the general population for asymptomatic carriers of some of the more common genetic diseases such as cystic fibrosis is also becoming possible, thus identifying couples who are at risk of having an affected child even before they have begun their families.

This chapter will describe how modern methods of DNA analysis have been applied to both direct and indirect testing for mutations in a variety of important inherited disorders. Most of the diagnostic assays make use of the polymerase chain reaction (PCR – see Chapter 3), which is already being recognized as one of the greatest scientific inventions of the twentieth century, and is now being used routinely in preference to Southern blotting by most DNA diagnostic laboratories.

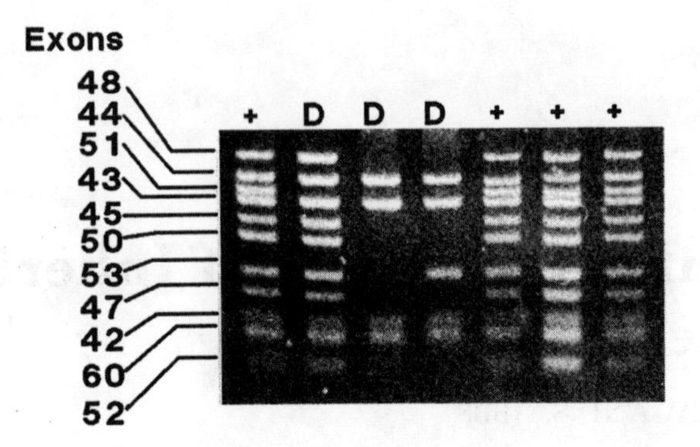

Figure 1 *Multiplex PCR amplification for the detection of deletions. Products obtained from simultaneous PCR amplification of 11 deletion-prone exons in the dystrophin gene. Tracks labelled '+' show products for all 11 exons, indicating that none of these exons are deleted in the samples. Deletions are detected in 3 samples (D) by failure to amplify exons 51, 45–53, and 45–52 respectively*

2 DIRECT DETECTION OF GENE MUTATIONS

The structural alteration of a gene which causes it to lose or alter its function may involve anything from mutation of a single nucleotide, usually referred to as a point mutation, to removal of a chunk of DNA millions of nucleotides in length. Detection of gross alterations is straightforward, whereas a variety of ingenious techniques have been developed to detect more subtle changes in gene structure.

2.1 Gene Deletions and Insertions

The causative mutation in a variety of inherited disorders is removal of part or all of the DNA coding for a particular gene product. In Duchenne muscular dystrophy (DMD), for example, which is a severe disorder causing profound muscle weakness in males, about 60% of cases are caused by deletion of part of the dystrophin gene on the X chromosome. These deletions can be detected as the absence of restriction fragments of particular size on a Southern blot (see Chapter 3) of the patient's DNA when it is probed with DNA from the coding regions (exons) of the gene. However, they can be detected much more rapidly by amplification of exons using PCR. Oligonucleotide primers are designed such that the PCR product from each exon is of a particular length. Deletion of an exon in a male DMD patient (who has only one X chromosome) is then detected by the absence of a band of the appropriate size when the products of the PCR are analysed by gel electrophoresis. Diagnosis is complicated by the fact that the dystrophin gene contains no less than 79 exons. However, certain exons are more prone to deletion than others, and eleven or more exons can be amplified simultaneously in a single PCR (see Figure 1). Thus it is possible to screen for almost all possible deletions associated with DMD in only two PCRs.[1,2]

[1] J. S. Chamberlain, R. A. Gibbs, J. E. Ranier, P. N. Nguyen, and C. T. Caskey, *Nucleic Acids Res.*, 1988, **16**, 11141.
[2] S. Abbs, S. C. Yau, S. Clark, C. G. Mathew, and M. Bobrow, *J. Med. Genet.*, 1991, **28**, 304.

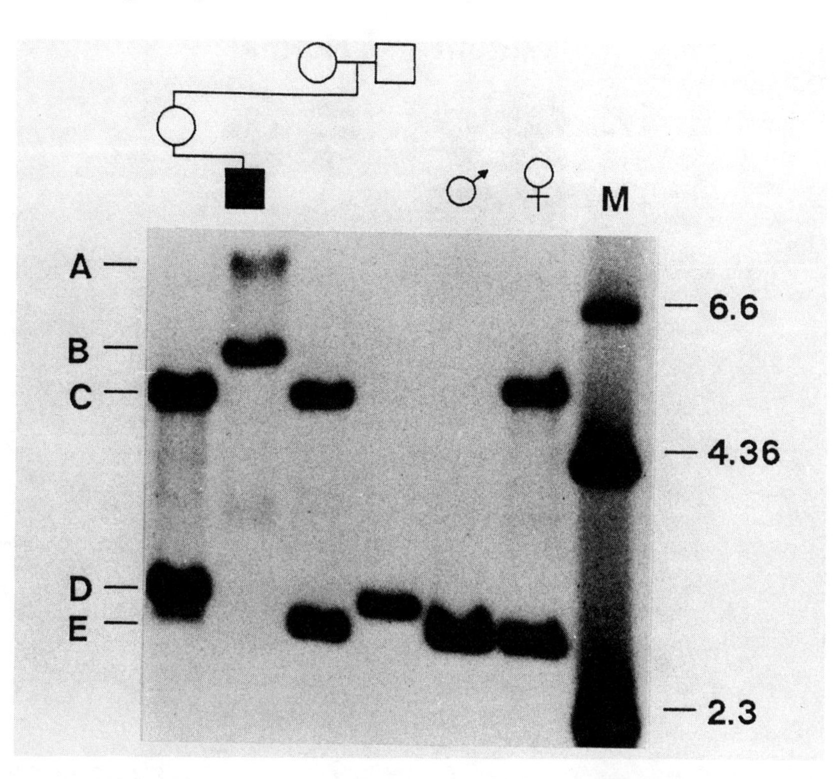

Figure 2 *Detection of insertional mutations by Southern blot analysis. Autoradiograph showing hybridization of the probe StB12.3 to DNA digested with both EcoR1 and the methylation-sensitive enzyme Eag1, from a family affected with the Fragile X syndrome. Control samples from a normal male and female are on the right (♂, ♀), showing the normal 2.8 kb fragment (E) from the normal, unmethylated X chromosomes, and in the female a 5.2 kb fragment (C) from the methylated, inactive X. The affected male (■) has an insertional mutation detected by hybridization to a larger, methylated fragment (B), which has mutated from the smaller insertional fragment (D) transmitted from his normal grandfather (□) through his mother (○). The affected male also has an additional larger mutation in some of his cells, as shown by the presence of an additional higher molecular weight band (A)*

Two major breakthroughs in genetic research during the past year have been the identification of the genes responsible for the fragile X syndrome (which causes mental retardation in males) and for the muscle disease myotonic dystrophy.[3] Both appear to be caused by excessive amplification of unstable trinucleotide repeat sequences such as CTG and CGG within the relevant gene. These are detected on Southern blots as restriction fragments of increased size (see Figure 2). PCR is not yet suitable for the analysis of these large insertions since segments of DNA that are several thousands of base-pairs in length are difficult to amplify. However, it can be used as an initial screen for the presence of DNA fragments of normal size that do not have the insertion.

[3] C. T. Caskey, A. Pizzuti, Y. Fu, R. G. Fenwick, Jn., and D. L. Nelson, *Science*, 1992, **256**, 784.

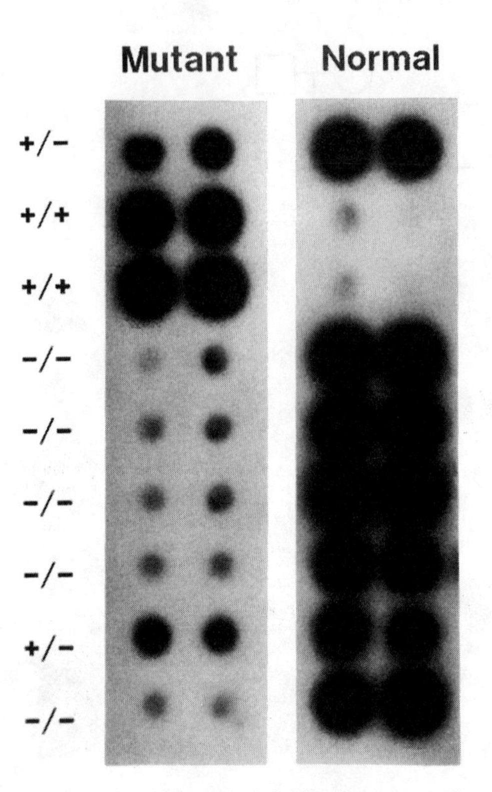

Figure 3 *Mutation analysis by ASO probe hybridization. Autoradiographs obtained after the products of amplification from exon 11 of the cystic fibrosis gene have been dot-blotted (in duplicate) onto a nylon membrane and hybridized with ASO probes which match the normal and mutant sequence of the G542X mutation. Individuals who are heterozygous for this mutation (*i.e. *they have one normal copy and one mutated copy) show strong hybridization with both ASOs (+ / −), whereas mutant (+ / +) or normal (− / −) homozygotes hybridize strongly with only the mutant or normal ASO*

2.2 Point Mutations

2.2.1 Allele-specific Oligonucleotides. An alteration of a single nucleotide in the sequence of a gene can be detected by synthesizing oligonucleotide probes which are complementary to either the normal or to the mutated sequence. Under carefully controlled conditions, a probe of 18–20 nucleotides will hybridize to a homologous sequence but not to a sequence which contains a single mismatch within it.[4] The region of the gene which contains the mutation is first amplified by PCR, and the products dotted on to a nylon membrane. This is then incubated with a radioactively labelled allele-specific probe complementary to the normal or mutated sequence. Unbound probe is then removed by washing the filter at a carefully controlled temperature, and the bound probe is detected by autoradiography (see Figure 3). Probes can also be labelled by non-radioactive methods, such as biotinylation.

[4] R. Saiki, T. L. Bugawan, G. T. Horn, K. B. Mullis, and H. A. Erlich, *Nature (London)*, 1986, **324**, 163.

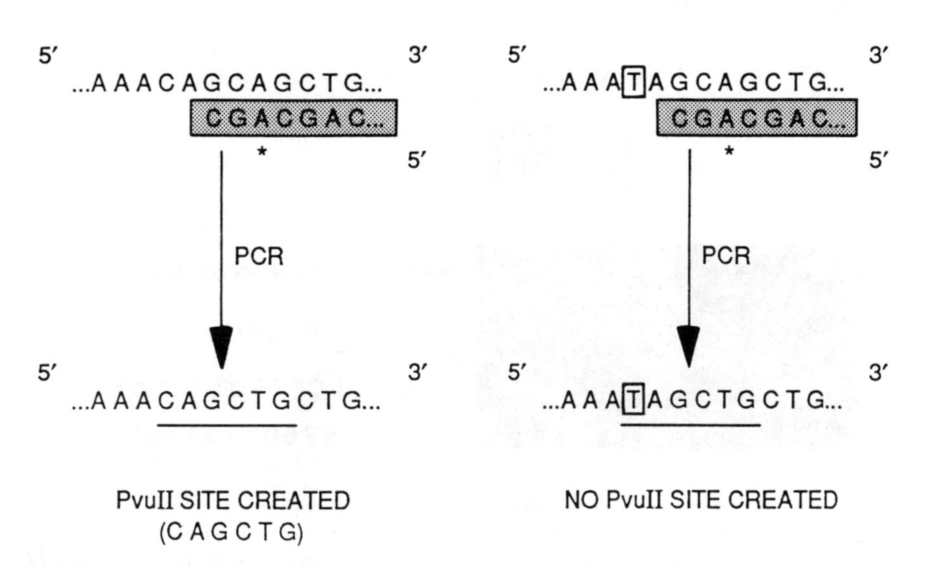

Figure 4a *Diagram illustrating how a restriction site can be engineered into the products of PCR amplification from a normal sequence (A), but not from the mutant sequence (B), in cases where the mutation does not naturally create nor destroy a known restriction site. See the text for a detailed explanation of the method. The results of typing a family with this assay for a point mutation causing Duchenne muscular dystrophy are shown in Figure 4b.*

2.2.2 Restriction Enzyme Site Analysis. Bacterial restriction endonucleases which recognize and cleave specific nucleotide sequences are an integral part of the cloning process. They have also been widely used in molecular diagnosis, since they can recognize mutations which happen to occur within their target recognition sequence. For example, sickle-cell anaemia is caused by a mutation in the gene which codes for the β-chain of haemoglobin. The DNA sequence is mutated from **CCTGAGGAG** to **CCTG*T*GGAG**. The restriction enzyme MstII cuts DNA at the sequence CCTGAGG, so will cut the normal β-globin gene sequence at this point, but not the mutated form. Thus if we amplify this region of the gene by PCR and digest the PCR products with MstII, a gene bearing the sickle-cell mutation will be detected as a larger (undigested) band on gel electrophoresis.

The advantage of this method is that it is very simple to perform. No radioactive or other probe labelling is required, and multiple samples can be tested within about 6 hours. Its limitation is that many mutations do not happen to fall within the recognition sites of any known restriction enzymes. However, a recent clever adaptation of the PCR technique has solved this problem for the

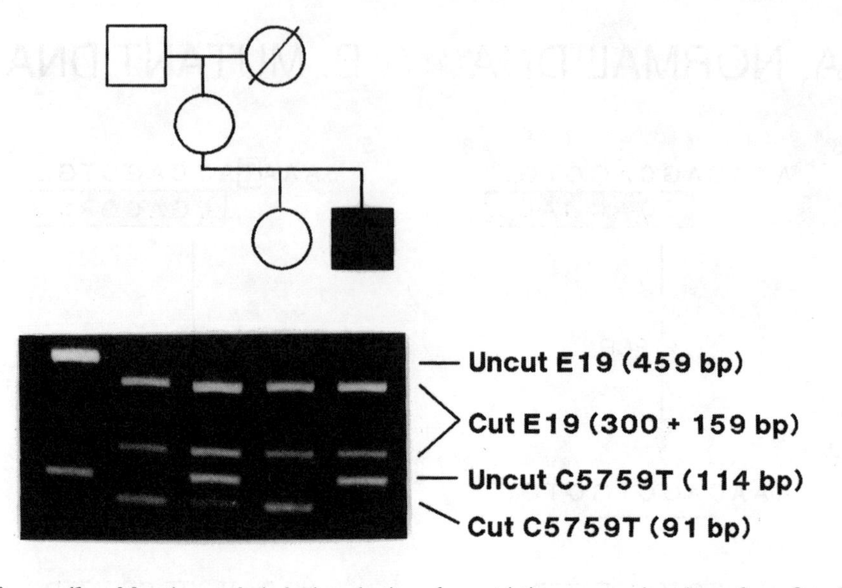

Figure 4b *Mutation analysis by introduction of a restriction enzyme site. Assay for a C to T substitution (C5759T) associated with a Duchenne muscular dystrophy family using a primer to introduce a restriction enzyme (PvuII) recognition seuqence in the absence of the mutation. Exon 19 (E19) of the dystrophin gene is used as an internal control for completeness of digestion, and undigested PCR products are shown in the first lane of the gel. The C5759T PCR product (114 bp) obtained from normal DNA digests to 91 bp and 23 bp (23 bp band not shown). Failure to digest this product, while obtaining complete digestion of the internal control indicates the presence of the mutation. Thus the affected male (■) shows the mutation in his X chromosome; his mother (○, lane 3) is a carrier of the mutation, having one normal and one mutant X; and his sister (○, lane 4) and grandfather (□) do not carry the mutation*

majority of cases.[5] If no restriction site is present, we can create one in the amplified PCR product by altering the sequence of one of the PCR primers by one or even two nucleotides. We have used this technique to devise an assay for a mutation identified in the dystrophin gene of a Duchenne muscular dystrophy patient. In this case, we altered the sequence of one of the PCR primers so as to engineer a PvuII site into the sequence of the PCR product obtained from the normal gene, but not from the mutant gene, as shown in Figure 4a.

The mutation involves the substitution of a cytosine (C) with thymine (T), as shown in the open box in B. The sequence of the primer (in the shaded boxes) is complementary to the normal sequence apart from the incorporation of an adenine (A) in place of a thymine at the asterisked position. Since the recognition sequence of PvuII is CAGCTG, the PCR products from normal DNA will be cleaved, but product from the mutant DNA will not contain a PvuII site and thus will not cut. The outcome of the experiment is shown in Figure 4b.

2.2.3 'ARMS'. The rather grand and cumbersome name of this technique is the **a**mplification **r**efractory **m**utation **s**ystem, or 'ARMS' for short.[6] It is based on

[5] E. J. Sorscher and Z. Huang, *Lancet*, 1991, **337**, 1115.

[6] C. R. Newton, A. Graham, L. E. Heptinstall, S. J. Powell, C. Summers, N. Kalsheker, J. C. Smith, and A. F. Markham, *Nucleic Acids Res.*, 1989, **17**, 2503.

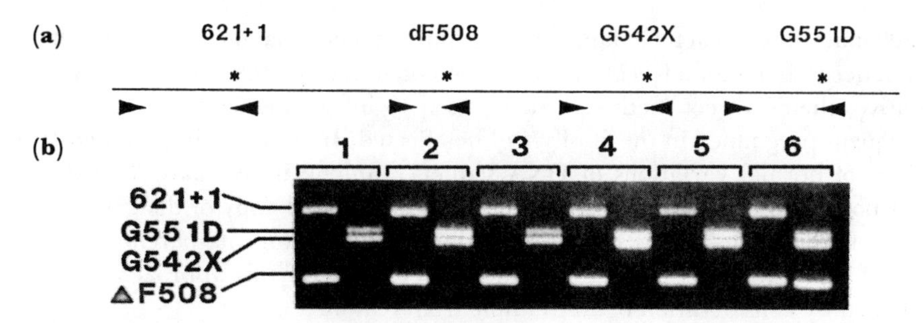

Figure 5 *Mutation analysis using the ARMS technique.*
(a) The region containing each mutation is amplified using an ARMS primer positioned at the mutation site (*) and a second primer positioned at a distance from the ARMS primer which is different for each of the four mutations. The size of the PCR product is therefore specific for each mutated region, with the 621 + 1 product being the largest and the ΔF508 the smallest.
(b) Products of PCR amplification of 6 samples that have been typed for four common mutations in the cystic fibrosis gene using ARMS primers. The first track for each sample shows products obtained with primers specific for the normal sequence at the 621 + 1 and ΔF508 mutations, and for the mutant sequence at the G551D and G542X mutations. Products in the second track of each sample were obtained with primers specific for the mutant sequences of 621 + 1 and ΔF508, and for the normal sequence of G551D and G542X. Thus samples 1–5 are typed as normal for all four mutations, whereas sample 6 is heterozygous for the ΔF508 mutation

the fact that the identity of the most 3′ nucleotide of a PCR primer is critical for the success of the amplification. If this nucleotide is not complementary to the target sequence, amplification is likely to fail. Thus if the PCR primers are designed so that the 3′ end of one of them falls right at the position of the gene where the mutation occurs, the presence of the altered nucleotide causes failure of the PCR, and no product is obtained. The disadvantage of the ARMS method is that a second PCR tube must be set up in which the primer contains a perfect match with the mutated sequence. Thus a PCR product is only obtained in the second tube if the mutation is present. However, once the assay has been devised, it is very easy to perform, and no digestion with restriction enzymes is required. Furthermore, multiple mutations can be tested for simultaneously by positioning the second, non-critical primer at a different distance from the mutation-specific primer for each of the mutations. Figure 5 shows the results of an ARMS assay developed by Cellmark Diagnostics Ltd. to test for 4 different mutations in the cystic fibrosis gene simultaneously. This multiple testing approach is especially important in a disease like cystic fibrosis, since more than 200 different mutations have been identified in patients with this condition!

3 INDIRECT DIAGNOSIS WITH LINKED GENETIC MARKERS

In cases where the gene which caused the disease is known and the particular mutation identified, accurate diagnosis for other family members using one of the methods described above is straightforward. However, the mutation which causes the disease in a particular family is not always known. We have pointed out that the condition cystic fibrosis (CF) is genetically very complex since it can be caused by any of at least 200 mutations. At present, it is not practical to test for all 200 mutations in each CF family who requires genetic diagnosis. What is done in practice is to test for a few of the most common mutations first. If these tests are negative, then indirect methods can be used to infer whether the foetus in a subsequent pregnancy in the family will be affected. Indirect testing is based on the use of normal variations of DNA sequence among individuals, known as DNA polymorphisms, to 'mark' the normal and mutated copy of the gene in a family. If the foetus inherits the same polymorphisms from its parents as the affected child, it too is likely to be affected. The use of this approach is best explained by consideration of an example from real life.

Spinal muscular atrophy (SMA) is a severe inherited neuromuscular disorder in which affected children often die within the first year of life. Although the gene responsible for the condition is unknown, it has been localized to a particular region of chromosome 5. We can use DNA polymorphisms from this region to perform predictive testing in families with SMA. There are many different kinds of polymorphisms, but the most widely used are those composed of CA repeats. These are regions of DNA between genes or in their non-coding parts which are composed of a series of tandem repeats of cytosine (C) and adenosine (A) nucleotides. They are polymorphic because the number of CA units is variable from one individual to another. The number of CA repeats at a particular position in an individual can be determined by PCR amplification of the region with primers specific for the unique DNA sequences flanking that particular CA repeat, followed by gel electrophoresis of the PCR products. Usually, one of the PCR primers is radioactively labelled, and the PCR products separated by electrophoresis under conditions which allow bands which differ by only two base pairs (*i.e.* one CA repeat) to be resolved from each other. The bands are then detected by autoradiography of the gel for 2–3 hours.

In the example shown in Figure 6, a family with SMA has been typed with a CA repeat close to the gene on chromosome 5. SMA is known as a recessive genetic disorder because offspring must inherit a defective gene from both parents in order to be affected with the disease. In this family, the affected son has inherited a type 3 chromosome from both of his parents. Unfortunately, the foetus has inherited the same parental genotypes, and is therefore also likely to be affected. This approach is known as linkage analysis because we are using a genetic marker linked to the disease gene to make the diagnosis.

The disadvantage of diagnosis by linkage analysis is that the genotype of the affected child must be known in order to identify the parental chromosomes which carry the mutation. Also, there is a small risk of error as a result of separation of the actual disease gene and the marker onto different chromosomes during the exchange of genetic material which occurs at the formation of the sperm and egg in meiosis. However, most families take the view that a diagnosis with an error rate of 2–5% is better than no diagnosis at all.

4 THE FUTURE

The pace at which new disease genes are being localized, isolated, and their mutations identified means that genetic diagnosis by DNA analysis will become available for an ever increasing number of inherited disorders. The number of

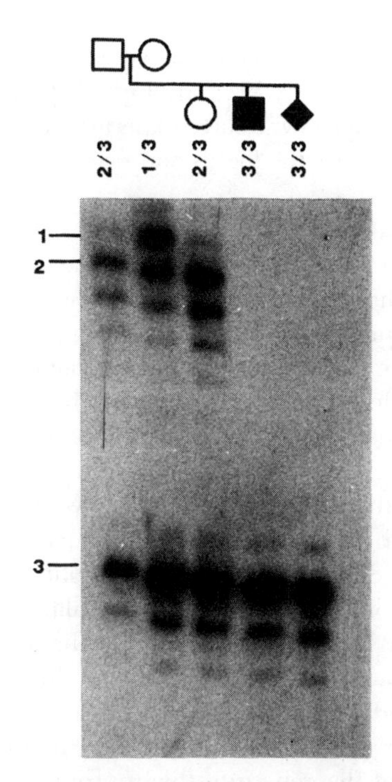

Figure 6 *Diagnosis by linkage analysis with DNA polymorphisms. Autoradiograph showing typing of the CA repeat polymorphism D5S127 for the prenatal diagnosis of spinal muscular atrophy (SMA). Since each of the parents have differently sized CA repeats, their copy of chromosome 5 which carries the mutant SMA gene can be identified by inspection of the genotype of the affected child (■). A tissue sample from the foetus (♦) has the same type (3,3) as the affected child, and is therefore likely to be affected with SMA. The multiple fainter bands seen below each strong band are thought to be caused by 'stuttering' of the Taq DNA polymerase during PCR of the CA repeats, which produces PCR products which are 2, 4, and 6 bp shorter than the main band*

individuals affected by these disorders will therefore decline, as will the suffering and trauma of patients and their families. The demand for DNA testing will create the need for automated analysis, as happened to clinical chemistry in the 1960s. The advent of the PCR technique and of sophisticated hardware for the automated fluorescent analysis of multiple DNA samples is the beginning of the development of high throughput, automated DNA analysers. The identification of the genes for some of the most common genetic disorders such as cystic fibrosis and the fragile X syndrome will lead to widespread population screening in order to identify couples who are at risk of having children affected with these disorders before they have begun to have children. Pilot studies of population screening for cystic fibrosis have already been carried out.[7] Methods for prenatal diagnosis will become increasingly sophisticated. The staggering sensitivity of the PCR method means that it is possible to analyse the DNA from a single cell for the presence of a particular mutation. Thus pre-implantation diagnosis, in which a single cell is

[7] Editorial, *Lancet*, 1992, **340**, 209.

removed from an embryo that has been fertilized *in vitro* and tested for a mutation should be feasible, and has already been attempted.[11] This technique would be particularly suitable for couples who have experienced the trauma of terminating several pregnancies because the foetus had been shown to be affected with a severe disorder.

One of the most exciting new areas of genetic research is the investigation of inherited susceptibility to common disorders such as heart disease and cancer. The detection of several thousand DNA polymorphisms scattered throughout the human genome, and the progress in the development of detailed genetic maps of all human chromosomes means that it is now possible to search for genes which predispose to these disorders by searching for co-inheritance of genetic markers with disease genes in large numbers of families and individuals. Recently, a gene associated with early onset breast cancer has been located in this way.[12] This discovery should lead to the early identification of women at risk in some of the families with a history of breast cancer, and to the identification of the causative gene itself. Tracing such genes could have major implications for the screening of the general population for predisposition to this alarmingly common disease. Although great progress has been made in the diagnosis and prevention of genetic disease, the ultimate goal of the research is to find a cure for patients who suffer from them. The identification of the affected gene for many of these disorders has created the possibility of correcting the defect by introducing a normal functional copy of the gene into the appropriate tissue or organ of the patient. The most promising approach is to splice the normal gene into a harmless virus, which is then allowed to infect cells from the patient. The first successful example of gene therapy has been reported for a rare immunodeficiency disorder, which has been corrected by the introduction of functional copies of the adenosine deaminase gene.[13]

A consequence of the rate at which new advances in molecular genetics have been reported during the past 8 years is that this chapter has to be entirely rewritten for each new edition of this volume! There is every indication that the pace of progress is quickening, helped in part by substantial international investment in the Human Genome Project,[14] which aims to map and sequence every gene in our genetic repertoire.

[8] A. H. Handyside, E. H. Kontogianni, K. Hardy, and R. M. L. Winston, *Nature (London)*, 1990, **344**, 768.

[9] J. M. Hall, M. K. Lee, B. Newman, J. E. Morrow, L. E. Anderson, B. Huey, and M. C. King, *Science*, 1990, **250**, 1684.

[10] W. F. Anderson, *Science*, 1992, **256**, 808.

[11] J. C. Stephens, M. L. Cavanaugh, M. I. Gradie, M. L. Mador, and K. K. Kidd, *Science*, 1990, **250**, 237.

CHAPTER 12

Vaccination and Gene Manipulation

M. MACKETT

1 INFECTIOUS DISEASE – THE SCALE OF THE PROBLEM

Vaccination against viral and bacterial diseases has been one of the success stories of human and veterinary medicine. Probably the most outstanding example of the effectiveness of vaccination is the eradication of smallpox. In 1967 between 10 and 15 million cases of smallpox occurred annually in some 33 countries. By 1977 the last naturally occurring case was reported in Somalia. Polio too has been controlled in developed countries, for example the number of cases in the USA was reduced from over 40 000 per year in the early 1950s, before a vaccine was available, to only a handful of cases in the 1980s. Diphtheria is now almost unheard of yet over 45 000 cases in 1940 led to 2480 deaths from diphtheria in the UK (similar numbers to those who died from AIDS in the UK in the entire 1980s). This has been reduced in the UK to only 13 cases and no deaths from the bacterium between 1986 and 1991. Despite this tremendous progress infectious disease is still a major global issue. The scale of the problem is enormous (see Table 1)[1] – over 10 million deaths worldwide per year are due to infectious disease. UN development programme figures[2] suggest that cancers, circulatory problems, and injuries cause fewer deaths in developing countries than infectious disease.

The International Community has responded vigorously to the challenge and the WHO expanded programme of immunization (EPI) aims to immunize greater than 90% of the world's newborn against a number of viral and bacterial diseases (Table 2) by the year 2000.

If the WHO expanded programme of immunization were completely successful only half of the diseases in Table 1 would have been dealt with. Thus there is also a pressing need for new vaccines to viral, bacterial, and parasitic pathogens (see Table 3). Even in developed countries new vaccines are being introduced,

[1] S. B. Halstead, *World J. Microbiol. Biotechnol.*, 1991, **7**, 121.

[2] J. A. Walsh, 'Establishing Health Priorities in the Developing World', United Nations Development Programme, New York, 1988.

Table 1 *Illness and mortality burden for selected infectious diseases*

Condition	*Episodes* (× 1000)	*Deaths* (× 1000)
Respiratory disease	15 000 000	10 000
Diarrhoea	29 000 000	4300
Measles	67 000	2000
Malaria	150 000	1500
Tetanus	1800	1200
Tuberculosis	7000	900
Hepatitis B	3700	800
Pertussis	51 000 000	600
Typhoid	35 000 000	600
Shistosomiasis	10 000 000	250
HIV	250	200

Table 2 *Vaccines included in the WHO expanded programme of immunization*

Tuberculosis	Diphtheria	Pertussis	Tetanus
Poliomyelitis	Measles	Yellow fever in parts of Africa	
Japanese encephalitis in parts of Far East			

Table 3 *Selected diseases or pathogens requiring vaccines*

Dengue	Filariasis	HIV-1 and 2	Leishamiasis	Malaria
Mycobacterium leprae		Rotavirus	Schistosomiasis	Trypanosomiasis

Table 4 *Selected vaccines that require improvement*

Influenza A and B	Japanese encephalitis	Meningococcus A and C	
Mycobacterium tuberculosis	Pertussis	Poliovirus	Rabies virus
Salmonella typhi	*Streptococcus pneumoniae*	Yellow fever	Cholera

for example from the end of 1992 all children under 6 months old in the UK will receive a course of vaccinations to prevent small numbers of cases of meningitis associated with infection by *Haemophilus influenzae B*.

Not only are new vaccines required but more effective and safer vaccines than those currently used are also still needed (see Table 4). For example a more effective vaccine against cholera is desirable. The current vaccine is at most effective in only 50% of vaccinees and the duration of immunity is relatively short. Even the very low levels of vaccine associated poliomyelitis $(< \times 10^{6})$ has prompted some vaccine companies worried about litigation to stop manufacture

Table 5 *Vaccines intended for universal use in the UK. A whole series of other vaccines are also available for special circumstances,* e.g. *Cholera vaccine, Yellow fever 17D vaccine, or Japanese encephalitis virus vaccine are widely available for travel to parts of Africa or South America and the Far East where the pathogens are endemic. Another reason to be vaccinated may be due to high risk occupations such as individuals that come into contact with human blood; in this instance a hepatitis B virus vaccine should be used*

Vaccine	*Type*	*Age given*	*Route*
Haemophilus influenzae B	Capsular polysaccharide conjugated to protein	Three doses at 2, 3 and 4 months	Deep subcutaneous (SC) or intramuscular (IM)
Diphtheria (D)	Toxoid		
Tetanus (T)	Toxoid		
Whooping cough (P-pertussis)	Killed whole organism		
Oral poliovirus (OPV)	Live attenuated (three serotypes)		Oral
Measles Mumps Rubella	Live attenuated	12–18 months	Deep SC or IM
Booster D, P, T	as above	4–5 years	as above
Rubella	as above	10–14 years (girls only)	as above
BCG	Live attenuated	15–18 years	Intradermal
Booster P, T	as above	15–18 years	as above

of the polio vaccine. In recent years a considerable effort has gone into producing a more stable Sabin type 3 attenuated vaccine strain (see Section 3.3.3).

Admittedly vaccines are not the only way of tackling this problem nor are they the only health measure that would be required (see Section 6) however, they are the simplest, most cost effective way of reducing the burden of infectious disease. Genetic manipulation offers new technology to tackle the problems in a variety of ways. To best appreciate what can be done a brief summary of the principles of vaccination follow.

2 VACCINE STRATEGIES

There are two classical strategies for vaccination. One involves vaccination with either killed pathogenic organisms or subunits of the pathogenic organism. The other utilizes live attenuated viruses or bacteria that do not cause disease but have been derived from the pathogenic parent organism.

Table 5 summarizes the standard vaccines given to everybody in the UK[3] and

[3] Compiled from 'Immunisation against Infectious Disease', HMSO, London, 1992.

illustrates the widespread use of both live attenuated and killed or subunit vaccines. These are examples from human infectious disease, however, there are equally successful vaccines from the point of view of veterinary medicine. Leptospirosis, rinderpest, and foot and mouth in cattle, clostridial diseases in sheep, cattle, and pigs, diarrhoea caused by enterotoxic *E. coli* in piglets, and Newcastle disease and Mareks disease in poultry are all examples of diseases in animals for which highly effective vaccines are available.

2.1 Inactivated Vaccines

Inactivated vaccines are made from virulent pathogens by destroying their infectivity usually with β-propiolactone or formalin to ensure the retention of full immunogenicity. Vaccines prepared in this way are relatively safe, and stimulate circulating antibody against the pathogens surface proteins thereby conferring resistance to disease. Two or three vaccinations are usually required to give strong protection and booster doses are often required a number of years later to top up flagging immunity.

Subunit vaccines can be seen as a sub-category of inactivated vaccines because similar considerations apply to subunits and whole organisms. Doses, routes, duration of immunity, and efficacy of these vaccines are all very comparable. In this case a part of the pathogen, such as a surface protein, is used to elicit antibodies that will neutralize the infectivity of the pathogenic agent. The widespread use of hepatitis B virus surface antigen purified from the blood of carriers[4] (or more recently from recombinant yeast[5]) shows that this can be a very effective way to immunize. Hepatitis B virus surface antigen, the product of a single gene, assembles into a highly antigenic 22 nm particle which if used in three 40 μg doses at 0, 1, and 6 months gives virtually complete protection against infection with hepatitis B virus.

Another example that can be included in the subunit vaccine class is the use of bacterial toxoids. Many bacteria produce toxins which play an important role in the development of the disease caused by a particular organism. Thus, vaccines against some agents, for example tetanus and diptheria, consist of the toxin inactivated with formaldehyde conjugated to an adjuvant. Immunization protects from disease by stimulating antitoxin antibody which neutralizes the effects of the toxin.

A further type of vaccine included in the subunit category is the capsular polysaccharide vaccines, for example those against *Haemophilus influenzae* and meningococcal meningitis. In this case an extract of the polysaccharide outer capsule of the bacterium is used as a vaccine and is sometimes conjugated to protein to improve immunogenicity. Antibody persists for several years and is able to protect against the bacterium.

[4] 'Hepatitis B Vaccine', INSERM Symposium 18, ed. P. Maupas and P. Guesry, Elsevier, Amsterdam, 1981.

[5] P. Valenzuela, A. Medina, W. J. Rutter, G. Ammerer, and B. D. Hall, *Nature (London)*, 1982, **298**, 347.

Table 6 *Relative merits of live versus killed vaccines*

		Live	*Killed/subunit*
Production	Purification[a]	Relatively simple	More complex
	Cost	Low[b]	Higher
Administration	Route	Natural or injection	Injection
	Dose	Low, often single	High, multiple
	Adjuvant	None	Required[d]
	Heat lability	Yes	No
	Need for refrigeration[c]	Yes	Yes
Efficacy	Antibody response	IgG; IgA	IgG
	Duration of immunity	Many years	Often less
	Cell mediated response	Good	Poor
Safety	Interference	Occasional OPV only[e]	No
	Reversion to virulence	Rarely[f]	
	Side effects	Low level[g]	No

[a] Increasing safety standards mean that for new vaccines some of the older methodologies would not be acceptable. [b] The price for new vaccines will approach that of killed subunit vaccines as safety standards are increased. [c] The need for refrigeration increases the costs significantly. [d] Very few adjuvants for human use are acceptable. [e] Especially in the Third World. [f] At very low levels (less than 1 case per 10^6 vaccinations). [g] This varies from occasional mild symptoms with rubella and measles vaccines to possible brain damage with pertussis vaccine.

2.2 Live Attenuated Vaccines

Table 5 shows that over half of the standard vaccines in the UK are live attenuated mutants of parent pathogenic organisms. In effect live vaccines mimic natural infection, yet produce subclinical symptoms and elicit long lasting immunity often giving rise to resistance at the portal of entry. Most of today's attenuated vaccine strains have been derived by a tortuous often empirical route involving passage in culture until the pathogen is found to lose its virulence. This loss of virulence is tested in animal model systems before being tested in human volunteers. For example the vaccine used to immunize against tuberculosis was derived after 13 years' passage in bile-containing medium by Calmette and Guerin (hence the name BCG – bacille Calmette – Guerin).

2.3 The Relative Merits of Live *Versus* Killed Vaccines

There has been much debate over the past 40 years as to the relative merits of live and killed vaccines often generating more heat than light! The evidence is that both routes will give adequate vaccines that can be used to protect against disease under the appropriate conditions. Table 6 shows some of the major points of debate which are discussed more fully in Reference 6. Many factors including

[6] C. A. Mimms and D. O. White, 'Viral Pathogenesis and Immunology', Blackwell Scientific Publications, Oxford, 1984.

cost, safety, number of immunizations, ease of access to vaccines, politics, and social acceptance will determine whether there is a high uptake of a particular vaccine and whether it is ultimately successful in eradicating the target disease. Even if a perfectly viable, relatively safe vaccine is available uptake may be limited. For example it has been estimated that vaccination against measles within the WHO EPI has prevented over 60 million cases and 1.37 million deaths. Despite these efforts there are still some 70 million cases of measles annually resulting in nearly 1.5 million deaths, consequently a recent WHO congress[7] adopted the following goals:

(i) Increasing immunization coverage.
(ii) Improving surveillance.
(iii) Developing laboratory services and improving vaccine quality.
(iv) Training.
(v) Promoting social mobilization.
(vi) Developing rehabilitation services.
(vii) Research and development.

This again also serves to illustrate the importance of factors other than the efficacy of the vaccine itself in disease prevention.

An example of the fact that many approaches to a vaccine can be taken is that of the current typhoid vaccine. In the UK three different vaccines have been licensed.[3] One is a killed whole cell vaccine, a second is based on a capsular polysaccharide extract of typhoid, and the third is a live attenuated strain of *Salmonella typhi* (Ty21a). The most recently introduced, the capsular polysaccharide vaccine, (1992) requires the least number of doses and is likely to become the preferred vaccine.

The single most important issue in developed countries is the safety of a vaccine, a single death in a million vaccinations for a new vaccine would be unacceptable (except possibly if it were an effective AIDS vaccine). While this is obviously important in a third world country, other issues such as cost and how to deliver the vaccine are of paramount importance.

3 THE ROLE OF GENETIC ENGINEERING TECHNIQUES IN CONTROLLING INFECTIOUS DISEASE

3.1 Identification, Cloning, and Expression of Antigens with Vaccine Potential

Many pathogens are virtually impossible to culture outside their natural host and this makes it unlikely that conventional approaches to vaccination would be successful. For example hepatitis B virus, the agent of human syphilis (*Treponema pallidum*) and the bacterium that causes leprosy (*Mycobacterium leprae*) have never been grown *in vitro* although they can be propagated in animal models. Consequently it is not possible to generate live attenuated or inactivated vaccines by culturing the agents. Recombinant DNA technology allows the transfer of

[7] 42nd World Health Assembly, Expanded Programme on Immunisation, WHO 42.32, 1989.

genetic information from these fastidious organisms to more amenable hosts such as *E. coli*, yeast, or mammalian cells.

Not all protective antigens are as simple to identify, clone, and express as the surface antigen gene of hepatitis. The entire sequence of the hepatitis B virus genome became available[8] and as it is less than 10 kb it was relatively simple to establish which open reading frame to express. It has been known for many years that irradiated malarial sporozoites protect against malaria.[9] As the sporozoite stage in the life cycle of the malarial parasite can only be grown in small quantities it was left to recombinant DNA technology to identify, clone, and express components of the sporozoite that might be of use in vaccine production. The genome of the malarial parasite is many thousands of times larger than the genome of HBV and therefore provides a different scale of problem, not only was there little sequence data available but there was also no idea of which gene products may be protective. See Section 3.1.3 for detail of the initial cloning of the malarial sporozoite surface antigen.

The starting point of any recombinant DNA work is to generate a library of DNA in *E. coli* which is representative of the organism under study. Once having a cDNA bank or a genomic library there are three basic ways of identifying and isolating a gene of interest.

3.1.1 DNA/Oligonucleotide Hybridization. If there is some pre-existing knowledge of the nucleic acid sequence, or where purified mRNA is available, it is possible to detect recombinant clones by hybridization of ^{32}P labelled DNA or RNA to bacterial colonies or bacteriophage plaques. Often a protein has been purified and some amino acid sequence is available which allows a corresponding nucleic acid sequence to be synthesized. Due to the degeneracy of the genetic code a complex mixture of oligonucleotides is required to ensure that all possible sequences are represented. Labelling this mixture of oligonucleotides yields a probe that can be used to screen a cDNA (or possibly genomic) library that might be expected to contain the gene of interest.

3.1.2 Hybrid Selection and Cell Free Translation. A second approach is to use hybrid selection of mRNA coupled with cell free translation. DNA clones from a library, either individually or in pools of clones, can be immobilized by binding to a solid support and mRNA hybridized to them. Only the mRNA that corresponds to the clones will bind and this can then be eluted and translated to protein in a cell free system. The protein can then be immunoprecipitated with antisera to the gene product of interest or assayed for activity. An example that encompasses both this approach and the sequence route is in the development of a vaccine for Epstein–Barr virus. It had been known since 1980[10] that antibody to the major membrane antigen of the virus (gp350/220) would neutralize the virus. Around 1983 a fragment of the virus genome was cloned[11] and sequenced; using computer predictions

[8] Y. Ono, H. Onda, R. Sasada, K. Igarashi, Y. Sugino, and K. Nishioka, *Nucleic Acids Res.*, 1983, **11**, 1747.
[9] V. Nussenzweig and R. S. Nussenzwieg, *Cell*, 1985, **42**, 401.
[10] J. R. North, A. J. Morgan, and M. A. Epstein, *Int. J. Cancer*, 1980, **26**, 231.
[11] M. Biggin, P. J. Farrell, and B. G. Barrell, *EMBO J.*, 1984, **3**, 1083.

the gp340/220 gene was identified. The experimental evidence that confirmed this prediction was published in 1985[12] and came from experimental work that managed to hybrid select EBV mRNA using genomic DNA clones. This was followed by cell free translation of the eluted mRNA and immunoprecipitation of gp340/220 with a high titre antibody. The DNA clone that hybridized with the gp340/220 mRNA was the one predicted to encode the gp340/220 gene by computer analysis. The hybrid selection approach is rather labour intensive and has for the most part been superseded by one of the forms of expression cloning.

3.1.3 Expression Cloning. This approach is invaluable when the only means of identification is an antisera against the protein or pathogen of interest.

Probably the most laborious form of this approach is its use in conjunction with a biological assay. cDNA libraries are cloned into a plasmid that will allow expression in eukaryotic cells, *e.g.* SV40 or EBV vectors. Clones or pools of clones are then transferred to appropriate cell types, *e.g.* COS cells for SV40 vectors and cell extracts or cell supernatant is assayed for biological activity. If a pool of clones gives the biological activity then the individual clones can be reassayed and the desired cDNA clone identified. This methodology although tedious has allowed many of the interleukin genes to be cloned probably because the assays for these proteins are very sensitive.

Other gene products or vaccine antigens may require an enrichment step. For example many genes expressed on the cell surface, *i.e.* receptors, adhesion molecules, *etc.* have been cloned by 'panning' techniques where the cells expressing the gene of interest are selected out either with antibody or by interaction with other cells. cDNA libraries are constructed in *E. coli* and the library is transferred to eukaryotic cells. Those cells expressing the gene of interest are enriched for and the library transferred back to *E. coli*, this can be done for several rounds of expression and eventually individual clones conferring the selected phenotype will be isolated.

The most extensively used form of expression cloning involves the use of plasmid or bacteriophage vectors in *E. coli* and identification of DNA clones using antisera to the protein of interest. Here a vector such as the bacteriophage λgt11 is set up so that when cDNA fragments are cloned into sites adjacent to the β-galactosidase gene bacteria will express a β-galactosidase fusion protein containing epitopes present in the cDNA. Recombinant phage are detected with antisera.[13] The cDNA insert is then sequenced and the whole gene can then be isolated in a more traditional way. The antisera used can be monoclonal antibodies, polyclonal monospecific antisera, or even polyclonal antisera with many antibody specificities present. A variation on this method allowed the initial cloning of the malarial sporozoite surface antigen.[14] Malarial sporozoite stage cDNAs were introduced into the ampicillin resistance gene of the plasmid

[12] M. Hummel, D. Thorley-Lawson, and E. Keiff, *J. Virol.*, 1984, **49**, 413.

[13] C. K. Stover, V. F. de la Cruz, T. R. Fuerst, J. E. Burlien, L. A. Benson, L. T. Bennett, G. P. Bansal, J. F. Young, M. H. Lee, G. F. Hatfull, S. B. Snapper, R. G. Barletta, W. R. Jacobs, and B. R. Bloom, *Nature (London)*, 1991, **351**, 456.

[14] J. Ellis, L. S. Ozaki, R. W. Gwadz, A. H. Cochrane, V. Nussenzweig, R. S. Nussenzweig, and G. N. Godson, *Nature (London)*, 1983, **7**, 302, 536.

pBR322. Low levels of expression of the sporozoite surface antigen were detected by solid phase radioimmunoassay using a monoclonal antibody specific for the protein. In this way a cDNA clone coding for the antigen was isolated and subsequently sequenced. This information was then used to design peptide vaccines which have already been tested in humans.

The λgt11 system is a more sophisticated bacteriophage version of the plasmid system described above and has been used to isolate many different antigens from various stages in the life cycle of the human malarial parasite using human immune sera as well as antigens from other pathogens.

3.1.4 Expression of Potential Vaccine Antigens. Other chapters in this book (see 2, 4, and 6) refer to cloning and expression of proteins in prokaryotes and cell culture. In general, in the future eukaryotic cell culture is likely to be the method of choice for the production of subunit vaccine antigens (see Section 3.4) where the organism to be vaccinated against replicates in eukaryotic cells. *E. coli* are unable to post translationally modify some vaccine candidates, for example bacterial systems cannot add carbohydrate which is important in the antigenicity and structure of many protective antigens.

3.2 Analysis of Vaccine Antigens

The structural analysis of a potential vaccine antigen can yield valuable information in the development of a vaccine, for example a knowledge of the location of epitopes against which neutralizing antibody can be raised will allow the suitability of peptide vaccines, poliovirus chimeras, and anti-idiotype vaccines to be investigated. Epitopes are usually referred to as continuous or discontinuous. Continuous epitopes are peptides that are recognized in their random coil form so that antisera to the epitope will react with the whole molecule from which the peptide sequence is derived. Discontinuous epitopes are made up of molecules that are brought together due to secondary structure of a protein or arise from constraints imposed by the structure of a particular infectious agent. Some neutralizing epitopes are continuous while others are discontinuous.

An example of the variety of techniques that can be used to analyse vaccine antigens is the VP1 protein of foot and mouth disease virus (FMDV). As long ago as 1973 it was shown that of the four virus structural proteins only VP1 could induce FMDV neutralizing antibody.[15] It was not until 1982 that immunogenic epitopes were identified on VP1 and this was achieved by the classical method of cyanogen bromide (CNBr) cleavage.[16] CNBr cleaves at methionines and fragments the VP1 protein. Comparison of CNBr cleaved isolated VP1 and enzymatically digested virus particles suggested that the immunogenic epitopes were contained within amino acid sequences 146–152 and 200–213.

A second and more indirect approach worked on the basis of sequence analysis of four isolates of FMDV. It was argued that the most variable regions of VP1

[15] J. Laporte, J. Grosclaude, J. Wantyghem, S. Bernard, and P. Rouze, *C.R. Acad. Sci., Ser. D*, 1973, **276**, 399.
[16] K. Strohmaier, R. Franze, and K. H. Adam, *J. Gen. Virol.*, 1982, **59**, 295.

would be those subject to the most immunological pressure for mutation and would thus be the sites of greatest antigenicity. The sequence showed 80% conservation and three highly variable regions amino acids 42–61, 138–160, and 193–204. As antigenic sites are likely to be on the surface of the virus particle and therefore hydrophilic in nature it was assumed (and later shown) that the 42–61 hydrophobic sequence was not antigenic.[17]

Subsequently the Pepscan method was used to identify immunogenic sites on VP1. This method is a 'sledgehammer' approach and involves synthesis of overlapping hexapeptides corresponding to the amino acid sequences 1–6, 2–7, 3–8, 4–9, *etc.* These peptides are synthesized on polyethylene rods in a microtitre plate format and each pin is reacted with neutralizing antisera in a microtitre plate. Those peptides to which antibody attached are detected by an anti-species secondary antibody conjugated to an indicator enzyme. Such an analysis of the protein identified the 146–152 region of the molecule as antigen. Several groups went on to show that the 200–213 region of the molecule was also antigenic and if coupled to the 141–160 peptide gives an improved antibody response compared to either of the two individual peptides.[18]

A further approach that proved not to be successful with this particular protein was the use of bacteria to express fragments of the protein and analyse the reactivity of antibody to VP1. VP1 proved to be extremely difficult to produce in bacterial expression systems but other virus and bacterial antigens have been analysed in this way.

3.3 IMPROVEMENT AND GENERATION OF LIVE ATTENUATED VACCINES

Molecular biological techniques allow the analysis of virulence and antigenicity at the molecular level. This enables a more rational approach to the generation of attenuated organisms eventually enabling engineering of live vaccines with desired properties. DNA containing viruses and other micro-organisms including parasites can be engineered directly in some cases where the DNA is infectious, *e.g.* adenoviruses and herpes simplex virus and indirectly in other cases where plasmids are used to transfer information by recombination into the genome. RNA viruses are somewhat more problematical and although there has been some success with poliovirus (see Section 3.5.4) and influenza virus the compactness of the virus genome and packaging constraints of virus particles makes it unlikely that vaccines to many RNA viruses will be achieved by engineered attenuation.

It should be noted that as the incidence of immune suppression due to HIV infection increases the number of vaccine associated complications from live vaccines is also likely to increase. Vaccine manufacturers are by nature very cautious particularly when litigation from vaccine associated complications is possible. It is therefore likely that they will only favour the live approach under

[17] J. L. Bittle, R. A. Houghten, H. Alexander, T. M. Shinnick, J. G. Sutcliffe, R. A. Lerner, D. J. Rowlands, and F. Brown, *Nature (London)*, 1982, **298**, 30.

[18] H. M. Geysen, R. H. Meloen, and S. J. Barteling, *Proc. Natl. Acad. Sci. USA*, 1984, **81**, 3998.

well defined circumstances particularly if viable subunit or peptide vaccines are available.

3.3.1 New Vaccines for Pseudorabies Virus.[19] The first live viral vaccine on the open market produced by recombinant DNA technology was licensed in the USA in January 1986, for use in pigs to combat pseudorabies virus (PrV), a member of the herpesvirus family. The disease is a serious problem for swine production because of reproductive problems, death of piglets, and increased secondary respiratory disease. Vaccination is used to reduce economic losses and aid in preventing reactivation and shedding of the virus. The recombinant vaccine was generated by introducing a plasmid with 148 bp deletion in the PrV thymidine kinase (TK) gene into cells infected with PrV. At a low level homologous recombination takes place between PrV genomic DNA at the TK locus and the deleted TK gene carried by the plasmid. The recombinant virus is TK negative and can be distinguished from parental virus by its ability to grow unimpaired in the presence of 5-bromodeoxyuridine (BdUR). The TK gene lesion reduces the ability of the virus to establish neuronal replication leading to a latent infection. Since the initial system was worked out other gene-deleted pseudorabies vaccines have been licensed and used in both the USA and Europe.

Use of these live attenuated vaccines can lead to problems in diagnosis of natural infections. Is any particular PrV outbreak in pigs due to wild type virus or reversion of the vaccine? To answer this question ELISAs have been developed that are specific for the wild-type protein which is absent from the vaccine strain. Genetic deletion of genes for specific glycoproteins has also enabled the development of diagnostic kits to distinguish an antibody response to the vaccine from the response to concomitant infection with field strains. A further sophistication has also been incorporated in some vaccines where added DNA sequences have been used to trace and distinguish vaccine virus strains. Such markers in any recombinant vaccine would be useful for clear identification of the vaccines should they be suspected of circulating in the environment or reverting to virulence. Recombination of live vaccines with field viruses could also be followed if modified viruses were genetically marked in this way.

3.3.2 Improving Attenuation – Vibrio cholerae. As mentioned previously the currently available cholera vaccine is fairly ineffective. A great deal of time and effect has been expended in analysing the molecular basis for the pathogenicity of *Vibrio cholerae* and in engineering specific deletions for testing as potential vaccines.

Cholera is a severe diarrhoeal disease of humans caused by infection of the small intestine by virulent strains of *V. cholerae.* Although these virulent strains of *V. cholerae* may possess a number of virulence determinants, the clinical manifestations of cholera are primarily due to the action of the holotoxin on the intestinal epithelia. Cholera toxin is comprised of two distinct subunits, termed A and B, which are encoded by genes located on the bacterial chromosomes. The B subunit specifically binds to the GM1 ganglioside located on the surface of

[19] S. Kit, *Vaccine*, 1990, **8**, 420.

eukaryotic cells, whereas the A subunit possesses ADP-ribosylating activity resulting in the ribosylation of the G_s protein of the adenylate cyclase system in host cells. Virulent strains of *V. cholerae* are noninvasive, and, therefore, stimulation of humoral immunity by the currently available inactivated *Vibrio* strains is likely to provide only limited protection against the disease. Because the B subunit of the holotoxin is immunogenic, it has been investigated as a vaccine candidate to stimulate mucosal immunity at the site of infection. The production of specific immunoglobulins at the site of infection may lead to neutralization of cholera toxin activity by inhibiting adsorption to target cells.

Recently a recombinant plasmid in which the ctxB determinant was placed under the control of the tac promoter has been constructed resulting in over-expression of the B subunit by strains possessing this plasmid.[20] The introduction of these plasmids into nontoxigenic strains of *V. cholerae* results in transformants analogous to a current B-subunit/whole cell vaccine that has undergone field trails in Bangladesh.[21]

The use of recombinant DNA technology has made it possible to develop live attenuated strains of *V. cholerae* that can be used as oral vaccines.[22,23] This was achieved by an *in vivo* marker exchange procedure which involved the use of a plasmid in which the toxin A and B subunits were replaced by a mercury resistance gene. This gene and DNA flanking the toxin genes were mobilized into *V. cholerae* on a plasmid belonging to the P incompatibility group. Homologous recombination occurred with the net result of transferring the mercury resistance gene to the genome and eliminating the toxin genes. The recombination was detected by introducing a second IncP plasmid containing sulfur resistance and selecting for both mercury and sulfur resistance. The requirement for sulfur resistance would maintain the second plasmid while eliminating the now incompatible mercury resistance plasmid. If recombination has taken place then a dual resistant *V. cholerae* will have the toxin genes deleted.

The advantages of using attenuated strains of *V. cholerae* as vaccine material relies upon the observation that nontoxigenic strains of *V. cholerae*, when ingested by volunteers, produce a mild diarrhoea.[24] These results indicate that the extreme purging observed in severe cases of cholera is due to the action of the cholera toxin but that other virulence factors produced by the bacteria contribute towards infection. Such a hypothesis is strengthened by the fact that attenuated strains of *V. cholerae* that also possess no ctx genes can be used to stimulate a significant degree of protection in volunteers challenged with virulent strains.[24] Virulence determinants, other than the holotoxin, that have been implicated in mediating infection by *V. cholerae* include colonization antigens and

[20] J. Sanchez and J. Holmgren, *Proc. Natl. Acad. Sci USA*, 1989, **86**, 481.
[21] J. Clemens, D. A. Sack, J. R. Harris, J. Chakrbarti, M. R. Kahn, B. F. Stanton, B. A. Kay, M. U. Kahn, M. D. Yunus, W. Atkinson, A-M. Svennerholm, and J. Holmgren, *Lancet*, 1986, **ii**, 124.
[22] J. B. Kaper, H. Lockman, M. M. Baldini, and M. M. Levine, *Nature (London)*, 1984, **308**, 655.
[23] M. M. Levine, J. B. Kaper, D. Herrington, J. Ketley, G. Losonsky, C. O. Tacket, B. Tall, and S. Cryz, *Lancet*, 1988, **ii**, 467.
[24] M. M. Levine, J. B. Kaper, D. Herrington, G. Losonsky, J. G. Morris, M. L. Clements, R. E. Black, B. Tall, and R. Hall, *Infect. Immun.*, 1988, **56**, 161.

secondary toxins. Therefore, the strategy for the development of a safe, efficient cholera vaccine is currently focused upon the use of live, attenuated strains of *Vibrios* that possess a functional ctxB gene but lack the ctxA determinant.

3.3.3 Improving Stability – Poliovirus. The possibility of producing engineered polioviruses is dependant on the observation that infectious viruses can be rescued from cDNA clones when the appropriate sequence is put under the control of a strong eukaryotic promoter and transfected into susceptible cells.[25] Thus by standard molecular technology it is possible to introduce defined mutations or alterations into the cDNA and rescue a recombinant virus. This technology coupled with the knowledge of the crystal structure of the virus[26] and a knowledge of the major antigenic epitopes of the virus will allow more stable polioviruses to be generated and opens the door on generating polioviruses expressing foreign gene epitopes (see Section 3.5.4).

The 3 Sabin attenuated strains of poliovirus have been used successfully for many years to protect against paralytic polio. The strains are very stable but at low frequency it has been shown that the type 2 and 3 strains can revert to virulence and cause vaccine associated paralytic polio. Sabin type 1 however is much more stable. Two possible approaches to the generation of a more stable viruses have been investigated. The first is to replace the type 1 major virus neutralizing epitopes with type 3 epitopes (as in Section 3.5.4); this would hopefully give as stable a virus as Sabin type 1 and have some of the antigenic characteristics of type 3.[27] A second approach was based on the observation that virulence was in part associated with the 5′ non-coding region of the virion RNA. The 5′ non-coding region of the type 1 strain was used in place of the 5′ non-coding region of the type 3 strain. Here the chimeric virus RNA codes for a completely type 3 virus particle but hopefully will have a more stable phenotype.[28] These theoretical improvements may not be taken up by vaccine manufacturers for a variety of reasons not the least of which is being able to show conclusively that a complication rate in humans is reduced from 1 in 10^6 vaccinations to an even smaller number!

3.3.4 Improving Immunogenicity – Examples From Vaccinia Virus. A number of possible approaches to improving immunogenicity have been analysed using vaccinia recombinants (see Section 3.5.1). Vaccinia virus has many genes which appear to be involved in controlling the host response to the virus. There are four genes which are involved in evasion of the antiviral effects of γ-interferon as well as genes that interfere with complement fixation and processing of antigens. Deletion of any of these genes may result in an increased immunogenicity due to a more vigorous response to the virus. They may also become less attenuated but at least in one situation this seems not to be the case. Insertion into the serine protease inhibitor (Serpin) genes of vaccinia improve the immune response to a

[25] B. L. Semler, A. J. Dorner, and E. Wimmer, *Nucleic Acids Res.*, 1984, **12**, 5123.
[26] J. M. Hogle, M. Chow, and D. J. Filman, *Science*, 1985, **229**, 1358.
[27] K. L. Burke, G. Dunn, M. Ferguson, P. D. Minor, and J. W. Almond, *Nature (London)*, 1988, **332**, 81.
[28] G. Stanway, P. J. Hughes, G. D. Westrop, D. M. A. Evans, G. Dunn, P. D. Minor, G. C. Schild, and J. W. Almond, *J. Virol.*, 1986, **57**, 1187.

foreign gene incorporated into the virus.[29] This may be due to an absence of the Serpin gene product which if present might mute the immune response by preventing processing of antigens and subsequent presentation in conjunction with MHC class II. It is conceivable that other pathogens have adapted to their host and acquired the ability to modify the host immune response. Identification of these types of genes and deletion of them may improve immune responses in other pathogens.

A second possible route to improve immunogenicity is to engineer the organism to express an appropriate cytokine. IL2 has been expressed in vaccinia virus and has been shown to markedly attenuate the virus in immune compromised hosts.[30,31] Although the immune response is not dramatically enhanced small improvements were seen in some vaccinated mice. It will be interesting to see if there are any effects of other cytokines such as IL6 or IL10.

Another possible approach is to fuse a structural gene of the pathogen with the vaccine antigen under study.[32] This has been done in pilot experiments with vaccinia virus where a marker gene (β-galactosidase) was fused to a structural gene of the virus. A recombinant virus was generated which expressed this fusion and it was shown that the fusion product was also part of the virus particle. Presentation as part of the particle should enhance response to the vaccine antigen. In principle this may be extended to any organism, indeed recombinant fimbrae which are expressed on the surface of the bacteria have been expressed in *E. coli* (see Section 3.5.7).

3.4 Generation of Subunit Vaccines

The best example of the power of recombinant DNA technology is seen in the development of the current hepatitis B virus (HBV) subunit vaccine. About 300 million people worldwide have been infected with the virus. Infection in adults leads to a short acute phase associated with viral replication. In about 10% of cases patients develop a carrier state and are at a high risk from liver cirrhosis and hepatocellular carcinoma. It has been estimated that 800 000 deaths per year are due to hepatitis B virus infection and its sequelae. In order to eradicate the virus large quantities of an efficient and affordable vaccine will be required, however, traditional approaches were not tenable because, as mentioned previously, the virus cannot be grown *in vitro*. The first vaccine for HBV was licensed in 1981[4] and consisted of HBV surface antigen protein, which self assembles into 22 nm particles, purified from the blood of hepatitis B virus carriers. With the advent of HIV vaccine manufacturers turned to recombinant DNA technology for a source of HBV surface antigen to avoid using blood products. Table 7 shows some of the systems used to express the HBsAg gene and when they were described in the literature.

[29] J. Zhou, L. McClean, X-Y. Sun, M. Stanley, N. Almond, L. Crawford, and G. L. Smith, *J. Gen. Virol.*, 1990, **71**, 2185.
[30] C. Flexner, A. Hugin, and B. Moss, *Nature (London)*, 1987, **330**, 259.
[31] I. A. Ramshaw, M. E. Andrew, S. M. Phillips, D. B. Boyle, and B. E. H. Coupar, *Nature (London)*, 1987, **329**, 545.
[32] C. Haung, W. A. Samsonoff, and A. Grzelecki, *J. Virol.*, 1988, **62**, 3855.

Table 7 *Expression of the hepatitis B virus surface antigen in a variety of systems*

Gene	Vector/host	Year
HBsAg-β-galactosidase fusion protein	Plasmid/*E. coli*	1980[a]
HBsAGg-β-lactamase fusion protein	Plasmid/*E. coli*	1979[b]
HBsAGg	Plasmid/*E. coli*	1983[c]
HBsAGg	Plasmid/*S. cerevisiae* (yeast)	1982[d–f]
HBsAGg-HSV gD fusion	Plasmid/*S. cerevisiae* (yeast)	1985[g,h]
HBsAGg	Adenovirus/mammalian cells	1985[i]
HBsAGg	Herpes virus/mammalian cells	1985[j]
HBsAGg	Vaccinia virus/mammalian cells	1983[k]
HBsAGg	Varicella Zoster virus (Oka strain)	1992[l]
HBsAGg	SV40/Cos cells	1984[m,n]
HBsAGg	NIH 3T3 or C127/Bovine papillomavirus	1983[o,p]
HBsAGg	Mouse LMTK	1982[n,q]
HBsAGg	Vero	1984[r]
HBsAGg	Rat1 cells	1982[q]
HBsAGg + preS region	CHO cells	1986[s]

[a] P. Charnay, M. Gervais, A. Louise, F. Galibert, and P. Tiollais, *Nature (London)*, 1980, **286**, 893. [b] M. Pasek, T. Goto, W. Gilbert, B. Zink, H. Schaller, P. Mackay, G. Leadbetter, and K. Murray, *Nature (London)*, 1979, **282**, 575. [c] Y. Fujisawa, Y. Ito, R. Sasada, Y. Ono, K. Igarashi, R. Muramoto, V. Kikuchi, and Y. Sugino, *Nucleic Acids Res.*, 1983, **11**, 3581. [d] P. Valenzuela, A. Medina, W. J. Rutter, G. Ammerer, and B. D. Hall, *Nature (London)*, 1982, **298**, 347. [e] K. Murray, S. A. Bruce, A. Hinnen, P. Wingfield, P. M. C. A. van Erd, A. de Reus, and H. Schellekens, *EMBO J.*, 1984, **3**, 645. [f] A. Miyanohara, A. Toh-e, C. Nozaki, F. Hamada, N. Ohtomo, and K. Matsubara, *Proc. Natl. Acad. Sci. USA*, 1983, **80**, 1. [g] P. Valenzuela, D. Coit, M. A. Medina-Selby, C. H. Kuo, G. van Nest, R. L. Burke, P. Bull, M. A. Urdea, and P. V. Graves, *Biotechnology*, 1985, **3**, 323. [h] P. Valenzuela, D. Coit, M. A. Medina-Selby, C. H. Kuo, G. van Nest, R. L. Burke, M. A. Urdea, and P. V. Graves, 'Antigen Engineering in Yeast: Synthesis and Assembly of Hybrid HbsAg-HSV1 gD Particles', in 'Vaccines 85', ed. R. A. Lerner, R. M. Chanock, and F. Brown, Cold Spring Harbor Laboratory Press, New York, 1985, p. 285. [i] A. Ballay, M. Levrero, M.-A. Buendia, P. Tiollais, and M. Perricaudet, *EMBO J.*, 1985, **4**, 3861. [j] M. Arsenakis, and B. Roizman, 'Genetic engineering of Herpes Simplex Virus Genomes', in 'High Technology Route to Virus Vaccines', American Society for Microbiology, Washington DC, 1986. [k] G. L. Smith, M. Mackett, and B. Moss *Nature (London)*, 1983, **302**, 490. [l] R. S. Lowe, P. M. Keller, B. J. Keech, A. J. Davison, Y. Whang, A. J. Morgan et al., *Proc. Natl. Acad. Sci. USA*, 1987, **74**, 3896. [m] H. Will, R. Cattaneo, E. Pfaff, C. Kuhn, M. Roggendorf, and H. Schaller, *J. Virol.*, 1984, **50**, 335. [n] C. Nozaki, A. Miyanohara, A. Fujiyama, F. Hamada, N. Ohtomo, and K. Matsubara, *Gene*, 1985, **38**, 39. [o] K. J. Denniston, T. Yoneyama, B. H. Hoyer, and J. L. Gerin, *Gene*, 1984, **32**, 357. [p] Y. Wang, C. Stratowa, M. Schaefer-Ridder, J. Doehmer, and P. H. Hofschneider, *Mol. Cell. Biol.*, 1983, **3**, 1032. [q] N. M. Gough and K. Murray, *J. Mol. Biol.*, 1982, **162**, 43. [r] G. Carloni, Y. Malpiece, M.-L. Michel, A. L. Patezour, E. Sobczak, P. Tiollais, and R. E. Sheek, *Gene*, 1984, **31**, 49. [s] M.-L. Michel, P. Tiollais, D. R. Milich, F. V. Chisari, P. Pontisso, E. Sobczak, Y. Malpiece, and R. E. Streeck, 'Synthesis in CHO Cells of Hepatitis B Surface Antigen Particles Containing the Pre-S2 Region Expression Product', in 'Vaccines 86', ed. F. Brown, R. M. Chanock, and R. A. Lerner, Cold Spring Harbor Laboratory Press, New York, 1986.

The most appropriate system particularly for scale up proved to be expression of the surface antigen in yeast (*S. cereviseae*[5]). The production of surface antigen lipoprotein particles allowed the development of the first licensed recombinant vaccine for human use. Highly purified preparations of such yeast derived HBsAg particles were shown to be innocuous and have a high protective efficacy in humans and a license for general use of the vaccine was granted in 1986 in the USA.

The subunit approach is probably the most used current method for the development of new vaccines. This is because there are much less complications from vaccination and current technology has enabled large quantities of the appropriate molecules to be produced (see chapters on expression in bacteria, cell culture, *etc.*).

3.5 Recombinant Live Vectors

This strategy has been pioneered by vaccinia virus recombinants and uses currently available live attenuated vaccines as hosts for foreign genes. With careful choice of the protective antigen gene the immune response to the carrier vaccine and to the foreign gene product can be sufficient to protect against the original target of the host vaccine vector and the pathogen the foreign gene is derived from.

3.5.1 Vaccinia Virus Recombinants. Vaccinia virus has been used for over 150 years as a live attenuated vaccine for the control of smallpox. The cheapness and simplicity of the vaccine to manufacture and administer, its stability without refrigeration, potency as a single inoculation, and stimulation of both cell mediated and antibody responses are all advantages traditionally associated with vaccinia. These advantages should also be enjoyed by recombinants that express foreign genes.

Over 100 different vaccinia recombinants expressing genes from viral, bacterial, and parasitic pathogens have been described.[33] Many of them have been shown to protect in animal model systems against challenge with the appropriate pathogen. A vaccinia recombinant expressing the HIV1 envelope glycoprotein gp160 has already been tested in humans and shown to induce immunological responses to gp160.[34] However complications associated with vaccination and ever increasing number of individuals with immunodeficiencies (a contraindication for vaccination with vaccinia) may limit the usefulness of recombinants for human vaccination.

Plans for use of two vaccinia based vaccines in animals are, however, well advanced. One vaccine will protect cattle against rinderpest, the other protects wildlife against rabies.[35] The closest to extensive use is the vaccinia recombinant

33 'Recombinant Poxviruses', ed. M. M. Binns and G. L. Smith, CRC Press, Boca Raton, Florida, 1992.

34 E. L. Cooney, A. C. Collier, P. D. Greenberg, R. W. Coombs, J. Zarling, D. E. Arditti, M. C. Hoffman, S-L. Hu, and L. Corey, *Lancet*, 1991, **337**, 567.

35 P-P. Pastoret, B. Brochier, J. Blancou, M. Artois, M. Aubert, M-P. Keiny, J-P. Lecocq, B. Languet, G. Chappuis, and P. Desmettre, in 'Recombinant Poxviruses', ed. M. M. Binns and G. L. Smith, CRC Pres, Boca Raton, Florida, 1992, p. 163.

expressing the rabies virus glycoprotein. It has been shown to induce neutralizing antibody and cytotoxic T-cells in vaccinated animals. More impressively it protects foxes, fox cubs, skunks, and raccoons against challenge with wild type rabies virus even when the vaccine is presented as baited food. Field trials in Belgium dropping baited food from the air over large areas have already demonstrated that wildlife can be protected from rabies virus and that it appears to be safe with very little spread of the recombinant virus in the environment.[36]

Poxviruses have been found in many species of animal and they often have a limited host range. For instance fowlpoxvirus will only replicate in avian species. Thus it is well suited for a vector for expression of vaccine antigens to immunize poultry against pathogens such as Newcastle disease virus.[37] Indeed recombinants expressing genes from Newcastle disease virus will protect poultry against the disease. Other poxvirus such as capripoxviruses are being engineered to express antigens relevant for vaccination of sheep and goats.

3.5.2 Recombinant BCG Vaccines. BCG is an avirulent bovine tubercle bacillus that is the most widely used vaccine in the world. Since 1948 over 2.5 billion vaccinations have been carried out. Within the last few years it has been possible to introduce foreign DNA into BCG[13,38] to express antigens from other organisms, for example the envelope glycoprotein of Human immunodeficiency virus (HIV).[39] Recombinant BCG has a number of distinct advantages over other approaches for multivalent vaccines primarily due to experience gained with the parent BCG vaccine. Advantages include the fact that BCG and oral poliovaccine are the only two vaccines WHO recommend to be given at birth, the younger the age at which vaccination can begin the better the chances of success in vaccination programs. A single immunization with BCG gives long lasting cell mediated immunity to tuberculosis, it can be given repeatedly, is very safe with less than 1 complication per million vaccinations, and is a highly potent adjuvant in its own right. Although phage and plasmid vectors have been used with some success a significant amount of development is still required both to achieve higher levels of expression and to allow the system to be more readily manipulated.

3.5.3 Attenuated Salmonella Strains as Live Bacterial Vaccines. It is possible to introduce totally defined mutations or deletions in a variety of bacterial strains in order to attenuate them. These rationally designed attenuated vaccines can also be used as carriers for antigens cloned from other pathogenic organisms. Attenuated salmonella strains seem good candidates for this approach because they can be used as oral vaccines to stimulate secretory and cellular immune responses in the host. For example the gene for heat labile B subunit of enterotoxic *E. coli* was introduced into the attenuated *AroA* strain of salmonella. This recombinant salmonella was able to induce IgG and IgA antibodies to the enterotoxic B

[36] B. Brochier, M-P. Keiny, F. Costy, P. Coffens, B. Bauduin, J-P. Lecocq, B. Languet, G. Chappuis, P. Desmettre, K. Afiademomvo, R. Libois and P-P. Pastoret, *Nature (London)*, 1991, **354**, 520.
[37] J. Taylor, C. Edbauer, A. Rey-Senelonge, J-F. Bouquet, E. Norton, S. Goebel, P. Desmettre, and E. Paoletti, *J. Virol.*, 1990, **64**, 1441.
[38] W. R. Jacobs, Jr, M. Tuckman, and B. R. Bloom, *Nature (London)*, 1987, **327**, 532.
[39] A. Aldovini and R. A. Young, *Nature (London)*, 1991, **351**, 497.

subunit (as well as salmonella) in vaccinated animals.[40] A further modification of this strategy is to incorporate peptides into the flagellin gene of salmonella.[41] A potential hepatitis B virus (HBV) vaccine was constructed by incorporating synthetic oligonucleotides coding for sequences from the HBV surface antigen and from the pre S2 antigen into the flagellin gene followed by introducing the hybrid gene into a flagellin negative salmonella strain.[42] The recombinant salmonella expressed the hybrid flagellin gene and when used to vaccinate mice, guinea pigs, or rabbits induced antibodies that reacted with native hepatitis B virus surface antigen. In addition isolated T-cells, from immunized mice, proliferated in response to the hepatitis peptide contained in the flagellin gene, showing that T-cell mediated immune response can also be generated by recombinant salmonella. The antibody responses were greater in mice immunized by intramuscular inoculation rather than those vaccinated by the oral route. However continued efforts to improve oral immunization are important, not only because of the ease of administration (syringes and needles are not required) and the reduced costs but also because mucosal immune responses to antigens after oral vaccination may offer more protection against pathogens that have their initial replicative cycle on similar mucosal surfaces.

3.5.4 Poliovirus Chimaeras. The live attenuated poliovirus type 1 Sabin strain has proved to be a very safe and effective vaccine stimulating good secretory and circulating antibody responses. A knowledge of the crystal structure of the virus together with the ability to generate virus from cDNA molecules has allowed antigenic domains from other pathogens to be incorporated precisely into the virus particle at the most antigenic sites. For example the DNA coding for the major antigenic site of the Sabin type 1 strain was replaced by a peptide sequence from HIV1. Antisera to the peptide recognized the recombinant poliovirus particle and it was found that in immunization studies the recombinant virus could induce broadly neutralizing anti-HIV antibody.[43]

These new poliovirus *chimaeras* suffer some of the same limitations as peptide vaccines but do not require potent adjuvants to work.

3.5.5 Cross-species Vaccination – 'Live–dead' Vaccines. Debate over the safety of vaccinia virus has led to the suggestion that poxviruses of other species might be used to immunize humans. Canarypoxvirus replicates in cells of avian origin but is blocked in its ability to replicate in human cells. The idea is to express a vaccine antigen in a recombinant canarypoxvirus and use this as an immunogen. As it cannot replicate in the vaccinee there is no danger of virus spread, however the virus does enter cells and produce the antigen. This 'live–dead' vaccine may prove to have the advantages of live vaccines with authentic antigenic presentation without the possible complications associated with live viruses. Indeed a

[40] G. Dougan and J. Tite, *Semin. Virol.* 1990, **1**, 29.

[41] S. M. C. Newton, C. O. Jacob, and B. A. D. Stocker, *Science*, 1989, **244**, 70.

[42] Y. J. Wu, S. Newton, A. Judd, B. Stocker, and S. W. Robinson, *Proc. Natl. Acad. Sci. USA*, 1989, **86**, 4726.

[43] D. J. Evans, J. McKeating, J. M. Meredith, K. L. Burke, K. Katrak, A. John, M. Ferguson, P. D. Minor, R. A. Weiss, and J. W. Almond, *Nature (London)*, 1989, **339**, 385.

canarypoxvirus recombinant expressing the rabies virus glycoprotein induces rabies virus neutralizing antibody in vaccinated animals.[44]

3.5.6 Other Virus Vectors. A number of other virus vectors which are appropriate for vaccination have been described in the past few years. These include both replication deficient and replication competent recombinant Adenoviruses. Adenovirus types 4 and 7 have been used to immunize military personnel for some time and as it is now possible to engineer Adenoviruses to express vaccine antigens they have some potential as live recombinant vaccines. Poliovirus and hepatitis virus antigens have been expressed in these viruses and recombinants can induce significant levels of antibody to the foreign gene.[45,46] At present there are reservations about these viruses from several standpoints, large amounts of virus are needed to achieve any effective vaccination. For replication competent viruses there are concerns over the E1a and E1b gene products which have oncogenic potential and for the replication defective variants there is concern over the cell-lines that are required for growth of the recombinants.

The Oka strain of varicella zoster virus (chicken pox virus) is highly attenuated and can be engineered to express foreign vaccine antigens, for example the hepatitis B virus surface antigen has been expressed and immunization with the recombinant was shown to induce antibody to hepatitis B surface antigen.[47] Herpes simplex virus has also been used to express foreign genes with vaccine potential.

3.5.7 Recombinant E. coli *Strains.* Enterotoxigenic *E. coli* strains (ETEC) cause diarrhoeal diseases in young pigs and under some circumstances in man. These bacteria adhere to the intestine of the host via surface-associated fimbrae and secrete toxins which can be classified into heat stable (ST-toxin) and heat labile (LT-toxin). The fimbrae are highly antigenic and the first vaccines against ETEC consisted of whole cells or acellular extracts enriched for fimbrae.

Vaccines prepared from ETEC strains gave significant levels of adverse reactions due to high levels of lipopolysaccharide and capsular antigens on the surface of the wild type ETEC strains. *E. coli* K12 gave far fewer adverse reactions and was used as a vector for plasmid constructs that expressed one or more different antigenic types of fimbrae.[48] However to produce a vaccine with a wider spectrum of protection an anti-toxin component was introduced. Plasmid vectors were constructed using a strong prokaryotic promoter that expressed the LT toxin B-subunit at high levels.[49] Cetus corporation now market a pig vaccine which consists of an *E. coli* K12 strain that expresses high levels of the LT-B subunit and contains fimbrae from ETEC strains. This engineered *E. coli* was the

[44] J. Taylor, C. Trimarch, R. Weinberg, B. Languet, F. Guillemin, P. Desmettre, and E. Paoletti, *Vaccine*, 1991, **9**, 190.

[45] R. Dulbecco, International Patent Number PCT/US83/00015, 1983.

[46] A. Ballay, M. Levrero, M-A. Buendia, P. Tiollais, and M. Perricaudet, *EMBO J.*, 1984, **4**, 3861.

[47] R. S. Lowe, P. M. Keller, B. J. Keech, A. J. Davison, Y. Whang, and A. J. Morgan *et al.*, *Proc. Natl. Acad. Sci. USA*, 1987, **84**, 3896.

[48] M. Kehoe, M. D. Winther, P. Morrisey, G. Dowd, and G. Dougan, *FEMS Microbiol. Lett.*, 1982, **14**, 129.

[49] S. Attridge, J. Hackett, R. Marona, and P. Whyte, *Vaccine*, 1988, **6**, 387.

first licensed vaccine produced by recombinant DNA technology to be used in the USA.

4 OTHER APPROACHES TO VACCINES

4.1 Peptide Vaccines

This approach is the ultimate in the reductionist approach to vaccines. As mentioned in Section 3.2 it is possible to identify the epitopes within a protein that can induce neutralizing antibody or epitopes that are important in T-cell responses to vaccines. Chemical synthesis of these epitopes is relatively straightforward and with appropriate adjuvant or conjugation to carrier proteins they can induce antibody or T-cell mediated responses to the synthesized epitope. These immune responses are in some cases sufficient to give protection against the organism the protein epitope was derived from. For example 100 μg of a 20 amino acid synthetic peptide to amino acids 141–160 of VP1, the major coat protein of foot and mouth disease virus (FMDV), will protect guinea pigs against a severe challenge with FMDV.[50]

The cloned sporozoite surface antigen of the malarial parasite (see Section 3.1.3) was shown to contain a repeat sequence within the molecule. This repeat seemed an ideal target for the development of an antimalarial peptide vaccine. Indeed several human clinical trials of vaccines based on this sequence have been carried out with some success.[50]

Modifications to the basic technique of coupling a peptide to a protein and using it as a vaccine include; (a) incorporation of a helper T-cell epitope within the peptide and use of the peptide on its own, (b) cyclization of the peptide to improve antigenicity, and (c) incorporation of the peptide into antigenic regions of other proteins such as hepatitis B virus surface (HBsAg) or core (HBcAg) antigens. The advantage of incorporation into HBsAg or HBcAg rather than other proteins is that they both form particles which by their very nature are highly immunogenic.

The advantages of this approach include the fact that the product is stable and chemically defined without the presence of an infectious agent. Many other approaches require large scale production plants and complex down stream processing, peptide vaccines are relatively simple and require limited work up and purification. The major disadvantage often cited for peptide vaccines is based on the fact that often only one peptide is used. Many pathogens are characterized by the fact that there is extreme variation in the antigenic proteins of the agent. Could a single epitope or even multiple epitopes be found that protect in all cases in the face of the extreme variation? Despite these reservations the ease of production, stability, and safety of peptide vaccines make them an approach that will receive much attention in the future.

4.2 Anti-idiotypes

It has been suggested that anti-idiotype (anti-Id) antibodies may make effective vaccines. This is based on the finding that antibodies themselves can act as

[50] F. Brown, *Semin. Virol.*, 1990, **1**, 67.

immunogens. An immune response raised against the unique antigen combining site of an antibody is termed an anti-idiotypic response and may bear a structural resemblance to the original antigen. When this occurs the anti-idiotypic antibody (monoclonal or polyclonal) may be able to induce an antibody response that recognizes the original antigen and hence act as a vaccine. Anti-idiotypes have been shown to give protection in a vareity of animal model systems; probably the best demonstration of the potential of this approach was the protection of chimpanzees from hepatitis B virus associated disease by previous immunization with anti-Id.[51]

There are several advantages of anti-Id's over traditional approaches, most of which are true of other subunit vaccine approaches. These advantages are apparent:

(1) Where antigen is difficult to obtain, *i.e.* when the infectious agent is hazardous or cannot be grown *in vitro*.
(2) Where attenuated vaccines have high reversion frequencies or possess genes that may be involved in oncogenesis. Any problems vaccinating immunocompromised individuals with live vaccines would also be avoided.
(3) Where a single epitope can confer protection but other epitopes of the whole molecule might induce autoimmunity.
(4) When organisms display wide genetic diversity but a single cellular receptor. An anti-Id response could theoretically induce a serological response that mimics the receptor and binds the infectious agent at its receptor binding site.
(5) A significant advantage is the ability of anti-Id to mimic non-proteinaceous epitopes such as carbohydrate, lipid, or glycolipid. All of which cannot yet be produced easily to act as subunit vaccines.

There are a number of disadvantages to this approach in particular the restriction of the vaccine to a single epitope or a few epitopes administered together which may not be enough to protect against some organisms and the limitations associated with multiple use of anti-Id preparations. Over time it is assumed that with repeated anti-Id immunization, antibody to constant region immunoglobin determinants will arise and this might then prejudice subsequent immunizations with anti-Id.

Despite these limitations and although this approach offers little advantage over other recombinant sources of antigen other than for non-proteinaceous epitopes it may prove to be an important adjunct to other strategies for immunization.

5 SUMMARY AND CONCLUSIONS

The increasing knowledge of basic immunological processes and the contribution of different types of immune responses in the prevention and control of infectious diseases has clarified the immunological requirements for a vaccine to give long

[51] R. C. Kennedy, J. W. Eichberg, R. E. Lanford, and G. R. Dreesman, *Science*, 1986, **232**, 220.

lasting immunity. The ideal vaccine should generate large numbers of memory T and B lymphocytes, be capable of being processed to give peptides which associate with MHC antigens and induce T-cell responses to a sufficient number of T-cell epitopes to overcome genetic variability between hosts, and result in the persistence of antigen so that B memory cells are continually recruited to produce circulating antibody.

Not only do these immunological criteria need to be taken into account but also a series of other basic criteria need to be considered when designing new vaccines. For example, target epitopes must be clearly identified and characterized if possible and these target epitopes must remain conserved across any variant population and expressed to sufficient levels to allow immune mechanisms to function. Immune effector mechanisms should be identified and the direct involvement of the target epitope confirmed. New immunostimulating agents (Adjuvants) need to be identified and approved for use. The safety of the immunogen must be established, *i.e.* minimize side effects such as immunosuppression, auto-immunity, or excessive inflammation.

In practice this is a difficult set of criteria to meet for many organisms. For example the identification of a protective antigen can be difficult. It is not enough to assume that a major protein on the surface of the pathogen will be a good vaccine antigen because of antigenic variability of the molecule concerned and the possibility that antibody will not protect. It is also very clear that even if protective antigens can be identified new adjuvants will need to be used to improve immune responses to antigens. At present aluminium hydroxide gel (alhydrogel) is the only widely licensed adjuvant for human use and it is not effective enough for some antigens. For example although there are a number of difficulties with the HIV envelope protein gp160 as a vaccine for HIV (antigenic variation being the most serious problem) it has been used in human trials. These trials have indicated that alhydrogel is not a very effective adjuvant for the glycoprotein. Consequently a large trial has been set up to assess a series of adjuvants in conjunction with HIV gp160 produced by recombinant DNA technology in mammalian cells.

Once a candidate antigen has been identified then it is pertinant to ask which of these multitude of approaches should be taken? The simple answer is that in some cases one approach will be appropriate and in others a different strategy would be appropriate. A simple guide for the developed countries is that subunit vaccines produced by eukaryotic cells will probably receive the widest acceptance because of their perceived safety. Peptide vaccines if shown to be effective will also be valuable and a distinctly viable approach. In the third world many more secondary requirements come into play such as those listed below.

Secondary Requirements for Effective Vaccination

(1) Cost. The six childhood vaccines supplied to the WHO global programme of immunization are supplied through UNICEF for a cost of about 20 pence per schedule of administration. Any new vaccines must cost a similar amount for them to be afforded on a global scale. There is a difference

between what developed countries feel is appropriate and what third world countries can afford.

(2) Many vaccines are administered parenterally which adds to the cost of administration requiring skilled personnel, needles, and syringes. Delivery via mucosal surfaces preferably orally offers many advantages.

(3) Thermal stability, particularly for those vaccines used in tropical countries, is vital if the cost of setting up a cold chain is to be avoided.

(4) A single immunization, if effective, would also reduce administration costs and improve success rates. It is often difficult in a third world country to complete a course of immunizations.

(5) Long lived immunity. Live vaccines tend to be more effective at stimulating long lasting immunity than non-infectious vaccines without adjuvant.

For the successful application of this new technology and in particular for the choice of which route to use another important consideration is the portal of entry of the pathogen. It may be that the most effective immunity to a particular agent will be generated by immunity at the site of entry of that pathogen. The disease would then dictate to some degree the most effective means of vaccination. It is probable that the immune response required for protection against a blood borne pathogen such as a virus or a parasite would be quite different from a virus or bacterium that infects at a mucosal surface. Part of the overall assessment requires an answer to the question of whether protection is best achieved by circulating antibody or secretory antibody or a vigorous T-cell response.

It should also be borne in mind that neither immunization nor recovery from natural infection always protects a person against infection with the same organism. This principle holds true for diseases that have been successfully dealt with by immunization, for example polio, measles, and rubella. Control is achieved not by inducing sterile immunity, *i.e.* no infection, but rather by limiting replication and spread of the organism; for example polio virus is prevented from reaching nervous tissue but a limited replication does occur. Immunity to respiratory infections is often poor following natural infection and it is probably unrealistic to expect any vaccine to produce sterile immunity. However it may be possible to protect against serious pulmonary disease yet not prevent rhinitis. Thus limiting serious disease but not completely eliminating the pathogen may be a valuable approach to decreasing the burden of infectious disease.

Ultimately medical, political, economic, and social considerations will determine the use of any vaccine. It is too soon to say if any one avenue of vaccine research will provide a global answer but with the plethora of new approaches to vaccines it is to be hoped that before too long many of the problems associated with infectious disease will be tackled.

6 FURTHER READING

1. Seminars in Virology, Volume 1, Issue 1, 'Modern Approaches to Vaccines', ed. Fred Brown, Saunders Scientific Publications/W. B. Saunders Company, Philadelphia, Jan 1990.

2. 'Immunization against Infectious Disease', HMSO, London, 1992. (Written for health workers and lists all licensed vaccines in the UK with basic information on them.) (ISBN 0-11-321515-0)
3. 'Recombinant DNA Vaccines: Rationale and Strategy', ed. R. E. Isaacson, Marcel Dekker Inc, New York, 1992. (ISBN 0-8247-8699-8)
4. Modern Vaccines Series in the Journal Lancet (1990). G. L. Ada, 'The Immunological Principles of Vaccination', *Lancet*, 1990, **335**, 523–526; F. Brown, 'From Jenner to Genes–New Vaccines', *Lancet*, 1990, **335**, 587–590; R. M. Anderson and R. M. Hay, 'Immunization and Herd Immunity', *Lancet*, 1990, **335**, 641–645; A. R. Hinman and W. A. Orenstein, 'Immunization Practice in Developed Countries', *Lancet*, 1990, **335**, 707–710; A. J. Hall, B. M. Greenwood, and H. Whittle, 'Practice in Developing Countries', *Lancet*, 1990, **335**, 774–777; A. J. Beale, 'Polio Vaccines: Time for a Change in Vaccination Policy?', *Lancet*, 1990, **335**, 839–842; F. Shann, 'Pneumococcus and Influenza', *Lancet*, 1990, **335**, 898–901; M. Levine, 'Enteric Infections', *Lancet*, 1990, **335**, 958–961; P. E. M. Fine and L. C. Rodrigues, 'Mycobacterial Diseases', *Lancet*, 1990, **335**, 1016–1020; G. C. Schild and P. D. Minor, 'Human Immunodeficiency Virus and AIDS: Challenges and Progress', *Lancet*, 1990, **335**, 1081–1084; A. Eddleston, 'Hepatitis', *Lancet*, 1990, **335**, 1142–1145; K. G. Nicholson, 'Rabies', *Lancet*, 1990, **335**, 1201–1205; J. H. L. Playfair, J. M. Blackwell, and H. R. P. Miller, 'Parasitic Diseases', *Lancet*, 1990, **335**, 1263–1266; E. R. Moxon and R. Rappuoli, '*Haemophilus influenzae* Infections and Whooping Cough', *Lancet*, 1990, **335**, 1324–1329; D. Isaacs and M. Menser, 'Measles, Mumps, Rubella, and Varicella', *Lancet*, 1990, **335**, 1384–1387; A. Robbins, 'Progress to Vaccines We Need and Do Not Have', *Lancet*, 1990, **335**, 1436–1438.

Acknowledgements. I would like to acknowledge the patience and encouragement of my wife and family as well as generous financial support from the Cancer Research Campaign.

CHAPTER 13

Transgenesis

L. J. MULLINS AND J. J. MULLINS

1 INTRODUCTION

Transgenesis may be defined as the introduction of exogenous DNA into the genome, such that it is stably maintained in a heritable manner. Over the last decade, the introduction of transgenes into the mammalian genome has become a routine experimental tool, and is gaining increasing importance in the biotechnology industry. Traditionally, DNA (the transgene) is introduced into the one-cell embryo by micro-injection, and surviving embryos are subsequently reimplanted into a pseudopregnant female and allowed to develop to term. In a proportion of the embryos – provided that the DNA integrated into the genome prior to the first cell division – the transgene will be passed on to subsequent generations through the germline.

Transgene techniques have far-reaching research applications. At the molecular level they allow the identification of *cis*-acting DNA sequences important in directing developmental and/or tissue-specific gene expression, and the specific manipulation of gene expression *in vivo*. One can equally ask broad developmental questions concerning cell or organ function with a degree of finesse not possible with other techniques. With the advent of embryo-stem (ES) cell technology and the development of strategies for achieving homologous recombination, the researcher now has the ability to question the function of specific genes and to ascertain the *in vivo* effects of precise alterations to gene function. These innovations have important implications for many areas of biomedical research including the design of disease models, the use of mammals as bioreactors for the production of human therapeutic proteins, and ultimately, the correction of inborn errors of metabolism by gene targeting.

Although there is now a growing body of literature regarding transgenesis in plants, and several of the strategies discussed here are equally applicable to plant biotechnology, we will be restricting the scope of this review to the application of transgenesis in mammalian systems.

2 THE PRODUCTION OF TRANSGENIC ANIMALS BY MICRO-INJECTION

The first transgenic mice were produced more than ten years ago, and both for scientific and practical reasons, the mouse is still the animal of choice in the majority of transgenic experiments. During recent years, however, transgenic techniques have been extended to other species including the rabbit, the rat, and also a range of commercially important animals – notably the cow, the pig, the goat, and the sheep.

2.1 Transgenic Mice

The technique by which transgenic mice are produced is schematically outlined in Figure 1. The first stage in development of transgenic animals is the isolation of sufficiently large numbers of fertilized eggs for micro-injection. This is achieved by the superovulation of young virgin females (approximately 4–5 weeks of age), which are injected with a source of follicle-stimulating hormone (pregnant mare's serum gonadotrophin). Then 48 h later, the females are given an artificial leutinizing hormone surge by administration of human chorionic gonadotrophin, and are paired with proven stud males. The following day, females which have mated (as identified by the presence of a vaginal plug) are sacrificed, and fertilized eggs are removed from the swollen ampullae of the fallopian tubes, by dissection. Using such a protocol, up to 30 zygotes can be isolated per female, depending on the strain used. The zygotes are freed from attached cumulus cells by brief incubation in the presence of hyaluronidase, transferred to appropriate medium, and are stored in a CO_2 incubator at 37 °C prior to micro-injection.

Typically 20–25 fertilized eggs are processed at a time. One by one, they are picked up by gentle suction onto a holding pipette, and injected with the micro-injection needle. The movement of both holding and micro-injection pipettes is controlled by micromanipulators (Figure 2) which are either hand or pneumatically controlled. Suction to the holding pipette is applied through oil-filled tubing via a micrometer-controlled syringe.

The micro-injection needle, which has an internal tip diameter of approximately 1 μm and contains DNA at a concentration of approximately 1 $\mu g\ cm^{-3}$, is manipulated gently but firmly until both the zona pellucida and the nuclear membrane of one of the pronuclei have been pierced. Care must be taken to ensure that the highly elastic nuclear membrane is punctured and that the needle does not touch the nucleoli, causing blockage of the needle and damage to the egg. DNA is injected into the pronucleus using a pneumatic pump or a hand operated syringe/micrometer and successful injection is indicated by swelling of the pronucleus prior to removal of the pipette tip.

Eggs that have been successfully injected are returned to the incubator and may be left to develop to the two-cell stage overnight. This allows one to check that the eggs are still viable, following injection. (A proportion of the injected eggs do not survive the ordeal.) Whether at the one-cell or two-cell stage, embryos are reimplanted into the oviduct of anaesthetized pseudopregnant

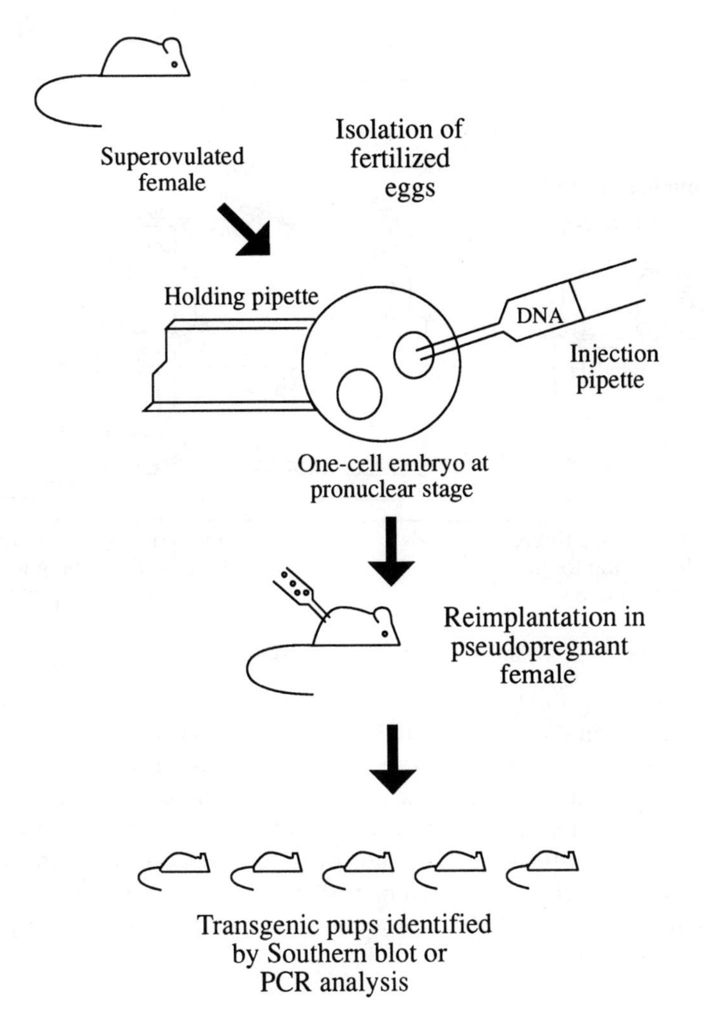

Figure 1 *Generation of transgenic animals by micro-injection of the one-cell embryo*

females (experienced mothers which have been mated the previous night with vasectomized or genetically infertile males). The females are allowed to recover from the anaesthetic and the pregnancy is continued to term. At weaning, progeny are tested for incorporation of the transgene, by polymerase chain reaction (PCR)[1] or Southern blot analysis[2] of genomic DNA isolated from tail biopsies or by PCR analysis of whole blood.[3]

Provided that the foreign DNA was incorporated into the genome prior to the first cell division, the transgene should be present in every cell of the resultant pup, including those of the germline. A suitable breeding strategy can be

[1] K. B. Mullis and F. Faloona, *Methods Enzymol.*, 1987, **155**, 335.
[2] E. M. Southern, *J. Mol. Biol.*, 1975, **98**, 503.
[3] A. J. Ivinson and G. R. Taylor, in 'PCR – A Practical Approach', ed. M. J. McPherson, P. Quirke, and G. R. Taylor, Oxford University Press, New York, 1991, Chapter 2, p. 15.

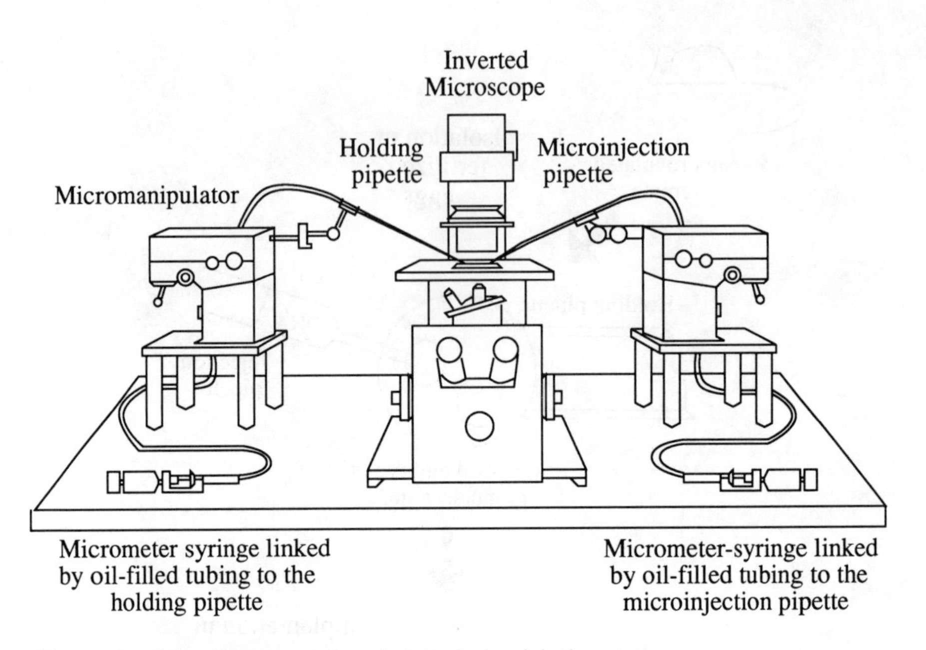

Figure 2 *Schematic representation of a typical micro-injection set-up*

initiated to maintain the transgene, in heterozygous and eventually homozygous form, as a unique transgenic line. Sometimes, where transgene integration occurred after the first cell division, the founder is chimaeric, and depending upon the representative proportion of cells populating the germline, a fraction of its progeny may carry the transgene, allowing it to be rescued. If the transgene is not represented in cells of the germline, then it is not possible to generate a transgenic line from the founder. For a more technically detailed description of transgenesis see Hogan *et al.*[4]

2.2 Transgenic Rats

Transgenic rats are generated using the same strategy as outlined for transgenic mice, but with some important modifications. Firstly, virgin females (approximately 30 days of age), have been found to be most responsive to superovulation. On that day, a highly purified source of follicle-stimulating hormone ('Foltropin', Vetrepharm, Canada), is administered, using a subcutaneously implanted osmotic 'minipump'.[5] Following induction with human chorionic gonadotrophin 48 h later, and mating, between fifty and one hundred fertilized eggs can be isolated from a single female (depending on the strain used). The pronuclei take several hours longer to develop in the rat embryo than those of the mouse, and it

[4] B. Hogan, F. Costantini, and E. Lacy, 'Manipulating the Mouse Embryo – A Laboratory Manual', Cold Spring Harbor Laboratories Press, 1986.
[5] D. T. Armstrong and M. A. Opavsky, *Biol. Reprod.*, 1988, **39**, 511.

has been found that the zona pellucida and pronuclear membranes are much more elastic, making them slightly more difficult to micro-inject.

2.3 Application of Micro-injection Techniques to Other Animals

Transgenic technology has been extended to include a number of agriculturally important animals. Details regarding superovulation vary from species to species and the reader is referred to other publications for specific information (*e.g.* production of transgenic sheep,[6,7] goats,[8] and pigs).[9] In some species, the pronuclei of the fertilized embryo have to be visualized by centrifugation, prior to micro-injection.[8,9]

3 GENERAL CONSIDERATIONS

3.1 Characterization of Construct

Correct tissue-specific or developmental expression of the transgene requires the incorporation of all necessary flanking control elements within the construct, but this is not always sufficient to ensure expression. If feasible, constructs should be transfected into appropriate cell-lines prior to introduction into the germline, to determine whether the construct can be expressed *in vitro*. This can often give valuable information about the extent of promoter sequences necessary for expression, prior to *in vivo* studies. One should consider the relative benefits of linking promoter sequences to a reporter gene if expression might be masked by the endogenous copy of the gene.

3.2 Choice of Animal

As previously stated, the mouse is traditionally the animal of choice for transgenic research. A transgenic animal programme requires significant animal breeding facilities, firstly to ensure the regular production of large numbers of eggs for micro-injection and pseudopregnant females to receive the injected eggs, and secondly to maintain breeding colonies for the various transgenic lines generated from each micro-injection series. For some physiological studies, such as cardiovascular research, neurobiology, and pharmacology, however, rat transgenics may be preferable, because of size constraints in the mouse[10] or the historical use of this species within a particular discipline.

From mouse studies, it is apparent that F1 cross-bred animals are superior to

[6] J. P. Simons, I. Wilmut, A. J. Clark, A. L. Archibald, J. O. Bishop, and R. Lathe, *Bio/Technology*, 1988, **6**, 179.

[7] G. Wright, A. Carver, D. Cottom, D. Reeves, A. Scott, P. Simons, I. Wilmut, I. Garner, and A. Colman, *Bio/Technology*, 1991, **9**, 830.

[8] K. M. Ebert, J. P. Selgrath, P. DiTullio, J. Denman, T. E. Smith, M. A. Memon, J. E. Schindler, G. M. Monastersky, J. A. Vitale, and K. Gordon, *Bio/Technology*, 1991, **9**, 835.

[9] R. E. Hammer, V. G. Pursel, C. E. Rexroad, R. J. Wall, D. J. Bolt, K. M. Ebert, R. D. Palmiter, and R. L. Brinster, *Nature (London)*, 1985, **315**, 680.

[10] J. J. Mullins, J. Peters, and D. Ganten, *Nature (London)*, 1990, **344**, 541.

inbred strains because they produce higher numbers of eggs on superovulation, have larger litter sizes, and are generally better mothers. Factors such as characteristics of endogenous gene *versus* the transgene, however, may play a part in the decision to use inbred rather than F1 hybrids.

If larger animals are to be used, it is practical to carry out initial transgenic studies with a given construct in rodents prior to costly trials in large domestic animals. It must be noted, however, that the response of one species to a gene construct may vary from that of another.[10,11]

3.3 Transgene Expression

The transgene integrates into the genome in an entirely random fashion, often as a head–tail concatomer. Expression of the transgene is not necessarily copy number related, however, and may be very low despite the presence of a high copy number. This is due to the fact that expression can be affected by sequences flanking the site of integration, for example if it integrates into an area of the genome that is actively repressed/silenced. Alternatively, expression of the transgene may be so high that it causes unexpected phenotypic alterations to the organism. There have been examples in the literature where chronic expression of a transgene led to tissue ablation.[12] Finally, the transgene may integrate within an endogenous gene, altering or insertionally inactivating that gene function and presenting as a totally unexpected phenotype.[13–15] Despite all these cautions, transgenesis is a very powerful tool with many applications, some of which are outlined below.

4 DESIGN OF THE TRANSGENIC EXPERIMENT

4.1 Investigating Gene Expression

Increasing the expression of a gene is often informative in evaluating the role that the gene product plays in normal development. It is possible to determine the effect of over-expression of a gene by placing it under the control of a strong, heterologous promoter.[16] Alternatively, expression can be rendered constitutive by linking the transgene to a housekeeping gene promoter, such as that of the phosphoglycerate kinase gene (PGK), or inducible by using, for example, the metallothionine promoter.[12]

In order to identify and define the control elements in and around a gene which effect its tissue-specific and developmental pattern of expression, one can

[11] R. E. Hammer, S. D. Maika, J. A. Richardson, J-P. Tang, and J. D. Taurog, *Cell*, 1990, **63**, 1099.
[12] R. R. Behringer, R. L. Cate, G. J. Froelick, R. D. Palmiter, and R. L. Brinster, *Nature (London)*, 1990, **345**, 167.
[13] R. P. Woychik, T. A. Stewart, L. G. Davis, P. D'Eustachio, and P. Leder, *Nature (London)*, 1985, **318**, 36.
[14] A. K. Ratty, L. W. Fitzgerald, M. Titeler, S. D. Glick, J. Mullins, and K. W. Gross, *Mol. Brain Res.*, 1990, **8**, 355.
[15] U. Karls, U. Muller, D. J. Gilbert, N. G. Copeland, N. A. Jenkins, and K. Harbers, *Mol. Cell Biol.*, 1992, **12**, 3644.
[16] M. E. Steinhelper, K. L. Cochrane, and L. J. Field, *Hypertension*, 1990, **16**, 301.

design a series of constructs with nested deletions around the promoter region, the 3′ end of the gene, and, if necessary, within introns. By analysing expression patterns of the transgenes in the resultant transgenic lines – with ontogeny studies and tissue surveys – one can build up a map of the sequences which are absolutely required for correct gene expression.

If expression of the transgene is likely to be masked or affected by expression of its endogenous counterpart then it is possible to link the promoter sequences to a reporter gene, such as the SV40 T-Antigen,[17] CAT,[18] or β-galactosidase (*lacZ*).[19] One can then ascertain whether or not the promoter directs expression to the appropriate cell types. In the former example, the resultant tumours arising in the target cells have proved to be invaluable in generating new cell lines which retain highly differentiated phenotypes.[20,21] In the latter case, X-gal staining of the embryo/tissue slices readily identifies any promoter-directed sites of expression.

By such strategies, regulatory elements well upstream of certain genes have been identified, which play important roles in controlling gene expression. These are the so-called dominant or locus control regions, which have been characterized for the human globin gene locus,[22] the chicken β-globin gene,[23] the chicken lysozyme gene,[24] and recently, the red-green visual pigment genes.[19] Importantly, these elements appear to confer position independent, tissue-specific expression on the transgene. Some of the transcription factor binding sites within these locus control elements are currently being identified.[25,26] Identification of other similar regulatory elements, may give the researcher a greater degree of control over construct design – limiting expression of the transgene to specific cells, and yet ensuring good expression levels.

4.2 Reduction of Gene Function

Abolishing or reducing the function of a gene can be equally informative with respect to the role which the gene plays *in vivo*. There are a number of strategies by which this can be achieved. The first is the introduction into the organism, of a gene encoding antisense RNA. If the transgene is driven by the promoter sequences of its endogenous counterpart, then expression is limited to those sites where it can reduce the amount of targeted gene product. The mechanism by which antisense inhibition occurs remains obscure, but it is probably brought

[17] C. D. Sigmund, K. Okuyama, J. Ingelfinger, C. A. Jones, J. J. Mullins, C. Kane, U. Kim, C. Wu, L. Kenny, Y. Rustum, V. J. Dzau, and K. W. Gross, *J. Biol. Chem.*, 1990, **265**, 19916.

[18] M. L. Lui, A. L. Olson, W. S. Moye-Rowley, J. B. Buse, G. I. Bell, and J. E. Pessin, *J. Biol. Chem.*, 1992, **267**, 11673.

[19] Y. Wang, J. P. Macke, S. L. Merbs, D. J. Zack, B. Klaunberg, J. Bennett, J. Gearhart, and J. Nathans, *Neuron*, 1992, **9**, 429.

[20] J. J. Windle, R. I. Weiner, and P. L. Mellon, *Mol. Endocrinol.*, 1990, **4**, 597.

[21] P. L. Mellon, J. J. Windle, P. C. Goldsmith, C. A. Padula, J. L. Roberts, and R. I. Weiner, *Neuron*, 1990, **5**, 1.

[22] P. Collis, M. Antoniou, and F. Grosveld, *EMBO J.*, 1990, **9**, 233.

[23] M. Reitman, E. Lee, H. Westphal, and G. Felsenfeld, *Nature (London)*, 1990, **348**, 749.

[24] C. Bonifer, M. Vidal, F. Grosveld, and A. E. Sippel, *EMBO J.*, 1990, **9**, 2843.

[25] S. Philipsen, D. Talbot, P. Fraser, and F. Grosveld, *EMBO J.*, 1990, **9**, 2159.

[26] D. Talbot, S. Philipsen, P. Fraser, and F. Grosveld, *EMBO J.*, 1990, **9**, 2169.

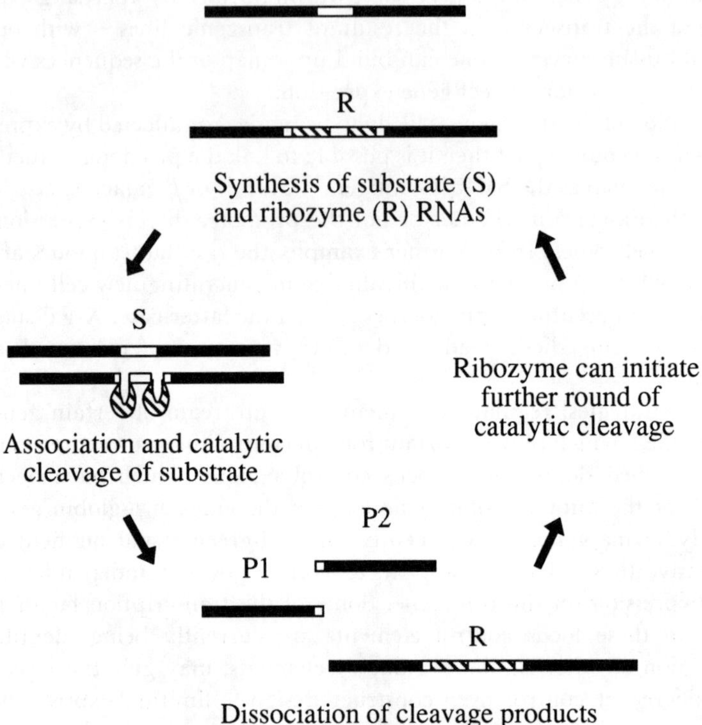

Figure 3 *Schematic representation of the mechanism by which the ribozyme construct interacts with and cleaves its target substrate*

about by RNA–DNA interactions interfering with transcription, general interference with RNA processing and transport, inhibition of translation, and/or rapid degradation of sense–antisense RNA hybrids. The general efficacy of antisense inhibition has yet to be proven, although there are a growing number of examples of its application in the literature in both plants [27,28] and mammals.[29,30]

The second strategy involves the uses of a ribozyme – an RNA molecule with enzymic activity which is capable of cleaving specific target RNA molecules. If such a sequence is placed within an appropriate antisense sequence, one would predict a much more efficient inhibition of targeted gene expression (see Figure 3). To date, no transgenic animals carrying such a construct have been gener-

[27] J. Stockhaus, M. Hofer, G. Renger, P. Westhoff, T. Wydrzynski, and L. Willmitzer, *EMBO J.*, 1990, **9**, 3013.
[28] A. J. Hamilton, G. W. Lycett, and D. Grierson, *Nature (London)*, 1990, **346**, 284.
[29] M. I. Munir, B. J. F. Rossiter, and C. T. Caskey, *Somat. Cell and Mol. Genet.*, 1990, **16**, 383.
[30] M-C. Pepin, F. Pothier, and N. Barden, *Nature (London)*, 1992, **355**, 725.

ated, but ribozyme-mediated destruction of RNA has been demonstrated in tissue culture.[31–34]

4.3 Cell Ablation

To answer questions about the lineage, fate, or function of a cell, it can be informative to observe the effects of removing that cell.[35] By introducing genes encoding cytotoxins, such as the catalytic subunits of Diphtheria toxin (DT-A)[36,37] or ricin (RT-A),[38] under appropriate cell-specific promoters, one can selectively ablate cell types which might be difficult or impossible to remove by physical means. Toxigenic ablation is potentially very powerful, since the toxic gene products may act at very low concentrations. However, it is important that the toxin should be confined to the cells where it is expressed or damage to neighbouring cells may occur. To overcome this problem, toxigenes often lack the signal sequences of their native counterparts. Obviously, some promoter–toxigene constructs are likely to be lethal to the developing embryo, if the ablated cells are essential for viability. To give the researcher more control over the degree of cell ablation or timing of the event during development, a number of strategies have been devised. The first is the use of an attenuated DT-A gene.[39] Although the degree of penetration of such a gene may be variable, it should prove more versatile in achieving cell ablation with a broad range of cell- and tissue-specific promoter sequences.

An alternative strategy makes ablation dependent on the administration of drugs. This has been achieved by introducing the herpes simplex virus-1 thymidine kinase (HSV-1-*tk*) transgene under the control of appropriate promoter elements. Cells expressing the gene are rendered susceptible to drugs such as gancyclovir.[40] The power of this strategy is that the timing and degree of ablation is controlled by the investigator. Evidence suggests that both activity dividing and non-dividing cells may be susceptible to drug-induced ablation indicating a wide application for this strategy. (RAS personal communication.) With these more sophisticated methods of regulation, it should be possible to use promoters that are active in cells essential to development.

31 M. Cotten and M. L. Birnstiel, *EMBO J.*, 1989, **8**, 3861.

32 F. H. Cameron and P. A. Jennings, *Proc. Natl. Acad. Sci. USA*, 1989, **86**, 9139.

33 N. Sarver, E. M. Cantin, P. S. Chang, J. A. Zaia, P. A. Ladne, D. A. Stephens, and J. J. Rossi, *Science*, 1990, **247**, 1222.

34 B. Dropulic, N. H. Lin, M. A. Martin, and K-T. Jeang, *J. Virol.*, 1992, **66**, 1432.

35 C. J. O'Kane and K. G. Moffat, *Current Opinion Genet. Dev.*, 1992, **2**, 602.

36 R. D. Palmiter, R. R. Behringer, C. J. Quaife, F. Maxwell, I. H. Maxwell, and R. L. Brinster, *Cell*, 1987, **50**, 435.

37 M. L. Breitman, S. Clapoff, J. Rossant, L-C. Tsui, L. M. Globe, I. H. Maxwell, and A. Bernstein, *Science*, 1987, **238**, 1563.

38 C. P. Landel, J. Zhao, D. Bok, and G. A. Evans, *Genes Dev.*, 1988, **2**, 1168.

39 M. L. Breitman, H. Rombola, I. H. Maxwell, G. K. Klintworth, and A. Bernstein, *Mol. Cell Biol.*, 1990, **10**, 474.

40 E. Borrelli, R. A. Heyman, C. Arias, P. E. Sawchenko, and R. M. Evans, *Nature (London)*, 1989, **339**, 538.

4.4 Biopharmaceuticals in Transgenic Livestock

Transgenic livestock hold the promise of being able to produce large quantities of important therapeutic proteins. The important distinction between 'pharming' and protein production by large scale mammalian or microbial cell culture is that proteins produced *in vivo* can be post-translationally modified in a manner identical to that of the native produce.

Early attempts to generate transgenic farm animals were hampered by low frequencies of transgene integration, low numbers of animals expressing the recombinant protein, and reproductive and physiological problems. Some of these problems can be by-passed by directing expression of the transgene to the mammary gland, an exocrine organ in which the expressed transgene remains separate from the animal's blood circulation. Typically, the transgene is fused to the regulatory sequences of a milk protein.[8,41,42] The yield of heterologous protein has been found to be extremely variable, but can account for a significant proportion of the total milk proteins. The best example, to date, is the transgenic sheep which expresses 35 g dm^{-3} (50% of the total milk protein) of glycosylated human α-1-antitrypsin under the direction of the ovine β-lactoglobulin promoter.[7] Taking into account the milk yield of goats, sheep, and cows, such levels of transgene expression would represent a substantial yield per year, compared to that from mammalian or microbial cell culture. The range of proteins which can be produced in milk may not be limitless, however, since high expression of certain proteins may adversely affect the physiology of the mammary gland.[41]

5 EMBRYO STEM CELL TECHNOLOGY

An alternative and more powerful strategy for transgenesis involves the introduction of foreign DNA into embryonic stem (ES) cells (Figure 4a.). Initially, cells are removed from the inner cell mass of the developing blastocyst, and are passaged on feeder layers, or in the presence of differentiation–inhibiting activity (DIA),[44] to maintain their undifferentiated state. Foreign DNA can be introduced into the ES cells by a number of means – electroporation, transfection, or micro-injection. The cells are then reintroduced into a blastocyst and are reimplanted into a pseudopregnant female and allowed to develop to term. An important distinction between pups obtained in this way, and those resulting from micro-injection of the one-cell embryo, is that they will, by definition, be chimaeras, since cells harbouring the transgene only constitute a proportion of the inner cell mass of the blastocyst. However, providing transgene-contain-

[41] A. L. Archibald, M. McClenaghan, V. Hornsey, J. P. Simons, and A. J. Clark, *Proc. Natl. Acad. Sci. USA*, 1990, **87**, 5178.

[42] A. J. Clark, H. Bessos, J. O. Bishop, P. Brown, S. Harris, R. Lathe, M. McClenaghan, C. Prowse, J. P. Simons, C. B. A. Whitelaw, and I. Wilmut, *Bio/Technology*, 1989, **7**, 487.

[43] R. J. Wall, V. G. Pursel, A. Shamay, R. A. McKnight, C. W. Pittius, and L. Henninghausen, *Proc. Natl. Acad. Sci. USA*, 1991, **88**, 1696.

[44] A. G. Smith, J. Nichols, M. Robertson, and P. D. Rathjen, *Dev. Biol.*, 1992, **151**, 339.

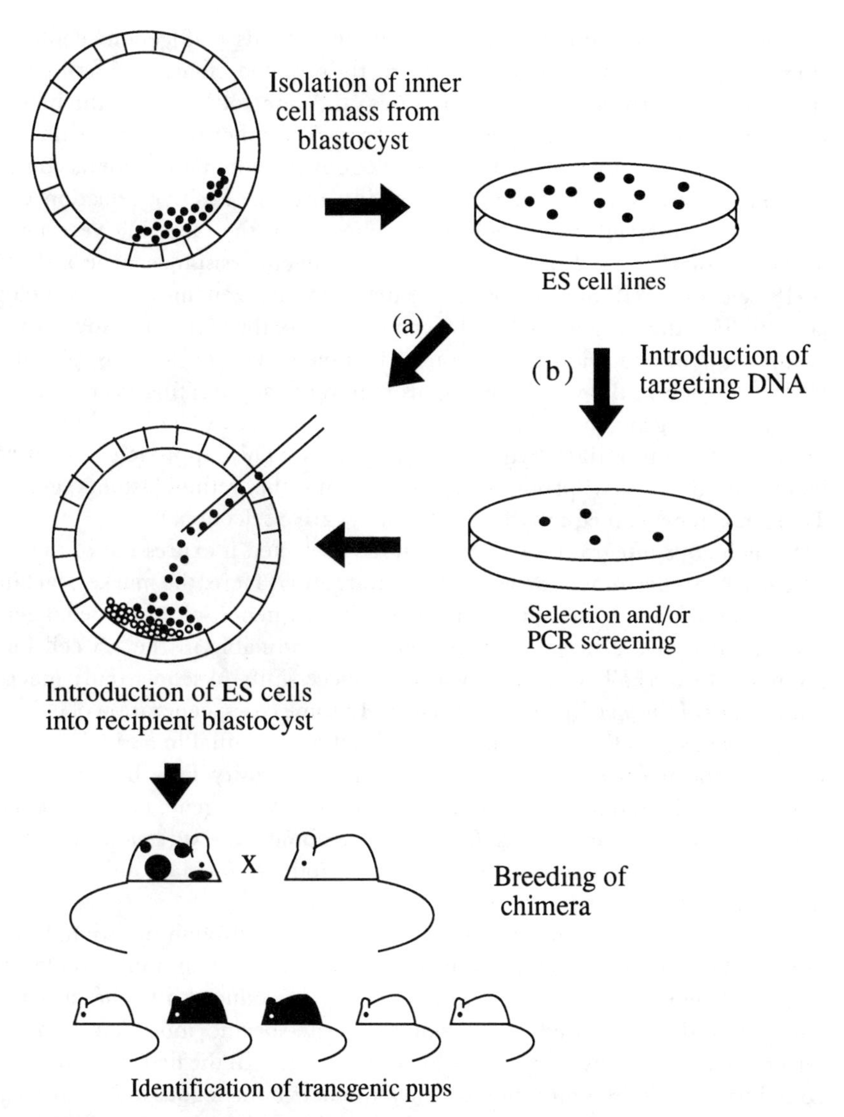

Figure 4 *(a) Generation of transgenic animals by means of ES cell manipulation. (b) Following introduction of DNA, ES cells carrying the homologous recombination event can be selected prior to introduction into the blastocyst*

ing cells have contributed to the germline, then a suitable breeding strategy will allow the establishment of a transgenic line.[45]

6 HOMOLOGOUS RECOMBINATION AND TRANSGENESIS

The introduction of foreign DNA into the ES cell has significant advantages over micro-injection of the one-cell embryo, by virtue of the fact that the foreign DNA

[45] J. Nichols, E. P. Evans, and A. G. Smith, *Development*, 1990, **110**, 1341.

can be targeted to homologously recombine with its endogenous counterpart (Figure 4b).[46] Following selection, correctly targeted clones can be identified, either by Southern blot[2] or PCR[1] analysis, and reintroduced into the blastocyst to generate the desired transgenic chimaeras. A number of strategies have been devised for the positive selection of homologous recombination events combined with negative selection against random integration. Positive selection can be achieved by interrupting homologous sequences in the targeting vector with a selectable marker, such as the bacterial neomycin-resistance gene *neo*^{r} (using G418 selection). If the vector integrates into the genome in a homologous fashion, then the *neo*^{r} gene will be incorporated into the germline. Any additional vector sequences outside the region of homology will be lost. By placing the HSV-1-*tk* gene in the non-homologous region of the targeting vector, any cells retaining this gene through random integration events, will be killed in the presence of appropriate synthetic nucleosides. This approach was used to inactivate the *Wnt*-1 proto-oncogene[47,48] and the retinoblastoma gene.[49–51] Toxigenic genes can equally be used as a negative selection.[52]

Homologous integration can be indirectly selected if expression of the selectable marker is dependent on correct gene targeting. Here, the marker, lacking its translational start, is fused, in frame, to coding sequences of the targeted gene. It is essential that the target gene be active, or inducible in the ES cell for this approach to work. The strategy was used successfully to sequentially inactivate both alleles of the *pim*-1 proto-oncogene.[53] In some cases, generation of chimaeras using homozygous ES cells lacking gene functions essential to early development may be informative where breeding to homozygosity fails because of early lethality. To investigate expression patterns of a target gene, one can replace it with a reporter gene such as *LacZ*.[54] Developmental expression can then be followed by X-gal staining for the gene product, β-galactosidase, which is potentially more sensitive than *in situ* hybridization.

All the above techniques generate null mutations through insertional inactivation of the target gene. By using the so-called hit and run or double replacement strategies, more subtle gene alterations can be introduced into the target gene. The hit and run procedure introduces a site-specific mutation into a non selectable gene by a two step recombination event.[55] In the first step, the vector – containing the desired mutation within sequences homologous to the target gene, together with selectable markers for monitoring the integration and reversion

[46] M. J. Evans, *Mol. Biol. Med.*, 1989, **6**, 557.
[47] K. R. Thomas and M. R. Capecchi, *Nature (London)*, 1990, **346**, 847.
[48] A. P. McMahon and A. Bradley, *Cell*, 1990, **62**, 1073.
[49] E. Y-H. P. Lee, C-Y. Chang, N. Hu, Y-C. J. Wang, C-C. Lai, K. Herrup, W-H. Lee, and A. Bradley, *Nature (London)*, 1992, **359**, 288.
[50] T. Jacks, A. Fazeli, E. M. Schmitt, R.T. Bronson, M. A. Goodell, and R. A. Weinberg, *Nature (London)*, 1992, **359**, 295.
[51] A. R. Clarke, E. R. Maandag, M. van Roon, N. M. T. van der Lugt, M. van der Valk, M. L. Hooper, A. Berns, and H. te Riele, *Nature (London)*, 1992, **359**, 328.
[52] T. Yagi, Y. Ikawa, K. Yoshida, Y. Shigetani, N. Takeda, I. Mabuchi, T. Yamamoto, and S. Aizawa, *Proc. Natl. Acad. Sci. USA*, 1990, **87**, 9918.
[53] H. Riele, E. R. Maandag, A. Clarke, M. Hooper, and A. Berns, *Nature (London)*, 1990, **348**, 649.
[54] H. L. Mouellic, Y. Lallemand, and P. Brulet, *Proc. Natl. Acad. Sci. USA*, 1990, **87**, 4712.
[55] P. Hasty, R. Ramirez-Solis, R. Krumlauf, and A. Bradley, *Nature (London)*, 1991, **350**, 243.

events – integrates into the target gene by single reciprocal recombination. The resultant duplication is resolved, by single intrachromosomal recombination, to yield clones restored to wild type, or carrying the desired mutation. The double replacement strategy[44] requires two targeting constructs. The first introduces a functional *hprt* minigene into the target gene,[56] whilst the second removes the *hprt* gene and replaces it with a subtly altered target gene. Selection both for and against the incorporation of *hprt* can be achieved by selection on HAT and 6-thioguanine, respectively.

Though gene targeting strategies are becoming increasingly elegant, the design of replacement vectors is somewhat empirical, since factors underlying the frequency of gene targeting are still poorly understood. Progress towards defining the extent of homology required between targeting vector and target locus, for high-fidelity recombination to occur, is now being made.[57,58] Additionally, recent data suggest that for optimal targeting efficiency the transgene should be isogenic, *i.e.* isolated from the same strain as that from which ES cells were derived.[58,59] It appears that heterology in nonisogenic DNA sequences significantly reduces recombination efficiency. One major limitation in the widespread application of homologous recombination, is that ES cells have, as yet, only been isolated from the mouse.

Interestingly, homologous recombination has recently been employed in murine zygotes, to reconstruct a large functional gene from micro-injected DNA fragments.[60] The size of DNA fragment micro-injected into the one-cell embryo has been a limiting factor in successful generation of transgenics because of potential problems with fragment preparation and handling. Pieper *et al.* injected three overlapping genomic DNA fragments, which together constituted the human serum albumin (hSA) gene. A significant proportion of resulting transgenic mice contained the correctly reconstituted hSA gene – the transgenic transcript and hSA protein being indistinguishable from the native product. The three fragments recombined to yield a 33 kb segment of DNA spanning the hSA structural gene. The authors report a high frequency of homologous recombination in murine zygotes, with transgenes up to 56 kb being reconstructed *in situ*. The limits of this co-injection procedure have yet to be determined, but the method should produce transgenic animals having genotypes which were previously unattainable.

7 FUTURE PROSPECTS

Transgenesis is having and will continue to have far-reaching effects on the fields of animal model production, gene therapy strategies, new therapeutic drug treatments, and the commercial production of biologically important molecules.

[56] R. Ratcliff, M. J. Evans, J. Doran, B. J. Wainwright, R. Williamson, and W. H. Colledge, *Transgen. Res.*, 1992, **1**, 177.
[57] K. R. Thomas, C. Deng, and M. A. Capecchi, *Mol. Cell Biol.*, 1992, **12**, 2919.
[58] C. Deng and M. A. Capecchi, *Mol. Cell Biol.*, 1992, **12**, 3365.
[59] H. te Riele, E. Robanus Maandag, and A. Berns, *Proc. Natl. Acad. Sci. USA*, 1992, **89**, 5128.
[60] F. R. Pieper, I. C. M. de Wit, A. C. J. Pronk, P. M. Kooiman, R. Strijker, P. J. A. Krimpenfort, J. H. Nuyens, and H. A. de Boer, *Nucleic Acids Res.*, 1992, **20**, 1259.

One should not ignore the emergence of plant transgenesis which will have an equally significant impact on large scale production of such products. With improvements in gene targeting frequency, and the extension of ES cell technology to include commercially important animals, the potential for subtle gene alteration promises to yield exciting future developments.

CHAPTER 14

Enzyme Engineering

J. M. WALKER

1 INTRODUCTION

Industrially used organic catalysts generally lack specificity for the reaction they are catalysing and often have to be used under extreme conditions of temperature and pressure. In comparison, enzymes show high specificity for the reactions that they catalyse and are able to function under extremely mild conditions and at very low concentrations. It is not surprising therefore that enzymes have found a number of applications as catalysts in industrial processes, particularly in the fine chemical, food, and pharmaceutical industries and in clinical analysis. However, although over 2000 enzymes have been described in the scientific literature, only a handful, mainly the thermostable proteases and glycosidases, have found large scale industrial applications. The reasons for the industrial under-use of enzymes is fairly clear. Enzyme structures have evolved in response to metabolic demands found *in vivo*, and are consequently well suited to their *in vivo* role. However, when the same enzymes are considered for use as industrial catalysts, they are invariably exposed to unnatural (non-physiological) environments (*e.g.* the presence of organic solvents, elevated temperatures, pH values outside their normal *in vivo* value, *etc.*) which can denature the enzyme with consequent loss of activity. One of the major goals of the enzyme technologist therefore is to enhance the stability of enzymes so that they may function effectively under non-physiological conditions. This will provide enzymes more suitable (*i.e.* with longer half-lives) for the conditions found in industrial processes and should also improve the yield of active enzyme following immobilization processes. (See Chapter 16 for a discussion of enzyme immobilization.) Long-term enzyme stability is often a more important characteristic of an industrially-used enzyme than enhanced catalytic activity. Both the chemical modification of proteins, and the manipulation of solvent media to enhance enzyme stability have found some success (see Chapter 15 and references 1–4), but the chemical modification of enzymes is harsh and rather non-specific. The recent introduction of the technique of site-directed

[1] K. E. Noet, A. Nanci, and D. E. Koshland, Jn., *J. Biol. Chem.*, 1968, **243**, 6392.
[2] D. Kowalski and M. Laskowski, *Biochemistry*, 1976, **15**, 1300 and 1309.
[3] A. Freeman, *Trends Biotechnol.* 1984, **2**, (6), 147.
[4] G. E. Means and R. E. Feeney, 'Chemical Modification of Proteins', Holden-Day, 1971.

mutagenesis has provided a far more subtle approach to modifying enzyme structures. This approach, described in detail below (Section 2), allows us to specifically alter a chosen base (or bases) in the gene for a given enzyme. This results in the replacement of a specific amino acid by another of one's own choice in the polypeptide chain of the enzyme. Recombinant DNA technology then allows us to produce large quantities of the purified, re-designed enzyme (see Chapter 2 for a discussion of the general procedures for cloning a gene). Using site-directed mutagenesis we are therefore in a position to alter, in a predictable manner, specific amino acids in an enzyme structure. In this way it should be possible to engineer enzymes with much improved stabilty characteristics by making appropriate amino acid replacements. Such changes could include the introduction of disulfide bridges, increasing the number of salt bridges in the molecule, or increasing the number of internal hydrogen bonds. These changes will be made based on our understanding of the factors responsible for enzyme stability and a detailed knowledge of the three-dimensional structure of the enzyme.

While enzyme engineering offers us the potential to increase enzyme stability, many other exciting potential applications are also apparent. These include:

1. The possibility of enhancing catalytic activity and increasing substrate affinity.
2. The modification of substrate specificity, thus constructing novel enzymes from pre-existing enzymes (*i.e.* creating enzymes not found in nature to catalyse new reactions).
3. Producing an enzyme with an altered pH-activity profile, thus allowing the enzyme to function at non-physiological pH values.
4. The design of enzymes stable to oxidizing agents by replacing easily oxidizable residues such as Cys, Trp, or Met by sterically similar non-oxidizable amino acids such as Ser, Phe, or Glu respectively (*e.g.* producing washing powder enzymes that will function in the presence of bleach).
5. Improving stability to heavy metals by replacing Cys and Met residues and by removing surface carboxyl-groups.
6. The design of enzymes resistant to proteolytic degradation. (Many reaction processes have proteases present, usually as contaminants, which can degrade the enzyme of interest.)
7. Producing enzymes that are stable and active in non-aqueous solvents.
8. The elimination of allosteric sites involved in feedback inhibition. (The activities of some enzymes are inhibited in the presence of excess product by the product binding to an allosteric site on the enzyme which causes conformational changes resulting in an inactive enzyme. Such effects are undesirable if one is aiming to produce a high yield of product.)
9. The fusion of enzymes involved in a particular reaction pathway, so that a multienzyme process might be carried out using one protein.

By introducing only some of the above improvements, it should be possible to transform many traditional high pressure and high temperature industrial processes into enzymatic processes that can be operated with a low energy consumption. Mutagenesis to produce enzymes with altered activity is of course not new. As discussed in Chapter 1, the search for more suitable enzymes for industrial use has, in the past, included extensive searches for improved enzymes by mutation and selection programmes to enhance the properties of the wild-type enzyme. The random mutation of micro-organisms to produce strains with enhanced characteristics is a well tried technique, particularly within the pharmaceutical industry. A classic example of this general approach is the bacterial production of penicillin which has been improved about 10 000-fold over the last 40 years. There are many cases where the activities of bacterial enzymes have been directed toward novel substrates *in vivo* by selecting spontaneous mutants of the enzyme which allows the organism to grow on the new substrate.[5–8] Enzymes lacking allosteric inhibition have also been isolated using this approach.[9] However, these conventional mutagenesis techniques are generally limited to producing very minor (usually single amino acid) changes in the enzyme structure. Should it prove necessary to change several specific amino acids throughout the protein chain to make a particular improvement, one is unlikely to detect such an event in a mutant population because of the extremely small probability of its occurrence. The introduction of such multiple alterations *is* possible, however, using site-directed mutagenesis. Also screening for random mutations is a highly time-consuming procedure which lacks the subtlety and design aspect provided by site-directed mutagenesis. However, such mutation studies have shown us that enzyme properties *can* indeed be improved, and it is anticipated that enzyme engineering by site-directed mutagenesis will build on this knowledge to provide even greater improvements. The ultimate goal of enzyme engineering must be the design and construction of enzymes to order such that they will catalyse *any* desired reaction, if necessary even under the most extreme conditions.

Although the potential for enzyme engineering is enormous, at present our ability to apply the technique to its full potential is limited by a number of factors. These include the following:

1. A lack of *X*-ray crystallographic data for many enzymes means that the three-dimensional structure of the majority of enzymes is not yet known. Such structural data are essential for each enzyme to be modified, since they provide the structural basis upon which appropriate changes in enzyme structure are decided. For those enzymes where the three-dimensional structures are presently available, the application of computer graphics, which depict the translation and rotation of an enzyme and its substrate, in real-time, is providing a powerful tool in helping to design appropriate changes to enzyme structure. The co-ordinate sets for a wide variety of macromolecules are provided by the Brookhaven Databank of Protein

[5] B. G. Hall, *Biochemistry*, 1981, **20**, 4042.
[6] C. Turberville and P. H. Clarke, *FEMS Microbiol. Lett.*, 1981, **10**, 87.
[7] A. Hall and J. R. Knowles, *Nature (London)*, 1976, **264**, 803.
[8] C. Scazzocchio and H. M. Sealy-Lewis, *Eur. J. Biochem.*, 1978, **91**, 99.
[9] M. Pabst, J. Kuhn, and R. Sommerville, *J. Biol. Chem.*, 1973, **248**, 901.

Structures and the Cambridge Structural Database.[10] Recent developments in Laue *X*-ray methodology, and the application of 2-D NMR to the determination of protein structure suggests that the rate at which three-dimensional structures are produced will increase in future. However, at present this general lack of detail on the three-dimensional structures of enzymes is likely to prove a rate-limiting step in the progress of enzyme engineering.

2. We lack the knowledge of the exact interactions (hydrogen bonding, electrostatic interactions, and hydrophobic interactions) made between an enzyme and its substrate at the active site. Without this information it is not possible to predict the effect of altering specific amino acids at the active site. Many of the initial experiments in enzyme engineering therefore involved investigating the effect of removing residues suspected of being involved with substrate binding and observing the consequent effect on substrate binding (see Section 3).
3. We lack knowledge of the precise factors involved in conferring stability to proteins. However, it is clear that salt bridges and other electrostatic interactions confer thermostability, as do amino acid changes that stabilize secondary structures and interactions between secondary structures. Again, many initial experiments in enzyme engineering have involved the investigation of the effect on stability of substituting residues thought to be involved in enzyme stability (see Section 3).

A number of exciting examples of enzyme engineering have appeared in the literature in recent years. Given the above limitations, it is not surprising that initial studies were made on enzymes with a known or suspected relationship between structure and function. Some examples of enzyme engineering are described in Section 3 and provide a good indication of the potential of this technique. Although this article is restricted to a description of *enzyme* engineering, it should be stressed that the basic approach is equally applicable to the improvement of the stability or function of *any* protein, *e.g.* clinically useful compounds such as interferon or growth hormone, antibodies, veterinary and agricultural products, *etc.* References to proteins that have been modified by site-directed mutagenesis are given at the end of Section 3.

2 SITE-DIRECTED MUTAGENESIS

The term 'site-directed mutagenesis' refers to any technique that allows one to specifically (site-directed) change (mutate) a base in a length of DNA. There are in fact a number of different technical approaches to selectively mutating DNA, and these have been well reviewed in references 11–14. Three commonly used

[10] F. H. Allen and M. F. Lynch, *Chem. Brit.*, 1989, **25**, 1101–1108.
[11] M. Smith, *Ann. Rev. Genet.*, 1985, **19**, 423–462.
[12] D. Shortle and D. Botstein, *Science*, 1986, **229**, 1193–1201.
[13] J. G. Williams and R. K. Patient, 'Genetic Engineering', IRL Press, Oxford, 1988.
[14] J. C. Murrell and L. M. Roberts, 'Understanding Genetic Engineering', Ellis Horwood, Chichester, 1989.

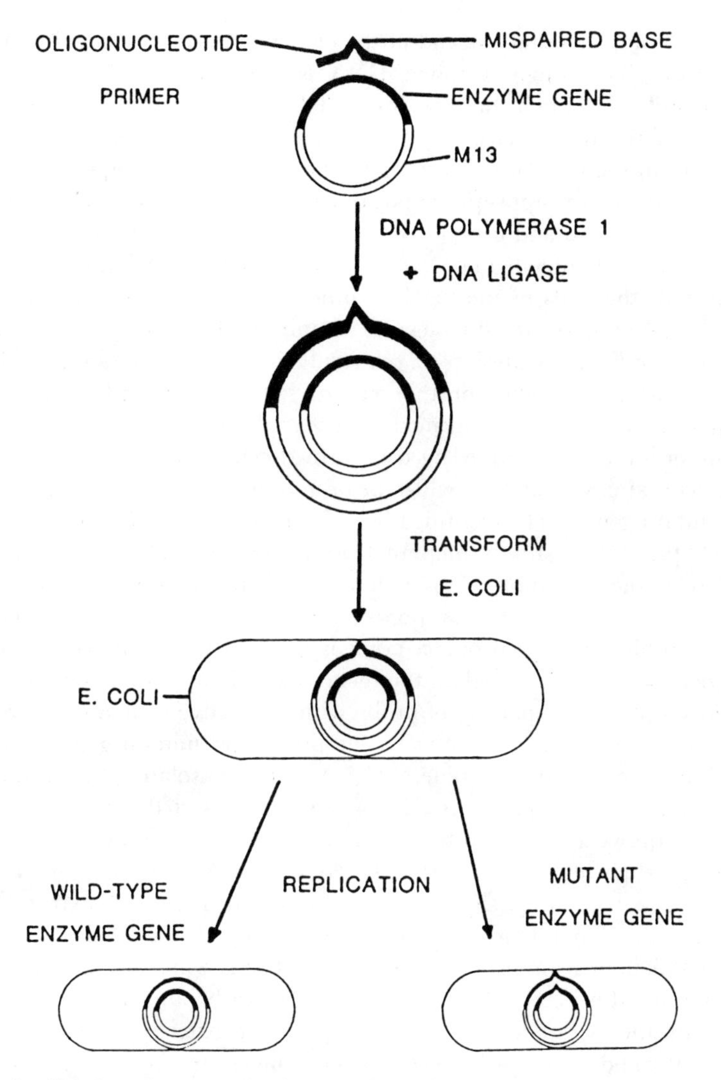

Figure 1 *Site-directed mutagenesis using a synthetic oligonucleotide (see text for details)*

methods will be described briefly here. It should be stressed that all the site-directed mutagenesis techniques require the cloning of the gene for the enzyme under study and its incorporation into a suitable carrier such as a bacteriophage vector. This, however, should not be a problem since genetic engineering technology has now reached the point where we can now clone the gene for essentially any protein found in nature.

The technique of oligonucleotide-directed mutagenesis requires the synthesis of a short (15–20) oligonucleotide that is complementary (*i.e.* base pairs) to the gene around the site to be mutated, but which contains a mismatch at the base that we wish to mutate.

The basic procedure is shown in Figure 1. We start with a single-stranded clone of the complementary strand of the enzyme molecule, carried in an M13 phage vector. This single-stranded DNA is then mixed with the synthetic oligonucleotide. Although there is a mismatch, as long as this mismatch is near the centre of the oligonucleotide, and as long as the mixing is done at low temperature in the presence of high salt concentration, the oligonucleotide will hybridize (bind) to the appropriate position on the enzyme gene. DNA polymerase is then introduced and uses the oligonucleotide as a primer to synthesize the remainder of the complementary strand of the DNA. DNA ligase is also introduced to join the ends of the newly synthesized DNA to the oligonucleotide primer. This double-stranded molecule, containing the mismatch, is then introduced into *E. coli*. Replication of *E. coli* results in bacteria containing either the original wild-type sequence or the mutant sequence. Transformed cells are differentiated from non-transformed cells by plating out and observing the formation of inhibition plaques by transformed cells. These plaques are caused by phage containing either the wild-type or mutant gene. Those phages containing the mutant gene can be identified by hybridizing their DNA (on a nitrocellulose blot of the plate), with the oligonucleotide (now radiolabelled) that was used to introduce the mutation. Although at room temperature this probe will base-pair with both the native and mutant gene, by increasing the temperature at which probing is carried out, a point is reached where the complete base-pairing between the probe and mutant gene results in hybridization, but the less stable, incomplete base-pairing of probe with the native gene does *not* result in hybridization. Hybridization between the probe and mutant gene is therefore detected by autoradiography. The M13 DNA is then isolated, the mutant gene excised using an appropriate restriction enzyme, inserted into an expression vector, and the vector used to transform an appropriate organism (*e.g.* yeast). The gene product (enzyme) can then be obtained either by rupturing the cells, or directly from the culture supernatant if the enzyme is secreted extracellularly.

The method of oligonucleotide directed mutagenesis can also be used to introduce deletions or insertions of sequences. For example, an oligonucleotide containing an additional codon in its sequence can be annealed to the gene of interest with the three additional bases looping out. After replication the newly synthesized strand contains the extra codon. Conversely, a primer, *e.g.* 30mer, is synthesized where the first 15 residues are complementary to one region of the gene, and the other 15 to a region some bases beyond this region. When this primer is annealed to the gene, the intervening section loops out, and after replication the synthetic strand lacks this looped-out section.

Another technique for site-directed mutagenesis is the method of 'cassette mutagenesis'. This method is used if the segment of gene to be mutated lies between two close-spaced, unique, restriction enzyme cleavage sites. The intervening sequence is excised and replaced by a chemically synthesized oligonucleotide containing the required mutation.[15] Oligonucleotides containing a mixture of substitutions at a particular site can be used in this method to generate a large family of mutant enzymes with different replacements at a particular codon (see,

[15] J. A. Wells, M. Vasser, and D. B. Powers, *Gene*, 1985, **34**, 315–323.

for example, reference 38). Methods have also been developed that allow the generation of random mutant libraries, containing all-possible single base substitution mutations within a complete gene.[16,17] In one such method, the single-stranded gene to be mutated is incubated with an oligonucleotide primer, DNA polymerase, three deoxynucleotides in excess, and one at limiting concentrations. Elongation stops at different points along the sequence where nucleotide at limiting concentration is then required. Misincorporation to the 3′ ends is then introduced using reverse transcriptase and the three wrong nucleotides. Finally, all four nucleotides, a polymerase, and ligase are added to give covalently closed double-stranded DNA. Site-directed mutagenesis is now a well-tried and proven technique. However, each experiment introduces its own specific problems and a number of modifications to these basic approaches exist. For further details the interested reader is referred to the methods sections of the papers cited below.

3 SPECIFIC EXAMPLES OF ENZYME ENGINEERING

3.1 Tyrosyl tRNA Synthetase

In the cell, synthesis of proteins occurs on cytoplasmic structures known as ribosomes. When a specific amino acid is required for incorporation into the growing polypeptide chain it is brought to the ribosome by a specific transfer RNA molecule (tRNA) to which it is covalently attached. This specific amino acid–tRNA attachment has been earlier achieved by a specific synthetase enzyme. (The process of attachment is often referred to as the 'charging' of tRNA.) The enzyme tyrosyl tRNA synthetase catalyses the reaction between the tRNA for tyrosine ($tRNA^{tyr}$) and the amino acid tyrosine, in a two-stage reaction. In the first step tyrosine is activated by ATP to give enzyme-bound tyrosyl adenylate. In the second step this complex is attached by the tRNA to give the final product tyrosine–tRNA.

$$\text{Tyrosine} + \text{ATP} \rightarrow \text{tyrosyl adenylate} + \text{pyrophosphate} \quad (1)$$

$$\text{Tyrosyl adenylate} + \text{tRNA}^{\text{tyr}} \rightarrow \text{tyrosine–tRNA} + \text{AMP} \quad (2)$$

The enzyme tyrosyl tRNA synthetase from *B. stearothermophilus* has been purified, crystallized, its three-dimensional structure determined by *X*-ray crystallography, and the active site identified.[18,19] With the aid of computer graphics, the possible effects of changing amino acid side-chain contacts between the enzyme and substrate, or of distorting the polypeptide chain at the active site have been predicted, and then these theoretical predictions have been tested in practice using site-directed mutagenesis. This enzyme probably represents the most

[16] D. W. Leung, E. Chen, and D. V. Goeddel, *Technique*, 1989, **1**, 11–15.
[17] P. M. Lehtovaara, A. K. Koivula, J. Banford, and J. K. C. Knowles, *Protein Eng.*, 1988, **2**, 63–68.
[18] T. Bhat, D. Blow, P. Brick, and J. Nyborg, *J. Molec. Biol.*, 1982, **158**, 699.
[19] J. Rubin and D. Blow, *J. Molec. Biol.*, 1981, **145**, 489.

detailed study to date of enzyme engineering of a given enzyme.[20–24] The following summarizes some of the information that has been obtained by site-directed mutagenesis of this enzyme.

(1) *X*-Ray studies showed that the enzyme residues cysteine-35, histidine-48, and threonine-51 all appear to form hydrogen bonds at the active site with the ribose moiety of tyrosine adenylate. To confirm the existence of these contacts, and to determine their relative importance, a series of mutants were constructed in which histidine-48 was altered to glycine, threonine-51 to alanine, and cysteine-35 to glycine respectively. Each of these changes eliminates the particular hydrogen bond involved. From kinetic data on each of these mutants it was deduced that the imidazole side chain of histidine-48, and the sulfydryl side-chain of cysteine-35 were indeed involved in binding to the substrate, each contributing about 1 Kcal mol^{-1} to the stability of the transition sate in tyrosine activation. (For example, the replacement of cysteine-35 with glycine reduced the catalytic activity of the enzyme by about 70%, mainly by lowering the strength of tyrosyl adenylate binding.) Surprisingly, removal of the hydrogen bonding by threonine-51 actually led to an *increase* in the binding of substrate by about a factor of two.[22] The possible reasons for this increased binding is discussed below. However, the above observations confirm the feasibility of using site-directed mutagenesis to examine predictions made from structural studies on the active site of an enzyme.

(2) Comparison of the structure of tyrosyl tRNA synthetase from *B. stearothermophilus* and *E.coli* showed that threonine-51 was replaced by proline in the *E. coli* enzyme. This was surprising since the presence of proline should disrupt the α-helical structure of the polypeptide backbone found in this region. A mutant of the *B. stearothermophilus* enzyme was therefore constructed with proline at residue 51 (mutant TP51).[22] Although this would indeed have disrupted the backbone of the enzyme, the binding of the enzyme to ATP was surprisingly found to have increased by a factor of 100. In an attempt to explain this observation, the crystal structure of TP51 was determined,[24] and revealed that the mutation Thr 51 → Pro 51 causes only a local effect on the structure of the enzyme. The authors conclude that the increased activity of the TP51 mutant probably resulted from the replacement of the polar threonine residue by a non-polar group. In the wild-type (Thr 51) enzyme substrate binding is disfavoured by the displacement of solvent from the vicinity of Thr 51. This unfavourable effect is absent in the TP51 mutant. The importance of this observation is that it shows that enzyme affinities *can* be improved by *in vitro* manipulation, an observation which has important consequences for the commercial applica-

[20] A. J. Wilkinson, A. R. Fersht, D. M. Blow, and G. Winter, *Biochemistry*, 1983, **22**, 3581.
[21] G. Winter, A. R. Fersht, A. Wilkinson, M. Zoller, and M. Smith, *Nature (London)*, 1982, **299**, 756.
[22] A. J. Wilkinson, A. R. Fersht, D. M. Blow, P. Carter, and G. Winter, *Nature (London)*, 1984, **307**, 187.
[23] G. Winter and A. R. Fersht, *Trends Biotechnol.*, 1984, **2**, (5), 115.
[24] K. A. Brown, P. Brick, and D. M. Blow, *Nature (London)*, 1987, **326**, 416–418.

tions of enzyme engineering. In addition, the studies on TP51 have shown that allowance must be made for changes in surrounding solvent as well as changes in main-chain structure and side-chain character when interpreting changes observed as a result of protein engineering.

(3) In the tyrosine binding-site of the enzyme, the phenolic hydroxyl of the tyrosine substrate appears to hydrogen bond to the side-chains of tyrosine-34 and aspartic acid-176 in the active site 'pocket'. The enzyme discriminates phenylalanine, which also has an aromatic ring side-chain, from tyrosine by a factor of 10 000 in the charging of $tRNA^{tyr}$. The phenolic hydroxy of tyrosine-34 seems to be a major determinant in discriminating against the binding of phenylalanine at the active site. However, when tyrosine-34 was replaced by phenylalanine, relatively little effect was observed on the binding of tyrosine but a fifteen-fold decrease in the discrimination against phenylalanine was observed.[23] Although this is only a relatively small change, it does show that site-directed mutagenesis offers the potential to alter the *specificity* of an enzyme.

3.2 β-Lactamase

β-Lactamase is an enzyme produced by certain antibiotic resistant bacteria. The enzyme catalyses the hydrolysis of the amide bond of the lactam ring of penicillins and related antibiotics (*e.g.* cephalosporins) and thus confers resistance to these antibiotics in the bacterium. The catalytic pathway includes an acyl intermediate and residues serine-70 and threonine-71 have been implicated as being important residues at the active site of the enzyme. It has been suggested that the hydroxyl group of serine-70 adds nucleophilically to the carbonyl group of the β-lactam ring, in a mechanism somewhat analogous to that of serine proteases. The role of threonine-71 is less clear, but it seems essential for catalytic activity because a mutant (not genetically engineered), probably with isoleucine at this position, shows no catalytic activity. The role of this region of the molecule in the catalytic activity of the enzyme has been investigated by site-directed mutagenesis.[25–27]

(1) The conversion of serine-70 to threonine gave a product with no β-lactamase activity. This was not surprising since the conversion has the effect of adding a methyl group to the serine side-chain, thereby hindering access to the hydroxyl group. However, the role of serine-70 in the catalytic activity of the enzyme has been confirmed by this experiment.

(2) The serine-70 residue in the wild-type enzyme has been replaced by a cysteine residue.[25,26] This effectively replaces one nucleophilic group (–OH) by another somewhat bulkier one (–SH) producing a thiol β-lactamase. This thiol enzyme has a binding constant for penicillin about the same as that for the wild-type enzyme, but the rate of hydrolysis is only

[25] I. S. Sigal *et al.*, *J. Biol. Chem.*, 1984, **259**, 5327.
[26] I. S. Sigal, B. Harwood, and R. Arentzen, *Proc. Natl. Acad. Sci. USA*, 1982, **79**, 7157.
[27] S. C. Schultz and J. H. Richards, *Proc. Natl. Acad. Sci. USA*, 1986, **83**, 1588–1592.

about 1–2% of that of the wild-type enzyme. However, against certain cephalosporin antibiotics (which also contain β-lactam rings) the binding constant is more than ten-fold greater than that of the wild-type enzyme and the rate of hydrolysis at least as great as that of the wild-type enzyme. The mutation therefore has essentially changed the specificity of the enzyme. Additionally, the mutant enzyme is three times more resistant to trypsin digestion than the wild-type enzyme. This difference has been related to increased thermal stability of the mutant enzyme, one of the major goals of enzyme engineering.

(3) The Thr residue at position 71 is conserved in all Class A β-lactamases, suggesting an essential role for this residue. To investigate the role of Thr 71, site-directed mutagenesis has been used to construct mutants containing all 19 possible other amino acids at positions 71.[27] Surprisingly, cells producing 14 of the mutant β-lactamases still displayed appreciable resistance to ampicillin (*i.e.* the mutant β-lactamase is still active). Only cells with mutants having Tyr, Trp, Asp, Lys, or Arg at residue 71 had no observable resistance to ampicillin. However, all active mutants were less stable to cellular proteases than the wild-type enzyme. These results suggest that Thr-71 is not essential for binding or catalysis but is important for the stability of the β-lactamase enzyme. An apparent change in specificity of the active mutants suggests that residue 71 influences the region of the protein that accommodates the side-chain attached to the β-lactam ring of the substrate.

3.3 Dihydrofolate Reductase

Dihydrofolate reductase (DHFR) catalyses the reduction of 7,8-dihydrofolate to 5,6,7,8-tetrahydrofolate, which in turn has a major metabolic role as a carrier of one-carbon units in the biosynthesis of purines, thymidylate, and some amino acids. The inhibition of DHFR by synthetic folate analogues (antifolates) results in the depletion of the cellular tetrahydrofolate pool, with consequent cessation of DNA synthesis leading to stasis and cell death. Such an approach is especially useful in destroying dividing cells (such as tumour cells) which are active in DNA synthesis. Antifolates, especially methotrexate, have therefore found particular use in the chemotherapy of cancer. It is not surprising therefore that DHFR is an enzyme of considerable interest to pharmaceutical chemists and drug designers. The three-dimensional structure of DHFR has been determined[28] and site-directed mutagenesis has been used to answer a number of questions concerning the struture and function of DHFR.[29–31]

(1) The crystal structure suggests that aspartic acid-27, buried below the enzyme surface, may form a hydrogen-bonded salt linkage with the pteridine ring of the substrate. It would appear to ultimately act as a proton

[28] D. Filman, J. Bolin, D. Matthews, and J. Kraut, *J. Biol. Chem.*, 1982, **257**, 13663.
[29] J. Villafranca *et al.*, *Science*, 1983, **222**, 782.
[30] E. E. Howell, J. E. Villafranca, M. S. Warren, S. J. Oatley, and J. Kraut, *Science*, 1986, **231**, 1123.
[31] J. E. Villafranca *et al.*, in 'Design, Construction and Properties of Novel Protein Molecules', ed. D. M. Blow, A. R. Fersht, and G. Winter, Royal Society, London, 1986, p. 113–122.

donor and to stabilize this transition state by providing a negatively charged carboxylate counter ion to hydrogen bond with the resulting positively charged pteridine ring. To test this, aspartic acid-27 was replaced by both asparagine or serine, both changes leaving the geometry of the site unchanged, but eliminating hydrogen bonding and formation of a cation.[30,31] The mutant enzymes had only about 0.1% of the activity of the wild-type enzyme, supporting the postulated role of the aspartic acid-27 side-chain in catalysis. Both these mutations apparently force the enzyme to utilize protonated DHF directly as a substrate.

(2) The crystal structure suggests that glutamic acid-139 may bridge with histidine-14, which may stabilize the β-sheet structure in that region. Glutamic acid-139 was therefore replaced with lysine which would not form a salt bridge. This alteration did not affect the catalytic activity of the enzyme, but introduced a significant decrease in stability of the enzyme. This is an important observation since it shows that structural stability can be uncoupled from enzyme activity.

(3) DHFR contains two cysteine residues but these are not involved in the formation of a disulfide bridge. The crystal structure suggested that if proline-39 was replaced by a cysteine residue it could form a disulfide bridge with cysteine-85 which might improve the stability of the enzyme. When this change was made, the reduced form of the enzyme (no disulfide bridges) showed normal activity. However, when oxidized to form a disulfide, enzymic activity was significantly reduced, presumably due to a loss of dynamic flexibility in the molecule. Obviously in this particular experiment the attempt to improve the stability of the enzyme was detrimental to enzyme activity.

(4) All the DHFR enzymes studied to date (from a variety of sources) contain a glycine–glycine dyad at positions 95 and 96, the glycine residues being linked by an unusual *cis*-peptide bond. It is thought that this particular topography may play a role in the working of the molecule, possibly as a conformational switch of some kind. A minimal change which should nevertheless alter the geometry in this part of the molecule would be the replacement of glycine-95 with alanine. When this mutation was made the enzyme was found to be completely inactive, and the mutant enzyme also had a lower mobility on non-denaturing gels, suggesting that a change in the conformation of the enzyme had occurred. The essential nature of the glycine–glycine dyad for the functioning of DHFR has therefore been confirmed by this experiment.

3.4 Subtilisin

One of the main sources of protein instability is susceptibility to oxidation with subsequent inactivation or denaturation. This is especially true for proteins containing methionine, cysteine, or tryptophan residues in or around the active site. Oxidative inactivation is therefore a major problem in the industrial application of some enzymes.

(1) Subtilisin (from *Bacillus* spp.) is a major industrial enzyme owing to its use in enzyme detergents. However, methionine-222 in subtilisin, which is invariant in subtilisin from all *Bacillus* spp. lies adjacent to the active site serine residue which is necessary for catalysis, and oxidation of this Met residue can lead to inactivation of the enzyme by preventing substrate access to the active site. This therefore presently precludes the use of bleach with enzyme detergents, and even inactivation due to oxidation by molecular oxygen in solution can be a problem. However, removal of the Met residue should reduce the susceptibility of the enzyme to oxidation. Estelle and co-workers have therefore prepared subtilisin molecules containing all 19 possible substitutions at position 222.[32] This specific activity of all mutants was determined relative to the wild-type Met enzyme (wild-type specific activity = 100%). The Cys mutant in fact showed increased specific activity (138%), but like the wild-type enzyme, was inactivated in the presence of H_2O_2, although at a somewhat lower rate. Although all other mutants showed some activity, the Ala and Ser mutants showed the greatest specific activities (53% and 35% respectively) and were also stable to extensive exposure to H_2O_2. This study therefore shows that it is possible to make oxidatively stable amino acid substitutions which maintain enzyme function. The production of industrially useful enzymes that are resistant to oxidative inactivation is therefore a distinct possibility.

(2) Two disulfide bonds have been introduced into subtilisin molecule, as a possible means of increasing thermal stability.[33] Although the mutant enzyme was efficiently secreted from *Bacillus amyloliquefaciens*, and the required disulfide bonds were produced quantitatively *in vivo*, autolysis of these mutants (taken as a measure of thermal stability) occurred at the same rate as in the native enzyme.

(3) Subtilisin has a large hydrophobic binding cleft, and the side-chain of residue 166 (glycine in wild-type enzymes) is found in this cleft. This glycine residue has been replaced by all possible 19 substitutions, and substrate specificity of the mutant enzymes examined.[34] A range of different substrate specificities was identified for the mutant enzymes. For example, the Lys-166 mutant had a k_{cat}/K_m 800 times greater than that of the wild-type enzyme when the substrate introduced glutamic acid into the binding cleft. The changes in substrate specificity shown by the mutant enzymes were attributed to a combination of steric hindrance and enhanced hydrophobic interactions between enzyme and substrate within the cleft. In general, as the volume of the amino acid in position 166 increases, the volume of the substrate amino acid decreases, and *vice versa*. Also, if the hydrophobic amino acid isoleucine is in position 166 the enzyme increases in catalytic efficiency as the hydrophobicity of the substrate is increased, as long as steric factors are not a problem. This study confirms the possibility of making changes in the specificity of an enzyme by site-directed mutagenesis.

[32] D. A. Estell, T. P. Graycart, and J. A. Wells *J. Biol. Chem.*, 1985, **260**, 6518–6521.
[33] J. A. Wells and D. B. Powers, *J. Biol. Chem.*, 1986, **261**, 6564–6570.
[34] D. A. Estell *et al.*, *Science*, 1986, **233**, 659.

(4) Electrostatic effects are important in enzyme catalysis and play an important part in stabilizing charged transition states. The chemical modification of serine proteases has shown that the pH dependence of catalysis alters with changes in overall surface charge. This suggests a possible general method of modifying pH dependence by altering the electrostatic environment of the active site by enzyme engineering. Such studies have been carried out on subtilisin where the change of just one surface charge has been shown to have a significant effect on the pH dependence of the enzyme.[35] Asp-99, which is 14–15 Å from the active site, was replaced by an uncharged serine residue. This change was shown to affect both the ionization constant of the active site (the pK_a of His-64 was reduced by 0.29 units) and the charge transition state of the enzyme (k_{cat}/K_m decreased by 20%). This result shows that the modification of a single charge in the vicinity of the active site can have a significant effect in the pH dependence of the catalytic reaction. It should of course be possible to induce larger changes by modifying groups that are closer to either charged catalytic residues or charges that develop in transition-state strutures. Such electrostatic effects could also be enhanced by engineering multiple-charge changes. The introduction of large effects on the pH dependence and catalytic rate constants by engineering surface-charge changes therefore seems possible.

(5) Enzymes have much greater potential uses in chemical synthesis if they can function in polar organic solvents where organic substrates are far more soluble. Unfortunately enzyme activity is often dramatically reduced in polar organic solvents even under conditions where the folded structure is stable. Random mutagenesis of subtilisin has produced an enzyme mutant that has enhanced activity in the presence of dimethyl formamide (DMF).[36] Two separate mutations, one located near the substrate binding pocket and one in the active site both improve the catalytic efficiency (k_{cat}/K_m) for the hydrolysis of a peptide substrate. When both mutations are introduced into the enzyme together the effects are additive. The inclusion of a third mutation into the enzyme results in an enzyme that is 38 times more active than wild-type subtilisin in 85% DMF.

The above examples have been presented to indicate the scope offered by site-directed mutagenesis both to examine the relationship between enzyme structure and function, and to improve this relationship as well as improving enzyme stability. The technique of enzyme engineering is an expanding field and one can look forward to an increasing number of publications in this area over the next few years. Indeed, the above examples only represent a small part of the papers of enzyme and protein engineering published at the time of writing. Lack of space precludes the description of further examples. However, the following list is a collection of some further enzymes and proteins that have been investigated by site-directed mutagenesis and directs the interested reader to appropriate references.

[35] P. G. Thomas, A. J. Russell, and A. R. Fersht, *Nature (London)*, 1985, **318**, 375–376.
[36] K. Chen and F. H. Arnold, *Bio/Technology*, 1991, **9**, 1073–1077.

a_1-*Antitrypsin*. Mutants with therapeutic potential have been constructed as well as mutants with specificity towards different proteolytic enzymes.[37–40]

Insulin. Mutants of the interstitial C-peptide have been constructed to investigate the roles of these regions.[41]

Interleukin-2. The role of the cysteine residues in the functioning of the protein have been investigated by site-directed mutagenesis.[42]

Carboxypeptidase A. The role of active-site residues has been investigated.[43,44]

Trypsin. Active-site residues have been investigated, and substrate specificity has been modified.[43,45]

Lysozyme. A disulfide bridge has been introduced into the enzyme, producing a thermally more stable enzyme with unaltered catalytic activity.[46–48] Studies have also been made on the thermodynamic stability of the enzyme.[49,50]

Triosephosphate isomerase. Changes in free-energy profiles have been determined following changes to residues involved in the catalytic activity of the enzyme.[51,52]

Dihydrolipoamide acetyl transferase (part of pyruvate dehydrogenase multi-enzyme complex). The effect of deletions and mutations on the assembly, catalytic activity, and active site coupling in the complex have been investigated.[52,53]

Alcohol dehydrogenase. Mutants that result in enlarging the active site have a marked increase in reaction velocity towards a range of higher alcohols.[54]

Phosphofructokinase. A single mutation has been shown to lead to phosphoenol-pyruvate being an activator rather than an allosteric inhibitor as it is for the native enzyme. The mutant enzyme is more than 100 times more active than the wild-type enzyme.[55] The role of active-site residues have also been investigated.[56]

Lactate dehydrogenase. The role of a mobile arginine residue has been identified by

37 M. Courtney, S. Jallat, L-H. Tessier, A. Benavente, R. G. Crystal, and J.-P. Lecocq, *Nature (London)*, 1985, **313**, 149.

38 S. Rosenberg, P. J. Barr, R. C. Najarian, and R. A. Hallewell, *Nature (London)*, 1984, **312**, 77.

39 M. Courtney *et al.*, in 'Design, Construction, and Properties of Novel Protein Molecules', ed. D. M. Blow, A. R. Fersht, and G. Winter, Royal Society, 1986, p. 89–98.

40 S. Jallat *et al.*, *Protein Eng.*, 1986, **1**, (1), p. 29–36.

41 R. Wetzel *et al.*, *Gene*, 1981, **16**, 63.

42 A. Wang *et al.*, *Science*, 1984, 29 June, 1431.

43 E. T. Kaiser, *Nature (London)*, 1985, **313**, 630.

44 S. J. Gardell, C. S. Craik, D. Hilvert, M. S. Urdea, and W. J. Rutter, *Nature (London)*, 1985, **317**, 551–555.

45 C. S. Craik, C. Largman, T. Fletcher, S. Roczniak, P. J. Barr, R. Fletterick, and W. J. Rutter, *Science*, 1985, **228**, 291–297.

46 L. J. Perry and R. Wetzel, *Science*, 1984, **226**, 555.

47 R. Wetzel, *Protein Eng.*, 1986, **1**, (1), 257.

48 L. J. Perry and R. Wetzel, *Biochemistry*, 1986, **25**, 733–739.

49 T. Albers, K. Wilson, J. A. Wozniak, S. P. Cook, and B. W. Matthews, *Nature (London)*, 1987, **330**, 41–46.

50 M. Matsumura, W. J. Becktel, and B. W. Matthews, *Nature (London)*, 1988, **334**, 406–410.

51 R. T. Raines *et al.*, in 'Design, Construction, and Properties of Novel Protein Molecules', ed. D. M. Blow, A. R. Fersht, and G. Winter, Royal Society, London, 1986, p. 79–89.

52 D. Straus, R. Raines, E. Kawashima, J. R. Knowles, and W. Gilbert, *Proc. Natl. Acad. Sci. USA*, 1985, **82**, 2272–2276.

53 L. D. Graham *et al.*, in 'Design, Construction, and Properties of Novel Protein Molecules', ed. D. M. Blow, A. R. Fersht, and G. Winter, Royal Society, London, 1986, p. 99–112.

54 C. Murali and E. Creaser, *Protein Eng.*, 1986, **1**, (1), p. 55–57.

55 F. Tat-Kwong Lau and A. R. Fersht, *Nature (London)*, 1987, **326**, 811–812.

56 H. W. Hellinga and P. R. Evans, *Nature (London)*, 1987, **327**, 437–439.

site-directed mutagenesis.[57] The structural basis of the catalytic and substrate binding properties have been determined[58] and mutant enzymes produced, one being specific for a new substrate, and one lacking allosteric regulation.[59]

Cytochrome C. An invariant Phe residue has been shown not to be essential for the transfer of electrons but to be involved in determining the reduction potential.[60]

Signal peptides. An analysis of the function of signal peptides involved in protein secretion has been made using site-directed mutagenesis.[61]

Thermostability. The thermostability of glucose isomerase has been enhanced by protein engineering.[62] The subject of protein engineering has been reviewed.[63]

Superoxide dismutase. The introduction of changes of electrostatic charges has facilitated substrate diffusion rates thus increasing the enzyme reaction rate.[64]

Metalloenzymes. Protein engineering of metalloenzymes has been reviewed.[65]

Tissue plasminogen activator (t-PA). Attempts at increasing plasma half-life, increasing fibrin affinity and decreasing the rate of reaction with plasma inhibitors have been reviewed.[66]

Use of non-standard amino acids. In the past, protein engineering has been restricted to the use of the 20 'proteinogenic' amino acids. A method for the incorporation of non-standard amino acids into a polypeptide has been described.[67]

[57] A. R. Clarke, D. B. Wigley, W. N. Chia, D. Barstow, T. Atkinson, and J. J. Holbrook, *Nature (London)*, 1986, **324**, 699–702.

[58] A. R. Clarke, T. Atkinson, and J. J. Holbrook, *Trends Biochem. Sci.*, 1989, **14**, 101–105.

[59] A. R. Clarke, T. Atkinson, and J. J. Holbrook, *Trends Biochem. Sci.*, 1989, **14**, 145–151.

[60] G. J. Pielak, A. G. Mauk, and M. Smith, *Nature (London)*, 1985, **313**, 152–153.

[61] S. Lehnhardt, S. Inouye, and M. Inouye, *Protein Engineering*, 1986, **1**, (1), p. 157–191.

[62] W. J. Quax, N. T. Mrabet, R. Luiten, P. W. Schuurhuizen, P. Stanssens and A. I. Lasters, *Bio/Technology*, 1991, **8**, 738–742.

[63] Y. Nosoh and S. Takeshi, *TIBTECH*, 1990, **8**, 16–20.

[64] E. D. Getzoff, D. E. Cabelli, C. L. Fisher, H. E. Parge, M. S. Viezzoli, L. Bacni, and R. A. Hallewell, *Nature (London)*, 1992, **358**, 347–350.

[65] J. N. Higaki, R. J. Fletterick, and C. S. Craik, *Trends Biochem. Sci.*, 1992, **17**, 100–104.

[66] J. H. Robinson and M. J. Browne, *TIBTECH*, 1991, **9**, 86–90.

[67] J. D. Bain, C. Switzer, A. R. Chamberlin, and S. A. Benner, *Nature (London)*, 1992, **356**, 537–539.

CHAPTER 15

Stability in Enzymes and Cells

M. D. TREVAN, O. M. POLTORAK, AND E. S. CHUKHRAI

1 INTRODUCTION

Biotechnology is all about producing products. Biological products are produced as a result of the biotransformation of one substance into another. These biotransformations rely almost totally on the catalytic power of enzymes, whether they are present as purified proteins or acting in concert in whole cells. Thus one, albeit biased, viewpoint is that ultimately biotechnology depends on our understanding and ability to use biocatalysts. This chapter is concerned principally with one of the major limitations to the extension of the use of biocatalysts: their relative instability, our understanding of its causes, and our present remedies.

In this context the term 'biocatalyst' is used to describe any biologically derived material which exhibits some enzymic activity. This may range from the highly purified enzyme to a whole viable metabolizing cell. In between these two extremes come unpurified enzymes, crude cell homogenates, non-viable cells, and coupled enzyme systems. The choice of the most appropriate form of biocatalyst for use in any given system will depend upon the individual circumstances and what is required of the biocatalyst. No general rules can be laid down as a guide to choice, but some of the relevant considerations regarding immobilized biocatalysts are presented in Table 1.

Some 2500 different enzymes have been isolated and described to date and it is estimated that this probably reflects only 10% of the enzymes that exist in nature. With all their potential advantages, it is at first sight surprising that fewer than 200 enzymes and cellular biocatalysts find common use.

There are a number of factors which have limited the range of biocatalyst applications, particularly in their use as industrial catalysts, the principles being, availability, cost, and, perhaps most significantly, stability.

It is perhaps surprising that enzyme application should be limited by availability, given the vast range of enzymes in nature. The problem is both practical and legislative. When considering the manufacture of a particular enzyme, the producer must, in order to reduce costs, manufacture the enzyme in a quality which is acceptable to all of his customers. In practice this means that the enzyme will often be produced as a single grade to satisfy the most stringent legislative demands, as it will then perform satisfactorily where lower quality will suffice.

Table 1 *Comparisons of applications of various immobilized biocatalysts*

	Enzymes	*Microbial cells*	*Plant cells*	*Animal cells*
Immobilization method	Possible by wide range of techniques	Possible mostly by entrapment or adsorption	Entrapment in biogels: few other methods tested	Possible by adsorption
Inherent stability of unimmobilized form	Low	Medium → high	Medium	Low
Stabilization of biocatalyst upon immobilization?	Yes, but apparent reactions	Some stabilization	Real, marked stabilization	?
Single-step processes	Yes, mostly hydrolytic reactions	Yes, mostly hydrolytic reactions with non-viable cell	Possible but enzymes or microbial cells probably more efficient	Not efficient
Single-step reactions involving (expensive) coenzymes	Possible, but stil prohibitively costly	Possible, but fermentation may be cheaper	Possible, may be uniquely useful for processing waste plant products	May be possible but very costly
Multistep processes for biotransformation of added precursors	Perhaps technically feasible but prohibitively expensive	Possible, but traditional fermentation may be cheaper	Possible, (see above); immobilization probably more efficient than fermentation	May be possible
De novo synthesis from simple carbon sources	No	Possible, but fermentation almost certainly cheaper	Possible and probably cost effective if price + market volume is high enough	Essentially protein products
Major actual or potential application	Probably confined to single-step mostly hydrolytic reactions	Used as 'unpurified' enzyme or possibly for single-step process involving coenzymes	Multistep process or *de novo* synthesis	Production of desirable animal proteins

Table 1 *(continued)*

	Enzymes	*Microbial cells*	*Plant cells*	*Animal cells*
Type of product most likely to be produced	High volume organic compounds produced by single-step process	Medium to high volume organics possibly involving single-step or multi-step reactions	Medium to low volume high cost plant products, photosynthetic processes	As above
Alternative production method	Chemical processes immobilized microbial cells	Fermentation	Agricultural, fermentation but not for all products	Extraction from animal tissue, genetic engineering
Long-term prospects	Good but limited	Advanced fermentation* may be more cost effective	Immobilized cells may be more cost effective	Genetic engineering may be ultimately more cost effective

* Except in processes involving organic solvent soluble products/substrates when immobilized microbial cells may be uniquely useful
(Reproduced with permission from M. D. Trevan *et al.*, 'Biotechnology, the Biological Principles', Open University Press, Milton Keynes, UK, 1987)

The most stringent regulations are, of course, for food grade enzymes. Where the enzyme is derived from a microbial source, this immediately limits production to a food-compatible micro-organism.

Thus most industrial enzymes are produced from only eleven fungi, four yeasts, and eight bacteria. Any new enzyme required is first sought from these organisms. This approach has an additional advantage in that the producer has to hand the knowledge and technology to grow and extract enzymes from these microbes.

The disadvantage of this philosophy is, of course, that the isolated enzyme may not show ideal characteristics for its intended purpose; for example substrate specificity or stability may not be as high as that for a similar enzyme from an untried micro-organism.

So why use microbes to produce enzymes, why not use plant or animal materials? The answer lies in part in the ease with which fluctuations in demand for a particular enzyme can be catered for by altering fermentation capacity, partly in the wide range of reactions exhibited by most microbial cells, and in part the convenience and certainty of not having to rely on an external supply of raw material.

To the enzyme user all this may restrict the choice of enzyme, but will assure that the enzyme will be reliably available in bulk and of a known quality. What redress does the enzyme user have if a suitable enzyme is not available? He may attempt to alter the characteristics of an available enzyme, look for new reactions catalysed by existing enzymes, or select a new source and do it himself. All three approaches have been researched in recent years with some interesting results.

Selecting a new source of organism and alteration of the characteristic of available enzymes are discussed below in Section 3 and elsewhere in this volume.

The search for new reactions from old enzymes is a relatively new field of endeavour but already there are some surprises. For example, glucose oxidase (EC.1.1.3.4), normally associated with the catalysis of glucose and oxygen to gluconolactone and hydrogen peroxide, has been employed to convert benzoquinone into hydroquinone in the presence of glucose. Carboxypeptidase A (EC.3.4.17.1) has two activities, esterase and peptidase, and normally contains a zinc atom. If this is replaced by manganese or cadmium the peptidase activity is virtually lost whereas the esterase activity is enhanced. α-Thrombin (EC.3.4.21.5) also has two enzyme activities, amidase and esterase, the balance between which can be altered by adjusting the level of dimethyl sulfoxide. Proteases in general may be used to synthesize esters or peptides, reversing the normal hydrolytic reaction by performing the reaction in organic solvents. Thus α-chymotrypsin (EC.3.4.21.1) in chloroform will synthesize *N*-acetyl-L-tryptophan ethyl esters, with 100% yield, from ethanol and *N*-acetyl-L-tryptophan; use of carbon tetrachloride as a solvent lowers the yield to less than 60%. The fungal enzyme, cyanide hydratase (EC.3.5.5.1), produced by some plant pathogens as a defence against cyanide production by cyanogenic plants, has recently been used to convert acrylonitrile into acrylamide on an industrial scale (*ca.* 5 tonnes p.a.). The same enzyme has been proposed for the formation of L-alanine from aminoproprionitrile. This approach to extending the availability of biocatalysts has been reviewed by Neidleman (1984).[1]

Why is stability, or rather the lack of it, so important a consideration in the practical application of biocatalysis? The answer in a word is cost. In the case of soluble enzymes, the cost of recovering the enzyme at the end of a process is usually prohibitive; thus the enzyme can be used only once. This is fine just so long as the enzyme itself is cheap and does not contribute greatly to the total cost of the process. Even the cheapest of bulk enzymes, for example alkaline proteases used in washing powder formulations, may contribute a significant fraction of the cost of a product, in this case 5–10%. Where the enzyme is used as a soluble, dispensable reagent, too great a stability may be a disadvantage, as the enzyme will have to be separated from the product, and the easiest way to do this may be by heating the solution. For example, α-amylases are used in the processing of flour. If there is a residual enzyme activity left in the product, subsequent processes may be adversely affected. In cases where the cost of the enzyme is high, single use of the enzyme will be uneconomic.

Clearly a re-usable enzyme would be advantageous; the problem to overcome is the enzyme's solubility. The solution is to render the enzyme insoluble. This may be easily done by fixing or immobilizing the enzyme onto some inert insoluble 'polymer' (see Chapter 16 and reference 2). Thus the enzyme may be retained in the reactor, on the surface of the biosensor, or whatever. The same considerations apply equally to unpurified enzymes and non-viable cell preparations. Quite obviously, however, a re-usable enzyme will be useless unless it is stable enough to allow re-use!

When it comes to the use of viable cells as biocatalysts, the traditional approach has been that based upon fermentation processes. However, whereas microbial cells grow rapidly and easily, it takes a considerable expenditure of time and effort to grow a reasonable biomass of plant and animal cells. In many cases the desired product may be produced only during the stationary phase of growth, and thus the object is to increase the longevity of the cells in this phase. Apart from using viable cells for the *de novo* synthesis of natural (or even cloned) products, work on their potential to perform technical enzyme transformations on exotic, added precursor molecules dates back over ten years. In this latter case the cells are essentially being used as convenient bags containing the necessary (sequence of) enzymes to perform not only the biotransformation, but also to recycle and retain any coenzymes required by the reaction(s). In such cases in particular, the stability of the biocatalyst is of concern.

When considering the stability of biocatalysts we are concerned with the resistance of enzymes to denaturation or irreversible inhibition and, where appropriate, with factors that adversely affect the viability of cells. In the discussion that follows we shall first consider enzymes and non-living cells and then, separately, consider the viable cells.

[1] S. L. Neidleman, in 'Biotechnology and Genetic Engineering Reviews', Vol. 1, ed. G. E. Russell, Intercept, Newcastle upon Tyne, 1984, pp. 1–38.

[2] M. D. Trevan, 'Immobilized Enzymes: Introduction and Application in Biotechnology', John Wiley, Chichester, 1980.

2 ENZYME STABILITY

The first question we must ask is what is meant by the term stability? This may seem an obvious question, but to judge by the variety of implied definitions extant in the literature it would appear that the term stability has a variety of subtly different meanings. In this context, however, we shall define stability simply as the ability of the enzyme to retain its catalytic activity.

The next question is perhaps equally obvious. What factors make an enzyme lose its activity? There are many; microbial digestion, hydrolysis by protease (including autolysis), irreversible inhibition, for example by metal ions or oxidation reactions, aggregation and precipitation of the enzyme, for example by organic solvents, the dissociation of subunits of a polymeric enzyme, and denaturation, that is the unfolding of the molecular structure of the enzyme. The first four of these factors may be fairly simply countered by making the enzyme inaccessible to the inactivating agent. For example, immobilization will surround the enzyme with a cage that will physically prevent contact with micro-organisms and hydrolytic enzymes, or, in the case of an immobilized protease, prevent inter-molecular contact and hence autolysis. Alternatively the surface of the molecule may be modified chemically, for example by alkylation or glycosylation, thus covering up the possible sites of action of proteases on the enzyme's surface. Equally, the polymer matrix of an immobilized enzyme may well, for steric or electrostatic reasons, prevent the irreversible inhibition of an enzyme by excluding or preferentially binding the inhibitor, be it hydrogen or metal ions, organic molecules, or even oxygen. Immobilization may even prevent inactivation caused by the aggregation of enzyme molecules which may be brought about by, for example, organic solvents, either by physically holding the bound enzyme molecules apart or by negatively partitioning the solvent away from the enzyme (see, for example, the work of Blanco *et al.*[3] on the effect of water immiscible solvents on the activity and stability of both native and immobilized α-chymotrypsin). Molecular denaturation or subunit dissociation of the enzyme may be caused by physical factors, for example heat or pH, or by chemical factors, for example urea or organic solvents. In principle, however, their mode of action is similar: the disruption of the various electrostatic, hydrophobic, hydrogen bonds, *etc.* which hold the enzyme in its native shape, thus allowing the peptide chain to unfold and new associations to be made. This similarity of action of course means that a successful attempt to stabilize an enzyme against the action of, for example, urea may stand a good chance of stabilizing it to heat. The trick is in effect to reinforce the native structure of the enzyme, and there are a number of ways in which this can, in theory at least, be done.

The need for heat stability is self-evident: the more stable the enzyme the longer it will last in a reactor. In addition thermostable industrial enzymes allow the possibility of carrying out processes at elevated temperatures, reducing the risk of microbial contamination, and markedly elevating reaction rates. This latter point in particular aids process economics, as it allows either a smaller

[3] R. M. Blanco, P. J. Halling, A. Bastida, C. Cuesta, and J. M. Guisan, *Biotechnol. Bioeng.*, 1992, **39**, 75.

enzyme reactor or a greater productivity. However, there are other advantages in having a thermostable enzyme. Shelf life of the enzyme will be enhanced, enabling batch production (of enzyme) to be performed on a larger, more cost-effective scale, as well as promoting user convenience. Thermostable enzymes will be less prone to denaturation by the conditions under which they are produced, which may in turn lead to increased yields of enzyme or permit the use of simpler and cheaper production methods.

The rationale for organic solvent-stable enzymes has been hinted at above (reverse hydrolysis) but there is another, wider reason. Many interesting and valuable substances which could be produced from enzyme-catalysed processes are too insufficiently soluble in water to make process design feasible. The ability, therefore, to run the reaction in the presence of organic solvents would be valuable in the processing of molecules such as steroids or triglycerides.

One of the problems of studying enzyme stability has been the lack of suitable models of enzyme deactivation. Generally deactivation has been considered to be a first-order process, which although successfully applied to a number of enzymes could not explain the deactivation of many others. Equally, simple deactivation models also gave rise to a number of dubious claims for enhancement of enzyme stability by this or that procedure. This lack of understanding has probably hindered the application of rational approaches to enzyme stabilization, as has a lack of knowledge of the molecular forces responsible for enzyme stability. Thus, to this time, most attempts at enzyme stabilization have been empirical in nature and often with dubious results.

Happily, the situation is changing with a number of articles occurring in the mid-1980s on both these subjects.[4,5] Henley and Sadana[5] described a two-step model for deactivation, involving the irreversible interconversion of three conformations of the enzyme molecule:

$$\mathrm{E} \xrightarrow{k_1} \overset{a_1}{\mathrm{E}_1} \xrightarrow{k_2} \overset{a_1}{\mathrm{E}_2}$$

where k_1 and k_2 are first-order deactivation rate constants, E, E_1, and E_2 are specific activities of the three conformational states of the enzyme, and $a_1 = E_1/E$ and $a_2 = E_2/E$. Either a_1 or a_2 could be greater or less than 1 and k_1 and k_2 could vary independently > 0.

Assuming that the enzyme activity a measured at any time,

$$a = (\mathrm{E} + a_1 \mathrm{E}_1 + a_2 \mathrm{E}_2)/\mathrm{E}_0$$

where E_0 is the specific activity at time = 0, by integration this model yields an equation relating the total enzyme activity to the values of a_1, a_2, k_1, k_2, and time. They then went on to produce model curves (by altering these values which fitted known experimental data), and found that the experimental data could be divided into 14 cases, broadly classified into two categories, the first where the activity a at any time never exceeds the initial activity a_0 and the second,

[4] V. V. Mozhaev and K. Martinek, *Enzyme Microbiol. Technol.*, 1984, **6**, 50.
[5] J. P. Henley and A. Sadana, *Enzyme Microbiol. Technol.*, 1985, **7**, 50.

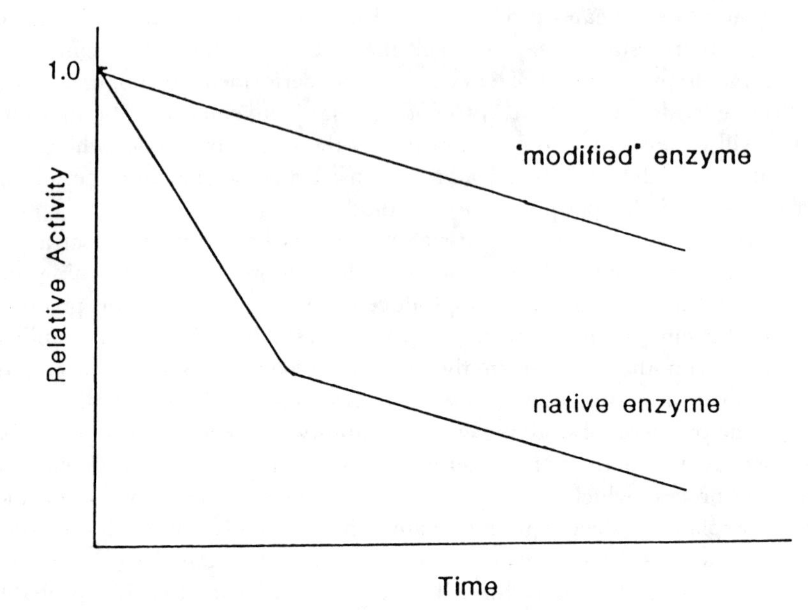

Figure 1 *Artefactual stabilization by differential denaturation of coexisting enzyme conformations*

intriguingly, where a can exceed a_0; that is, the enzyme, at least transiently, becomes more active.

The lesson to be drawn from these and similar studies is that enzymes which have been modified may indeed show enhanced stability over the native enzyme by virtue of the fact that the modification may have altered the balance between co-existing conformations of the enzyme. Consider the case where $k_1 > k_2$ and $a_1 < 1$ but $> a_2$. The native enzyme will exhibit rapid initial deactivation followed by a slower deactivation rate (Figure 1). If modification of the enzyme substantially denatures E but not E_1 then the modified enzyme will apparently display greater stability. This mechanism is seen in the deactivation of trypsin (EC.3.4.21.4) which exists in two forms, α and β, differing only in the extent of cleavage of the peptide chain. The rate constant of deactivation of the α-form is 100 times that of the β-form.[6]

Figure 2 illustrates the kind of experimental data that Henley and Sadana's theory was designed to explain. It shows the effect of pH on the stability of both the free and immobilized forms of 3-phosphoglycerate kinase (EC.2.7.1.2) at 25 °C. Quite clearly the change in activity with time for the immobilized enzyme is anything but a first-order process. For the free enzyme it can be easily seen that the pH has a significant effect, for only at pH 7.0 does the enzyme activity at first increase. In terms of Henley and Sadana's theory, this can be explained by postulating an increase in the value of k_2 with increasing pH; hence only at low pH value does the (more active) E_1 form of the enzyme accumulate to any great extent. Why should, however, the immobilized enzyme respond in such a

[6] R. A. Beardslee and J. C. Zahnley, *Arch. Biochem. Biophys.*, 1973, **153**, 806.

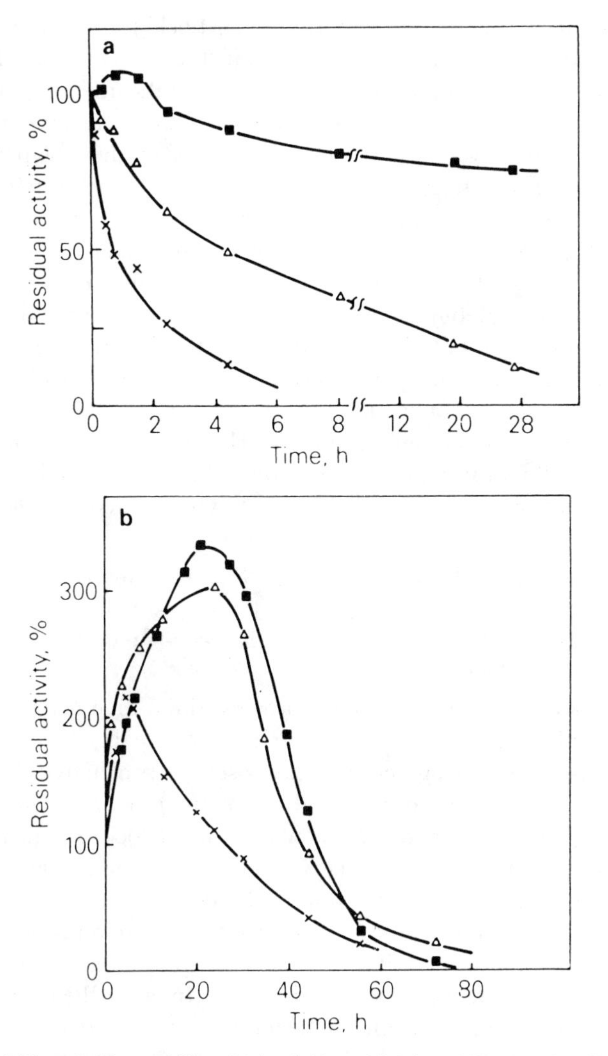

Figure 2 *Effect of pH on stability of* (a) *soluble and* (b) *immobilized 3-phosphoglycerate kinase at* 25°C *and* ■, *pH* 7.0; △, *pH* 8.0; ×, *pH* 9.0. *Experiments were carried out in* 0.1 mol dm^{-3} *triethanolamine/HCl buffer. Enzyme concentrations used were* 20 μg cm^{-3} *for the soluble enzyme and* 12 mg *solid* cm^{-3} *for the immobilized enzyme*
(Reproduced with permission from Simon *et al.*, *Enzyme and Microbiol. Technol.*, 1985, **7**, 357.)

dramatically different way? There are a number of reasons which are discussed below, but suffice it to say at this point that it may be a result of hydrogen ion partitioning towards the polymer matrix, lowering the effective pH in the microenvironment of the enzyme.

Even the experimentally verifiable models such as those of Henley and Sadana do not deal with the potential effects of dissociation on the stability of polymeric or multi-subunit enzymes. A number of approaches have been described, and one of the most interesting is the 'Conformational Lock' theory of Poltorak and his co-workers.[7-13] Their theory is based on the concept of Shaknovich and Finkelstein,[14,15] who proposed that the first step in the unfolding process of a peptide chain is the destruction of the densely packed surface layers of the protein molecule, caused by movements of the side-chains of the polypeptide, but without the penetration of water molecules into the core of the protein molecule. The subsequent step involves ingress of water into the 'pores' formed in the protein molecule. This unpacked structure was termed a 'molten globula'. If there are a number of intermediates in the process of conversion of the native protein into the 'molten globula' and these intermediates are catalytically active, but the 'molten globula' is not, then the kinetics of the denaturation process will show a lag phase or induction period before loss of activity is observed. The existence of this 'molten globula' state has been observed during the denaturation of cyctochrome c and apomyoglobin at pH 2 in the presence of Cl^-.[16]

In their studies Poltorak and Chukhrai describe the thermodegradation of oligomeric enzymes by a scheme which for catalytically active dimers may be written as;

E_2	$\rightleftharpoons E_2x$	$\rightleftharpoons 2E_1x(\mathcal{N})$	$\rightarrow 2E_1(R_m)$	$\rightarrow 2E_d(R_s)$
Native stable dimer	Active labile dimer	Inactive monomer	Inactive 'molten globula'	Denatured protein

In the initial stages of the denaturation process, the native active dimers are converted into labile (less stable) dimers which are able to dissociate more easily. This will again give rise to a lag period in the onset of loss of activity. It has been suggested[7] that the native dimers are stabilized by the process of conformational lock in protein–protein contact. This conformational lock occurs when total complementarity in inter-chain contact is achieved and the contacting parts of the two sub-units form one, common, quasi-solid domain. Raising the temperature causes, through changes in the surface structure, these quasi-solid domains to become quasi-liquid, diminishing the energy of protein–protein interaction and thereby facilitating the process of dissociation. In effect the conformational lock melts. Evidence for this concept can be found in the observations that many native oligomers are very stable but have negligibly small constants of dissociation, but, in the course of thermodegradation, dissociation occurs after the

[7] O. M. Poltorak and E. S. Chukhrai, *Vestinik Moskov. Univ., Khimia*, 1986, **27**, (3), 237.
[8] O. M. Poltorak and E. S. Chukhrai, *Vestnik Moskov. Univ., Khimya*, 1979, **20**, (3), 195.
[9] O. M. Poltorak, E. S. Chukhrai, V. N. Tashlitskii, and R. S. Svanidze, *Vestnik Moskov. Univ., Khimya*, 1980, **21**, (3), 224.
[10] O. M. Poltorak, E. S. Chukhrai, and A. N. Pryakhin, *Zhur. Fiz. Khim.*, 1985, **59**, (7), 1585. [*Russ. J. Phys. Chem.*, 1985, (7)].
[11] E. S. Chukhrai, *Vestnik Moskov. Univ., Khimya*, 1985, **59**, (7), 1585.
[12] O. M. Poltorak, A. N. Pryakhin, E. A. Arens, and G. Kh. Goldshtein, *Vestnik Moskov. Univ., Khimya*, 1982, **23**, (3), 249.
[13] O. L. Voronina, E. A. Tarasova, V. I. Telepneva, E. S. Chukhrai, and O. M. Poltorak, *Zhur. Fiz. Khim*,m, 1991, **65**, (12), 3293. [*Russ. J. Phys. Chem.*, 1991 (12)].
[14] E. I. Shaknovich and A. V. Finkelstein, *Biopolymers*, 1989, **28**, (10), 1667.
[15] A. V. Finkelstein and E. I. Shaknovich, *Biopolymers*, 1989, **28**, (10), 1680.
[16] I. Goto, N. Takashi, and A. L. Fink, *Biochemistry*, 1990, **29**, (14), 3480.

temperature has been raised by only a few degrees. This is difficult to reconcile with the temperature dependence of the thermodynamic constant of dissociation (because the energy of dissociation of oligomeric proteins is small), but is consistent with the properties of a melting-like process in the inter-protein contact domains. For example, native muscle lactate dehydrogenase and aldolase and the native dimeric yeast glucose-6-phosphate dehydrogenase under *in vivo* conditions do not apparently easily dissociate, but rapidly undergo dissociative thermodegradation for only a small increase in temperature.[17]

In some instances the conformational lock may occur as a result of interaction between catalytically active subunits and inactive (small) binding subunits. For example, aspartate carbamoyltransferase is a complex of type C_6R_6, where catalytically active trimers (C_3) are connected by smaller binding proteins (R_2).[18] The dissociation constant of this complex is less than 10^{-14} (*i.e.* it shows little tendency to dissociate), but removal of the R_2 subunit leads to thermodegradation of the complex accompanied by its dissociation.

Miller has proposed[19] typical structures for protein interaction based on a consideration of 9 dimeric and 9 tetrameric proteins. Stability is achieved because of the presence of large hydrophobic sites surrounded by charged amino acid side-chain residues. The hydrophobic contact is not flat but is formed by long β-sheets, contacting parts of helix–helix and β-chain–β-chain; formation of this contact is the final step in the process of protein–protein interaction.

Thus the conformational lock in stable oligomers may be achieved in a number of ways, of which three are of prime importance.

(1) Hydrophobic interaction accompanied by conformational restructuring of the contact sites. This results in the liquid half domains of protein subunits forming quasi-solid domains.
(2) The quasi-solid contact may occur only in the presence of a low molecular mass effector entering into the inter-peptide interaction site and changing its structure and properties. For example, stable tetramers of threonine desaminase are formed only in the presence of L-valine or L-leucine.[20]
(3) The quasi-solid binding between dimers may appear only after adsorption of a low molecular mass allosteric effector such as a metal ion.[21,22]

The formation of the types of structure proposed by the conformational lock hypothesis is relatively slow, and thus the time taken to reach the equilibrium position in a dissociation process is long. For example, for mannose-6-phosphate specific receptor it has been shown[23] that at 4°C and pH 7.5 the equilibrium

[17] L. Goldstein and S. M. Gartler, *Exp. Cell. Res.*, 1979, **122**, 185.
[18] H. K. Schachman and S. J. Edelstein, *Biochemistry*, 1966, **5**, 2681.
[19] S. Miller, *Protein Eng.*, 1989, **3**, (2), 77.
[20] G. W. Hatfield and S. O. Borns, *Science*, 1970, **167**, 75.
[21] H. Paulus and J. B. Alpers, *Enzymes*, 1971, **12**, 385.
[22] B. I. Kurganov, 'Allosteric Enzymes. Kinetic Behaviour', John Wiley and Sons Ltd., Chichester, 1982.
[23] A. Waheed, A. Hille, U. Junghans, and K. Von Figura, *Biochemistry*, 1990, **29**, (10), 2450.

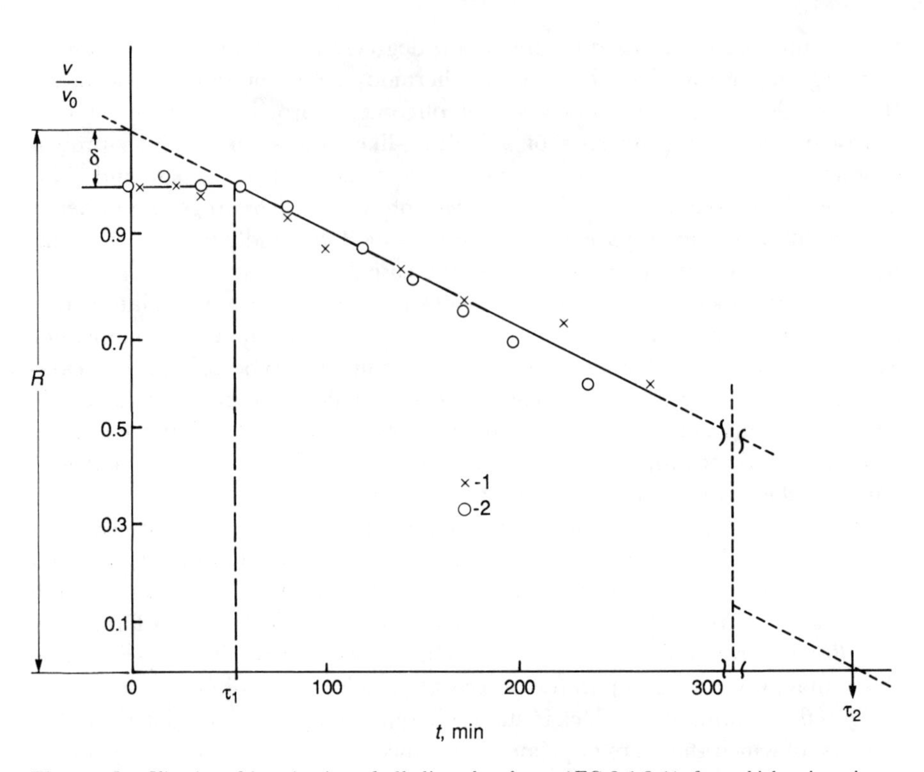

Figure 3 *Kinetics of inactivation of alkaline phosphates (EC.3.1.3.1) from chicken intestine at pH* 8.5 *and* 45°C *in presence of* 1 M *urea. Initial concentrations of enzyme* 1.4×10^{-8} M (1) *and* 4.3×10^{-8} M (2). *Enzymic activity was measured in reaction hydrolyse of* p-*nitrophenyl phosphate at* 400 nm. *(R, τ_1 and τ_2 – see explanation in text)*

between the dimeric and tetrameric forms takes 6–8 days to reach. Uchida *et al.*[24] showed that ATP, ADP, and fructose-6-phosphate, in addition to their effect on the activity of phosphofructokinase also influence its stability. The effect of Zn^{2+} ions which enhance both the activity and stability of trimers of ornithine carbamoyltransferase is mediated through slow structural changes.[25,26] Catalase can exist in a homogenous tetrameric form which is stabilized by peptide 'arms' which penetrate adjacent subunits forming stable inter-subunit connections, thus preventing rapid dissociation.[27]

All these examples illustrate that many oligomeric proteins *in vivo* are stable complexes, but during the course of thermodegradation they transform into labile forms capable of rapid dissociation. This process may be regarded as the destruction (melting or opening) of the conformational lock which of course will occur during the lag phase or induction period in the kinetic studies of loss of activity (see Figures 3–5).

24 I. Uchida, T. Koyama, and A. Hachimori, *Compar. Biochem. Physiol. B.*, 1990, **96B**, (2), 399.
25 S. Lee, W.-H. Shen, A. W. Miller, and L. C. Kuo, *J. Mol. Biol.*, 1990, **211**, (1), 271.
26 S. Lee, W.-H. Shen, A. W. Miller, and L. C. Kuo, *J. Mol. Biol.*, 1990, **211**, (1), 255.
27 B. K. Vainshtein, W. R. Melik-Adamyan, V. V. Barynin, A. A. Vagin, A. I. Grebenco, V. V. Borisov, K. S. Bartels, I. Fita, and M. G. Rossman, *J. Mol. Biol.*, 1986, **207**, (1), 49.

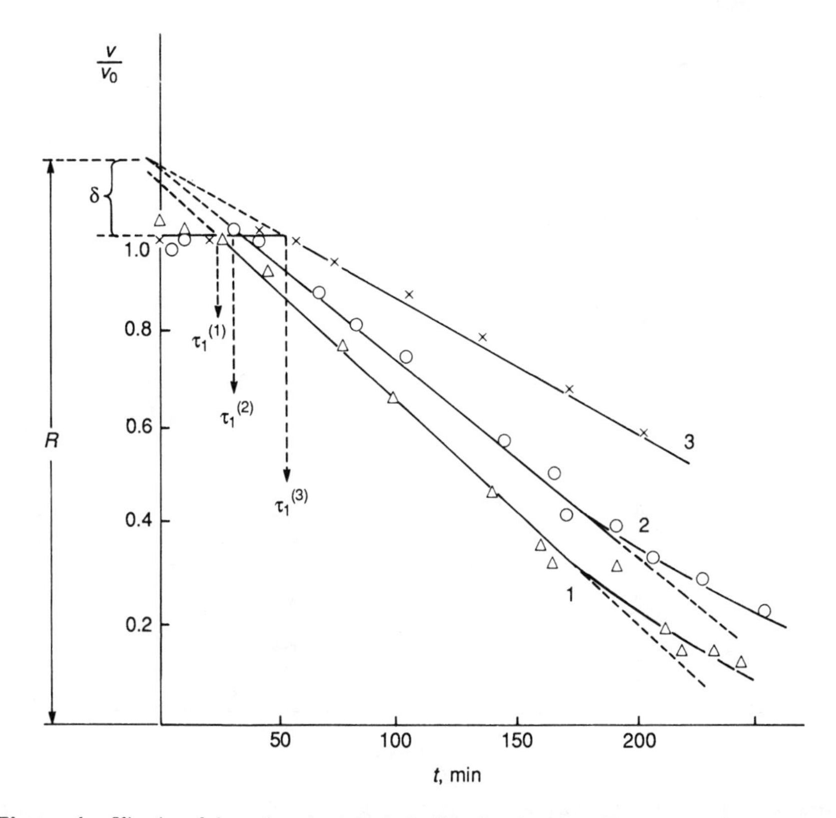

Figure 4 *Kinetics of thermoinactivation of alkaline phosphates at pH* 7.5 *and* 58 °C *in presence of* 0.03 M $MgCl_2$. *Initial concentrations of enzyme* 1.4×10^{-8} (2) *and* 4.3×10^{-8} M (3) *(δ – see explanation in text)*

One of the major and significant practical consequences of these observations and hypotheses is that it provides an explanation for the often observed differences in stability of different batches of the same enzyme. If labile oligomers arise from the native enzyme during storage, then the precise time point at which the enzyme is used, whilst not affecting the actual level of enzymic activity displayed, may have a considerable effect upon the stability characteristics of the enzyme. In other words, how stable the enzyme appears to be will depend upon from which point along the induction or lag phase portion of the denaturation curve that the enzyme sample is taken. Thus we have the potential situation that during this lag phase, which under the recommended storage conditions of any one enzyme may in practise be a matter of months, the enzyme's initial activity does not change but its stability potential declines; catalytic activity cannot be taken in isolation as a measure of the 'nativity' of an enzyme. A much more pertinent measure of 'nativity' would be the number of intermediate labile forms of an oligomeric enzyme occurring between original native enzyme and

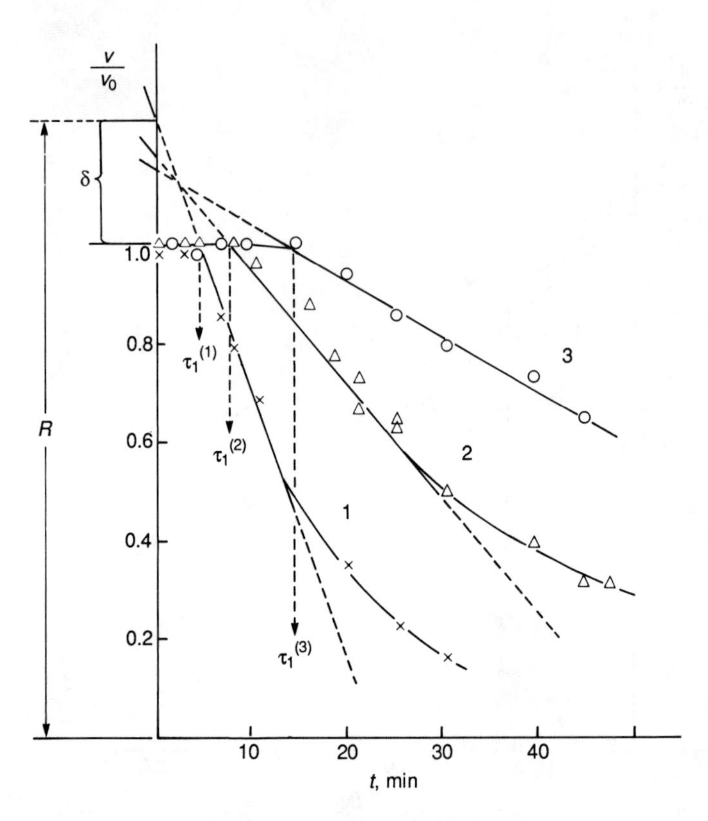

Figure 5 *Kinetics of thermoinactivation of glucose-6-phosphate dehydrogenase (EC.1.1.1.49) at pH* 8 *and* 46°C. *Activity* (v) *was measured on extinction of NADPH at* 340 nm *after cooling to* 20°C. *Initial concentration of enzyme* 0.65×10^{-8} (1), 1×10^{-8} (2), *and* 2×10^{-8} M (3)

the finally denatured form. Bednarek *et al.*[28] have proposed a kinetic method for the estimation of the minimal number of such labile intermediate forms.

It can be seen from Figures 3–5, which are plots of relative rate (v/v_0) against time (t), where v_0 is the rate at $t = 0$ the period of induction before the rate declines (τ), as a result of internal changes in the structure of the protein subunits and thus the appearance of an unknown number (n) of catalytically active but less stable intermediate forms. Degradation in this case may be written as the reaction sequence:

$$E^{(n)} \underset{k_{-n}}{\overset{k_n}{\rightleftharpoons}} E^{(n-1)} \rightleftharpoons \ldots \underset{k_{-1}}{\overset{k_1}{\rightleftharpoons}} E^{(1)} \xrightarrow{k_d} E_d \quad (1)$$

For active dimers with inactive subunits this takes the form of;

$$E_2^{(n)} \underset{k_{-n}}{\overset{k_n}{\rightleftharpoons}} \ldots \underset{k_{-2}}{\overset{k_2}{\rightleftharpoons}} E_2 \underset{k_{-1}}{\overset{k_1}{\rightleftharpoons}} 2E_1 \xrightarrow{k_d} E_d \quad (2)$$

[28] P. Z. Bednarek, V. P. Kaberdin, V. N. Tashlitskii, O. M. Poltorak, and E. S. Chukhrai, *Vestnik, Moscov. Univ., Khimya*, 1987, **28**, (6), 523.

Complete theoretical solution of the kinetics is not possible, but by computer calculation it is possible to evaluate the minimum value of n. It is interesting to note that the minimum period of lag or induction is reached when in scheme (2) all steps are irreversible and all specific rates are equal.

$$\underbrace{E_2^{(n)} \rightarrow E_2^{(n-1)} \rightarrow \ldots \rightarrow E_2^{(1)}}_{\text{Active forms}} \rightarrow \underbrace{2E_1 \rightarrow}_{\substack{\text{inactive}\\ \text{forms}}} \tag{3}$$

Conversely, the existence of any reversible steps of inequality of rate constants inexorably leads to an increase in τ if the activities of the different forms of E_2 are assumed to be equal.[28] Although it is impossible to determine the true value of n, its lower limit is of interest as a characteristic of the process of thermodenaturation, and as a practical parameter of stability.

To determine the minimum value of n from kinetic analysis, it is convenient to introduce a dimensionless ratio (δ) for the period of induction (lag) τ_1 and the period of inactivation τ_2 shown in Figure 3.

$$\delta = \frac{\tau_1}{\tau_2 - \tau_1} = R - 1$$

where R is shown on the ordinates of Figure 3.

For scheme 3 above, accurate analytical analysis shows that the value of δ is dependent only on the number of steps through which the protein must pass before loss of activity (n).

$$\delta = e^{-(n-1)}\left[1 + \frac{(n-1)^{n-2}}{(n-2)!} + \sum_{m=1}^{n=2} \frac{(n-1)^m}{m!} - 1\right] \tag{4}$$

However in order to render this equation soluble it must be approximated to:

$$\delta = \frac{0.13(\tilde{n} - 1)}{1 - 0.05\tilde{n}}$$

or

$$\tilde{n} = \frac{0.13 + \delta}{0.13 - 0.05\delta} \tag{5}$$

In equation (5) $\tilde{n}$ is written instead of n and as such gives only the limiting minimal value of in equation (3), but not the true number of intermediate forms in equation (2). In evaluating $\tilde{n}$ from equation (5) the error connected with the use of this equation instead of the accurate version (4) is less than 4% when n lies between 3 and 11, and reaches 12% when $n = 2$. One consequence of the approximations is that $\tilde{n} < n$.

Figures 3–5 show examples of the thermodegradation of chicken intestine alkaline phosphatase and yeast glucose-6-phosphate dehydrogenase; both enzymes are dimers under the experimental conditions used. The tangents to the kinetic curves at their inflection points (*i.e.* when the maximum rate of enzymatic degradation has been reached) gives the values of τ_1 and τ_2. From Figure 1, the degradation of alkaline phosphatase at 40°C in the presence of 1 M urea, it can

Table 2 *Values of ñ calculated by equation* (5) *for the thermodegradation of alkaline phosphatase*

	pH			
Temp °C	7.2	7.5	8.5	9.0
50	stable	stable	2.1	2.1
52	stable	stable	1.9	1.7
55	2.4	2.4	1	1
58	2.0	2.1	1	1
60	1	1.75	1	1
63	0	0	0	0

be determined that $\delta = 0.1$ which corresponds to a value for $n = 1.84 \sim 2$. In other words, under the given experimental conditions the decomposition of alkaline phosphatase is occurring in at least 2 steps, through one labile but active dimeric form. Monomers of alkaline phosphatase are inactive or at least much less active. The transition to the inactive form of the enzyme in the presence of urea is irreversible. The period of induction (lag) $\tau = 50$ minutes and is independent of the protein concentration. The decrease in enzymic activity after the induction period may be described by the first order rate constant $k_d = 1.4 \times 10^{-5}\ s^{-1}$.

This is significantly different to the pattern presented by Figure 2 which shows the thermodegradation of alkaline phosphatase at 58°C in the absence of urea. Here, in contrast, it is quite clearly shown that the period of induction, τ, is dependent on the concentration of the enzyme. The values of $\delta = 0.11$, 0.13, and 0.15 correspond to a value of $\tilde{n}$ between 1.93 and 2.2. The simplest kinetic scheme therefore for this process is;

$$\underset{\text{Active}}{E_2} \longrightarrow E_2x \underset{k_{-1}}{\overset{k_1}{\rightleftarrows}} \underset{\text{Inactive}}{2E_1} \xrightarrow{k_d} \ldots$$

because only the reversible dissociation of labile dimers can explain the dependence of the observed rate of degradation on the concentration of protein. The irreversible nature of the transition from E_2 to the labile form E_2x was shown experimentally.

Table 2 summarizes the values of $\tilde{n}$ for the deactivation of alkaline phosphatase in solutions of varying values of temperature and pH.

The values $\tilde{n} = 0$ correspond to the absence of any period of induction on the kinetic plots. However, this means only that the limiting stage of the process gives no information about the properties of any active intermediate, but it does not mean such intermediates do not exist. It is also clear that if any useful information is to be extracted in this way about the degradation process, then experiments must be carried out under carefully controlled non-extreme conditions.

The data in Table 2 show that alkaline phosphatase dimers are most stable at pH 7.5. At this pH the induction period in the degradation process is present up

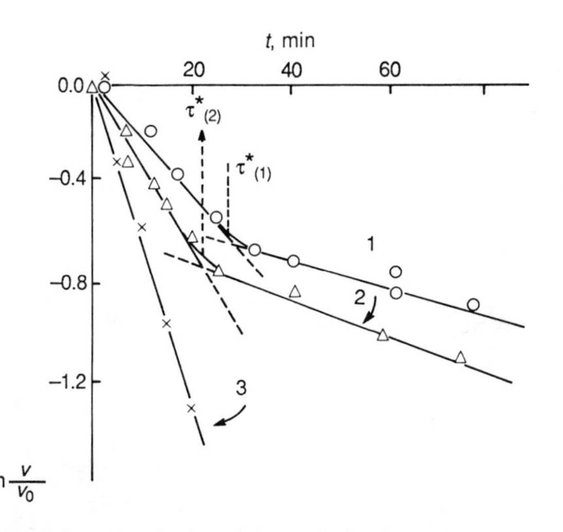

Figure 6 *Kinetics of thermoinactivation of glucose-6-phosphate dehydrogenase at* 40 °C (1), 45 °C (2), *and* 48 °C (3) *at pH* 8. $[E_0] = 2.10^{-9}$ M

to temperatures of 63 °C. In contrast, at pH 8.5 and 9.0 the induction period disappears at temperatures above 55 °C. At pH 7.2 and 7.5 and a temperature of 52 °C the enzyme is stable for the duration of the experiment. This dependence of the stability of alkaline phosphatase on pH over such a narrow range, is probably the result of ionization of the molecule:

$$E_2H_2^+ \overset{H^+}{\rightleftharpoons} E_2H \overset{H^+}{\rightleftharpoons} E_2^-$$

where E_2H is the most stable form.

In the optimum pH range for catalytic activity (pH 8.5–9.0), inactivation of the form E_2^- starts at 50 °C. At 55 °C the period of induction of deactivation disappears and there are no observable kinetic differences between the E_2H and E_2^- forms. Inactivation under these conditions may be described simply by the step of reversible dissociation of the labile dimer and will depend upon the total protein concentration:

$$\underset{\text{Active}}{E_2} \underset{k_{-1}}{\overset{k_1}{\rightleftharpoons}} \underset{\text{Inactive}}{2E_1} \xrightarrow{k_d} 2E_d \qquad (6)$$

and it is this part of the process to which we shall now turn.

As we have seen the degradation of an oligomeric protein occurs via many intermediates, each with different properties. The description of the full scheme by one equation is impossible. Therefore, the only pragmatic way to investigate the kinetics is to assume that one part of the degradation process dominates all others and is therefore in practise rate limiting.

So far we have considered only the properties of the catalytically active intermediates, undergoing transformations during the induction period τ. If we now take the simple case described by equation (6), we can see that the

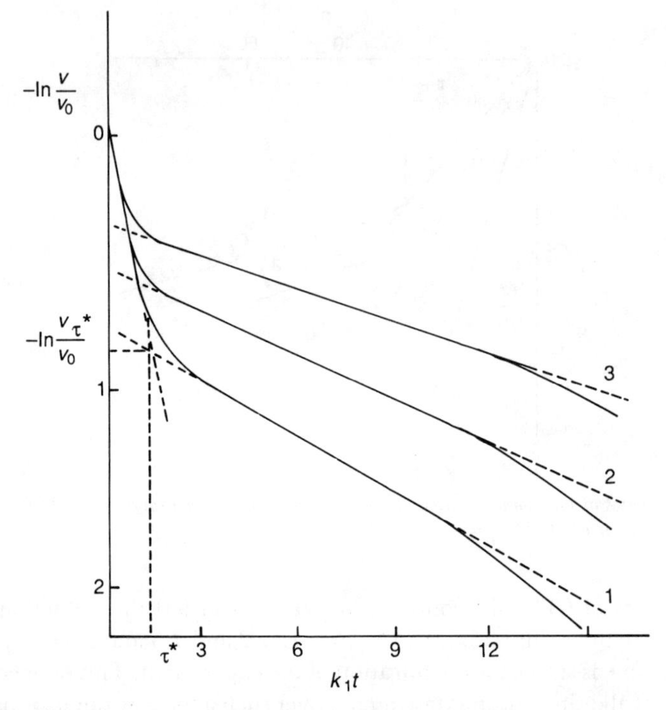

Figure 7 *Theoretical curves for drastic inactivation by scheme (9), obtained computer calculation for* $K_1/k_d = 10$ *and* $[E]_0/k_{dis}$ *values of* 0.3 (1), 1.0 (2), *and* 3.0 (3)

inactivation of the enzyme occurs simultaneously with the dissociation of the labile dimer. In practise, of course, the dissociation may produce less active rather than totally inactive subunits, which may then undergo further degradation. However experimental evidence suggests that this inactivation process occurs in at least two steps, and that under selected conditions these steps can be observed. Figure 6 shows the thermo-inactivation of glucose-6-phosphate dehydrogenase at three different temperatures. At the two lower temperatures, where inactivation is proceeding slowly, the kinetic plot is non-linear. At the highest temperature the kinetics are simple, first order. These effects can be considered from the point of view of the kinetic scheme given in equation (6), using the rate constants k_1, k_{-1} (or K_{dis}), and k_d.

Two factors are pertinent to the discussion. The first is that the dissociation of the labile, but active, form of the oligomer results in deactivation of the enzyme through the production of inactive (less active) subunits. This labile oligomer can be viewed as having an open conformational lock. The second factor is that the dissociation and concurrent inactivation of the labile oligomer is reversible. It is this reversibility which creates the inflection points on the curves in Figures 6 and 7 and creates the relationship between the point of inflection and the protein concentration (Figure 7). Such effects will not be observed if the dissociation/inactivation step is irreversible, nor will any such effects be produced by any first-order process, reversible or not, occurring within the oligomer itself.

Thermodynamically, the equilibrium concentrations (and therefore rate at which equilibrium is approached), for any reversible process in which there is a net change in the number of reactants – in this case an association or dissociation, will depend upon the concentration of the reactants.

In general the denaturation stage in the degradation process is more temperature dependent than any other process, including the dissociation of the oligomers. Thus it is likely that at low temperatures the limiting process will be the denaturation step, but that as the temperature rises the rate of dissociation increases faster than the rate of denaturation and therefore it is the dissociation process which becomes rate limiting at high temperatures. Looked at from the point of the rate constants, at low temperatures,

$$k_d \ll k_1 \tag{7}$$

and therefore the $2E_1$ form of the enzyme is produced faster than it is removed and an equilibrium in the dissociation process is established. This equilibrium will influence kinetically observable data when the protein concentration is varied at or near the value of K_{dis} as shown in equation (8),

$$0.5 < [E]_0/K_{dis} < 30 \tag{8}$$

At higher temperatures, regardless of the concentration of the enzyme $[E]_0$, the dissociation process becomes effectively irreversible because now, $k_d \gg k_1$ and $2E_1$ is converted into $2E_d$ faster than it can be produced. Thus in this case the dissociation process becomes the rate-limiting step.

This also means that the inflection points on the kinetic curves will only be seen over a relatively narrow temperature range, usually no more than 5–7 °C, that is when $0 < k_d < k_1$.

When the conditions in equations (7) and (8) are met then, for a plot of $\ln(v/v_0)$ *versus* time, the slopes of the tangents to the curves and their point of intersection (the inflection point) will yield the values of k_1, k_{-1}, and K_{dis} for the part of the curve where $t < \tau^*$, and k_d for observations made at $t > \tau^*$. The co-ordinates of the tangents' intersection (inflection point) are $\ln(v_\tau/v_0)$ and τ^* (see Figure 6). Computer simulation of the kinetics of equation (6) under conditions of equations (7) and (8) yield patterns of the kind shown in Figure 7. It can be clearly seen that the position of the inflection point in the curves is dependent on the enzyme concentration. It is only under these conditions that the concept of dissociative inactivation is valid.[8–11]

Precise solution of the kinetics of the kinetic scheme of equation (6) is not possible, but an approximate resolution has been proposed. Computer simulation[15] shows that, within the limits of $0.5 < [E]_0/K_{dis} < 30$ and $k_1/k_d > 2$, the error in evaluation of the parameters does not exceed 8%. Even at $k_1 = k_d$ the error estimation of k_d is less than 10%.

Various assumptions simplify the calculation: for example, if at time $t = 0$ all the protein is in the form of dissociable labile dimer, then the concentration of the active dimers is given by $[E_2x]/[E_2x]_0 = v/v_0$ and then according to[9] the dissociation constant of the labile dimers is

$$K_{dis} = k_1/k_{-1} = [E_1]^2/[E_2x] = 4[E]_0(v_0 - v_\tau)^2/v_0 v_\tau \tag{9}$$

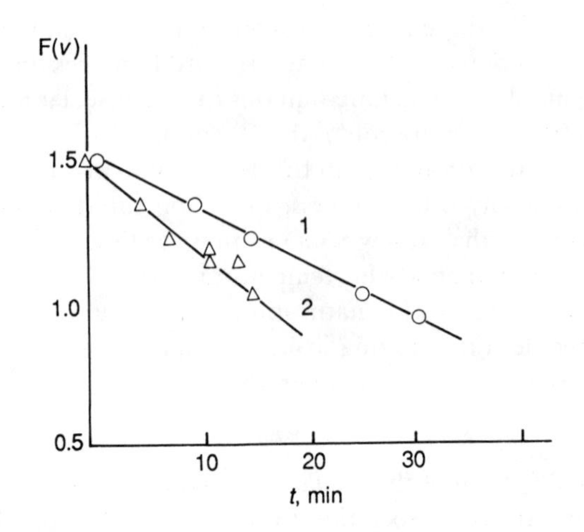

Figure 8 *Determination of* k_1 *by Equation (11) for glucose-6-phosphate dehydrogenase at pH* 8, *concentration of enzyme* 2×10^{-9} M, 40°C (1), *and* 45°C (2)

where v_τ is the rate of reaction at the inflection point of the kinetic curve and v_0 is the initial rate at $t = 0$.

k_1 Can be determined by one of two procedures. The first is based on the uses of equation (10) and is valid under prestationary conditions where $0 < t < \tau^*$:

$$(1/t)\ln(v/v_0) = -k_1 + (k_{-1}k_1^2[E]_0)4/3 \qquad (10)$$

Although in theory this equation should yield both the values of k_1 and k_{-1}, in practice it yields only k_1. Thus it is more convenient to use the simplified form of (10).

$$F(v) = 2(v/v_0) - 0.5(v/v_0)^2 = 3/2 - k_1 t \qquad (11)$$

The transformation of the kinetic curve (at $t < \tau^*$) into a plot of $F(v)$ *versus* t allows k_1 to be calculated with sufficient accuracy. This procedure is exemplified in Figure 8.

The determination of the value of k_d is possible at $t > \tau^*$, when the quasi-equilibrium of dissociation of the labile intermediate is reached

$$E_2x \rightleftarrows 2E_1$$

and the inactivation rate is relatively slow [*i.e.* the conditions in equation (7) apply to scheme (6)].

Here it must be remembered that the final denatured form of the enzyme E_d is unable to re-associate under the prescribed experimental conditions. The equation for this part of the kinetic curve (*i.e.* at $t > \tau^*$) has been shown[9] to take the form:

$$\frac{8[E_2]^{0.5}}{K_{dis}} + \ln[E_2] = \text{const} - 2k_d t$$

with the slope of the semi-logarithmic curve

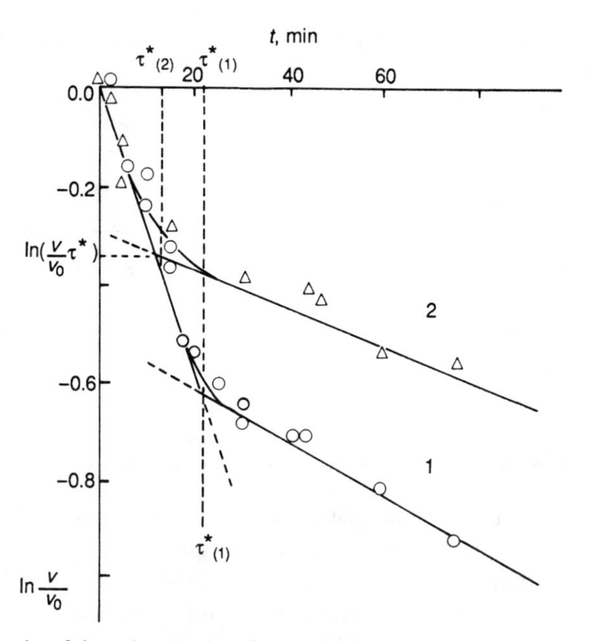

Figure 9 *Kinetics of thermoinactivation of glucose-6-phosphate dehydrogenase at pH* 8 *and* 40 °C. *Initial concentration of enzyme* 2×10^{-8} M (1) *and* 4×10^{-8} M (2). $v_{\tau*}/v_0$ *and* τ^* – *co-ordinates of break point*

$$\frac{d\ln[E_2]}{dt} = I = \frac{2k_d(K_{dis})^{0.5}}{4[E_2]^{0.5} + K_{dis}^{\ 0.5}}$$

This, of course, may be determined experimentally. When all the protein at $t = 0$ is in catalytically active form, this equation is soluble with respect to k_d, which may be determined from the slope of the semi-logarithmic plot to the right of the inflection point (I_τ), *i.e.* at $t > \tau^*$.

$$k_d = I_{\tau*} \frac{v_0 + v_{\tau*}}{2(v_0 - v_{\tau*})} \qquad (12)$$

Figures 6 and 9 show the inactivation of glucose-6-phosphate dehydrogenase, from which can be calculated the values of K_{dis}, k_1, and k_d. Figure 6, curves 1 and 2, show clear inflection points, whereas curve 3 corresponds to the case where the different steps in the degradation process are kinetically indistinct. Figure 9 also shows that the position of the inflection point depends upon the concentration of the enzyme. The temperature range employed is approximately 8 °C. At temperatures below 40 °C, even after 10 hours of incubation, there is insufficient dissociation of the labile oligomer to allow its detection. At temperatures above 48 °C apparently $k_d \gg k_1$ and it is not possible to observe distinct events in the thermodegradation process. Table 3 summarizes the kinetic parameters for the thermodegradation of glucose-6-phosphate dehydrogenase. It can be seen from this that k_1 and k_d are very similar; Figure 9 clearly shows inflection points on the kinetic plots.

Table 4 gives similar kinetic parameters, but this time for alkaline phosphatase

Table 3 *Kinetic parameters for the final steps of the thermodegradation of glucose-6-phosphate dehydrogenase at pH* 8.0 *and* 40 °C

$[E]_0$ 10^8 M	v_τ/v_0	$k_1 \times 10^4$ s^{-1}	$k_d \times 10^4$ s^{-1}	$K_{dis} \times 10^8$ M
2	0.54	3.1 ± 0.6	1.5 ± 0.3	3.1 ± 0.6
4	0.71	3.1 ± 0.6	1.9 ± 0.4	1.9 ± 0.4

Table 4 *Kinetic parameters for the thermoinactivation of alkaline phosphatase at* 60 °C. $[E]_0 = 4.3 + 10^{-8}$ M

pH	v_τ/v_0	$k_1 \times 10^4$ s^{-1}	$k_d \times 10^4$ s^{-1}	$K_{dis} \times 10^8$ M
7.2	0.79	1.3 ± 0.2	3.7 ± 0.3	1.0 ± 0.2
9.0	0.46	9.3 ± 0.4	9.7 ± 0.4	10.4 ± 2.2

at a fixed enzyme concentration but at two different values of pH. The kinetic plots do not display an induction period in this case, but it is supposed that this is because the formation of the labile dimers of the enzyme was complete before the experiment was started. In other words the sample of alkaline phosphatase used was old and not in the best of health!

This shows that at pH 9.0 the dimers of alkaline phosphatase are less stable and the values of K_{dis} and k_1 are much larger than at pH 7.2. Thus we can see that destabilization is mostly connected with a rise in the value of k_1, with little change in k_{-1} (K_{dis} increases by approximately the same factor as k_1).

Kinetic studies, together with other methods of investigation, are useful to investigate the function of quaternary structure and the kinetic mechanisms in large oligomeric enzymes. Approaches based on those described above were used to study the hexamer of glutamate dehydrogenase (EC.1.4.1.3).[16] Biphasic kinetic curves for the thermodeactivation of the enzyme were observed at temperatures between 35–48 °C in the presence of 2-oxoglutarate (one of the substrates). Intermediate products of the thermodegradation process were investigated by HPLC and it was shown that the primary intermediates were trimers. These trimers were relatively stable but of lower enzymatic activity than the native hexamer; inactive dimers and monomers were also observed in the final stages of degradation. Until this time the role of catalytically active trimers in the degradation of glutamate dehydrogenase was unknown.

Taking the approaches outlined above, it is possible from this form of kinetic analysis to determine, under the appropriate conditions, the minimal number of catalytically active intermediates in the degradation process ($\tilde{n}$), the value of the dissociation constant of the labile oligomeric form of the enzyme (K_{dis}), the rate constant of dissociation (k_1), and the real rate constant of irreversible denaturation (k_d). Simple studies of the degradation of oligomeric enzymes at high temperatures, where the process is apparently a first-order reaction, provide little information by comparison. However, the real virtue of an understanding of the approaches described above is that they can provide an explanation for three otherwise difficult to rationalize, but common, observations:

(i) that the process of loss of enzymic activity is rarely first-order;
(ii) that stability of enzymes is often proportional to the enzyme's concentration;
(iii) that different samples of the same enzyme prepared by the same method from the same source may display different operational stabilities.

These theories of Poltorak and Chukrai, taken together with the observations of, for example Henley and Sadana, illustrate one fundamental point in the use of enzymes as practical catalysts and that is that there is only one way to be certain as to how an enzyme will behave in practice, and that is to measure it oneself. However, even having done so, do not expect the same sample, showing the same initial activity some several months later, to behave in the same fashion!

3 ENZYME STABILIZATION

There are a number of approaches which may be taken in the search for more stable enzymes, and each of these will be discussed in turn. Which approach is to be considered the most suitable will obviously depend upon particular circumstances and will finally depend upon considerations of cost effectiveness. Obviously, a combination of approaches may be used. Some of the approaches described are based upon empirical observation, some are based on primordial logic, but few are based on a fundamental grasp of the true nature of enzyme stability. This is one of those areas of scientific endeavour where it is not only difficult to separate out the applied research from the basic, but also one which refuses to obey the principle that basic research leads onto the applied; if anything the converse is true.

3.1 Selection and Genetic Engineering

Based on the dictum that if you do not want to do something yourself, get somebody else to do it for you, the simplest approach to stable enzymes must be to select them from a suitable micro-organism. In this respect the increasing body of knowledge of thermophilic organisms is proving invaluable. Thermophiles, by virtue of their ability to survive at high temperatures, invariably seem to contain enzymes with a high thermal stability. At this time the only major areas in which this approach has been successfully applied is the production of thermostable proteases for inclusion in 'biological' washing powders. One of the limitations of this approach is, unlike mesophilic organisms, our present inability to persuade thermophilic organisms to overproduce a particular enzyme. This problem may, theoretically, be overcome by transferring the gene for a thermostable enzyme from a thermophile into an appropriate mesophile. However, such studies are still largely undeveloped and few successful reports exist in the literature. Ohshima and Soda[29] have reviewed the characteristics of amino acid dehydrogenases from the thermophiles *Bacillus stearothermophilus* and *B. sphearicus* in the context of the use of the enzymes in the production of L-amino acids (particularly L-leucine), and attempts at the gene cloning of these enzymes. Saha and Zeikus[30]

[29] T. Ohshima and K. Soda, *Trends Biotechnol.*, 1989, **7**, 210.
[30] B. C. Saha and J. G. Zeikus, *Trends Biotechnol.*, 1989, **7**, 234.

have reviewed the characteristics of pullulanase and α-amylase produced from various thermophiles, and also briefly examine some potential novel uses of these enzymes.

The reason why enzymes from thermophiles are often more stable than the same enzyme isolated from a mesophile is not, at present, well understood. However, it seems unlikely that it is due to additional covalent intramolecular bonds (that is disulfide bonds) as thermostable enzymes invariably seem to contain fewer disulfide bonds than their less stable counterparts. Indeed, it is almost as if the introduction of disulfide bonds into a protein's structure is a last ditch attempt by nature to introduce a modicum of stability into an unstable structure! It could be suggested that thermostable enzymes are intrinisically stable without the need for extra covalent bonding. In this it seems that the enzyme may rely more on internal hydrophobic bonding, than on more heat labile hydrogen bonds.

3.2 Protein Engineering

The uncertainty concerning the precise reasons for thermostability (or any other stability) is the major limiting factor hindering attempts to increase the stability of enzymes using the techniques of protein engineering (that is, the use of site-directed mutagenesis – see Chapter 14). In the longer term this approach may generate a new breed of super-enzymes. However, it must be reiterated that first we need to know a great deal more about the relationship between protein structure and function. An increasing number of reports now exist, for example describing the use of site-directed mutagenesis to stabilize subtilisin E for use in organic solvents.[31]

3.3 Reaction Environment

There are many examples of instances where the presence of substrate molecules or specific metal ions (*e.g.* Ca^{2+} or Zn^{2+}) have a marked effect on an enzyme's stability. For example, α-amylase (EC.3.2.1.1) from *Bacillus licheniformis* is stabilized by the presence of 4 p.p.m. Ca^{2+} ions to such an extent that it can retain 100% of its activity after six hours incubation at 70°C. In the absence of Ca^{2+} ions it is totally denatured after four hours at this temperature. This approach cannot be applied to all α-amylases; that from *Bacillus amyloliquefaciens* shows no increased stability in the presence of Ca^{2+} ions. Such effects may be explained by the contribution that these additional bonds make to the stability of the protein's structure or indeed to its native conformation. For example, removal of the haem group from myoglobin causes a decrease in total α-helix content of the protein from 75% to 60%.

Addition of high molecular weight hydrophilic polymers (*e.g.* dextrans or polyethyleneglycols) to an enzyme solution can stabilize the enzyme[32] (see also reference 49) either by increasing the viscosity of the solution or by lowering the water activity around the enzyme removing its hydration shell, and thus restrict-

[31] P. Martinez, M. E. Van Dam, A. C. Robinson, K. Chen, and F. H. Arnold, *Biotechnol. Bioeng.*, 1992, **39**, 141.
[32] R. D. Schmidt, *Adv. Biochem. Eng.*, 1979, **12**, 41.

ing the enzyme's ability to alter its conformation. Presumably, by removing the ability of the enzyme's hydrophilic groups to interact with water they have no option but to interact with each other, thus preventing an unfolding of the molecule. This form of stabilization is illustrated by α-amylase which in the presence of 80% dry solid starch 'solution' can be made to operate effectively at a temperature of 110°C.

Recently the use of trehalose (α-D-glucopyranosyl-α-D-glucopyraoside) to stabilize dried preparations of enzymes has been investigated.[33] A remarkable degree of stabilization to heat of nuclease enzymes was observed, which was unique to trehalose amongst a wide range of mono-, di-, tri-, and polysaccharides investigated. The investigators concluded that neither of the two current theories of this form of stabilization, that is the water replacement or the glassy state theory, adequately explained the specificity of effect observed. The water replacement theory could not explain the greatly increased stabilization effect specific to trehalose compared to, for example glucose or maltose; the glassy state theory, which postulates that trehalose undergoes a glass transformation resulting in a vitreous ice like amorphous phase which inhibits molecular motion, cannot explain the stabilization observed at temperatures above the measured glass transition temperatures as measured by differential scanning calorimetry. They concluded that the relative chemical stability and non-reducing nature of trehalose may be the key factors in this effect.

The water concentration of an enzyme solution can of course be lowered by other means, often with the same stabilizing effect. Zaks and Klibanov[34] have reported the stabilization of porcine pancreatic lipase (EC.3.1.1.3) by decreasing the water content of the system. In an aqueous medium the enzyme instantaneously loses activity upon heating to 100°C. However, when the dry powdered enzyme was placed in a mixture of 12 mol dm^{-3} n-heptanol in tributyrylglycerol (suitable substrates for the lipase) the half-life of the enzyme at 100°C was more than twelve hours. In this case the water content of the system was only 0.015%. Increasing the water content to 0.8% caused a reduction in half-life to 15 minutes.

However, these examples also illustrate the problem associated with attempts to restrict conformational changes in enzymes. In both cases the specific activity of the enzyme was reduced, substrate specificity altered, and undesirable by-products formed. In the case of α-amylase various trisaccharides were produced. The stabilized lipase, unlike the native lipase, was less able to utilize bulky secondary and tertiary alcohols, and its specific activity at 100°C was no greater than that for the native enzyme in water at 20°C.

One final form of reaction environment modification which may successfully increase enzyme stability is to increase the protein concentration. This may function in one of three ways. First, if proteolysis or autolysis is the cause of deactivation, then providing a high concentration of an alternative substrate will reduce loss of the enzyme. Secondly, physical interactions between proteins are often dependent upon the concentration. Thus high concentrations whilst apparently reducing the tendency of protein molecules to aggregate and precipitate,

[33] C. Colaco, S. Sen, M. Thangavelu, S. Pinder, and B. Roser, *Biotechnology*, 1992, **10**, 1007.
[34] A. Zaks and A. M. Klibanov, *Science*, 1984, **224**, 1249.

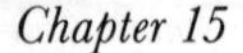

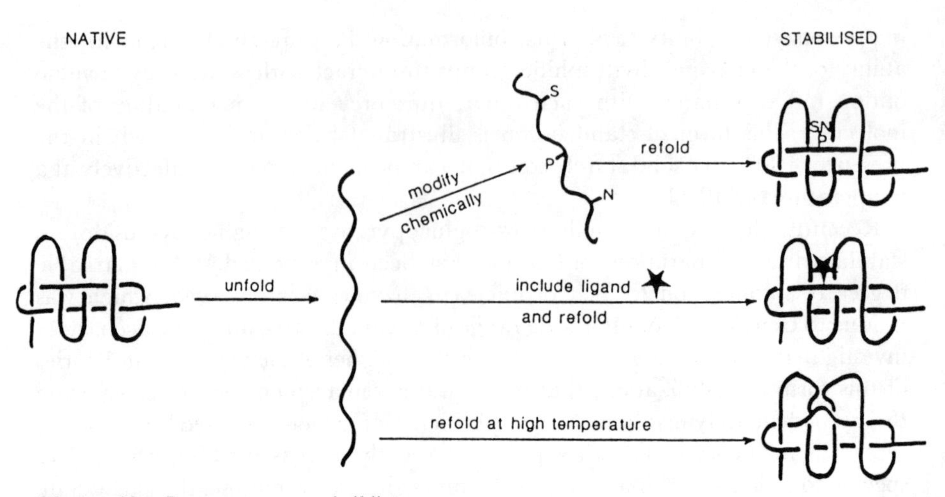

Figure 10 *Routes to enzyme rebuilding*

may well allow specific, stabilizing interactions to occur (see Section 3.7). Forniani *et al.*[35] showed that human glucose-6-phosphate dehydrogenase (EC.1.1.1.49) retains 90% of its activity at 37 °C for 90 days at a concentration of 1 unit cm^{-3} but lost 80% of its activity under the same conditions at 0.06 units cm^{-3}. Thirdly, the stabilization may be the result of the kind of kinetic mechanism proposed above for oligomeric proteins (Section 2).

3.4 Rebuilding

On the theoretical basis that thermostability in enzymes is largely a result of hydrophobic interactions in the core of the enzyme,[36,37] Mozhaev and Martinek[4] have proposed a route to enzyme stabilization by enhancing these interactions. They suggested that in order to gain access to the hydrophobic groups the enzyme is first unfolded (by treatment with urea and a disulfide) and then refolded after chemical modification, in the presence of low molecular weight ligands which would be incorporated into the centre of the enzyme molecule, or refolded in non-native conditions (*e.g.* at a higher temperature) (Figure 10). When they tried the last of these approaches on trypsin, refolding it at 50 °C, the subsequent active enzyme was five times as stable at 80 °C as either the native enzyme or the unfolded enzyme refolded at 20 °C. Presumably the enhanced enzyme stability is a result of strengthened hydrophobic interactions brought about by the increased temperature during refolding.

3.5 Chemical Modification

Chemical modification of the surface of an enzyme molecule as a means of achieving stabilization is a well practised art. In theory any modification to an

[35] G. Forniani, G. Leoncini, P. Segni, G. A. Calabria, and M. Dacha, *Eur. J. Biochem.*, 1969, **7**, 214.
[36] C. Tanford, *Science*, 1978, **200**, 1012.
[37] D. J. Merkler, C. K. Farrington, and F. C. Wedler, *Int. J. Pept. Protein Res.*, 1981, **18**, 430.

enzyme's primary structure may result in conformation effects which will alter stability. The trick is to ensure that these alterations are beneficial. By virtue of the lack of understanding of the physical relationship between structure and stability this, and other, approaches have been largely empirical, many modifications causing either total loss of activity or reduced stability. However, some encouraging results have been reported.

For example, Tuengler and Pfleiderer[38] enhanced the heat and alkali stability of lactate dehydrogenase (EC.1.1.1.27) by treating the enzyme with methyl acetimidate. Interestingly this treatment also increased the enzyme's resistance to digestion by trypsin; presumably the modification of the enzyme surface obscured the normal points of attack of trypsin.

Torchillin *et al.*[39] describe the stabilization of α-chymotrypsin by the modification of surface amino groups. The 15 available amino groups were either alkylated or acylated with acrolein or succinic anhydride respectively. Modification of up to 5 amino groups with acrolein gave no significant stabilization; with between 5 and 13 amino groups modified there was a marked, exponential increase in stability up to 120 times that of the native enzyme. Modification of all 15 amino groups produced an enzyme with no more stability than the native enzyme. Torchillin postulated therefore that there were just a few (2–5) amino-groups whose modification might lead to stabilization. Acetylation gave a similar pattern but without such striking stabilization.

Lipases and their stability (or lack of it) have attracted increased attention in recent years, as has the application of the concept of chemical modification to enhance the stability of this group of enzymes. For example, Kawase *et al.*[40] reported the improvement in heat stability of yeast lipases produced as a result of modification with hetero-bifunctional photogenerated reagents; Kaimal *et al.*[41] demonstrated the enhancement of the catalytic activity of porcine pancreatic lipase by reductive alkylation.

3.6 Intramolecular Cross-linking

The modification of an enzyme's surface structure may be taken one stage further to the cross-linking of groups on the surface. By linking rigid molecular brackets onto and around the enzyme molecule, the potential for conformational change should, in theory, be reduced and the enzyme's stability therefore enhanced. Stabilization of oligomeric proteins may be achieved by cross-linking the subunits together. Dialdehydes, in particular glutardialdehyde, di-isocyanates, and bisdiazonium salts have been most commonly used as these molecular brackets. The stabilizations achieved have been somewhat variable, often being a result of single-point chemical modification, largely because of the lack of any attempt to match the size of the bracket molecule to the distance between reactive groups on the enzyme's surface.

[38] P. Tuengler and G. Pfleiderer, *Biochim. Biophys. Acta*, 1977, **484**, 1.
[39] V. P. Torchillin, A. V. Mak, I. V. Berezin, A. M. Klibanov, and K. Martinek, *Biochim. Biophys. Acta*, 1979, **567**, 1.
[40] M. Kawase, K. Sonomoto, and A. Tanaka, *J. Ferment. and Bioeng.*, 1990, **70**, 155.
[41] T. N. B. Kaimal and M. Saroja, *Biotech. Lett.*, 1989, **11**, 31.

Table 5 *Effect of alkyl chain length of diaminoalkyl cross-linking reagent on the stability of native and succinylated α-chymotrypsin. Stability is inversely related to thermodeactivation rate constant (from Ref. 40)*

Chain length of cross-linking alkyldiamine	*Rate constants of thermodeactivation* min^{-1}	
	Native chymotrypsin	*Succinylchymotrypsin*
Non-crosslinked	0.25	0.25
0	0.48	0.05
2	0.10	0.01
4	0.08	0.04
5	0.15	0.05
6	0.24	0.09
12	0.27	0.07

By taking the surface geometry of the enzyme into account Torchillin *et al.*[42] found that α-chymotrypsin was markedly stabilized when tetramethylenediamine was covalently linked to carbodi-imide activated carboxy-groups on the enzyme's surface. Hexamethylenediamine, however, produced no stabilization, whilst single-point attachment of aminopropanol actually destablized the α-chymotrypsin. When the surface content of carboxylic acid groups was increased by treating the enzyme with succinic anhydride, ethylenediamine was found to be the most effective stabilizing agent. In addition the stability was even greater than for the native, tetramethylenediamine-treated enzyme. Presumably the distance between modified carboxylic acid groups was decreased, requiring a shorter bracket for optimal stabilization, and the number increased allowing more brackets to be applied (see Table 5).

Another approach, reviewed by Shami *et al.*,[43] is the only use of poly- and monoclonal antibodies to stabilize various enzymes with little or no loss in activity.

3.7 Immobilization

If ever there was to be a goose with a golden egg in enzyme technology, immobilization of enzymes must be the number one candidate. However, there is little general agreement about the precise date of ovulation. Nevertheless, one of the most frequent claims for the benefits of immobilization is the potential for enhancement of the enzyme's stability. As we shall see, however, the field is positively mined with artefacts and the wise tread warily.

Immobilization may affect an enzyme's activity in a number of ways (Figure 11). The process of immobilization may in itself alter the *intrinsic* characters of the enzyme. Partitioning effects of the polymer may separately give rise to a change in the enzyme's *inherent* parameters and, in addition, diffusional constraints

[42] V. P. Torchillin, A. V. Maksimenko, V. N. Smirnov, I. V. Berezin, A. M. Klibanov, and K. Martinek, *Biochim. Biophys. Act*, 1977, **522**, 277.

[43] E. Y. Shami, A. Rothstein, and M. Ramjeesingh, *Trends Biotechnol.*, 1989, **7**, 186.

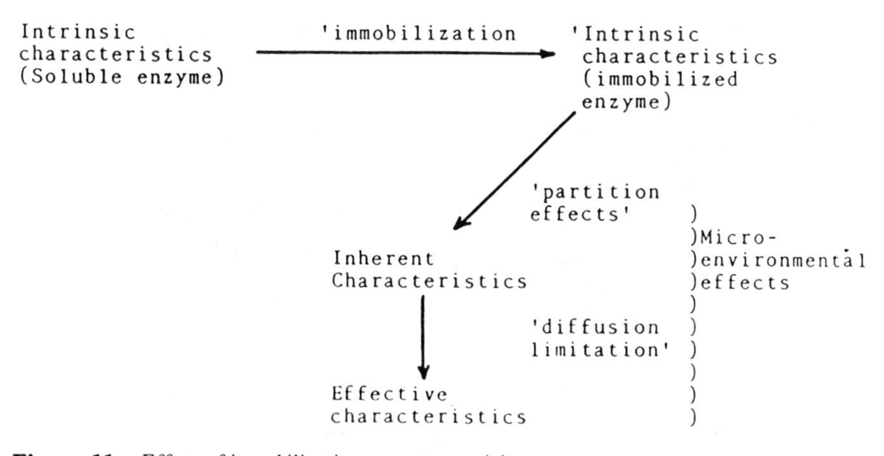

Figure 11 *Effects of immobilization on enzyme activity*

imposed by the polymer matrix may alter the inherent parameters to give the experimentally observed *effective* parameters.

The most obvious alteration of intrinisic characteristics would be brought about by total or partial inactivation of the enzyme, caused by gross conformational changes, or reaction of some essential group at the enzyme's active site. More subtle conformational changes induced in the enzyme could cause destabilization or alteration of allosteric effects or kinetic parameters. Some enzymes demonstrate enhancement of effective stability when immobilized, the reasons for which are discussed below.

The microenvironment provided by the polymer matrix affects the activity of the immobilized enzymes by causing heterogeneity of distribution of solute. Microenvironmental effects may be subdivided into *partitioning* effects and *diffusion limitation* (Figure 12).

Partitioning of solutes by the polymer matrix is a result of hydrophobic or electrostatic interactions and affects reacting and non-reacting solutes alike.

Limitation to free diffusion of solute molecules by the physical presence of the polymer matrix will alter the microenvironmental concentrations of solutes taking part in the reaction. The interaction between diffusion and reaction produces non-linear solute distribution throughout the polymer matrix. Diffusion limitation may be either external (*i.e.* up to the enzyme–polymer surface through the unstirred Nernst layer) or internal (*i.e.* diffusion within the polymer matrix). Where external diffusion limitation is present reaction occurs after diffusion whereas when internal diffusion limitations are present diffusion and reaction occur concurrently.

Thus immobilization provides a number of scenarios by which an enzyme may be stabilized, and many claims have been made for enzyme stabilization as a consequence of immobilization. With systems of such complexity two questions arise. Are these stabilization effects real or merely apparent? Does it matter?

To take the second question first the answer is, in many instances, probably not. Where an immobilized enzyme is to be used as, for example, an industrial catalyst or as a biosensor then the important consideration is operational stability

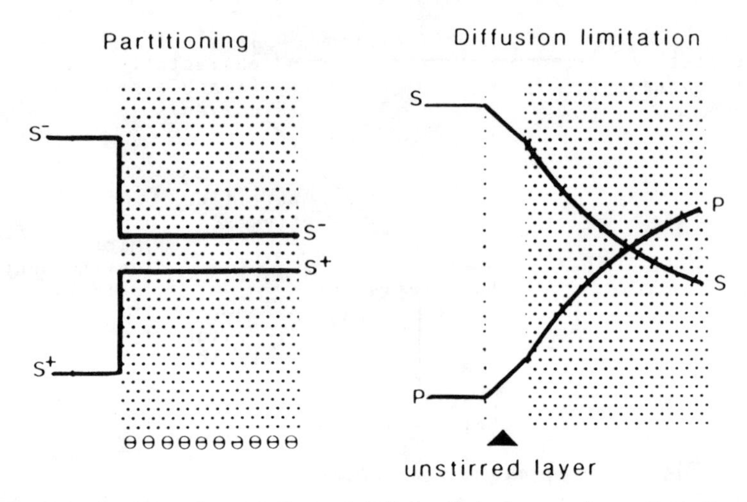

Figure 12 *The effect of partitioning and diffusion limitation on solute concentration profiles in immobilized enzymes*

of the whole immobilized enzyme and the cause and reality of the stability are largely irrelevant. However, if the objective is to understand how immobilization may stabilize an enzyme then the real must be disentangled from the artefactual. The rest of this discussion then will concern itself with such distinctions, starting with real intrinsic stabilization.

In retrospect, and in view of work of the nature discussed above on chemical modification and intramolecular cross linking of enzymes, it now seems evident that multipoint binding of an enzyme molecule to a polymer surface might bring about some degree of intrinsic stabilization. The emphasis here is on multi-point binding; obviously single-point binding is unlikely to constrain conformational changes in the protein. Most usually the binding will be covalent, although electrostatic interactions may be involved. For example, trypsin attached to ethylene maleic anhydride copolymer is active in 8 mol dm^{-3} urea;[44] α-chymotrypsin entrapped in 50% polymethacrylic acid has a theoretical half-life at 60 °C of several million years.[45] However, unguarded optimism for this approach must be tempered by four cautions. First, the surfaces of the enzyme and polymer are unlikely to be complementary, and unless the polymer is built up around the enzyme only part of the enzyme's surface may be rigidified with little consequent stabilization. Second, stabilization might be due to chemical modification of the enzyme. Third, the enzyme may be destabilized because its conformation is locked in an unstable state. Fourth, conformational change is important to the catalytic function of an enzyme, and restricting such changes may well cause inactivation or alter its kinetic characteristics (*e.g.* substrate specificity).

The second way in which real stabilization of an immobilized enzyme may be achieved is through the effect of protein concentration on stability as discussed

[44] L. Goldstein, *Biochemistry*, 1964, **3**, 1913.
[45] K. Martinek, A. M. Klibanov, A. V. Tchernyshera, V. V. Mozhaev, I. V. Berezin, and B. O. Glotov, *Biochim. Biophys. Acta*, 1977, **485**, 13.

above. Obviously, when an enzyme is immobilized, the local concentration of enzyme within or around the polymer matrix is far greater than the average value throughout the reaction medium.

We can now turn to those stabilizing effects of immobilization which are more apparent than real; that is, the intrinsic stability of the enzyme is not altered, but the effective stability of the immobilized enzyme, when compared to the free native enzyme, is enhanced. One such artefact, the differential denaturation of multiple forms of an enzyme with varying stabilities as a result of the immobilization procedure, has already been discussed (Section 2).

The presence of a microenvironment around an immobilized enzyme may prevent access by denaturing agents (*e.g.* H^+ ions, organic solvents) to the enzyme molecule. For example, denaturation at unfavourable pH may be prevented by coupling an enzyme to a polyelectrolyte. The buffering capacity of the polymer in the locality of the enzyme will be far greater than could be achieved with a solution of buffer. If a polycation is used the positively charged polymer will partition H^+ ions away from the enzyme. In either case the pH around the enzyme will be different, and possibly less harmful to the enzyme, to that in the bulk phase. Dixon *et al.*[46] have reported such stabilization of lactate dehydrogenase (EC.1.1.1.27) covalently attached to porous glass. Similarly, stabilization of enzyme to amphipathic polymers, such as polyethylene glycol, has been proposed.[47,48] In a recent review of studies using this approach, Inada *et al.*[47] summarize the effect the attachment of soluble 2,4-(*O*-methoxypolyethylene glycol)-6-chloro-*s*-triazine to a wide range of enzymes has on their activity in organic solvents. The activity of the enzyme depended, in general both on the degree of substitution with the activated PEG – the more substitution the greater the activity, and on the nature of the solvent. In practice, the enzyme was protected from water-immiscible solvents but not water-miscible ones. Presumably, the PEG either excludes the organic solvent from the enzyme's microenvironment and/or itself replaces the enzyme's hydration shell. This approach towards enzyme stabilization has also resulted in a number of patent applications.[49]

Clostridial hydrogenase has been stabilized to oxygen deactivation by immobilizing it to polyethyleneimine cellulose, when it has a half-life of 1 week in air-saturated water, compared with 4 minutes for the free enzyme.[50] Oxygen solubility in water is reduced as ionic strength is increased, and it seems likely that oxygen would be virtually insoluble in the highly ionic microenvironment provided by the polymer matrix. Proteolytic enzyme molecules constrained within or by a polymer matrix are unlikely to come into contact with each other, and thus immobilized proteases rarely suffer from autolysis.

Perhaps the most common cause of stabilization by immobilization is diffusion

[46] J. E. Dixon, F. E. Stolzenbach, J. A. Berenson, and N. O. Koplan, *Biochim. Biophys. Res. Commun.*, 1973, **52**, 905.
[47] Y. Inada, K. Takahashi, T. Yoshimoto, A. Ajima, A. Matsushima, and Y. Saito, *Trends Biotechnol.*, 1986, **4**, 190.
[48] M. W. Baillargeon and P. E. Sonnet, *Fed. Proc. Fed. Am. Soc. Exp. Biol.*, 1986, **45**, 1539.
[49] German Pat. No. DE 3508-906.
[50] A. M. Klibanov, N. O. Nathan, and M. D. Kamen, *Proc. Natl. Acad. Sci. USA*, 1978, **75**, 3640.

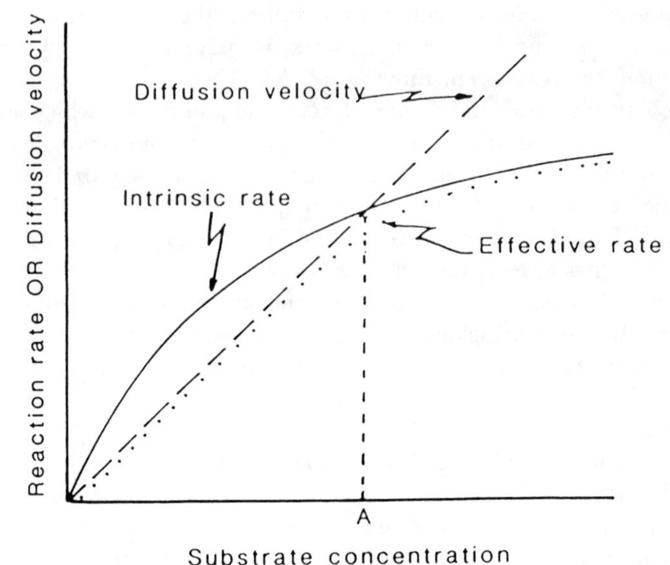

Figure 13 *Relationship between intrinsic reaction rate* (——), *effective reaction rate* (. . .), *and diffusion velocity* (---) *for an immobilized enzyme*

limitation.[2,51] As we have seen the polymer matrix may present a barrier to the free diffusion of substrate molecules to the enzyme. In effect the conversion of substrate into product becomes a sequence of events, diffusion of substrate up to and into the polymer, catalysis, and diffusion of product back into the bulk phase. The overall rate will be governed by the slowest step which may be substrate diffusion. This is illustrated by Figure 13, which schematically shows the relationship between the diffusion rate of substrate, the potential intrinsic rate of reaction, and the actual effective rate of reaction of an immobilized enzyme when the substrate concentration in the bulk phase is varied. At substrate concentration below A the rate is controlled by the diffusion rate, whereas above A it is under the kinetic control of the enzyme reaction. Clearly a higher enzyme concentration would lead to an increase in the value of A and result in the actual effective reaction rate being a function of diffusion rate for most substrate concentrations. The converse will be true for a reduced enzyme activity. It can also be clearly seen that raising the substrate concentration will overcome diffusion limitation of the effective rate of reaction. Thus if substrate diffusion is limiting, events which might cause a change in the intrinisic activity of the enzyme, *e.g.* partial thermal denaturation of changes in pH, will have little or no effect upon the effective catalytic activity of the immobilized enzyme particle, until such time that the intrinisic enzyme activity falls below the rate of substrate diffusion and itself becomes rate limiting. To put it another way, if the polymer matrix is highly loaded with enzyme, all the substrate molecules diffusing into the particle may be converted into product by the enzyme con-

[51] M. D. Trevan, *Trends Biotechnol.*, 1987, **5**, 7.

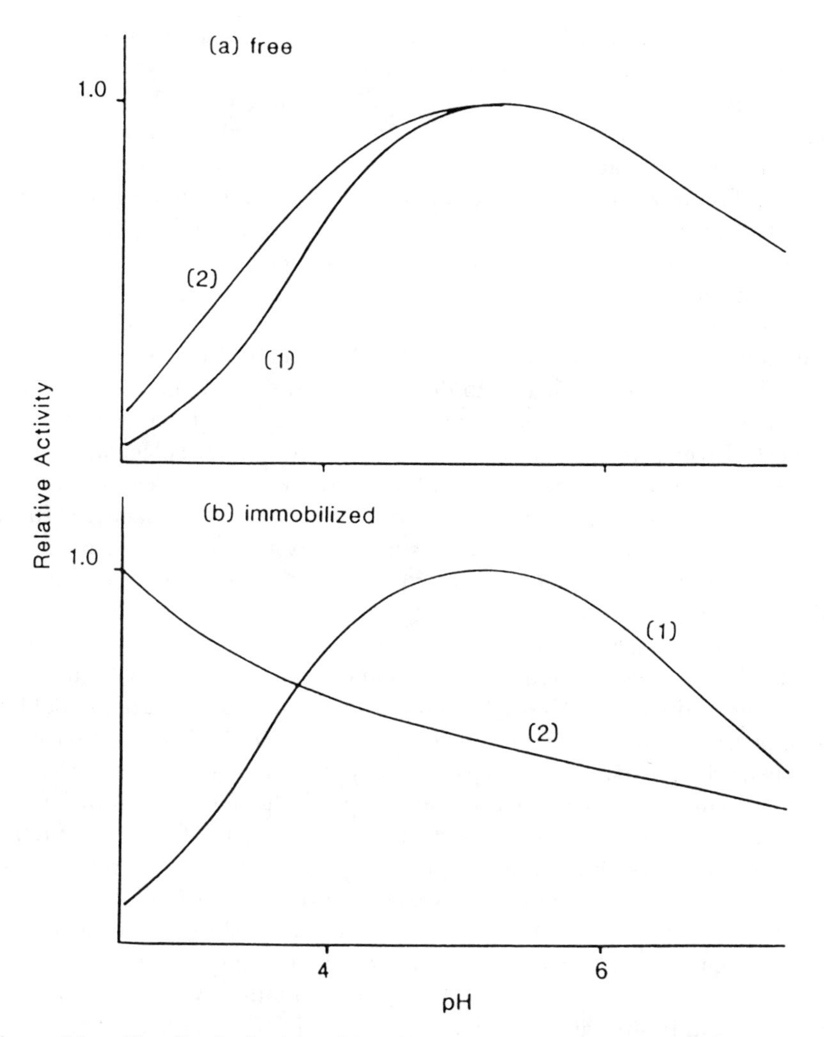

Figure 14 *pH profiles for free* (a) *and immobilized* (b) *glucose oxidase at* 0.1 M (1) *and* 0.1 M (2) *glucose concentrations*

tained within the outer shell of the polymer particle. The particle therefore has a large reserve of unused enzyme to replace any enzyme molecules inactivated by heat, pH, *etc.* It is interesting to note at this point that immobilized enzymes subject to severe diffusional limitation may not only appear very stable, but will also be largely insensitive to normal effects of their kinetic characteristics; inhibitors will appear ineffective or the effective reaction rate may show almost no response to pH.

For example, Figure 14 shows the response of glucose oxidase both in the free form (a) and when immobilized by entrapment in polyacrylamide (b) at different substrate concentrations. Quite clearly, the pH profile of the free enzyme is unaffected by substrate concentration, whereas the immobilized enzyme at high substrate concentrations behaves like the free enzyme, but at low substrate concentrations becomes more or less insensitive to pH. This is, of course, in keeping with the fact that substrate diffusion limitation will become more pronounced as the substrate concentration is lowered. Just about the only thing which will affect the rate of reaction will be the substrate concentration and this in a linear fashion over several orders of magnitude. (Were it not for effects such as this the construction and use of biosensors would be much complicated.)

However, no matter how logical potential explanations may seem, there are always enigmatic cases. Lenders and Crichton[52] studied the effect of immobilization to dextran of two enzymes, amylase and pullulanase. The striking feature of this study is that there appears to be little coherence or generality to the results. Thus binding of amylase to dextran (to give a conjugate of molecular mass $2–6 \times 10^5$) shows a significant degree of stabilization to both heat and urea, the effect of which is directly related to the degree of intermolecular bonding between enzyme and support polymer, but the value of K_m decreases.

We shall conclude this section with two tales of enzyme stabilization. In the first case immobilization was used as means both to stabilize the enzyme and to allow its use in a continuous reactor. The enzyme in question, fumarase (EC.4.2.1.2), was present as a non-living cell preparation of either *Brevibacterium ammoniagenes* or *B. flavum*. In their studies Takata *et al.*[53] used two basic methods of immobilization, entrapment in polyacrylamide or κ-carrageenan, both with or without the addition of polyethyleneimine. The various preparations could be ranked in order of increasing stability: free cells < free cells plus polyethyleimine < polyacrylamide-entrapped cells < polyacrylamide-entrapped cells plus polyethyleneimine < κ-carrageenan-entrapped cells and most stable κ-carrageenan-entrapped cells treated with polyethyleneimine. This latter preparation was some 22 times more stable than the polyarylamide preparation. They concluded that this high stability towards heat, urea, high pH, and ethanol (see Table 6) was a result of a three-way interaction between the cell membrane, polyethyleneimine, and the κ-carrageenan. They did not speculate how such an interaction could stabilize an individual enzyme contained with the cell, nor is it easy to explain. It is possible that some of the stabilization effect could have been real and due to the presence of the polyethyleneimine, as polyamines are known to stabilize cells and are used as cryoprotective agents. However, much of the observed stabilization could have been due to diffusion limitation. Certain indicators of diffusion limitation were present, for example broadening of pH profiles, less than expected increase of catalytic activity with increases in temperature, high catalyst loading. The differences observed between the different immobilization matrices could simply reflect variations in polymer structure and hence diffusional resistance; the inclusion of polyethyleneimine changed some of the physical characteristics of the κ-carrageenan gel. In addition, the initial loading of cells/enzyme in the preparations could have been affected by the immobilization method. Acrylamide monomers, and the free radicals produced by the polymerization reaction in the formation of polyacrylamide gels, are

[52] J. P. Lenders and R. P. Crichton, *Biotech. Bioeng.*, 1984, **26**, 1343.
[53] I. Takata, K. Kayoshima, T. Tosa, and I. Chibata, *J. Ferment. Technol.*, 1982, **60**, 431.

Table 6 *Comparison effect of immobilization method on stability of fumarase of* Brevibacterium flavum *(from Ref. 53)*

Cells and immobilization method	*Initial fumarase activity (units/*ml *gel) at* 37°C	*Half-life days at* 37°C	*Relative remaining activity (%) after heat treatment*	
			at 55°C *for* 60 min	*at* 60°C *for* 15 min
Free cells				
Native	n.d.	n.d.	18	0
Immobilized cells				
Polyacrylamide	10.2	94	60	0
Polyacrylamide (*B. ammoniagenes*)	8.8	53	n.d.	n.d.
Polyacrylamide polyethyleneimine	n.d.	n.d.	88	37
κ-Carrageenan	15.0	160	100	59
κ-Carrageenan + polyethyleneimine	16.3	243	100	100

known to be toxic and it is thus likely that these preparations contained a lower active concentration of cells/enzyme than the κ-carrageenan.

The second tale of enzyme stabilization is told in a series of recent papers by Evans *et al.*[54–57] It concerns the enzyme phenylalanine ammonia lyase (EC.4.3.1.5) – PAL. Normally this enzyme undertakes the deamination of phenylalanine to *trans*-cinnamic acid and ammonia. However, at high values of pH (> 10) and ammonia (4–7 M) and with *trans*-cinnamic acid concentrations 200 mM the reaction will run backwards, thus producing (expensive) L-phenylalanine from (cheap) substrates. Interest in routes for the production of L-phenylalanine has been growing since the introduction of the dipeptide sweetener aspartame (α-L-aspartyl-L-phenylalanine-α-methyl ester). However, although a large-scale process using PAL has been introduced[58] by the Genex Corporation, serious problems of PAL instability in the harsh reaction conditions are experienced. Evans *et al.* used a permeabilized yeast *Rodotorula rubra* as the source of the enzyme. The enzyme is easily inducible, but undergoes rapid deactivation (< 2 h) under normal culture conditions. The reason for this deactivation seems to be either specific breakdown by a cellular protease and/or inherent instability of the molecule. Whatever the reason, isoleucine and phenylalanine retard this inactivation, whereas oxygen and chloride ions potentiate it. Not content with

[54] C. T. Evans, D. Conrad, K. Hanna, W. Peterson, C. Choma, and M. Misawa, *Appl. Microbiol. Biotechnol.*, 1987, **25**, 399.
[55] C. T. Evans, K. Hanna, C. Payne, D. Conrad, and M. Misawa, *Enzyme Microbiol. Technol.*, 1987, **9**, 417.
[56] C. T. Evans, C. Choma, W. Peterson, and M. Misawa, *J. Ind. Microbiol.*, 1987, **2**, 53.
[57] C. T. Evans, C. Choma, W. Peterson, and M. Misawa, *Biotechnol. Bioeng.*, 1987, **30**, 1067.
[58] B. K. Hamilton, H.-Y. Hsiao, W. Swann, D. Anderson, and J. Delente, *Trends Biotechnol.*, 1985, **3**, 64.

using just one method, Evans *et al.* employed a whole battery of different approaches. Thus the cells producing the enzyme were grown initially in fairly conventional aerobic media, until a maximum of enzyme had been induced, when the medium was deaerated by gassing with nitrogen. All subsequent stages were also operated under nitrogen. Chloride ions are replaced by sulfate ions in the reaction mixture. The polyhydric alcohols, sorbitol, and/or glycerol (used for many years to stabilize enzymes) were added to the reaction mixture, as was polyethyleneglycol (PEG) and the cells were cross-linked with glutaraldehyde. The relationship between these last two factors in their ability to stabilize the enzyme was complex, and illustrates the essentially empirical nature of much work of this type. Untreated cells (*i.e.* no glutaraldehyde) in the absence of PEG lost most of their activity in 2 h; in the presence of either 1% or 15% PEG most activity was lost after 19 h. Cells treated with 0.1% glutaraldehyde in the presence of 1% PEG lost most of their activity in 28 h, but in the presence of 15% PEG retained full activity for this period. Concentrations of glutaraldehyde >0.1% resulted in inactivation of the PAL. By using this combination approach up to 50 g dm^{-3} L-phenylalanine could be produced repeatedly in fed-batch systems, with 80% yield even after 50 days operation. The process was adapted to produce stabilized cells co-immobilized with PEG in a glutaraldehyde-hardened calcium alginate beads, which were then placed in a continuous, fluidized bed reactor. This reactor was operated continuously for 120 h, at conversion rates of between 0.45 and 0.3 g dm^{-3} h^{-1} L-phenylalanine. Although the approach was entirely empirical, this is not to imply any criticism. The important question when faced with an unstable enzyme is why do you want to stabilize it? The answer is a matter of identifying priorities. If the enzyme is to be used in an important industrial role, then there probably will not be time to undertake several years of fundamental physico-chemical studies on the enzyme's stability, so you use every known weapon in the arsenal of stabilization until it is stable.

4 STABILITY OF VIABLE CELLS

The discussion above has been largely confined to enzyme stabilization; what then of viable cells? Early mythology in this field seems to tell the story that immobilization increases the longevity of cells, in particular plant cells. What follows is an attempt to divine the truth of the situation.

There are two reasons why it might be advantgeous to have stable viable cell preparations. The first is obvious: the more stable they are the longer you can go on using them. The second reason is the desire to find ways of storing cell populations. Many cell types can be stored easily under liquid nitrogen. However, some, notably plant cells, are not so easy to store, and thus for some years investigations have been carried out searching for suitable cryoprotective agents. It is on the extension of longevity that we shall concentrate. However, it must be noted that in the many reports of the effects of immobilization on the behaviour of living cells, the key interest appears to be in productivity rather than longevity.

Amongst the early writings in the field in 1981, Bucke and Wiseman[59] reported

[59] C. Bucke and A. Wiseman, *Chem. Ind. (London)*, 1981, 234.

that increases in the longevity of plant cells, in particular, was an unexpected finding. They cite a number of examples, including the production of alkaloids by immobilized plant cells, and postulate that immobilization might by some means divert cell metabolism from growth towards secondary product formation. That is to say, it might induce a prolonged stationary phase in cells, under conditions where nutrient supply is not limiting, and thus cell death does not follow. Of the various immobilization matrices they mention, however, only alginate seems to have this effect on longevity. It must be noted, however, that, with plant cells in particular, caution must be exercised in interpreting such results, because the comparison of longevity is usually made between free-suspension culture cells and immobilized cells whose morphology usually more closely resembles that of callus culture.[60] This complication is avoided in the lower plants, the algae, many of which are naturally free living unicellular or small colonial forms. In a review of immobilized algae[61] a number of papers are cited which tend to indicate that growth rates are indeed depressed by immobilization and that some significant changes to the metabolic behaviour of the cells may be taking place. For example, respiration rates are generally depressed, whereas photosynthetic oxygen liberation rates are often enhanced. This latter observation would be in concordance with other reports that the protein–chlorophyll complexes of various algae are markedly stabilized by immobilization (in alginate). These and similar findings have led to the suggestion that immobilization of cells in alginate may be an easier method of storage of cell lines than cryopreservation.

There are an increasing number of reports of the use of co-immobilized cell preparations, which whilst not directly enhancing the stability of the producer cells, enhance the quantity and longevity of production. For example, Khang *et al.*[62] demonstrated that by co-immobilizing fungal and algal cells the β-lactam production by the fungal cells was increased.

Few hypotheses have been advanced, however, to explain these effects. The most notable, by Mattiason and Hahn-Hagerdal,[63] suggests that most of these effects may be the result of the immobilization matrix lowering either the water activity or oxygen supply to the immobilized cells. It is well known that insoluble polymers, such as calcium alginate, can effectively bind and structure the water surrounding them and may, by virtue of their polyionic nature, reduce the oxygen concentration in the vicinity of the immobilized cells. Thus, they suggest, the lowered water activity surrounding the immobilized cells leads to an increase in the maintenance metabolism, with subsequent slowing of growth and enhancement of secondary metabolite production. By way of example, they cite the effect of lowered water activity on lactobacilli, which is to decrease lactate but increase diacetyl production; immobilized yeasts can produce increased yields of ethanol from glucose in excess of the theoretical 51% maximum yield. In

[60] A. Dainty, K. H. Goulding, P. M. Robinson, I. Simkins, and M. D. Trevan, *Trends Biotechnol.*, 1985, **3**, 59.
[61] P. K. Robinson, A. Mak, and M. D. Trevan, *Process Biochem.*, 1986, August, 122.
[62] Y. H. Khang, H. Shanakar, and F. Senatore, *Biotech Lett.*, 1989, **10**, 867.
[63] B. Mattiason and B. Hahn-Hagerdal, *Eur. J. Appl. Microbiol. Biotechnol.*, 1982, **16**, 52.

relation to the effect of lowered oxygen availability, they cite the example of the reduction of cell growth in κ-carrageenan of immobilized *Bacillus amyloliquifaciens* coupled to an increase in α-amylase production, when it is known that the cells are operating under limiting oxygen supply.

5 CONCLUSION

We have seen in the course of this discussion how enzyme stability may be enhanced and why this is desirable. Many of these methods may indeed yield enzyme preparations of enhanced intrinsic stability, albeit at some cost, whereas for others, notably immobilization, enhanced stability may be apparent rather than real. To confuse the picture, enzyme deactivation is not necessarily a simple monomolecular process but may be a complex event. It should be clear, therefore, that great caution must be exercised when claiming superior intrinsic stability for an enzyme, as real stabilization as a result of manipulating the enzyme is probably the exception.

In practice, however, operational stability of a biocatalyst is the goal and the exact cause of explanation of enhanced stability is of secondary importance. He cares not for the means who profits by the ends.

Acknowledgements. We are grateful to all those who have provided help and advice in the preparation of this chapter, in particular to Dr. Christopher Evans for his indispensable help in the provision of information and critical evaluation of the section on phenylalanine ammonia lyase.

CHAPTER 16

Immobilized Biocatalysts

A. ROSEVEAR

1 INTRODUCTION

Most chemical reactions involving organic materials can be catalysed by specific enzymes, derived from living cells. A substantial number of these very specific biocatalysts are now commercially available or can be isolated from biological material using straightforward techniques. In order to take full advantage of the benefits of improved processing or more specific analysis that might arise from using these catalysts, the engineer or scientist has to consider how best to bring the components of reaction together in the most efficient manner. Immobilization is one of the most efficient mechanisms to facilitate this.

2 IMMOBILIZATION

2.1 The Concept

Immobilization is the means by which enzymes and cells are transformed into heterogeneous catalysts. The biocatalyst is confined to a restricted region through which the substrate solution is passed, emerging as a catalyst-free product. This approach makes it easier to control the environment in the immediate vicinity of the reaction and can dramatically increase the process intensity by concentrating all activity into a small reaction volume. Fortunately a decision to adopt the term 'immobilization' was taken by the pioneers of this technology and this has simplified the task of those wishing to survey the area. FEBS have attempted to unify the way in which data on immobilization is recorded,[1] and this further facilitates the intercomparison of relevant information. It is to be hoped that newcomers to the topic will continue to observe these conventions.

2.2 General Practice

There are numerous ways in which biocatalysts can be confined so as to create a heterogeneous reaction system. Successful immobilization ensures that there is

[1] FEBS Working Party, *Enz. Microb. Technol.*, 1983, **5**, 304; *ibid*, **9**, 315.

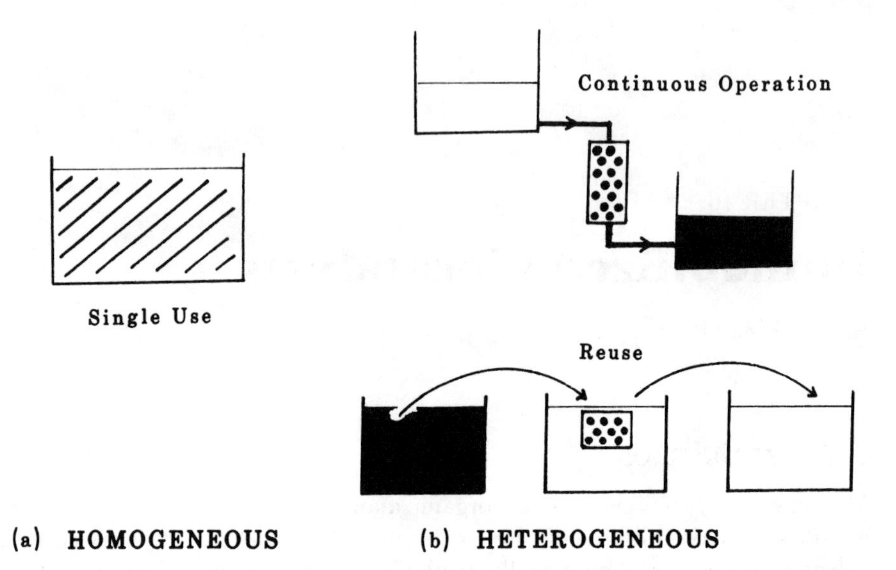

Figure 1 *The Basic Concept of Immobilization. Retaining the heterogeneous biocatalyst makes possible controlled, continuous operation or repeated use*

efficient contact between the phase containing the catalyst and the phase carrying substrate in and product away from the biologically active region. The most common means of achieving this is to have substrate in a continuous, aqueous phase and catalyst in a solid phase, retained within a reactor vessel (Figure 1). However, the same principle applies if the substrate is in a gas phase[2] or in a water immiscible phase.[3] The catalyst does not even have to be fully attached to support material so long as the complex is retained in the reactor vessel. For instance, enzymes and cells confined by means of a semi-permeable membrane will act as immobilized biocatalyst systems.

Methods for immobilizing biocatalysts and the types of reactor in which they can be employed are reviewed below. The problems which are created during immobilization will then be considered and finally ways in which the advantages of immobilization can be exploited will be discussed.

3 BIOCATALYSTS

The term biocatalyst covers a range of biological materials which are capable of transforming many substrate molecules into product, without themselves being changed. Most immobilized processes are concerned with aggregating these active materials on support particles which have larger dimensions than the free

[2] J. T. Enwright, J. L. Gainer, and D. J. Kirwan, *J. Environ. Sci.*, 1975, **9**, 586.
[3] I. V. Berezin and K. Martinek, *Ann. N.Y. Acad. Sci.*, 1984, **434**, 577.

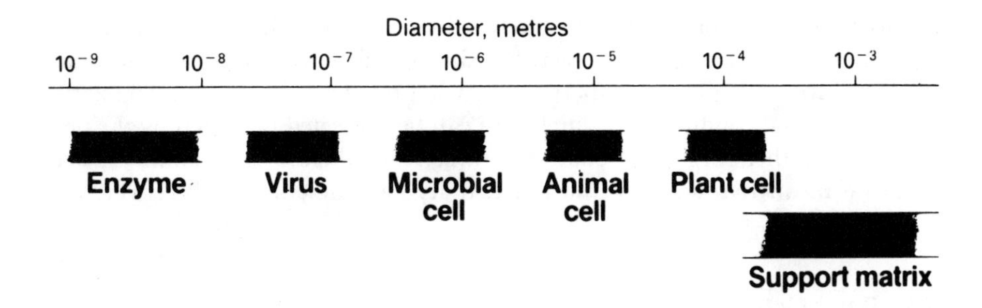

Figure 2 *Relative size of typical biocatalysts and the support matrices used for immobilization*

biocatalyst (Figure 2). Detailed reviews have been published on the immobilization of enzymes,[4] microbial cells,[5] algae,[6] plant cells,[7] and animal cells.[8]

3.1 Enzymes

The simplest biocatalytic unit is the enzyme. For the purposes of this discussion enzymes can be considered to be a protein chain, folded in a very specific manner to create an active site at which the reaction occurs. During immobilization this precise structure must be retained and steric hindrance of the active site avoided. Enzymic proteins are generally soluble in water so if they are not tightly bound to the support they will be swept away in the process fluids. However, the peripheral, less essential groups on the protein are relatively reactive chemically and so can be used to bind the whole enzyme to a support.

3.2 Live cells

The living cell is a much larger biocatalytic unit and unlike an isolated enzyme is not soluble in the process fluid. Cells contain a multitude of individual enzymes, held within a lipid bilayer membrane and the cell is capable of self-replication. Even the simplest prokaryotic cell has a diameter over 100 times that of an enzyme and depends for its activity on the co-ordinated functioning of organized sub-structures within the cellular membrane. The higher eukaryotes, such as plant and animal cells, are even larger than the simple bacteria. The *in vitro* cells used for biocatalysis are generally tissue cultures, derived from multicellular organisms which maintain homeostasis and rely on groups of differentiated cells performing specialist functions.

Cells are already heterogeneous catalysts in that they are solids, but in practice their neutral buoyancy and gelatinous nature makes it impossible to use them as such without immobilization of the particles. The outer membrane can be used as a site for attachment to supports but the cell's ability to change its structure

[4] I. Chibata, T. Tosa, and T. Sato, *J. Mol. Catal.*, 1986, Review Issue, 63.
[5] E. Corcoran, *Top. Enz. Ferment. Biotechnol.*, 1985, **10**, 12.
[6] P. K. Robinson, A. L. Mak, and M. D. Trevan, *Process Biochem.*, 1986, **21**, 122.
[7] A. Rosevear and C. A. Lambe, *Adv. Biochem. Eng.*, 1985, **31**, 37.
[8] A. Rosevear, 'The World Biotech Report 1985', Online, London, 1985, Vol. 1, p. 559.

means that it will only remain attached if provided with the correct micro-environment at all times. However, the size of cells makes it possible to retain them by entrapment in a semi-permeable structure. The range of immobilization and operational conditions to which cells can be subjected is limited by the need to maintain their organizational structure. Nevertheless, their ability to self replicate mean that some of the lost activity can sometimes be recovered by regrowth *in situ*.

3.3 Dead Cells

Non-viable cells are an intermediate group of biocatalysts. These materials have often been included in discussions of cell immobilization but in many respects they should be regarded as crude enzyme preparations. Dead cells have already lost the capacity to perform the co-ordinated multi-enzyme functions of the living cell so there are fewer restrictions on the choice of immobilization method and operation conditions. However, it may be necessary to permeabilize the cell membrane if the full enzymic activity is to be exhibited.

3.4 Choice of Biocatalyst

The actual reaction to be performed will determine the type of biocatalyst which can be considered but many single step processes can be performed by either isolated enzymes, dead cells or live cells. It is therefore wise to consider how pure an enzyme preparation should be in order to fulfil its desired function. Purification of enzymes can be justified either to remove undesirable contaminations (*i.e.* increase purity or specificity) or to minimize the amount of enzyme which must be added to achieve a result (i.e. increase specific activity). Over-purification is both costly and may even remove stabilizing components from the preparation.

This choice of catalyst purity will influence the type of immobilization method to be considered and will have an impact on the subsequent operating conditions. In general isolated enzymes are simpler to immobilize, give a higher process intensity and are more robust during subsequent use. However, where multistep processing with cofactor requiring enzymes is involved, it is often preferable to use whole cells which have an inbuilt system of cofactor recycling. Cell-free methods of regenerating cofactors such as NAD^+, have been used for some immobilized systems.[9]

Optimization of immobilization for a particular enzyme will almost certainly discriminate against some of the unwanted materials in the original biocatalyst mixture. Thus immobilization can act as a further stage of purification. Soluble contaminants are either washed away following immobilization or remain firmly bound to the immobilized conjugate and do not contaminate the product. In principle this means that less pure enzymes could be used. However, impurities, especially the gross impurities in dead cells, dilute the enzyme activity and

[9] C. R. Lowe, *Philos. Trans. R. Soc. London, Ser. B*, 1983, **300**, 335.

reduce the potential for greater process intensity, occupying valuable sites on a support matrix. Nevertheless, this loss is sometimes justified when the cellular debris increases enzyme stability by providing a beneficial microenvironment or protecting the enzyme from damage by reagents in the matrix or substrate solution.

4 METHODS OF IMMOBILIZATION

Biocatalysts can be retained in reactors by either entrapment or by attachment to a fixed structure or matrix. Several comprehensive reviews of techniques for immobilizing biocatalysts are available[10,11] and a number of systems can be purchased off the shelf. Methods of immobilization will be discussed briefly here, but it should be remembered that the type of reactor used to deploy the biocatalyst is just as important for the success of the process and should always be borne in mind when selecting immobilization methods.

4.1 Entrapment

This group of techniques involves creating a barrier through which substrate and product molecules will pass freely but which is impermeable to the biocatalyst. Since most biocatalysts are significantly larger than the substrates on which they act, there is very often scope to employ barriers which discriminate materials on the basis of size. Since this discrimination need only take place at the interface between the immobilized complex and the bulk phase containing the substrate, the biocatalyst can actually be in free solution within the confines of the entrapment system. As a result, it may be subject to only limited interference from the immobilization system, making entrapment the preferred approach for delicate biocatalysts such as cells. Its application to enzyme immobilization is limited by the size differential between substrate and catalyst, so that it is convenient for transformation of small molecules but is inappropriate when processing high molecular weight substrates.

Entrapment methods fall into two categories (Figure 3). In one, a single membrane or barrier encloses a defined area containing the catalyst. In the second case, the catalyst is dispersed within a three-dimensional gel.

4.1.1 Single Membranes. The membrane filtration units used to concentrate biochemicals are easily adapted to form membrane reactor. Although reverse osmosis, ultrafiltration, and microfiltration membranes can be employed, ultrafiltration modules are generally the most convenient compromise, achieving high flux at a low pressure-drop required while also giving sufficient size discrimination to ensure retention of the catalyst. Where these modules are operated in the mode recommended for concentration, the catalyst is essentially working in a homogeneous environment, suspended in the retentate. The system is well

[10] A. Rosevear, J. F. Kennedy, and J. M. S. Cabral, 'Immobilized Biocatalysts', Adam Hilgar, Bristol, 1985.

[11] K. Mosbach, 'Immobilized Enzymes', *Methods Enzymol.*, 1976, **44**.

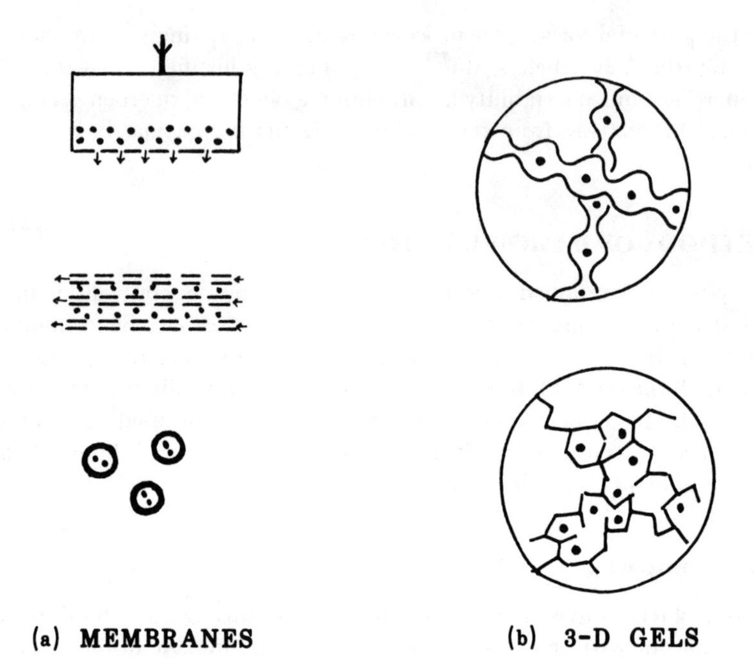

Figure 3 *Entrapment as a means of immobilization:* (a) *Membrane methods using sheets, hollow fibres, and encapsulation;* (b) *Three-dimensional gels by gelling prepolymers and polymerizing monomers*

stirred, ensuring that efficient reaction occurs in the fluid above the membrane. The natural tendency of these modules to suffer concentration polarization at the membrane surface can be used to advantage when considering them as membrane reactors. Formation of a concentrated layer of catalyst at the membrane creates a high activity region, ensuring that substrate molecules interact with the catalyst as they pass through the layer.[12] Protein–protein stabilization may also occur and the hold up volume of the system can be kept low, reducing contact times.

If the biocatalyst is itself large (*e.g.* an animal cell or plant cell aggregate), it may be possible to retain it in the culture vessel using just a coarse filter or even a simple sedimentation region[13] on the outlet of the vessel. The rotating filter[14] used to perfuse dense hybridoma cell cultures exemplifies this approach. The retention of important, small molecules such as cofactors can be improved by attaching them to soluble polymers which are retained more easily.[15] Cells are so much larger than the pores that there is less concern over leakage. When using hollow fibre membranes the cells can be placed on either the tube side or within the shell side, where they are perfused through the hollow fibres.[16] Damage to the

[12] G. Greco and D. Albanesi, *Eur.J.Appl.Microbiol.Biotechnol.*, 1979, **8**, 249.
[13] J. Feder and W. R. Tolbert, *Int. Biotech. Lab.*, 1985, June, 40.
[14] A. L. van Wezel, *Dev. Biol. Stand.*, 1983, **55**, 3.
[15] C. Wandrey, R. Wichmann, and A. S. Jandel, *Enzyme Eng.*, 1982, **6**, 61.
[16] J. P. Tharakon and P. C. Chau, *Biotechnol. Bioeng.*, 1986, **28**, 1064.

membrane may be caused by the pressure from growth of new biomass[17] or result from fluid being forced from shell side to lumen when there is least support to the delicate, macroporous face of the membrane.

An alternative approach is to create a film around individual droplets of concentrated biocatalysts. This can be achieved by forming a stable emulsion or by generating microcapsules (Figure 3). Emulsification has found widest application in the processing of solutes in water immiscible solvents. Even those enzymes which attack lipophilic substrates will generally have a greater affinity for the aqueous phase and often concentrate at the phase boundary. Thus they are effectively immobilized in the discontinuous phase of a water-in-oil emulsion, particularly when the emulsion is stabilized by a surfactant.[18] A solid phase introduced into the system acts as a stable support around which the biocatalyst aggregates and to which small quantities of water will bind. Lipase dried onto an inert support such as kieselguhr has been used to transesterify triglycerides[19] when contacted with a solution of a fatty acid in a hydrocarbon, pre-saturated with a small, catalytic amount of water. Two-phase aqueous systems have also been used to hold the biocatalyst in a distinct, easily separated region of a reactor. Such arrangements have facilitated continuous removal of inhibitory products such as ethanol from immobilized yeast cells.[20]

Microencapsulation has been employed to immobilize enzymes[21] although it has not proved to be a versatile technique. It is quite easy to form a thin membrane around an emulsion, but most methods of interfacial polymerization require quite corrosive reagents (*e.g.* sebacyl chloride and a diamine) which inactivate enzymes.

4.1.2 Three-dimensional Gels. A number of hydrogels have been used very successfully, particularly for the entrapment of cells. The approach involves gelation of an aqeous suspension of the cells so that the biocatalyst is distributed throughout the resulting sold mass. *In situ* polymerization of monomer mixtures such as acrylamide/bis-acrylamide, usually give stable gels with predictable physical properties, but the reagents inactivate delicate biocatalysts. Gels formed by cross-linking or precipitating an existing polymer such as the marine polysaccharides, generally have inferior physical properties but result in far less inactivation.

The physical shape of the resulting immobilized catalyst is determined by the form of the aqueous solution during gelation. Early studies with polyacrylamide merely allowed the bulk solution to gel as a single block which was subsequently broken up to give small particles. The product was ill-defined and the exothermic polymerization also resulted in thermal inactivation of enzymes in the centre of the block.[22] Emulsification of the pre-gel mixture by dispersion in

[17] D. S. Inloes, W. J. Smith, and D. P. Taylor, *Biotechnol. Bioeng.*, 1983, **25**, 2653.
[18] P. L. Luisi, P. Luthi, I. Tomka, J. Prenosil, and A. Pande, *Ann. N.Y. Acad. Sci.*, 1984, **434**, 549.
[19] A. R. Macrae, *J. Am. Oil Chem. Soc.*, 1983, **60**, 291.
[20] B. Hahn-Haegerdal and B. Mattiasson, *J. Chem. Technol. Biotechnol.*, 1982, **32**, 157.
[21] T. M. S. Chang, 'Biomedical Applications of Immobilized Enzymes and Proteins', Plenum, New York, 1977.
[22] M. A. Wheatley and C. R. Philips, *Biotechnol. Bioeng.*, 1983, **25**, 623.

organic solvents can overcome some of these problems. Solvents such as chloroform/toluene may inactivate biocatalysts and paraffin oil or tributyl phosphate may be preferred, particularly for the immobilization of delicate eukaryotic cells in thermosetting gels such as agarose. The cells are first suspended in a warm solution of the polymer, stirred vigorously in warm oil to create an emulsion, and the mixture then allowed to cool.[23] Residual oil can be floated off the beads, which have the perfectly spherical shape of the original emulsion.

An alternative approach is to add droplets of the suspension to the reagent which causes gelation. This has been used most successfully in the formation of calcium alginate beads.[24] A suspension of the cells in sodium alginate is added dropwise to a solution of calcium chloride. Gelation at the surface of the droplets is almost instantaneous and perfectly spherical beads can be obtained. Although this technique is extremely simple, the gel can deteriorate during subsequent use as a result of calcium loss to the mobile phase. The gel can be stabilized by the polymerization of small amounts of acrylamide within the sodium alginate solution. This facilitates formation of the gel as sheets reinforced with an open, woven cloth.[25]

4.2 Attachment

Binding the biocatalyst directly to a solid support can exploit almost every aspect of chemical and physical chemistry. In this review methods will be divided into those which involve a physical, often reversible interaction, and those in which a covalent bond is formed.

4.2.1 Adsorption. Figure 4a and b illustrate the types of adsorption which have been used to immobilize enzymes and cells. This list includes many of the adsorbents used for the chromatographic separation of biochemicals. The principal difference between the two situations is that during purification the enzyme must be released from the solid under mild conditions, while for a good immobilization support it is essential that the intearction is irreversible under operating conditions. The strength of the interaction increases down the series shown in Figure 4 (*e.g.* the dissociation constant for hydrophobic binding is around 10^{-6} M and that of the antibodies up to 10^{-12} M). However, the cost of the support follows a similar pattern so that the best compromise is often to be found towards the middle of the series, *e.g.* ion exchange or hydrophobic resins. Most of the ion exchangers sold commercially for purifying biochemicals have been used successfully as immobilized enzyme supports. Relatively low cost, non-ionic materials such as Duolite are available in grades specifically selected for enzyme immobilization.

The binding of amino acid acylase to ion exchange resins was the first example of the industrial use of immobilized enzymes. This simple adsorptive technique can be applied to whole cells.[26] It is only necessary to mix the adsorbent with the

[23] K. Nilsson, S. Birmhaum, and S. Flygare, *Eur. J. Appl. Microbiol. Biotechnol.*, 1983, **17**, 319.
[24] P. S. J. Cheetham and K. W. Blunt, *Biotechnol. Bioeng.*, 1979, **21**, 2155.
[25] A. Rosevear, European Patent Office 1981, 048109 A2.
[26] R. Bar, J. L. Gainer, and D. J. Kirwan, *Biotechnol. Bioeng.*, 1986, **28**, 1166.

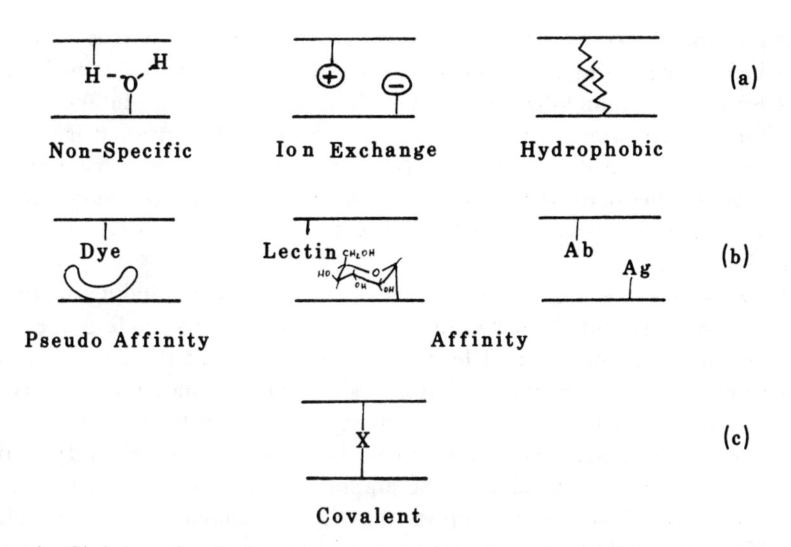

Figure 4 *Techniques for attaching biocatalysts to a support matrix:* (a) *General interaction;* (b) *Specific interaction;* (c) *Covalent binding. The strength of the attachment will depend on the individual interactions and the number of interactions per molecule*

enzyme solution for binding to occur but it is essential that the reagents and conditions used subsequently continue to favour the equilibrium with the biocatalyst on the support rather than in free solution. Salts will compete with the bound protein for sites on an ion exchanger resulting in loss of the catalyst.

Inorganic materials will form complexes with many enzymes and cells, titanium (IV) hydrated oxides giving particularly stable conjugates.[27] Many living cells will also adsorb non-specifically onto surfaces, particularly under nutrient limiting conditions.[28] Although the initial interaction may not be strong, subsequent synthesis of extracellular polysaccharides by the cell may eventually bind it very strongly to the surface. This mechanism underlies the build-up of microbial biofilms on almost inert supports such as sand in waste water trickle filters[29] and fluidized bed reactors. Bacteria attached to cotton cloth have been used as resident inocula in fermenters.[30] The phenomenon is not restricted to mixed populations or to prokaryotic cells. Microcarriers based on anionic exchangers and collagen beads are used extensively to bind animal cells such as fibroblasts.[31] Self immobilization of tissue cultured plant cells in open polyurethane foams[32] can be encouraged by careful selection of the pore size so as to provide a protected environment in which the relatively larger plant cells can aggregate and grow.

[27] C. A. Kent, A. Rosevear, and A. R. Thomson, *Top. Enz. Ferment. Biotechnol.*, 1978, **2**, 12.
[28] V. Bringi and B. E. Dale, *Biotechnol. Lett.*, 1985, **7**, 905.
[29] D. L. Oakley, C. F. Forster, and D. A. J. Wase, in 'Advances in Fermentation II', Turret-Wheatland Ltd., Richmansworth, UK, 1985, p. 20.
[30] S. Joshi and H. Yamazaki, *Biotechnol. Lett.*, 1985, **7**, 753.
[31] A. L. van Wezel, in 'Animal Cell Biotechnology', ed. R. E. Spier and J. B. Griffiths, Academic Press, London, 1985, p. 265.
[32] F. Mavituna and J. S. Park, *Biotechnol. Lett.*, 1985, **7**, 637.

Desorption of enzymes from a support can be prevented by cross-linking the adsorbed layer to form a network on the surface. The most widely used reagent used for protein immobilization is the difunctional reagent, glutaraldehyde.[33] This forms much more stable cross-links than might be expected from simple Schiff's base conjugates, and glutaraldehyde is uniquely effective within the homologous series of dialdehydes. The cross-linking can be even more successful if the protein is first concentrated on the surface or within a porous structure by precipitation.[34]

Immobilization using the more sophisticated affinity binding methods is restricted by cost and the instability of the proteinaceous binding agent. The application of an immobilized lectin such as Concanavalin A to bind glycoprotein enzymes[35] or the use of antibodies to bind enzyme antigens may offer some prospect of concentrating and purifying the enzyme during immobilization but have little practical application. It may be economic to consider derivatizing the enzyme rather than modifying the support. For example if the enzyme has a low affinity for a hydrophobic support, the interaction can be strengthened by attaching fatty acid chains to a protein.[36]

4.2.2 Covalent Binding. Formation of a covalent link between the biocatalyst and a support generates a stable conjugate which is unlikely to dissociate during subsequent use. This approach covers the widest range of novel methods of immobilization but, despite the elegance of the chemistry involved, very few of these techniques have found widespread use.

The conditions under which covalent links form often disrupt the delicate three-dimensional structure of the biocatalyst. This almost inevitably results in a loss of a significant proportion (*e.g.* up to 90%) of enzymic activity during coupling. This can only be justified when the superior stability of the conjugate is important, or the activated support is supplied in a convenient and reliable form. Thus covalent attachment is most frequently used where it is essential that no activity leaks into the product stream, or where the catalyst is to be used against a high molecular weight substrate. The use of pre-activated supports is also common in laboratories which are exploring new applications for an enzyme, when there is more concern with reliable immobilization rather than high enzyme recovery.

In order to minimize the exposure of the enzyme to corrosive chemicals, it is normal practice to pre-activate the support matrix under severe conditions (Figure 5). Any corrosive reagents are then removed and the activated material contacted with the target protein under mild conditions. An alternative, less frequently used approach, is to activate the enzyme (*e.g.* with acryloyl chloride) and after removing any inactive material, to react this with a support or form it into a copolymer. The thiolation of proteins and their subsequent oxidative

[33] P. J. Halling and P. Dunnill, *Biotechnol. Bioeng.*, 1979, **21**, 393.
[34] A. Rosevear, British Patent 1975, 1514707.
[35] H. Seligner, E. H. Tenfel, and M. Philips, *Biotechnol. Bioeng.*, 1980, **22**, 55.
[36] H. Okada and I. Urabe, in 'Immobilized Enzyme Technology', ed. H. H. Wheetall and S. Suzuki, Plenum, New York, 1975, p. 37.

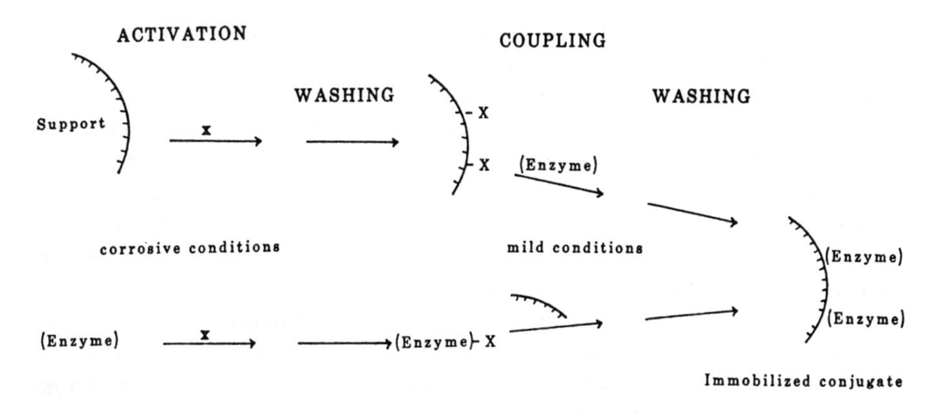

Figure 5 *Strategies covalently attaching an enzyme to a support*

coupling to thiolated supports[37] is a very mild method of applying this general approach.

Selection of a good support is as important as the chemistry of the linkage.[12] In principle it is possible to synthesize ready-activated polymers but producing a material with good physical characteristics at the same time is it not easy. Thus it is more common to activate a ready-made support which has the desired, physical properties. Natural polymers such as cellulose, cross-linked dextran, and agarose or collagen provide a suitable hydrophilic environment for the enzyme and commercially available forms of these gels can be activated by a variety of chemical methods. Synthetic polymers such as polyacrylates can be activated by chemical substituents while a range of inorganic matrices can be derivatized to give suitable pendant groups (*e.g.* by use of γ-aminopropyltriethoxy silane[27]). Many of the organic matrices are compressible and settle slowly, restricting both through-put and the size of vessel in which they can be employed. Composite materials (*e.g.* Macrosorb) combine the versatile chemistry of the organic polymers with the rigidity and good settling characteristics of the inorganic matrices and should be considered for any large-scale use.

Polyalcohol supports activated with cyanogen bromide (CNBr) are probably the most commonly encountered example of covalent coupling for enzymes. This popularity stems from the ready availability of CNBr-activated Sepharose rather than any inherent benefits in the chemistry of binding. Without this convenience, the method has few attractions, since the activation involves the use of a toxic, lachrymatory reagent and the final link between the support and the protein is far from stable.[38] There has been some debate over the chemistry of the reaction. It was originally thought that the activated group was a cyclic imidocarbonate. However, it now appears that cyanate esters are of more importance and the activation procedure can be modified to maximize formation of these.[39] By using

[37] J. Carlsson, R. Axen, and T. Unge, *Eur. J. Biochem.*, 1975, **59**, 567.
[38] L. Peng, G. J. Calton, and J. W. Burnett, *Enz. Microb. Technol.*, 1986, **8**, 681.
[39] J. Kohn and M. Wilchek, *Biochem. Biophys. Res. Commun.*, 1982, **107**, 878.

CNBr in acetone and triethylamine as a base, much less reagent is required to generate high degrees of substitution.

The use of chloroformates to give reactive carbonate groups provides an alternative method of activating polyalcohols.[40] The use of less volatile chloroformates is an improvement to this technique[41] and is said to give very stable conjugates. However, activation of polyols with tresyl chloride (2,2,2-trifluoroethane sulfonyl chloride)[42] is a very attractive technique. Although the matrix must be dehydrated prior to activation, the reagent is easy to use and appears to give at least the same degree of activation as CNBr.

Epoxides react efficiently with proteins under mild conditions and although the reagents needed to pre-activate gels are rather unpleasant, ready activated materials are available under the Trade name Eupergit. Other commonly used activation methods for polyols involve triazines or benzoquinone.[11]

The reaction of an aldehyde to form a Schiff base with amine groups on the protein, has been exploited in forming immobilized enzymes. The use of glutaraldehyde to interlink proteins has been noted above but it is also possible to use this reagent in excess to convert polyamine supports into polyaldehydes.[43] Alternatively the aldehydes can be generated by periodate oxidation of a polymer containing vicinal diols, *e.g.* dextrans.[44] There is some debate over whether reduction of the base is necessary to ensure stability of the link.[45] Although cyanoborohydride is relatively mild, the loss of activity encountered during the reduction rarely seems to justify this secondary treatment.

Many techniques for activating carboxylic acid and amino residues have been developed in protein chemistry and are applicable to immobilization. Most amino supports can be activated in the same way as the polyols. The most important of the activation techniques for acids is the use of various carbodiimides[46] and of *N*-hydroxysuccinimide activation which is used for the commercially available Affigel.

5 REACTORS

The overall productivity of the immobilized biocatalytic system is as much determined by the equipment as by the actual method of immobilization itself.[47] The main function of the reactor is to retain the immobilized complex and to ensure efficient and controlled contact between the catalyst and the feedstock in order to maximize product formation. Immobilization generally helps the process engineer since the biocatalyst is presented in a concentrated, easily handled form.

[40] J. F. Kennedy, S. A. Barker, and A. Rosevear, *J. Chem. Soc., Perkin Trans. 1*, 1973, 2293.
[41] T. Miron and M. Wilchek, *Appl. Biochem. Biotechnol.*, 1985, **11**, 445.
[42] K. Nilsson and P. O. Larsson, *Anal. Biochem.*, 1983, **134**, 60.
[43] C. C. Hon and P. J. Reilly, *Biotechnol. Bioeng.*, 1979, **21**, 505.
[44] J. F. Wright and W. M. Hunter, *J. Immunol. Methods*, 1982, **48**, 311.
[45] A. W. Miller and J. F. Robyt, *Biotechnol. Bioeng.*, 1983, **25**, 2795.
[46] J. A. Osborn and R. M. Ianniello, *Biotechnol. Bioeng.*, 1982, **24**, 1653.
[47] 'Process Engineering Aspects of Immobilized Cell Systems', ed. C. Webb, G. M. Black, and B. Atkinson, Institute of Chemical Engineering, Rugby, 1986.

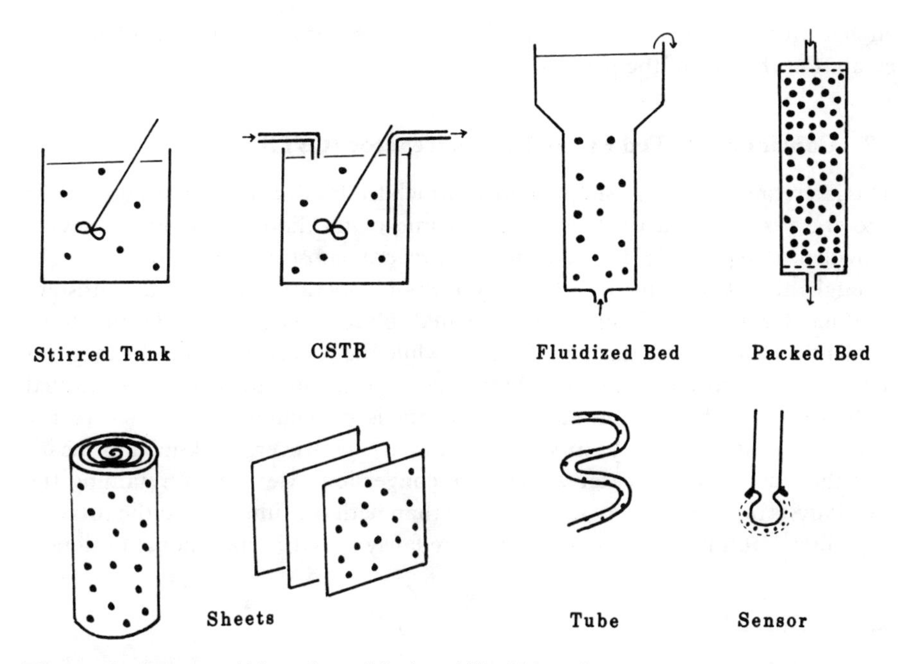

Figure 6 *Reactors which can be used for immobilized biocatalysts*

Reactors can be divided into those designs[48] which accept particulate biocatalysts and those which are intended to operate with fixed, specialist elements such as sheets, membranes, or tubes (Figure 6). Particle reactors are probably the more common and are in general more versatile, since they will easily accommodate minor changes in shape and loading of material. Non-particle reactors generally have more clearly defined flow patterns but often require expensive capital equipment and can only accept a fixed volume of catalyst.

5.1 Tank Reactors

These are the simplest type of reactor and span the range from the small, laboratory test-tube placed on a roller, through to large stirred vessels. A full batch of substrate is contacted with a single charge of immobilized catalyst and the reaction proceeds in a similar way to that with a free catalyst. It may be necessary to provide more efficient stirring to keep the particles suspended but the main difference is that at the end of the reaction the immobilized enzyme or cell can be removed rapidly and completely from the product solution. Transfer of the particles to a subsequent batch of substrate can be achieved either by letting them settle, or by retaining them in a coarse filter device, either in the vessel (*e.g.* like a tea bag) or on the outlet. Although efficient stirring is important to ensure good mass transfer, care should be taken not to grind up any particles as this may

[48] R. Miller and M. Melick, *Chem. Eng.*, 1987, 112.

change their apparent activity and will increase problems in recovering the catalyst at the end of the process.

5.2 Continuously Fed Stirred Tank Reactor (CSTR)

This is a refinement of the simple tank in which fresh substrate is continuously fed into the vessel and a corresponding volume of the liquid contents removed. Provided that mixing is sufficient to prevent substrate streaming straight through, this arrangement permits continuous operation to give a consistent product. The degree of conversion is related to the average retention time of an element of fluid passing through the tank, while the concentration of the reagents in the tank is constant at all times. Since the incoming substrate is rapidly diluted with previously formed product, the system is particularly useful where the substrate is inhibitory (*e.g.* in waste treatment). However, back-mixing means that the system never achieves complete conversion. Methods of retaining the biocatalysts need to be more sophisticated than with the simple tank: the rotating cage filter[14] and fluid settling systems[13] are fairly versatile operational options.

5.3 Fluidized Beds

Fluid beds keep the particles suspended by pumping liquid in at the base of the vessel at a velocity which prevents particles settling out. Like a CSTR the feed can be continuous but in the fluid bed, the flow is unidirectional through the bed, so that less back mixing occurs. Thus higher levels of conversion might be expected and in general the degree of conversion increases with bed length. Wash out of particles is prevented by a design such as a decoupling region at the top of the bed where the diameter of the vessel is increased so that fluid velocity drops and particles fall back into the bed.

5.4 Packed Bed

This is the most commonly used device for contacting an immobilized biocatalyst with the substrate solution. The particles are placed in a vessel fitted with inlet and outlet at opposite ends with a settled bed of particles between. Fluid flow may be upwards or downwards (side to side flow is less easy since settling may leave a void through which fluid will stream). The biocatalyst is packed at its maximum density within such a reactor and for a given set of reaction conditions, the conversion of the substrate to product increases with the length of the column. In principle, it is possible to achieve total conversion to product so that such reactors are ideal where total removal of a substrate is essential (*e.g.* detoxification). The main difficulty in operating a packed column is ensuring good flow throughout the bed. If the substrate contains any debris, or if gas is generated during the reaction, then the bed tends to act as a deep bed filter and it is advisable to use one of the other devices described above. Furthermore, if the particles are small or compressible, flow through the bed will be restricted. Although high flow with acceptable pressure drop across the bed is possible with short fat columns or trays, this may not give an adequate superficial velocity past the particles to ensure efficient mass transfer. Thus incompressible beads are a great advantage in packed beds.

5.5 Sheet Reactors

In principle sheets of material can be packed into a reactor space to give a biocatalyst density equivalent to a packed bed.[8] Parallel sheet and coaxially wound (Swiss Rolls) arrangements achieve the same end though care is needed to ensure distribution of flow to all channels. This is particularly important when flow is normal to the face of the sheet since the reactor then has the characteristics of a shallow packed bed where by-passing of fluid through low resistance channels will permit substrate solution to break through.

5.6 Membrane Reactors

In these devices, the reactor itself is the main means of immobilizing the biocatalyst. Stirred types of ultrafiltration module are easily adapted for enzyme immobilization and in principle hollow fibre modules can be used with little change for cell immobilization. However, adapted fibre and multiple flat sheet devices are now made specially for cell immobilization.[49] A critical feature of the design is to reduce the axial dimensions of the unit[50] so as to minimize the gradient of essential nutrients (especially oxygen) across the reactor and prevent selective growth of organisms at the inlet side.

5.7 Special Devices

Analytical applications often demand special consideration. Two commonly encountered arrangements are the reactive tube and the Enzyme Electrode.[51] The shape of the device is dictated by the detector but the reactor must be well mixed to give a consistent signal.

6 PROBLEMS IN USING IMMOBILIZED BIOCATALYSTS

Subjecting delicate enzymes and cells to new process condition gives rise to problems, but most of these can be minimized. These potential difficulties will be dealt with in two categories; firstly the loss of activity during immobilization and secondly the operational problems in using the immobilized conjugates.

6.1 Loss of Activity

Almost all the methods of immobilization result in a real loss of biological activity. The forces required to bind an enzyme to a surface are at least as strong as those, such as hydrogen bonds, which determine its three-dimensional struc-

[49] J. van Brunt, *Biotechnology*, 1986, **4**, 505.
[50] K. Ku, M. J. Kuo, J. Delente, B. S. Wildi, and J. Feder, *Biotechnol. Bioeng.*, 1981, **23**, 79.
[51] W. J. Aston and A. P. F. Turner, *Biotechnol. Genetic Eng.*, 1984, **1**, 79.

ture. Add to this the asymmetric nature of the binding force, *i.e.* only one side of the molecule is firmly held against a high energy surface, and it is to be expected that enzymatic function will be adversely affected. Although most covalent links will only form with peripheral, non-essential residues on the protein, our understanding of reactions with folded proteins is insufficient to predict the best way of avoiding damage to the active site. Empirical optimization of reaction conditions against specific activity is still the only effective way of minimizing loss. It may sometimes be possible to protect the critical groups around the active site by immobilizing the enzyme in the presence of substrate or a competitive inhibitor but the chemistry of binding often precludes this.

6.2 Apparent Changes

The poor recovery of activity following immobilization may not always be real but can arise due to changes in kinetics[52] in the immobilized conjugate (Figure 7). In the extreme, some large substrate molecules and almost all solid substrates will be prevented from reaching a catalyst held within the pores of a support matrix. Even when substrate molecules can reach the active site the diffusion rate of larger substrates in the pores will be greatly reduced compared with that in free solution so the enzyme may appear to have increased its specificity towards smaller molecules.

The very success of the immobilization may itself generate kinetic problems due to the very high concentation of enzyme within the restricted volume of the support matrix. In order to supply sufficient substrate to keep an enzyme working at its maximum velocity (V_{max}), there must be a substantial unidirectional flux of substrate into the matrix. Substrate passing from the well mixed bulk phase must move through a region close to the matrix where mass transport is diffusion controlled. At high reaction rates, *e.g.* near the optimum operating conditions, this restricted diffusion may become limiting. If the pH or temperature of the reaction mixture is varied around this optimum, it can appear to have far less effect on kinetics than is the case with the free enzyme. This apparent lack of response of the immobilized enzyme to reaction conditions is a direct result of physical diffusion rather than chemical kinetics being the determinant of the reaction rate.

Any factor which changes the depth of this diffusion controlled film will directly effect the overall rate of reaction. Increases in the superficial velocity of fluid past the beads, or of the stir speed, will thus appear to increase apparent activity (Figure 7) since they reduce the depth of the film. Thus efficient mixing is essential in order to measure the true activity. All substrate must pass through the surface layer so immobilized systems are vulnerable to anything which occludes the pores. Consequently it is important to remove all foulants from the system and avoid the generation of gas or precipitates which may blind surfaces and displace solution from the reactor core.

Enzyme in the outer pellicule of the matrix will deplete the substrate con-

[52] A. Bodalo, J. L. Gomez, E. Gomez and J. Bastida, *Enz. Microb. Technol.*, 1986, **8**, 433.

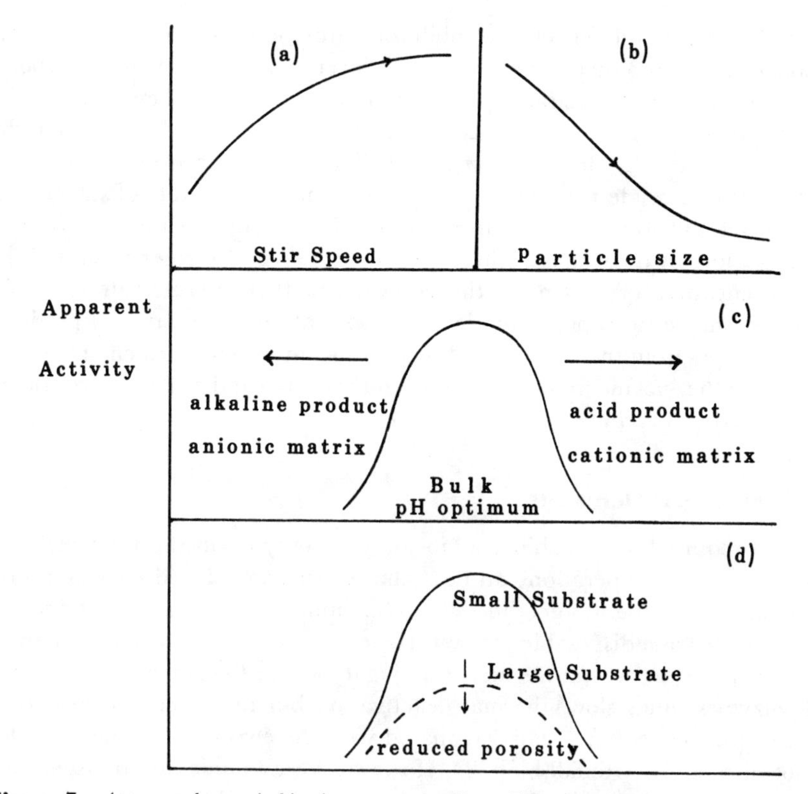

Figure 7 *Apparent changes in kinetics as a result of immobilization:* (a) *Due to film diffusion;* (b) *Due to internal pore diffusion;* (c) *Due to charged species in the matrix;* (d) *Due to restricted diffusion of large molecules*

centration so that enzyme at the centre sees a concentration which is lower than that in the bulk phase. Thus the bulk concentration required to ensure all the enzyme is working at V_{max} will be higher than that for the free enzyme and K_m will appear to increase. Since this effect is related to the depth of the internal pore diffusion path, changes in particle size and porosity will effect the apparent kinetics. Breaking up large catalyst particles will appear to increase activity (Figure 7), while the real loss of activity due to inactivation during use may be masked by the enzyme deep in the pores being used more efficiently as the surface bound enzyme is lost.

Restricted diffusion can have a further effect when the product is charged or is an inhibitor. At high reaction rates product concentrations within the pores will be higher than that in the bulk phase. If the product is charged (*e.g.* an acid generated by an esterase), the local pH will fall so that pH in the bulk phase needed to maintain efficient reaction must rise (*i.e.* the pH optimum of the enzyme may appear to rise). A similar effect will occur if the matrix itself is ionized (Figure 7) since the environmental pH within the beads will be dictated by charged groups close to the enzyme, rather than by the bulk fluid. These effects are most noticeable in weak buffers and can be totally absent in strong buffers. However, the effects of inhibitors are more difficult to suppress, especially if it is a product of the reaction and accumulates in the unstirred pores close to the catalyst. Undesirable products can be removed as they form (*e.g.* by co-immobilizing catalase in the same matrix as glucose oxidase so that the inhibitory peroxide is destroyed *in situ*[53]). The high concentration of intermediate product inside the porous catalyst will enhance the rate of subsequent reactions, relative to that seen in free solution. For instance in the hydrolysis of polysaccharides by dextranase, where the product is also a further substrate, the high concentration of enzyme in the pores means that intermediate sized oligomers are subject to rapid multiple attack so that only the terminal product, glucose, escapes from the matrix.[54] This might be interpreted as a change in the specificity of the enzyme from an *endo* to an *exo* hydrolase if the effects of restricted diffusion were not recognized.

6.3 Operational Demands

The effort required to immobilize a biocatalyst is only justified if the benefits of reuse or continuous operation can be realized. An immobilized enzyme represents a major investment, sometimes of similar value to a capital item, and so can not be regarded as a disposable catalyst. Hence a number of operational changes must be imposed to protect the active conjugate within the reactor.

All enzymes suffer slow, thermal deactivation but in normal homogeneous catalysis, process times are short compared with the enzymic half-life (time for the biological activity to fall by half). However, for an immobilized system, the operational half-life (as distinct from the storage half-life) is an important parameter which must be borne in mind in determining the application strategy. In a continuous reactor it is possible to compensate for the potential change in the output of the reactor by reducing the flow rate (thus increasing contact time) or by increasing the temperature. The latter policy not only raises the reaction rate of the remaining enzyme but also accelerates further deactivation. Alternatively, the catalyst can be constantly renewed on a cascade basis.

Superimposed on this fundamental decline in operational activity there is often a further loss attributed to the substrate solution. The fact that the immobilized catalyst treats a larger volume of liquid means that it is more vulnerable to low levels of irreversible inhibitors than is the free catalyst. Furthermore, any material which occludes the surface of the support matrix will seriously impair the operational activity of any immobilized enzyme or cell. These effects are highlighted by the dramatic fall in half-life from 39 days to only 7 days when the feedstock for an immobilized lactase was changed from 5% lactose solution to whole acid whey with an equivalent sugar content.[55]

The growth of contaminating micro-organisms in a reactor will also have a deleterious effect on performance either by direct attack of the catalyst and substrate or by coating of the surface. Fortunately the high concentration of

[53] P. H. S. Tse and D. A. Gough, *Biotechnol. Bioeng.*, 1987, **29**, 705.
[54] K. L. Smiley, J. A. Boundy, and D. E. Hensley, *Carbohydr. Res.*, 1982, **104**, 319.
[55] W. H. Pitcher and J. R. Ford, *Methods Enzymol.*, 1976, **44**, 792.

substrate and evaluated temperatures of industrial enzyme processes inhibit cell proliferation, but immobilized whole cells are particularly vulnerable to unintentional infection. It is advisable to immobilize delicate cells under aseptic conditions and it may be necessary to add an antibiotic to the substrate solution to avoid infection. The uncontrolled growth of the immobilized cells themselves can also cause problems. Fresh cells on the surface will interfere with free flow of liquid through the reactor or will become detached and contaminate the product stream. Growth also diverts valuable substrate away from product formation. Omitting a growth hormone or essential nutrient such as phosphate will prevent growth of eukaryotic cells after immobilization. This is less easy with microbial cells though it is reported that 0.03% azide will reduce the growth of immobilized yeast cells while maintaining production of ethanol.[56]

All these factors emphasize the need for a greater degree of control in immobilized systems. It is advisable to clarify solutions before they enter a reactor rather than to do this on the final product as is often the case when using soluble enzymes. The pH, temperature and fluid flow must be monitored and a fail-safe facility built into any control to protect the reactor. In addition greater levels of process hygiene are needed to prevent infection in the system. Finally it is important to realize that an immobilized cell reactor cannot be switched off in the same way as enzyme system. It is to be expected that in order to maintain cellular activity, the organism will probably need a minimum amount of oxygen and nutrients if it is to survive until it is next required.

7 APPLICATIONS

Immobilized glucose isomerase and penicillin acylase are already used successfully in very large scale manufacture of iso-sugar and semi-synthetic penicillin.[57] Novel applications such as the use of lipase for transesterification to increase the value of fats are commercially viable and the use of cyanide degrading enzymes has been evaluated for treating industrial wastes. At the other end of the size scale, enzymes immobilized on electronic devices now form a very important sector in the diagnostic sensors market (see Chapter 19).

Opportunities for using immobilized enzymes and cells arise from three important characteristics of such systems:

(i) the catalyst is retained in the reactor,
(ii) the catalyst is at high concentration, and
(iii) the reaction can be closely controlled.

7.1 Retention

The opportunity to reuse the catalyst many times has always been seen as a major advantage of immobilization. Thus it is possible to consider utilizing quite expensive enzymes in an immobilized form since the cost can be recouped by

[56] B. Hahn-Haegerdal and B. Mattiasson, *Eur. J. Appl. Microbiol. Biotechnol.*, 1982, **14**, 140.
[57] P. B. Poulson, Proceedings of the 3rd European Congress on Biotechnology, VCH, Weinheim, 1985, Vol. 4, p. 339.

repeated use. However, the second facet of retention, avoiding contamination of the product, is now seen as a great advantage. For instance, it is possible to control the hydrolysis of milk protein so that curd formation can take place in a second vessel.[58] Beer can be chillproofed, or wine conditioned with immobilized biocatalysts without the enzymes or cells being left behind in the product so that the catalyst becomes a process aid rather than an additive.[59] Avoiding release of biologically active proteins into patients is obviously an advantage in clinical treatments and immobilized enzymes such as heparinase have been used to advantage in extra-corporeal shunts to control the level of the anticlotting agent in blood.[60] Immobilization can also impose a steric restraint on a reaction. This has been used to advantage where immobilized neuraminidase only desialinates the outer surface of cells, in contrast to the free enzyme which also attacks intracellular material.[61]

Ensuring that the active enzymes do not enter the product stream will also prevent reactions continuing after the material leaves the reactor vessel. This reduces product loss but will also simplify purification processes. The benefits downstream are even greater when whole cells have to be used since retention of the biomass in the reactor obviates the need for primary separation of solid residues from the process stream. In practice some release of cells occurs due to slow growth in many immobilized cell reactors but the low load of solid in the process stream is a major advantage resulting from effective immobilization. Interestingly, this retention is often achieved by using modification of equipment and reagents (*e.g.* membranes and adsorbents) familiar to the separation scientist.

7.2 Concentration

Concentration of the activity into a smaller vessel and the resulting increase in process intensity greatly reduces the size and cost of equipment needed. The increased reaction rate which accompanies this is also important in producing radiochemicals which have a short half-life[62] or in rapid processing of unstable materials. For instance, compounds such as penicillin V can be deacylated with an immobilized enzyme before the lactam ring hydrolyses chemically.[63]

The shorter process time also means that there are fewer breakdown products produced by thermal degradation so that the product may require far less purification downstream. A further advantage may arise from the opportunity to move from batch operation to continuous processing. Immobilized systems are far better suited to continuous, automated operation, making better use of equipment and avoiding the need to hold large volumes of partially processed fluid.

[58] K. Ohmiya, S. Tanimura, and T. Kobayashi, *J. Food Sci.*, 1979, **44**, 1584.
[59] S. Gestrelius, *Enzyme Eng.*, 1982, **6**, 245.
[60] R. Langer, R. J. Linhardt, C. C. Cooney, and D. Tamper, *Enzyme Eng.*, 1982, **6**, 433.
[61] T. L. Parker and A. P. Cornfield, *Hoppe-Seyler's Z. Physiol. Chem.*, 1977, **358**, 789.
[62] A. J. L. Cooper and A. S. Gelbard, *Anal. Biochem.*, 1981, **111**, 42.
[63] S. Gestrelius, *Appl. Biochem. Biotechnol.*, 1982, **7**, 19.

7.3 Control

The microenvironment around an immobilized biocatalyst is strongly influenced by the nature of the matrix. Control over mechanical shear afforded by the matrix is particularly useful in protecting delicate cells. Control over the microenvironment around plant cells might also have a beneficial effect on the output of secondary metabolites.[7] The local concentration of reaction intermediates can be controlled to advantage as in the case of starch hydrolysis where pullalanase has been added to an immobilized amylase to debranch locally produced polymer fragments.[64] Contact time between enzyme and substrate can be precisely controlled by adjusting flow rate through immobilized enzyme devices and reaction conditions can be changed rapidly by employing sequential packed beds. These factors improve the efficiency with which the enzyme is used and makes it possible to isolate intermediates. Finally the fact that the enzymic reaction stops as soon as the substrate leaves the reactor has been exploited by employing immobilized carboxypeptidase Y to deprotect the growing peptide during peptide synthesis.[65]

This fine control should reduce the number of contaminants and by-products in the final product, thus reducing the need for multistage purification. However, control can only be exercised if the flow through the reactor is consistent. This further emphasizes the importance of the reactor design and monitoring of the system if the full benefits of immobilization are to be realized.

8 CONCLUSION

Immobilization is now an established technique for making most effective use of enzymes and cells. Where the substrate for the upstream process is itself soluble, immobilization of the biocatalyst can facilitate downstream operations. It simplifies downstream processing in general by a greater attention to process control upstream. Immobilization will certainly increase the initial costs of the upstream process. However, by making better use of the biocatalyst and simplifying downstream processing, it is likely to have a cost advantage overall, in a number of important industrial processes, therapeutic devices and analytical applications.

In many respects the characteristics of an immobilized biocatalyst are similar to those of any heterogeneous catalyst. The interesting feature of these biological examples is that the technologist is beginning to recreate the conditions of heterogeneity from which the scientist has laboured to remove enzymes by extraction and purification. The main improvement in this artifical environment is that the technologist can impose a level of outside control which has not been possible with the natural material.

[64] K. A. Ram and K. Venkatasubramanian, *Biotechnol. Bioeng.*, 1982, **24**, 355.
[65] G. P. Royer and G. M. Anantharmaiah, *J. Am. Chem. Soc.*, 1979, **101**, 3394.

CHAPTER 17

Downstream Processing: Protein Extraction and Purification

M. D. SCAWEN, T. ATKINSON, P. M. HAMMOND, AND R. F. SHERWOOD

1 INTRODUCTION

Enzymes are increasingly employed as reagents in clinical chemistry, as therapeutic agents in chemotherapy, and as catalysts in industrial processes. Furthermore, recent advances in molecular genetics have resulted in an increased awareness of the importance of protein recovery and purification.

Over the years, the increasing use of micro-organisms as a source of proteins, particularly enzymes, has led to improved efficiency in production and a more reproducible product. The great majority of enzymes in industrial use are extracellular proteins from organisms like *Aspergillus* sp. and *Bacillus* sp., and include α-amylase, β-glucanase, cellulase, dextranase, proteases, and glucoamylase.[1] Many of these are still produced from the original, wild-type strains of micro-organism. However, in the production of proteins for use in the fields of clinical diagnosis and for therapeutic applications, genetic and protein engineering are beginning to play an ever increasing role. Genetic engineering, besides permitting vast improvements in yields, has allowed the transfer of genetic material from animal to bacterial hosts. In this way, those products once only available in minute amounts from animal tissues can now be produced in virtually limitless quantities from easily grown bacteria. One such example is human growth hormone; this was once produced in tiny amounts from human pituitaries, until it was recognized to present a potential risk to the patient from contaminant virus which has been implicated in Creutzfeld–Jacob syndrome. Human growth hormone is now produced in far greater quantities from the bacterium *Escherichia coli*, and is completely free from the unwanted virus.

Enzymes or proteins produced by micro-organisms may be intracellular, periplasmic, or secreted into the culture medium. For extracellular enzymes, the degree of purification required is minimal, and large-scale processes may yield tonnes of protein product. It may also be acceptable for the intended application

[1] K. Aunstrup, in 'Applied Biochemistry and Bioengineering', ed. L. B. Wingard, E. Katchalski-Katzir, and L. Goldstein, Academic Press, New York, 1979, p. 27.

that the final product does not achieve absolute purity, particularly if the material is destined for industrial use.

Many other enzymes, however, may be produced in more complex intracellular mixtures and represent a greater challenge to the protein purification scientist. Those produced for therapeutic use must attain very high and exacting standards of purity and to achieve this it may be necessary to develop complex purification protocols. Again, genetic engineering has been able to help in this area. In the case of therapeutic proteins, there are great advantages if the protein of interest can be secreted, either into the periplasm or into the medium. This results in an enormous reduction in the level of contaminating proteins and other macromolecules, such that pure product can often be obtained in only 2 or 3 steps of purification. Genetic manipulation may also result in higher yields, thereby improving the 'quality' of starting material (in terms of raising the percentage of product in the total soluble cell protein). It is now also possible to add groups to a protein which aid its purification by conferring specific properties and subsequently removing these groups when they are no longer required.

2 CELL DISRUPTION

There are three main methods for the release of intracellular proteins from micro-organism; enzymic, chemical, or physical. Not all of the techniques available are suitable for use on a large scale. Perhaps the main example is sonication, which is frequently the method of choice for the small scale release of proteins. On a large scale it is difficult to transmit the necessary power to a large volume of suspension and to remove the heat generated.

2.1 Enzymic Methods of Cell Disruption

Lysozyme, an enzyme produced commercially from hen egg white, hydrolyses β-1,4-glycosidic bonds in the mucopeptide of bacterial cell walls. Gram-positive bacteria, which depend on cell wall mucopeptides for rigidity are most susceptible, but final rupture of the cell wall often depends upon the osmotic effects of the suspending buffer once the wall has been digested. In Gram-negative bacteria lysis is rarely achieved by the use of lysozyme alone, but the addition of EDTA to chelate metal ions will normally result in lysis. This technique is rarely used for the large scale extraction of bacterial enzymes, perhaps due to the relatively high cost of lysozyme. It has been used for the large scale release of an aryl acylamidase from *Pseudomonas fluorescens*.[2]

Other enzymes are infrequently used, but microbial glucanases have been used to hydrolyse the cell wall of yeasts, which contain β-1,3-glucan,[3] and lysostaphin may be used to release proteins from *Staphylococci*, for example Protein A from *S. aureus*.

[2] P. M. Hammond, C. P. Price, and M. D. Scawen, *Eur. J. Biochem.*, 1983, **132**, 651.
[3] R. Kobayashi, T. Miwa, S. Yamamoto, and S. Nagasaki, *Eur. J. Appl. Microbiol. Technol.*, 1982, **15**, 14.

2.2 Chemical Methods of Cell Lysis

2.2.1 Alkali. Treatment with alkali has been used with considerable success in small and large scale extraction of bacterial proteins. For example, the therapeutic enzyme, L-asparaginase, can be released from *Erwinia chrysanthemi* by exposing the cells to pH values between 11.0 and 12.5 for 20 minutes.[4] The success of this method relies on the alkali stability of the desired product. The high pH may inactivate proteases, and the method is of value for the combined inactivation and lysis of rDNA micro-organisms.

2.2.2 Detergents. Detergents, either ionic, for example sodium lauryl sulfate, sodium cholate (anionic), and cetyl trimethyl ammonium bromide (cationic), or non-ionic, for example Triton X-100 or X-450 or Tween, have been used to aid cell lysis. Ionic detergents are more reactive than non-ionic detergents, and can lead to the denaturation of many proteins.

The presence of detergents can affect subsequent purification steps, in particular salt precipitation. This can be overcome by the use of ion exchange chromatography or ultrafiltration, but obviously introduces additional steps. Nevertheless, detergents do have considerable uses in some extraction processes, for example Triton X-100 has been used for the large scale release of cholesterol oxidase from *Nocardia* sp.,[5] and sodium cholate used to solubilize pullulanase (pullulan-6-glucan hydrolase), a membrane-bound enzyme, from intact cells of *Klebsiella pneumoniae*.[6]

2.3 Physical Methods of Cell Lysis

2.3.1 Osmotic Shock. Osmotic shock has been used for the release of hydrolytic enzymes and binding proteins from the periplasmic space of a number of Gram-negative bacteria, including *Salmonella typhimurium* and *E. coli*.[7] The method involves washing the cells in buffer solution to free them from growth medium, and then suspending them in 20% buffered sucrose. After being allowed to equilibrate, the cells are harvested and rapidly resuspended in water at about 4°C. Only about 4–8% of the total bacterial protein is released by osmotic shock, and if the required enzyme is located in the periplasmic region it can produce a 14- to 20-fold increase in purification compared with other extraction techniques.[8]

The use of osmotic shock is gaining greater popularity with the increase in the number of heterologous proteins which are secreted into the periplasm. A typical example is the case of recombinant human growth hormone, produced in *E. coli*, where the product released by osmotic shock is 90% pure after the first purifi-

[4] T. Atkinson, B. J. Capel, and R. F. Sherwood, in 'Safety in Industrial Microbiology', ed. C. H. Collins and A. J. Beale, Butterworth, Oxford, 1992, p. 161.
[5] B. C. Buckland, W. Richmond, P. Dunnill, and M. D. Lilly, in 'Industrial Apsects of Biochemistry', ed. B. Spencer, North Holland Publishing Company, Amsterdam, 1974, p. 65.
[6] K. H. Kroner, H. Hustedt, S. Granda, and M-R. Kula, *Biotechnol. Bioeng.*, 1978, **20**, 1967.
[7] H. C. Neu and L. A. Heppel, *J. Biol. Chem.*, 1965, **240**, 3685.
[8] S. E. Charm and C. C. Matteo, *Methods Enzymol.*, 1971, **22**, 476.

cation step. In contrast, the cytoplasmic protein is only 10% pure after the same purification step.[9]

2.3.2 Grinding with Abrasives. Initially this technique was restricted to the grinding of cell pastes in a mortar with an abrasive powder, such as glass, alumina, or kieselguhr. This system has since been developed and mechanized using machines originally developed for the wet grinding and dispersion of pigments in the printing and paint industries. A typical product, the Dynomill (W. A. Bachofen, Switzerland) can be used to release proteins from a wide variety of micro-organisms. It consists of a chamber containing glass beads and a number of fixed and rotating impeller discs. The cell suspension is pumped through the chamber, and the rapid agitation is sufficient to break even the toughest of bacteria. The disintegration chamber must be cooled to remove the heat which is generated. A laboratory scale model, with a 600 cm^3 chamber can process up to 5 kg bacteria per hour, and production scale models are available with chambers of up to 250 dm^3 capacity.

Many factors influence the rates of cell breakage, such as the size and concentration of the glass beads, the type, concentration, and age of the cells, the agitator speed, the flow rate through the chamber, the temperature, and the arrangement of the agitator discs, and these have been investigated for yeasts[10] and bacteria.[11] This type of cell disrupter also has the advantage that it can be readily mounted in an enclosed safety cabinet when pathogenic or rDNA organisms are to be broken.

2.3.3 Solid Shear. Methods of cell disruption employing solid shear have long been used on a small scale. It involves the extrusion of frozen cell material through a narrow orifice at high pressure and an outlet temperature of about − 20 °C. It has found little application on an industrial scale, due to limitations on the amount of material which can be processed, although a semi-continuous process, based on the laboratory scale X-press has been described, which is claimed to have a throughput rate of 10 kg bacterial cell paste per hour at a pressure of 150 MPa, with a 90% efficient breakage rate.[12]

2.3.4 Liquid shear. Liquid shear is the principle choice for the large scale disruption of microbial cells, finding widespread application in both industrial processes and in research. It is particularly useful for the disruption of bacterial cells, but can also be effective in the breakage of yeast and fungal material. A recent publication describes the use of liquid shear combined with chemical treatment for the breakage of microbial cells.[13]

As with solid shear, the cells are passed through a restricted orifice under high pressure, this time in a liquid suspension. For smaller scale work, a continuous version of the French Press may be used (American Instrument Co., Silver Springs, USA). Larger scale work usually employs a homogenizer of the type

[9] L. Fryklund, *World Biotech. Report*, Part 3, 1987, 31.
[10] F. Marfy and M. R. Kula, *Biotechnol. Bioeng.*, 1974, **16**, 623.
[11] J. R. Woodrow and A. V. Quirk, *Enz. Microbiol. Technol.*, 1982, **24**, 385.
[12] K. E. Magnusson and L. Edebo, *Biotechnol. Bioeng.*, 1976, **18**, 975.
[13] S. T. L. Harrison, J. S. Dennis, and H. A. Chase, *Bioseparation*, 1991, **2**, 95.

developed for emulsification in the dairy industry. A temperature increase of at least 10°C in a single pass is not uncommon and it is necessary to pre-cool the cell suspension before homogenization.

For large scale work the Manton–Gaulin homogenizer (APV Ltd., Crawley, UK) is the most frequently used. It consists of a positive displacement piston pump with a restricted outlet valve, which can be adjusted to give the required operating pressure. Typically, pressures of up to 55 MPa can be attained.

The rate of cell breakage and of protein release is dependent on a number of factors, including cell type, concentration, and pre-treatment. The rate of protein release from yeast cells can be described by the empirical first order rate equation:[14]

$$\log(R_m/R_m - R) = K\, n\, p^{2.9} \tag{1}$$

where R_m = the theoretical maximum amount of soluble protein to be released, R = actual protein released, K = a temperature-dependent constant, n = number of passes, and p = operational back pressure.

The liquid shear homogenizer is normally operated at wet cell concentrations of about 20%, and the smallest Manton–Gaulin homogenizer, the 15M-8BA, has a throughput of about 50 $dm^3\ h^{-1}$ at a pressure of 55 MPa. A larger version, the MC-4, has a throughput of about 300 $dm^3\ h^{-1}$, again at a pressure of 55MPa.

There are many examples of the use of the Manton–Gaulin homogenizers for the large scale disruption of microbial cells. β-Galctosidase has been released from *E. coli*,[15,16] and carboxypeptidase from *Pseudomonas* spp.[17] A large number of enzymes have been isolated from the thermophilic bacterium *Bacillus stearothermophilus*,[18] including glycerokinase[19–21] and a glucose specific hexokinase.[22]

3 INITIAL PURIFICATION

3.1 Debris Removal

Following cell disruption, the first step in the purification of an intracellular enzyme is the removal of cell debris. The separation of solids from liquids is a key operation in enzyme isolation, and is normally accomplished by centrifugation or filtration. A variety of centrifuges are available for large scale enzyme purification.

[14] M. Follows, P. J. Hetherington, and M. D. Lilly, *Biotechnol. Bioeng.*, 1971, **13**, 549.
[15] P. P. Gray, P. Dunnill, and M. D. Lilly, in Proceedings of the IVth International Fermentation Symposium, ed. G. Terui, Society for Fermentation Technology, Osaka, Japan, 1972, p. 347.
[16] J. J. Higgins, D. J. Lewis, W. H. Daly, F. G. Mosqueira, P. Dunnill, and M. D. Lilly, *Biotechnol. Bioeng.*, 1987, **20**, 159.
[17] R. F. Sherwood, R. G. Melton, S. M. Alwan, and P. Hughes, *Eur. J. Biochem.*, 1985, **148**, 447.
[18] T. Atkinson, G. T. Banks, C. J. Bruton, M. J. Comer, R. Jakes, T. Kamalagharan, A. R. Whitaker, and G. P. Winter, *J. Appl. Biochem.*, 1979, **1**, 247.
[19] M. J. Comer, C. J. Bruton, and T. Atkinson, *J. Appl. Biochem.*, 1979, **1**, 259.
[20] M. D. Scawen, P. M. Hammond, M. J. Comer, and T. Atkinson, *Anal. Biochem.*, 1983, **132**, 413.
[21] P. M. Hammond, T. Atkinson, and M. D. Scawen, *J. Chromatogr.*, 1986, **366**, 79.
[22] C. R. Goward, R. Hartwell, T. Atkinson, and M. D. Scawen, *Biochem. J.*, 1986, **237**, 415.

3.2 Batch Centrifuges

Batch centrifuges are available with capacities ranging from less than 1 cm^3 up to several decimetres, and capable of applying a relative centrifugal force of up to 100 000 *g*. However, for the removal of bacterial cells, cell debris, and protein precipitates, fields up to 20 000 *g* are adequate. Many centrifuges of this type, and suitable for intermediate scale preparations are available, for example the DuPont Sorvall RC-3B has a maximum capacity of 6 dm^3 at 5000 *g*.

3.3 Continuous-flow Centrifugation

Because of the large volumes of liquid which need to be handled at the beginning of a large scale enzyme purification, it is preferable to use a continuous flow centrifuge to remove particulate matter. Three main types of centrifuge are available; the hollow bowl centrifuge, the disc or multi-chamber bowl centrifuge, and the basket centrifuge.

Hollow bowl centrifuges have a tubular rotor which provides a long flow path for the extract, which is pumped in at the bottom and flows upwards through the bowl. Particulate matter is thrown to the side of the bowl, and the clarified extract moves up and out of the bowl into a collecting vessel. As centrifugation proceeds, the effective diameter of the bowl decreases, so reducing the settling path and the centrifuge force which can be applied. The ease with which the bowl can be changed, and the possibility of using a liner to aid sediment recovery has contributed to the popularity of this type of centrifuge. The flow rate must be determined empirically as it will vary from one type of extract to another, but a rate of about 60 $dm^3 h^{-1}$ is generally satisfactory for the larger machines. Centrifuges of this type are produced by Pennwalt Ltd. (Camberley, Surrey, UK) and by Carl Padberg Gmbh (Lahr, W. Germany).

Disc centrifuges provide an excellent means of clarifying crude extracts, and in many cases the sediment may be discharged without interrupting the centrifugation process. The bowl contains a series of discs around a central cone. As the extract enters, particular matter is thrown outwards, impinging on the coned discs, and sedimented matter collects on the bowl wall. This provides a constant flow path, so there is little loss of centrifugal efficiency during operation. A disadvantge of these centrifuges is that some loss of activity may be experienced whilst discharging solids.

The rotors of instruments which do not have the facility to discharge sediment during operation are tedious to clean and this again may result in loss of product if the solids are required. These centrifuges achieve an RCF of about 8000 *g* and have a capacity of up to 20 kg sediment. As with hollow bowl centrifuges, the correct flow rate must be determined empirically. A variation of the disc-type centrifuge is the multi-chambered bowl centrifuge, in which the bowl is divided by vertically mounted cylinders into a number of interconnected chambers. The feed passes through each chamber from the centre outwards, before leaving the centrifuge. This type of arrangement also ensures a short and constant settling path as the bowl fills, and is easier to dismantle and clean than the disc-type of centrifuge.

Centrifuges of these types are produced by De Laval Separator Co. (New York, USA), Bird Machine Co. (S. Walpole, Ma, USA), and by Westfalia Separator Ltd. (Wolverton, UK). The development and trends in the construction of large scale centrifuge separators has been reviewed.[23]

A problem suffered by all types of centrifugation when applied on an industrial scale to enzyme recovery is that many homogenates produce a very sloppy precipitate. In some instances, this may be overcome by the addition of an agent such as the cellulose-based 'cell debris remover' (CDR) produced by Whatman Ltd. (Maidstone, UK). In many cases, however, the problem remains, leading to unacceptably high losses of valuable material. The influence of the physical characteristics of such protein precipitates on the performance of centrifuge separation has been reviewed.[24]

3.4 Basket Centrifuges

These are designed to operate at much lower *g* forces, perhaps only 1000 rev min^{-1}, and are basically centrifugal filters. The bowl is perforated and is normally lined with a filter cloth. The main use of these centrifuges is to collect large particulate material; in the context of enzyme purification this usually means ion exchange materials which have been used for the batch adsorption of the desired protein.

Perhaps the simplest example is the domestic spin-drier fitted with a suitable cloth liner. More powerful and robust machines are commercially available from Carl Padberg Gmbh (Hahr, W. Germany) and Thomas Broadbent (Huddersfield, UK).

3.5 Membrane Filtration

Filtration is often an alternative method of clarifying cell extracts. However, microbial preparations often tend to be gelatinous in nature, and are difficult to filter by traditional methods unless large filter areas are employed.

This can be overcome by using tangential or cross-flow filtration. In this method the extract flows at right-angles to the direction of filtration, and the use of a high flow rate tends to reduce fouling by a self-scouring action, although this action must be balanced against the possibility of losses due to shear effects. To date, there have been only limited reports of the large scale application of this method in downstream processing, although it may be expected to find increasing application in the future. Carboxypeptidase G, and aryl acylamidase from selected strains of *Pseudomonas* were successfully separated from broken cell debris by tangential microfiltration using a Millipore Pellicon system[25] [Millipore (UK) Ltd., London, UK]. This study concluded that the method was feasible,

[23] H. Hemfort and W. Kohlstette, *Chem. Ind.*, 1985, **108**, 412.
[24] M. Hoare, P. Dunnill, and D. J. Bell, *Ann. N.Y. Acad. Sci.*, 1983, **413**, 254.
[25] A. V. Quirk and J. R. Woodrow, *Biotechnol. Lett.*, 1983, **5**, 277.

leading to highly clarified extracts, but that the isotropic membranes used were prone to blocking due to the build up of protein material and debris.

Membranes with an asymmetric pore structure are less prone to blockage, and have been used to separate aryl acylamidase from *Pseudomonas* cell debris.[26] Membranes of the same type have also been used on a larger scale for the recovery of L-asparaginase from *Erwinia chrysanthemi*. A 1 m^2 membrane assembly was used to harvest the cells from 100 dm^3 of culture fluid in 2.5 h, when the solids concentration in the retentate increased from 0.55% to 22% dry weight. This same membrane assembly was then used to clarify the extract obtained by the alkali lysis of these bacteria. These data indicated that to harvest the cells from 500 dm^3 culture in 2.5 h would require 7.5 m^2 of membrane, and that the costs compared favourably with the costs of centrifugation.[27,28]

Membranes suitable for cross-flow filtration are now available as spirally wound cartridges, which offer the same surface area as flat membranes, but in a more compact space.

4 AQUEOUS TWO-PHASE SEPARATION

Aqueous two-phase systems, typically created by mixing solutions of polyethylene glycol and dextran or polyethylene glycol and specific salts such as potassium phosphate or ammonium sulfate, can be used both for the partitioning of enzymes during protein purification and also for the separation of proteins from cellular debris.[29–33] This phenomenon reflects a balance between the components involved, namely polymers, salts, proteins, and solvent (water).

The precise partitioning of a protein depends on parameters such as its molecular weight and charge, the concentration and molecular weight of the polymers, the temperature, pH, and ionic strength of the mixture, and the presence of polyvalent salts such as phosphate or sulfate.[29,30] The optimal conditions required for a particular protein must be found empirically. Although the conditions required to achieve satisfactory separation can often be precisely defined, the mechanism of partitioning is poorly understood, and the precise influence of the inorganic salts unknown. Moreover, polyethylene glycol, which is commonly used in such phase separations, can bind to proteins, and this may cause anomalous behaviour in subsequent conventional chromatographic steps.[34]

The phases can be separated in a settling tank, but a more efficient and rapid separation can usually be achieved by centrifugation. Since it is often easier to

[26] M. S. Le, L. B. Spark, and P. S. Ward, *J. Membrane Sci.*, 1984, **21**, 219.
[27] M. S. Le, L. B. Spark, P. S. Ward, and N. Ladwa, *J. Membrane Sci.*, 1984, **21**, 307.
[28] M. S. Le and T. Atkinson, *Process Biochem.*, 1985, **20**, 26.
[29] M. R. Kula, in 'Extraction and Purification of Enzymes, Applied Biochemistry and Bioengineering, Vol. 2', ed. L. B. Wingard jnr., E. Katchalski-Katzir, and L. Goldstein, Academic Press, New York, 1979, p. 71.
[30] M. R. Kula, K. H. Kroner, and H. Hustedt, *Adv. Biochem. Eng.*, 1982, **24**, 73.
[31] H. Walter, D. E. Brooks, and D. Fisher, 'Theory, Methods, Uses and Applications to Biotechnology: Partitioning in Aqueous Two Phase Systems', Academic Press, New York, 1985.
[32] G. Johansson, *J. Biotechnol.*, 1985, **3**, 11.
[33] H. Walter and G. Johansson, *Anal. Biochem.*, 1986, **155**, 215.
[34] J. Woodrow and A. V. Quirk, *Enz. Microb. Technol.*, 1986, **8**, 183.

separate liquids of different density than solids from liquids on the large scale, it has been suggested that this can be used to advantage in large scale enzyme preparations. Although the relatively low cost of the polyethylene glycol–salt system makes it attractive for large scale use, the more generally useful polyethylene glycol–dextran system can also have economic advantages in comparison with other purification methods, despite the high cost of purified dextran, providing the total processing costs are evaluated.[35] Cheaper substitutes for purified dextran are being investigated, such as crude dextran,[36] or hydroxypropyl starch.[37] A good overview of two-phase aqueous polymer systems is provided by Abbott *et al.*[38]

Two-phase separation has been used for the large scale separation and purification of pullulan-6-glucan hydrolase and 1,4-β-glucan phosphorylase from 5 kg quantities of cell paste of *Klebsiella pneumoniae*,[39] and RNA polymerase and glutamine synthetase of *E. coli*.[40] Its use is not restricted to materials of microbial origin, and the method has been successfully used to isolate materials from both plant[41] and animal[42] sources.

In affinity partition, various ligands are attached to the polymers in order to alter the partitioning of a protein.[29,32,33] All phase-forming polymers can have ligands attached covalently to them, and a wide range of such ligands have been investigated. The power of the technique is clearly demonstrated by the 58-fold purification of yeast phosphofructokinase that could be obtained in two steps using Cibacron Blue F-3GA immobilized on polyethylene glycol.[43,44] Besides the biomimetic dyes, a number of other ligands have been investigated. These include cofactors, such as the pyridine nucleotides used successfully in the affinity partitioning of a number of dehydrogenases.[45]

On the process scale, affinity partition has been used for the purification of formate dehydrogenase from 10 kg quantities of the yeast *Candida bodinii*, using the triazine dye, Procion Red HE-3B, immobilized on polyethylene glycol.[29] Aqueous two-phase separation is a method which can easily be scaled-up to a manufacturing level, although the cost of the polymers may prove to be a limiting factor.

5 PRECIPITATION

5.1 Ammonium Sulfate

Salting out of proteins has been employed for many years, and fulfils the dual purposes of purification and concentration. The most commonly used salt is

[35] K. H. Kroner, H. Hustedt, and M. R. Kula, *Process Biochem.*, 1984, **19**, 170.
[36] K. H. Kroner, H. Hustedt, and M. R. Kula, *Biotechnol. Bioeng.*, 1982, **24**, 1015.
[37] F. Tjerneld, S. Berner, A. Cajarville, and G. Johansson, *Enz. Microb. Technol.*, 1986, **8**, 417.
[38] N. L. Abbott, D. Blankschtein, and T. A. Hatton, *Bioseparation*, 1990, **1**, 191.
[39] H. Hustedt, K. Kroner, W. Stach, and M. R. Kula, *Biotechnol. Bioeng.*, 1978, **20**, 1989.
[40] T. Takahashi and Y. Adachi, *J. Biochem. (Tokyo)*, 1982, **91**, 1719.
[41] H. Vilter, *Bioseparation*, 1990, **1**, 283. [42] M. J. Boland, *Bioseparation*, 1990, **1**, 293.
[43] G. Koperschlager and G. Johansson, *Anal. Biochem.*, 1982, **124**, 117.
[44] G. Johansson, G. Koperschlager, and P. A. Albertsson, *Eur. J. Biochem.*, 1983, **131**, 589.
[45] A. F. Buckmann, M. Morr, and M-R. Kula, *Biotechnol. Appl. Biochem.*, 1987, **9**, 258.

ammonium sulfate, because of its high solubility, lack of toxicity towards most enzymes, and low cost. The precipitation of a protein by salt depends on a number of factors: pH, temperature, protein concentration, and the salt used.[46] The protein concentration is particularly important when scaling-up, because most large scale purifications are carried out at higher protein concentrations than laboratory-scale purifications. This can have a dramatic effect on the concentration of salt needed to precipitate a given protein.

5.2 Organic Solvents

The addition of organic solvents to aqueous solutions reduces the solubility of proteins by reducing the dielectric constant of the medium. Various organic solvents have been used for the precipitation of proteins, with ethanol, acetone, and propan-2-ol being the most important. Because proteins are denatured by organic solvents it is necessary to work at temperatures below 0 °C.

Because of their flammable nature and high cost, coupled with a low selectivity, organic solvents are not often used in large scale enzyme purification. The one notable exception is in the blood processing field, where ethanol precipitation is the major method for the purification of albumin; indeed it has been developed into a highly automated, computer-controlled system.[47]

5.3 High Molecular Weight Polymers

Other organic precipitants which can be used for the fractionation of proteins are water soluble polymers like polyethylene glycol. This has the advantage of being non-toxic, non-flammable, and not denaturing to proteins.[48] It is mainly used in the blood processing field.[49]

6 CHROMATOGRAPHY

The purification of proteins by chromatography has been a standard laboratory practice for many years. These same chromatographic techniques can equally well be applied to the isolation of much larger quantities of protein, although the order in which they are used must be considered carefully.

Although it is not generally considered applicable to the purification of low value/high volume products, chromatography has been used for the purification of whey proteins, for example.[50] For the purification of high value/low volume products, typically therapeutic or diagnostic proteins, chromatography is, perhaps, the most widely used method, even on the industrial scale. Chromato-

[46] M. C. Dixon and E. C. Webb, 'Enzymes', Longmans, London, 1979, p. 31.
[47] P. Foster and J. G. Watt, in 'Methods of Plasma Fractionation', ed. J. Curling, Academic Press, New York, 1980, p. 17.
[48] K. C. Ingham, *Methods Enzymol.*, 1984, **104**, 351.
[49] Y. L. Hao, W. Hoenig, and M. Wickerhauser, in 'Methods of Plasma Fractionation', ed. J. Curling, Academic Press, New York, 1980, p. 57.
[50] R. A. M. Delaney, in 'Applied Protein Chemistry', ed. R. A. Grant, Applied Science Publishers, Barking, UK, 1980, p. 233.

graphy is the only method with the required selectivity to purify a single protein from a complex mixture of proteins to a final purity of greater than 99.8%.

6.1 Scale-up and Quality Management

In analytical chromatography, as well as in many laboratory scale applications, the quantity of sample to be applied is small, and the overall aim is to achieve the maximum number of peaks or to produce a small amount of highly purified protein. The flow rates used are low, as ultimate resolution is of greater importance than throughput. In contrast, the aim in preparative chromatography is to purify the maximum amount of protein in the minimum amount of time. The flow rates used are high as throughput is often more important than resolution.

The greatest resolution and throughput will be given by the use of small particles at high flow rates. Unfortunately this combination results in high pressures, which can only be reduced by using a lower flow rate. Small particles are generally more expensive than larger ones, and high pressure equipment is more expensive than low pressure equipment. For these reasons, in process-scale chromatography it is common to use larger particles at the highest flow rate compatible with the gel, the resolution required, and the chromatography equipment.

The scale-up of a chromatographic separation is in principle simple, as chromatography theory shows that the column diameter has little effect on resolution, so any increase in scale can be accomplished by increasing the column diameter.

Before scaling-up it is important to ensure that the purification process is fully understood and optimized on the laboratory scale. The initial screening for adsorptive techniques can be conveniently carried out by the batch addition of different adsorbents under various conditions of pH and ionic strength. Suitable chromatographic methods can then be developed using either conventional, low pressure equipment, or by using higher pressure, high performance equipment.

Following this optimization process, the next step is to increase the sample load by a factor of between ten and twenty. The bed height of the column should be kept constant and the surface area increased in proportion to the sample load. The packing material should be the same, or at least have similar characteristics. The linear flow rate, the ionic strength, and the pH must all be held constant. If gradient elution is used the ratio of gradient volume to column volume must be the same. It is important that the protein load per unit column volume is held constant, rather than simple sample volume, because large scale production extracts normally contain a higher concentration of protein than the corresponding laboratory scale extracts, due to of the great differences in efficiency of laboratory and process scale centrifuges.

In principle this scaling-up process can be repeated until the desired scale of operation is reached. However, there are limitations. The maximum column diameter is governed by those available, and is currently around 120 cm, which for the bed heights of 15–20 cm commonly used for ion exchange chromatography is equivalent to 170–220 dm^3 of gel. As the column diameter is increased so does the cost, as well as the difficulty of ensuring even loading of the sample over the entire surface.

One of the prime considerations in scale-up and design of large scale purification protocols is the intended use of the final product. Following the above guidelines, it should be possible to scale-up most purifications to a scale suitable for manufacturing purposes. If the product is for industrial or research use, this may be all that is required. However, if the protein product is intended for human use, there are other factors besides the mechanics of scale-up to take into account. There may also be a requirement to demonstrate that any materials which would have come into contact with the product during manufacture do not have a residual presence in the final product. There needs to be a strict control over raw materials, including the chromatography matrices used, to ensure that they meet stringent criteria in terms of purity and acceptability. This in turn, may influence the choice of method.

In quality management, it is necessary to distinguish between contaminants deriving from the biological system or process raw materials and those introduced incidentally or by design, during processing. It is especially necessary to have suitable analytical methods for quantifying those impurities which may still be present in the final product. Contaminants introduced during processing can be wide-ranging in nature. Before the deliberate introduction of any component(s) during a purification protocol, it is necessary to ensure that their presence can be adequately monitored. It may also be necessary to demonstrate that they can be satisfactorily removed (using a positive removal step) at a later stage. All of these factors may have an influence on the design of the manufacturing protocol. For example, affinity chromatography may involve a biological ligand; in such an instance, it is necessary to show that there is no significant leakage of ligand into the product. Practically all affinity matrices leach ligand to some degree, and the purification scientist may therefore be faced with proving that the level of leakage is not significant. Analysis to a certain level of sensitivity may not be sufficient alone. It may merely reflect the fact that the level of leakage is below the sensitivity of the detection methodology; no detection does not equal no leakage of ligand!

Standards of purity must be extremely high, and this may rule out certain approaches to scale-up. The demonstration of purity is an important consideration of quality management and has led to the need for complex strategies for the purity analysis of proteins, not least where such protein pharmaceuticals have been produced by recombinant DNA technology.[51] The development of analytical methods for rapid structural characterization of such products is therefore a fundamental requirement in both research laboratories and in commercial production. Recently developed techniques include high performance capillary electrophoresis[52] and mass spectrometry.[53]

[51] V. R. Anicetti, B. A. Keyt, and W. S. Hancock, *Trends Biotechnol.*, 1989, **7**, 342.
[52] J. Frenz and W. S. Hancock, *Trends Biotechnol.*, 1991, **9**, 243.
[53] M. J. Geisow, *Trends Biotechnol.*, 1992, **10**, 432.

Table 1 *Chromatographic techniques for the large scale purification of proteins*

Molecular Property Exploited	*Chromatography Type*	*Characteristics*	*Application*
Size	Gel filtration	*Resolution* Moderate for fractionation. Good for buffer exchange. *Capacity* limited by volume of sample.	Fractionation is best left to later stages of purification. Buffer exchange can be used any time, although sample volume may be a limitation.
Charge	Ion exchange	*Resolution* can be high. *Capacity* is high and not limited by sample volume. *Speed* can be very high depending on matrix.	Most effective early in fractionation when volumes large.
	Chromatofocusing	*Resolution* can be high. *Capacity* can be high. *Speed* can be high.	Best used later in a purification, as matrices expensive.
Polarity	Hydrophobic interaction	*Resolution* can be high. *Capacity* is very high, and not limited by sample volume. *Speed* is high.	Can be used at any time, but best when ionic strength is high, after salt precipitation, or ion exchange.
Biological affinity	Affinity	*Resolution* can be very high. *Capacity* can be high, but may be low, depending on ligand. Not limited by sample volume. *Speed* is high.	Can be used at any time, but normally not recommended at an early stage.

6.2 Method Selection

All of the available chromatographic techniques, gel filtration, ion exchange, hydrophobic interaction, affinity, immunoaffinity, and chromatofocusing, can be used for the large scale isolation of proteins. Although all these techniques can be used, there are limitations to each technique, which must be taken into account when designing a process, as shown in Table 1. It is preferable to be able to proceed from one step to the next with the minimum of alteration to the conditions. Therefore, steps which concentrate the product, like ion exchange, hydrophobic or affinity chromatography, should precede steps which cause a dilution, like gel filtration. Hydrophobic interaction chromatography can conveniently follow an ion exchange step with the minimum change in buffer, because most proteins bind more strongly to a hydrophobic support at high ionic strength. Apart from its use as a means of exchanging buffers, gel filtration is probably best used as a final polishing step, when the volume of product is low. Affinity chromatography has provided a uniquely powerful method for the purification of proteins on the laboratory scale. Although it is used on a preparative scale the number of published examples are few, partly due to the problems of using expensive and delicate ligands on a large scale.

Table 2 *Examples of base matrices for large scale chromatography. This list is not meant to be exhaustive, but is intended to give some idea of the range of packing materials available*

Matrix	*Example*	*Manufacturer*
Cross-linked dextran	Sephadex	1
Cross-linked polyacrylamide	Biogel-P	2
Agarose	Sepharose	1
	Biogel-A	2
	Ultrogel-A	3
Cross-linked agarose	Sepharose CL	1
	Sepharose FF	1
	Superose	1
Composite of polyacrylamide and dextran	Sephacryl	1
Composite of polyacrylamide and agarose	Ultrogel AcA	3
Composite of dextran and agarose	Superdex	1
Hydroxylated acrylic polymer	Trisacryl	3
Ethylene glycol–methacrylate copolymer	Fractogel	4
Cellulose	Cellex	2
	DE-52, CM-52	5
	Sephacel	1
	Cellufine	6
Rigid organic polymers	Monobeads	1
	TSK-PW	4
	POROS	7
Porous silica	Zorbax	8
	Aquapore	9

[1] Pharmacia AB, Uppsala, Sweden. [2] BioRad Laboratories, Richmond, California, USA. [3] IBF Biotechnics, Villeneuve la Garenne, France. [4] TosoHass, Montgomeryville, Pennsylvania, USA. [5] Whatman Ltd., Maidstone, Kent, UK. [6] Amicon Ltd., Stonehouse, Gloucestershire, UK. [7] Perseptive Biosystems Inc., Cambridge, Massachusetts, USA. [8] Rockland Technologies, Wilmington, Delaware, USA. [9] Brownlee Laboratories Inc., California, USA.

6.3 Selection of Matrix

Perhaps the most important decision to be taken when designing a large scale purification process concerns the type of chromatography matrix to be used at each step. A matrix which is to be used for large scale chromatography should be hydrophilic, macroporous, rigid, spherical, chemically stable, yet easily derivatized, inert, and reuseable.

No single matrix can completely satisfy all of these criteria, and it is not surprising that many different types of matrix are available, as shown in Table 2. In many cases these are available in a variety of derivatized forms suitable for the different types of chromatography. All of the different types of matrix listed in Table 2 have advantages and disadvantages which must be taken into account, as shown in Table 3.

Some gels, such as those based on agarose or cellulose are natural products; others, such as those based on cross-linked dextrans or agarose, are modified natural products; yet others, such as those based on polyacrylamide, polyhydroxyethyl methacrylate, or polystyrene are wholly synthetic. The gels can be further

Table 3 *Properties of basic chromatographic matrices*

Matrix type	*Porosity*	*Non-specific adsorption*	*Rigidity*	*Stability*	*Ease of derivatization*	*Relative cost*
Cross-linked dextran	low	low	low	good	good	low/medium
Cross-linked polyacrylamide	low	low	low	good	good	medium
Agarose	high	low	low	poor	good	medium
Cross-linked agarose	high	low	medium	good	good	medium
Polyacrylamide/dextran composite	high	medium	medium	good	good	medium
Polyacrylamide/agarose composite	medium	low	medium	good	good	medium/high
Hydroxylated acrylic polymer	medium	medium	medium	good	good	medium
Ethylene glycol–methacrylate copolymer	low/medium	medium	medium	good	good	medium
Cellulose	medium/high	high	low	good	good	low
Porous silica	low/medium	high	high	poor	poor	high
Rigid organic polymers	medium/high	low	high	good	good	high

classified as macroporous, such as the agarose or cellulose gels, or microporous, such as the cross-linked dextran or polyacrylamide gels. The macroporous gels are most useful for ion exchange or affinity chromatography or for the size fractionation of very large molecules, such as viruses, large proteins, or glycoproteins. The microporous gels are most useful for the size fractionation of the majority of proteins. The rigidity of the gels varies widely. The earliest gels, based on cross-linked dextran, cellulose, polyacrylamide, or agarose were very soft and not readily suited to large scale chromatography. The newer generation of highly cross-linked, yet macroporous gels, based on agarose or on composites, such as polyacrylamide and agarose, are much more rigid and far more suited to large scale chromatography. In addition to their improved rigidity, these newer gels are also available in a smaller and more controlled particle size, ensuring that comparable resolution is obtained at the higher flow rates which are possible. The latest generation of hydrophilic, polymeric gels, for example Monobeads or Superose (Pharmacia) or the TSK-PW range (TosoHaas) are truly high performance gels, in that they are available as 10 μm particles that are able to withstand pressure of 3–10 MPa.

The truly inorganic materials, silica or porous glass, although ideal in terms of their rigidity and availability as small particles, are not suitable for the majority of biological applications. The hydrophobic nature of their surfaces often means that proteins are bound irreversibly, or can only be eluted under denaturing conditions. In addition silica particles offer a high resistance to the flow of aqueous solvents and are unstable at pH values > 8. These drawbacks can be alleviated to some extent by coating the particles with a hydrophilic polymer, which can then be further derivatized. The chemistry involved in this is very complex, and the resulting matrices are often too expensive for large scale use.

6.4 Gel Filtration

In gel filtration, separation is based on molecular size. The stationary phase consists of porous beads surrounded by a mobile solvent phase. When the sample is applied, the molecules in the mixture partition between the pores in the beads and the solvent. Large molecules are unable to enter the pores and so pass through the interstitial spaces and elute first. Smaller molecules, which can enter the pores are eluted later, in decreasing order of size.

The total volume of a column can be represented by:

$$V_t = V_o + V_i + V_m \tag{2}$$

where V_t = the total volume of the column, V_o = the volume of solvent external to the particles, V_i = the volume of solvent that occupies the interior of the particles, and V_m = the volume occupied by the matrix itself.

The elution volume of a protein can therefore vary between V_o, for one which cannot enter the pores in the gel, and V_i, for one which is fully able to enter the pores in the gel. Thus it is possible to calculate an effective partition coefficient, K_{av}, which can vary between zero and unity:

$$K_{av} = (V_e - V_o)/(V_t - V_o) \tag{3}$$

where V_e = the elution volume of the solute, V_o = the void volume of the column; V_t = the total volume of the column. For globular proteins it has been shown empirically that the value of K_{av} is inversely proportional to the logarithm of the relative molecular mass.

It is essential that there should be no interaction between the matrix and the solute, therefore the ideal gel filtration medium should be totally inert. For maximum capacity it should also be rigid and highly porous. For large scale working, rigidity is perhaps most important as it determines the highest flow rate that can be obtained.

The traditional gel filtration materials were based on cross-linked dextran, Sephadex (Pharmacia), or polyacrylamide (BioGel P). These materials are sufficiently inert, but, in the porosities suitable for the fractionation of most proteins, are too soft for easy application on the large scale. For this reason most applications of gel filtration are limited to desalting, using the low porosity, but rigid gels like Sephadex G-25 and G-50.

In recent years more rigid gel types, based on a variety of materials have been introduced, for example, Sephacryl (Pharmacia), based on dextran and polyacrylamide; Superdex (Pharmacia), based on dextran and agarose; Ultrogel AcA (IBF), based on agarose and polyacrylamide; Trisacryl (IBF), based on a hydroxylated acrylic polymer; Fractogel (TosoHaas), based on an ethylene glycol–methacrylate copolymer; Superose (Pharmacia), based on a highly cross-linked agarose; Cellufine (Amicon), based on cross-linked, beaded cellulose.

All of these materials are available in particle sizes which are smaller than their traditional counterparts, so that resolution is retained at the higher flow rates permitted by their increased rigidity. This increased rigidity also means that these materials are more suited to large scale chromatography.

6.5 Ion Exchange Chromatography

Traditionally ion exchange media for the fractionation of proteins have been based on cellulose substituted with various charged groups; see for example Table 4. Cellulose ion exchangers are ideally suited to batch-type operations at the early stages of a process. For example, in the purification of L-asparaginase from *Erwinia chrysanthemi*, a 6-fold purification and 100-fold reduction in volume can be achieved by batch adsorption and elution from CM-cellulose. For routine use in large scale columns cellulose cannot support the highest flow rates, and suffers the disadvantage that its volume changes with a change in pH or ionic strength, making it difficult to regenerate without unpacking the column, a laborious procedure. This volume change is less marked with the cross-linked, beaded forms.

Ion exchange Sephadex gels are prepared by introducing ion exchange groups into Sephadex G-25 or G-50 (Pharmacia). Those based on Sephadex G-25 are rigid, but have a low capacity for most proteins. Those based on Sephadex G-50 have a high capacity for proteins, but are very soft and show large changes in volume with variation in pH or ionic strength, making them unsuitable for large scale ion exchange chromatography. However, like the ion exchange celluloses, they are well suited to batch adsorption techniques.

Table 4 *Ion exchange substituents*

Cation exchange	
Carboxymethyl (CM)	$—O—CH_2—COO^-$
Sulfoethyl (SE)	$—O—CH_2—CH_2—SO_3^-$
Sulfopropyl (SP)	$—O—CH_2—CH_2—CH_2—SO_3^-$
Phosphate (P)	$—O—PO_3^-$
Sulfonate (S)	$—CH_2—SO_3^-$
Anion exchange	
Diethylaminoethyl (DEAE)	$—O—CH_2—CH_2—N^+—(C_2H_5)_2$
Quaternaryaminoethyl (QAE)	$—O—CH_2—CH_2—N^+—(C_2H_5)_3$
Quaternary amine (Q)	$—CH_2—N^+—(CH_3)_3$

The introduction of ion exchange groups into cross-linked agarose or the macroporous synthetic gels, such as Trisacryl or Fractogel, gives a range of ion exchange materials which are both rigid and of high capacity. The particles are small and spherical, supporting high flow rates with good resolution, and do not deform under the pressures normally encountered in large scale chromatography. In recent years the degree of cross-linking of the agarose-based gels has increased, giving rise to the 'Fast Flow' range of Sepharoses from Pharmacia, which can support even higher flow rates. These types of gel are particularly suited to the initial stages of a separation, when the volume of extract to be handled is largest. Because of their rigidity, all of these materials can be regenerated in the column with NaOH solutions.

A typical example of large scale ion exchange chromatography, using gradient elution from a 40 dm^3 column of DEAE Sepharose is the purification of glycerokinase from 20 kg *Bacillus stearothermophilus*. The column was 80 × 25 cm, and was eluted with a 200 dm^3 linear gradient of increasing phosphate concentration.[20] Recombinant *Erwinia* asparaginase has been purified by large scale ion exchange chromatography on S-Sepharose. The enzyme from 16 kg of bacterial cell was adsorbed onto a 30 dm^3 column at flow rate of 600 $dm^3\ h^{-1}$.[54]

6.6 Affinity Chromatography

Affinity chromatography can provide perhaps the most elegant method for the purification of a protein from a complex mixture. Although used extensively on a laboratory scale, it is only recently finding acceptance for industrial-scale purification.

Affinity chromatography relies on the interaction of a protein with an immobilized ligand. A ligand can be either specific for a particular protein, for example, a substrate, substrate analogue, inhibitor, or an antibody. Alternatively it may be able to interact with a variety of proteins, for example, AMP, ADP, NAD, dyes, hydrocarbon chains, or immobilized metal ions. Affinity chromatography using immobilized nucleotides is little used for process scale purification, perhaps

[54] C. R. Goward, G. B. Stevens, R. Tattersall, and T. Atkinson, *Bioseparation*, 1991, **2**, 335.

Table 5 *Examples of proteins purified by large scale dye affinity chromatography*

Enzyme	*Dye*	*Eluant*
Glycerokinase[a]	Procion Blue MX-3G	5 mM ATP
Glucokinase[b]	Procion Brown H-3R	2 mM ATP
Glycerol dehydrogenase[c]	Procion Red HE-3B	2 mM NAD
Methionyl tRNA synthetase[d]	Procion Green HE-4BD	Phosphate gradient
Tryptophanyl tRNA synthetase[d]	Procion Brown MX-5BR	50 mM tryptophan
3-Hydroxybutyrate dehydrogenase[e]	Procion Red H-3B	1 M KCl
	Procion Blue MX-4GD	1 M KCl + 2 mM NADH
Malate dehydrogenase[e]	Procion Red H-3B	1 M KCl + 2 mM NADH
	Procion Blue MX-4GD	0–0.7 M KCl gradient
Carboxypeptidase G_2[f]	Procion Red H-8BN	Bind in 0.2 mM Zn^{2+}, 0.1 M Tris HCl Elute with 10 mM EDTA pH 5.8 followed by 0.1 M Tris-HCl pH 7.3
Human serum albumin[g]	Cibacron Blue F3-GA	20 mM Na octanoate
Human serum albumin[h]	Cibacron Blue F3-GA	3 M NaCl, pH 8.6

[a] M. D. Scawen, P. M. Hammond, M. J. Comer, and T. Atkinson, *Anal. Biochem.*, 1983, **132**, 413. [b] C. R. Goward, R. Hartwell, T. Atkinson, and M. D. Scawen, *Biochem. J.*, 1986, **237**, 415. [c] P. Spencer, K. J. Brown, M. D. Scawen, T. Atkinson, and M. G. Gore, *Biochim. Biophys. Acta*, 1989, **994**, 270. [d] C. J. Bruton and T. Atkinson, *Nucleic Acids Res.*, 1979, **7**, 1579. [e] M. D. Scawen, J. Darbyshire, M. J. Harvey, and T. Atkinson, *Biochem. J.*, 1983, **203**, 699. [f] R. F. Sherwood, R. G. Melton, S. M. Alwan, and P. Hughes, *Eur. J. Biochem.*, 1985, **148**, 447. [g] J. E. More, A. G. Hitchcock, S. Price, J. Rott, and M. J. Harvey, in 'Protein-dye Interactions: Developments and Applications', ed. M. A. Vijayalakshmi and O. Bertrand, Elsevier, London, 1989, p. 265. [h] M. Allary, J. Saint-Blancard, E. Boschetti, and P. Girot, *Bioseparation*, 1991, **2**, 167.

because of their instability, expense, low capacity, and the difficulties of coupling them to a support matrix.

Immobilized dyes have been used for the large scale purification of many enzymes, where they offer the advantages of cheapness, ready coupling to a support matrix, stability, and high capacity.[55] Some proteins which have been purified by large scale dye affinity chromatography using Procion (ICI), or Cibacron (Ciba) dyes are shown in Table 5.

The reactive dyes are often anthroquinone-like structures, and are thought to bind proteins by interacting with the nucleotide-binding domain of dehydrogenases, for example. As so many different types of protein have been isolated by dye affinity chromatography,[56] the precise interaction must be more variable, and is in most cases unknown. The reactive dyes were originally made for the

[55] Y. D. Clonis, C. R. Lowe, T. Atkinson, and C. J. Bruton, 'Reactive Dyes in Protein and Enzyme Technology', Macmillan, London, 1987.

[56] M. D. Scawen and T. Atkinson, in 'Reactive Dyes in Protein and Enzyme Technology', ed. Y. D. Clonis, C. R. Lowe, T. Atkinson, and C. J. Bruton, Macmillan, London, 1987, p. 76.

textile dyeing industry, and are not ideally suited for the purification of proteins. Recently, the synthesis of ligands, loosely based on the structure of Cibacron Blue has been described,[57] which are better able to mimic the interaction between a natural cofactor and a protein. One such series of biomimetic dyes, in which a spacer group was inserted between the anthroquinone rings and the rest of the structure, and the means of attachment to the support could be varied, had a 3–4 fold higher affinity for alcohol dehydrogenase.[58] Other developments and applications of biomimetic dyes have recently been described.[59]

Immunoaffinity chromatography has been more used in recent years, because of the greater availability of monoclonal antibodies. For example recombinant leukocyte interferon[60] and three human pituitary hormones[61,62] have been purified by immunoaffinity chromatography.

Because of its high selectivity, affinity chromatography can be used to obtain extremely high purifications, in some cases up to several thousand fold in a single step. It is often possible to separate active from inactive forms of a protein, by using, for example an antibody or pseudo-substrate, or to remove a small amount of an impurity from an otherwise pure product. Examples of the large scale use of affinity chromatography have been given by Hill and Hirtenstein[63] and Clonis.[64]

Immobilized metal ions, such as Zn^{2+} or Ni^{2+}, can be used to separate proteins. This separation depends on the interaction between the metal ion and histidine residues on the surface of the protein. The metal ion is immobilized by chelation onto an iminodiacetate group attached to a suitable matrix, usually agarose. Bound proteins can be eluted with a competing ligand, such as imidazole.[65]

The criteria for the selection of a matrix for affinity chromatography are similar to those for ion exchange chromatography. As a result the most commonly used matrices are macroporous, such as Sepharose or Trisacryl. The matrix must be activated chemically to enable the ligand to be covalently coupled. A wide range of methods are available,[66] but one of the most generally used reagents is cyanogen bromide. For those who do not wish to activate their own matrix a range of activated matrices are available from the major suppliers of chromatography media. The reactive dyes have the advantage that they can be coupled directly to agarose with no prior activation.[56,67]

Methods for the elution of a bound protein may be either specific, using for example a substrate or cofactor, or non-specific, using for example salt or a

57 C. R. Lowe, S. J. Burton, J. C. Pearson, Y. D. Clonis, and C. V. Stead, *J. Chromatogr.*, 1986, **376**, 121.

58 S. J. Burton, C. V. Stead, and C. R. Lowe, *J. Chromatogr.*, 1990, **508**, 109.

59 C. R. Lowe, S. J. Burton, N. P. Burton, W. K. Alderton, J. M. Pitts, and J. A. Thomas, *Trends Biotechnol.*, 1992, **10**, 442.

60 T. Staehlin, D. S. Hobbs, H. F. Kung, C. Y. Lai, and S. Pestka, *J. Biol. Chem.*, 1981, **256**, 9750.

61 G. W. Jack and R. Blazek, *J. Chem. Tech. Biotechnol.*, 1987, **39**, 1.

62 G. W. Jack, R. Blazek, K. James, J. E. Boyd, and L. R. Micklem, *J. Chem. Tech. Biotechnol.*, 1987, **39**, 45.

63 E. A. Hill and M. D. Hirtenstein, in 'Advances in Biotechnological Processes', ed. A. Mizrahi and A. van Wezel, Alan R. Liss, New York, 1983, p. 31.

64 Y. D. Clonis, *Bio/Technology*, 1987, **5**, 1290.

65 E. Sulkowski, *Trends Biotechnol.*, 1985, **3**, 1.

66 M. Wilchek, T. Miron, and J. Kohn, *Methods Enzymol.*, 1984, **104**, 3.

67 C. R. Lowe and J. C. Pearson, *Methods Enzymol.*, 1984, **104**, 97.

Table 6 *Effects of ions on hydrophobic interactions*

	←——————— Increasing salting-out effect
Cations	NH_4^+, Rb^+, K^+, Na^+, Cs^+, Li^+, Mg^{2+}, Ca^{2+}, Ba^{2+}
Anions	PO_4^{3-}, SO_4^{2-}, CH_3COO^-, Cl^-, Br^-, NO_3^-, ClO_4^-, I^-, SCN^-
	Increasing chaotropic effect ———————→

change in pH. Non-specific elution is normally used with highly selective adsorbents which bind only one component; with a less selective adsorbent specific elution may give a greater degree of purification. Examples of available methods are, affinity elution with substrate or free ligand, change in pH or ionic strength, addition of a chaotropic or denaturing agent (*e.g.* KSCN, urea, guanidine–HCl), or change in solvent polarity, by the addition of an organic modifier, such as ethylene glycol.

The mildest method of elution is preferred, but must usually be found empirically, and in many cases cost considerations may be important. Although affinity steps can be used in batch or in columns, elution is best carried out in columns.

6.7 Hydrophobic Interaction Chromatography

Hydrophobic interaction chromatography was first developed following the observation that proteins were unexpectedly retained on affinity gels containing hydrocarbon spacer arms. This concept was extended and families of adsorbents prepared using an homologous series of hydrocarbon chains over the range C_2 to C_{10},[68] although in practice most proteins can be purified using agarose substituted with phenyl or octyl groups.

Hydrophobic interactions are strongest at high ionic strength, so adsorption can often be conveniently performed after salt precipitation or ion exchange chromatography, with no change in the salt concentration of the sample. Bound proteins can be eluted by altering the solvent pH, ionic strength, or by use of a chaotropic agent or an organic modifier such as ethylene glycol.

The various ions can be arranged in a series, depending on whether they promote hydrophobic interactions (salting-out effect) or disrupt the structure of water (chaotropic effect) and lead to a weakening of the hydrophobic interaction, as shown in Table 6. Those ions which promote the hydrophobic interaction, such as sulfate or phosphate, are useful in promoting binding. Those that are increasingly chaotropic are useful for strongly eluting proteins, which cannot be eluted by decreasing the ionic strength.

Chromatography on columns of Phenyl-Sepharose has been used for the purification of aryl acylamidase from *Pseudomonas fluorescens*. The enzyme from 2 kg bacteria was eluted from an ion exchange column in 0.3 M phosphate buffer, pH 7.6, and applied directly to a 500 ml column of Phenyl-Sepharose in the

[68] S. Shaltiel, *Methods Enzymol.*, 1984, **104**, 69.

same buffer. The enzyme was eluted by using a decreasing gradient from 0.1 M to 0.01 M Tris-HCl, pH 7.6.[2]

6.8 High Performance Chromatographic Techniques

One of the potentially most significant advances in chromatography has been the development of matrices of relatively small particle size capable of high resolution and operation under relatively high pressures. Originally developed for the separation of small organic molecules soluble in non-aqueous solvents, the technique has rapidly developed into a form suitable for the separation of proteins and enzymes in aqueous solvents, using all of the normally available chromatographic methods.[69,70] However, the majority of reported applications have been confined to the laboratory scale.

High performance matrices are highly efficient because of their small particle size (3 μm to 50 μm). Because of this small particle size, high pressures are needed to generate good flow rates. Thus, very rigid particles are necessary and two approaches have been taken to solve this problem. Silica-based matrices are sufficiently rigid, and can be extensively modified with monochloro- or monoalkoxy-silanes to give a hydrophilic surface which can be further derivatized.[71] Silica has the disadvantage of being unstable at pH values above pH 8, but this has been partially overcome either by using a polymer coating to reduce the availability of the inorganic particles to the solvent, or more specifically by surface stabilization with zirconium. The second approach has been the development of rigid, cross-linked polymeric supports such as Monobeads (Pharmacia) or TSK-PW (TosoHaas).

These true high performance matrices have been followed by higher performance derivatives of conventional packing materials, which have smaller particle sizes, for use in what is termed 'medium performance liquid chromatography' (MPLC). Thus Pharmacia have introduced Sepharose HR ion exchangers and a gel filtration matrix, Superdex, which have particle sizes of about 35 μm. These matrices have a particle size and a performance which is in between HPLC matrices (3 μm–20 μm) and conventional, low performance materials (>100 μm).

Another approach to the problem of flow rates and pressure in high performance separations has been the introduction of perfusion chromatography.[72] This technique differs from conventional chromatography in that the particles have large, interconnecting pores which pass through the particles as well as smaller, diffusive pores. This combination of pore sizes allows rapid access of molecules to the high surface area diffusive pores within the particles. The very short diffusive paths within the particles and the rapid bulk transport through the perfusive

[69] C. Horvath, 'High Performance Liquid Chromatography: Advances and Perspectives, Vol. 3', Academic Press, New York, 1983.
[70] J. F. Kennedy, Z. S. Rivera, and C. A. White, *J. Biotechnol.*, 1989, **9**, 83.
[71] R. E. Majors, in 'High Performance Liquid Chromatography: Advances and Perspectives, Vol. 1', ed. C. Horvath, Academic Press, New York, 1981, p. 2.
[72] N. B. Afeyan, S. P. Fulton, N. F. Gordon, I. Mazsaroff, L. Varady, and F. E. Regnier, *Bio/Technology*, 1990, **8**, 203.

pores makes separation and capacity largely independent of flow rate. The particles have a low resistance to flow, and are able to give high resolution performance using either HPLC-type systems or conventional pumping systems. The particles are composed of an organic polymer, and are available in ion exchange, hydrophobic, and reverse phase derivatives. They can be used at flow rates which are some ten fold higher than used for conventional high performance matrices; a 1 cm^3 column can give good resolution at 10 cm^3 min^{-1}. The material is commercially available under the trade name POROS (PerSeptive Biosystems, Cambridge, Massachusetts, USA).

There are relatively few examples of large scale high performance processes in the literature, mainly because they tend to be proprietary to specific products, although some examples of the use of conventional HPLC techniques and of high performance affinity techniques have been given.[73,74] Some specific applications that have been described are the purification of rabbit muscle lactate dehydrogenase using the dye, Procion Blue MX-R immobilized on silica. By using 3.3 dm^3 of matrix in a column 15 cm × 18.8 cm, 2 g of crude protein could be processed, to yield about 100 mg homogeneous enzyme with a cycle time of one hour, with a 46% recovery.[75] Lipoxidase, phosphoglucose isomerase, and lactate dehydrogenase have been purified by high performance hydrophobic interaction chromatography, using a 15 cm × 2.15 cm column of the TSK gel, Phenyl-5PW.[76] Lipoxidase, bovine erythrocyte superoxide dismutase, and human growth hormone have been purified to almost homogeneity by high performance ion exchange chromatography on 20 cm × 5.5 cm columns of the TSK gels, DEAE-5PW, and SP-5PW. With sample loading capacities of 240 mg to 1000 mg protein, and a 3 h cycle time, up to one gram of protein could be purified per day.[77] Recombinant human superoxide dismutase has been purified on a large scale by a process which takes advantage of the high performance offered by some of the newer chromatography materials.[78]

6.9 Maintenance of Column Packing Materials

The effective maintenance of the matrix in a large scale column is vital, so as to ensure both the integrity of the product and the longevity of the matrix. There are three main contributors to fouling of the gel bed: particulate matter in the sample, material which is non-specifically adsorbed, and microbial contamination.

The first of these can be dealt with by passing all solutions through filters with a pore size between 5 μm and 10 μm. Because large scale extracts are inevitably heavily contaminated with particulate matter, it is preferable to pass the sample and buffers through separate filters. Non-specifically bound material can be

[73] S. J. Brewer and B. R. Larsen, in 'Separations for Biotechnology', ed. M. S. Verrall and M. J. Hudson, Ellis Horwood, Chichester, 1987, p. 113.
[74] S. Ohlson, L. Hansson, M. Glad, K. Mosbach, and P-O. Larsson, *Trends Biotechnol.*, 1989, **7**, 179.
[75] Y. D. Clonis, K. Jones, and C. R. Lowe, *J. Chromatogr.*, 1986, **363**, 31.
[76] Y. Kato, T. Kitamura, and T. Hashimoto, *J. Chromatogr.*, 1986, **333**, 202.
[77] K. Nakamura and Y. Kato, *J. Chromatogr.*, 1985, **333**, 29.
[78] C. Scandella and T. Petersson, *Bioseparation*, 1990, **1**, 367.

removed by washing the gel with a variety of agents, either singly or in combination, such as 2 M NaCl, up to 6 M urea, non-ionic detergents, such as 1% Triton X-100, or up to 1 M NaOH. Most of the cross-linked matrices are stable in the presence of alkali, but the manufacturer's information should always be consulted.

In the case of many affinity media the problem is more complex because of the limited stability of many ligands. Some, such as hydrophobic and dye ligands are very stable, and can be subjected to the same harsh conditions used for ion exchange materials. Other ligands, such as nucleotides, lectins, or antibodies are much less stable, and washing procedures are limited to the use of high concentrations of salts, or perhaps alternating high and low pH, for example 8.5 and 4.5. For these reasons it is often best to apply affinity chromatography at a late stage in a purification, when the worst of the contaminants have been removed. One possible exception to this is in the purification of products from mammalian cell culture, where the medium is very clean in comparison to bacterial extracts.

The prevention and removal of bacterial contamination, or the pyrogens that can result from such contamination, can also be achieved by treating the column with 0.5–1 M NaOH, providing the matrix is sufficiently stable. Many gels can be autoclaved, but although this can sterilize it does not destroy all pyrogens and cannot be carried out without first removing the gel from the column, and in addition does not have any cleaning action.

For the long-term prevention of microbial contamination, during storage, for example, some form of antibacterial agent must be employed. In laboratory columns it is common practice to use agents like sodium azide or Merthiolate, but these are unsuitable for use on a large scale, particularly when proteins intended for therapeutic use are being purified. Suitable preservatives are 0.1 M NaOH or 25% ethanol, both of which can be readily removed from the column when required.

6.10 Equipment for Large Scale Chromatography

Columns for large scale chromatography should be constructed so as to have the minimum dead volume above and below their packing, and the end pieces should be designed so as to ensure an even distribution of sample over the entire surface area of the column, which may be 10 000 cm^2 or more.[79]

Large scale columns of glass or plastic construction are available from Amicon (Stonehouse, Gloucestershire, UK), Pharmacia (Milton Keynes, UK), and Whatman (Maidstone, Kent, UK). Amicon and Pharmacia also manufacture short, sectional, stack columns, either for adsorptive techniques, or for use with soft gels which cannot be used in long columns but where a long column is required. This type of configuration also has the advantages that not all of the column need be used at one time and that if one section becomes contaminated it can be removed without disrupting the remainder of the bed.

The Pharmacia 'stack' column is 37 cm in diameter, with a volume of 16 dm^3.

[79] J. C. Janson and P. Hedman, *Adv. Biochem. Eng.*, 1982, **25**, 43.

Up to ten of these columns can be connected in series, giving a column of 160 dm^3 capacity which has most of the flow characteristics of one only one-tenth that capacity. Amicon manufacture a range of sectional columns from 25 cm to 44 cm diameter, with capacities ranging from 10 dm^3 to 30 dm^3, and which have an adjustable end-piece to compensate for changes in bed height. Both of these manufacturers also supply columns in stainless steel. These columns have the disadvantage of being opaque, but in the industrial situation offer the advantages of robustness and ease of cleaning and sterilization. Stainless steel columns for high performance, high pressure operation are manufactured by Prochrom (Champignuelles, France).

The packing of large columns needs to be carried out with care so as to avoid stratifying particles of differing sizes or creating cavities within the bed. For packing the Sephadex gels, Pharmacia recommend that they be partially swollen in ethanol solution, and only be finally swollen once the column is fully assembled. The more rigid, cross-linked gels are supplied ready swollen, and the recommended procedure is to use an extension piece on the column and to fill it with a gel slurry containing the correct amount of gel, at a concentration of about 40% settled gel by volume in a solution containing 0.5 M NaCl. The solvent is then removed with a pump at a linear flow rate of about 5 cm min^{-1} until the top of the gel bed has dropped just below the top of the column. At this point the column outlet is closed and the top fitting attached as quickly as possible. Cellulose media can also be packed in this way, although both Amicon and Whatman recommend using a slurry packing technique, in which the column is filled with buffer, and a slurry consisting of about 25% settled gel pumped into the column at a high flow rate, until the column is filled with packing. Care must be taken during this operation not to exceed the maximum pressure limits of the column or packing.

Pumps used for process chromatography should be reliable, resistant to corrosion and able to operate at variable flow rates with minimum pulsation. They should not generate excessive heat or shear, and there must be no risk of contamination of the process stream with lubricants or seal materials. For the purification of therapeutic proteins a sanitary design is also important. The peristaltic pumps normally used in the laboratory are less widely used in process chromatography, because of the risk of loss of product if the tubing should split, although they do have the advantage that the pumping mechanism is separated from the liquid stream. The alternative is to use a lobe rotor type pump, fitted with a frequency controlled motor, for the control of flow rate. These pumps are readily available in a sanitary design, and cover a wide range of flow rates.

6.11 Control and Automation

Chromatographic separations can be readily automated, using either specialized microprocessor based controllers, or microcomputers with suitable programmes and interfaces, to operate valves at preset points in the process. Using such equipment it is possible to monitor the column effluent for pH, conductivity, flow rate, and pressure. The flow rate should be held constant by means of a feedback control between the pump and the flow meter, which will compensate for variations in the flow rate during the process. It is also convenient to be able to alter the flow rate during the process, perhaps high during equilibration, sample load, and washing, and low during elution and regeneration. Pressure transducers can be placed both before and after the column and be set to shut the system down should the pressures exceed the maximum and minimum values allowed. Sensors which can detect the presence of air in the liquid stream are also available, and these can be placed immediately before the column to protect it from running dry by shutting the system down should air enter for any reason. The column eluate should be passed through an ultraviolet monitor to detect protein, and this can be used to initiate the collection of product when the absorbance reaches a threshold value.

7 ULTRAFILTRATION

Ultrafiltration has become a standard laboratory technique for the concentration of protein solutions under very mild conditions. It can also be used as an alternative to dialysis or gel filtration for desalting or buffer exchange. By using affinity precipitants to increase the molecular weight of the desired protein it can also be used as a means of purification.

Ultrafiltration units are available as either stirred cells with a flat membrane, or as hollow fibres. These fibres have similar characteristics to the flat sheets, but for large scale processing give a much larger surface area for a given volume. For pilot scale operation, units are available with up to 6.4 m^2 of membrane area, which have ultrafiltration rates of up to 200 $dm^3\ h^{-1}$, depending on the protein concentration. Much larger units are available which have ultrafiltration rates of several hundred litres per hour, making this method applicable to almost any scale of operation.

8 DESIGN OF PROTEINS FOR PURIFICATION

Upstream factors can have a major impact on the development of enzyme and protein purification regimes, and recombinant DNA technology has had a major impact on protein purification. By fusing the gene of interest to an efficient promoter sequence a heterologous protein can be expressed in a host organism at 10–40% of the total soluble protein of the cell. This should be compared to the expression of many natural proteins, which may only constitute 0.01–4% of the total protein of the cell. As a result the subsequent purification of the protein is simplified. Such high levels of expression can lead to the protein being produced as dense, insoluble granules called inclusion bodies. These have been observed with many recombinant proteins, including urogastrone, interleukin-2, prochymosin, and interferons. After cell disruption, such granules can be sedimented at relatively low RCF, to yield insoluble material containing over 50% of the desired protein. The solubilization and renaturation of this material, particularly when disulfide bonds are involved, can be difficult, and requires closely controlled conditions. Solubilization requires the use of high pH, urea, or guanidinium chloride; the subsequent removal of these reagents can result in the desired protein precipitating in aqueous solution.

The reasons for the formation of inclusion bodies are not understood, since not all proteins expressed at high level form inclusion bodies. The host cell may also play an important role, because human growth hormone, which forms inclusion bodies in some *E. coli* strains, is freely soluble, with correctly formed disulfide bridges in *E. coli* RV308. The precise structure of the recombinant protein can also affect the formation of inclusion bodies. An extensive study using recombinant human interferon-γ showed that just a few amino acid changes could affect the transition between soluble and insoluble expression of the protein in *E. coli*.[80]

For proteins which are expressed in a soluble form, genetic techniques can be used to direct the newly synthesized protein into the periplasm of the cell, or even into the culture medium. This can increase the stability of the expressed protein, since only two of the eight known proteases of *E. coli* are wholly in the periplasm, and simplify the purification, because only about 8% of all *E. coli* proteins are periplasmic.

Another example of genetic design to aid protein purification is the concept of affinity tails. The gene for the protein of interest is fused to a DNA sequence that codes for some amino acid sequence which will simplify the purification of the protein, by modifying its properties in a predictable manner.[81,82] One of the first examples was the genetic fusion of several arginine residues to the *C*-terminus of urogastrone.[83] This made an unusually basic protein, which was strongly bound to a cationic ion-exchange matrix. Since this group of matrices binds only about 10% of all cellular proteins, a large purification can be obtained on elution. The polyarginine tail can then be removed by use of immobilized carboxypeptidase A; rechromatography of the liberated recombinant protein on the same matrix results in a major change in its elution position, but not in the elution position of the contaminants. Many other affinity fusion systems have been described, and several examples are shown in Table 7.

The problems of affinity fusion for purification are firstly the conditions needed for elution, and secondly the problems of cleaving the affinity tail from the purified protein. For example, fusions to protein A, although successfully applied to insulin-like growth factor, IGF-1,[84] the highly stable enzyme, alkaline phosphatase, and less successfully to galactosidase,[85] the technique suffers from two disadvantages. Extreme conditions of pH or high concentrations of chaotropic agents are required to elute protein A from immobilized IgG, which can lead to denaturation of the required product. Secondly there is the difficulty of proteolytically or chemically removing the protein A moiety without disrupting the structure of the desired product. This problem may be solved by the genetic introduction of highly specific points for protease cleavage or of acid-labile bonds

[80] R. Wetzel, L. J. Perry, and C. Veilleux, *Bio/Technology*, 1991, **9**, 731.
[81] H. M. Sassenfeld, *Trends Biotechnol.*, 1990, **8**, 88.
[82] R. F. Sherwood, *Trends Biotechnol.*, 1991, **9**, 1.
[83] H. M. Sassenfeld and S. J. Brewer, *Bio/Technology*, 1984, **2**, 76.
[84] B. Nilsson, E. Holmgren, S. Josephson, S. Gatenbeck, L. Philipson, and M. Uhlen, *Nucleic Acids Res.*, 1985, **13**, 1151.
[85] B. Nilsson, L. Abrahmsen, and M. Uhlen, *EMBO J.*, 1985, **4**, 1075.

Table 7 *Examples of protein purification methods using affinity tails*

Tail	*Ligand/matrix*	*Binding conditions*	*Elution conditions*
Polyarginine[a]	S-Sepharose	pH 4–8	NaCl gradient
Polyphenylalanine[b]	Phenyl Sepharose	1 M $(NH_4)_2SO_4$, pH 7	Ethylene glycol
Polyhistidine[c]	Nitrolotriacetate Sepharose (Ni^{2+})	pH 8 ± Guanidine HCl	Low pH gradient ± Guanidine HCl
His-Trp dipeptide[d]	Iminodiacetate Sepharose (Ni^{2+})	pH 8	Low pH gradient
Flag™ antigenic peptide[e]	Anti-Flag antibody pH 7.8	0.15 M NaCl, 1 mM $CaCl_2$	EDTA, pH 7.4
β-Galactosidase[f]	TPEG Sepharose	1.6 M NaCl, pH 7	0.1 M Borate
Chloramphenicol acetyl transferase[g]	*p*-aminochloramphenicol Sepharose	0.3 M NaCl, pH 7.8	5 mM chloramphenicol
Protein A[h]	IgG Sepharose	pH 7.6	0.5 M acetic acid
Glutathione *S*-transferase[j,k]	Glutathione Sepharose	pH 7.3	Thiol reducing agent

[a] S. J. Brewer and H. M. Sassenfeld, *Trends. Biotechnol.*, 1985, **3**, 119. [b] M. Persson, M. G. Bergstrand, L. Bulow, and K. Mosbach, *Anal. Biochem.*, 1988, **172**, 330. [c] E. Hochui, W. Bannwarth, H. Dobeli, R. Gentz, and D. Stuber, *Bio/Technology*, 1988, **6**, 1321. [d] M. C. Smith, T. C. Furman, T. D. Ingolia, and C. Pidgeon, *J. Biol. Chem.*, 1988, **263**, 7211. [e] T. P. Hopp, K. S. Prickett, V. L. Price, R. T. Libby, C. J. March, D. L. Urdal, and P. J. Conlon, *Bio/Technology*, 1988, **6**, 1204. [f] A. Ullman, *Gene*, 1984, **29**, 27. [g] J. A. Knott, C. A. Sullivan, and A. Weston, *Eur. J. Biochem.*, 1988, **174**, 405. [h] T. Moks, L. Abrahmsen, B. Osterlof, S. Josephson, S. Ostling, S-O. Enfors, I. Persson, I. B. Nilsson, and M. Uhlen, *Bio/Technology*, 1987, **5**, 379. [i] P. A. Nygren, M. Eliasson, L. Abrahmsen, M. Uhlen, and E. Palmcrantz, *J. Mol. Recognition*, 1988, **1**, 69. [j] S. Sankar and A. G. Porter, *J. Virol.*, 1991, **65**, 2993. [k] D. B. Smith and K. S. Johnson, *Gene*, 1988, **67**, 31.

Table 8 *Examples of methods for removing affinity tails*

Linker sequence	*Cleavage method*	*Conditions*
-Asn-↓-Gly-[a]	Hydroxylamine	pH9, 45°C
Asp-↓-Pro-[b]	Acid	10% acetic acid, 55°C
-Met-↓-Xxx-[c]	CNBr	70% formic acid, 20°C
-Trp-↓-Xxx-[d]	BNPS skatole	50% acetic acid, 20°C
-Xxx-↓-$(Arg)_n$[e]	Carboxypeptidase B	pH 8
-Xxx-↓-$(Lys)_n$[e]	Carboxypeptidase B	pH 8
-Xxx-↓-$(His)_n$[f]	Carboxypeptidase A	pH 8
-Pro-Xxx-↓-Gly-Pro-[g]	Collagenase	pH 7–8
-Gly-Val-Arg-Gly-Pro-Arg-↓-Xxx-[h]	Thrombin	pH 7–8
-Ile-Glu-Gly-Arg-↓-Xxx-[i]	Factor Xa	pH 7–8
-Asp-Asp-Asp-Lys-↓-Xxx-[j]	Enterokinase	pH 8

↓ Shows the bond that is cleaved. Xxx is any amino acid.
[a] T. Moks, L. Abrahmsen, E. Holmgren, M. Bilich, A. Olsson, M. Uhlen, G. Pohl, C. Sterky, and H. Hultberg, *Biochemistry*, 1987, **26**, 5239. [b] B. Nilsson, L. Abrahmsen, and M. Uhlen, *EMBO J.*, 1985, **4**, 1075. [c] B. Hammarberg, P-A. Nygren, E. Holmgren, A. Elmblad, M. Tally, U. Hellman, T. Moks, and M. Uhlen, *Proc. Natl. Acad. Sci. USA*, 1985, **86**, 4367. [d] C. W. Dykes, A. B. Bookless, B. A. Coomber, S. A. Noble, D. C. Humber, and A. N. Hobden, *Eur. J. Biochem.*, 1988, **174**, 411. [e] H. M. Sassenfeld and S. J. Brewer, *Bio/Technology*, 1984, **2**, 76. [f] M. C. Smith, T. C. Furman, T. D. Ingolia, and C. Pidgeon, *J. Biol. Chem.*, 1988, **263**, 7211. [g] J. A. Knott, C. A. Sullivan, and A. Weston, *Eur. J. Biochem.*, 1988, **174**, 405. [i] J. Shine, I. Fettes, N. C. Y. Lan, J. L. Roberts, and J. D. Baxter, *Nature (London)*, 1980, **285**, 456. [j] T. P. Hopp, K. S. Prickett, V. L. Price, R. T. Libby, C. J. March, D. L. Urdal, and P. J. Conlon, *Bio/Technology*, 1988, **6**, 1204.

at the junction of the desired recombinant product and the protein. Many methods for cleavage have been suggested, as shown in Table 8. Those that require strongly acid conditions are only suitable for proteins which are stable under such conditions, whilst enzymic methods rely on there being no susceptible bonds within the protein and on the chosen protease being free from all contaminating proteases.

9 FUTURE TRENDS

The future of large scale protein purification is assured, if only because of the biopharmaceuticals that will be coming onto the market over the next few years. Some of these, such as recombinant human haemoglobin,[86] will be required in very large amounts. The high value of such products will mean that high performance techniques will be used if the process is applicable. It is also likely that the problems currently associated with affinity tailing of proteins will be overcome, especially if recombinant proteases become widely available. This will hopefully eliminate the problem of non-specific contamination and unwanted cleavage of peptide bonds.

[86] M. Wagenbach, K. O'Rourke, L. Vitez, A. Wieczorek, S. Hoffman, S. Durfee, J. Tedesco, and G. Stetler, *Bio/Technology*, 1991, **9**, 57.

CHAPTER 18

Monoclonal Antibodies

M. WEBB

1 INTRODUCTION

Antibodies, or immunoglobulins (Igs), are protein molecules produced by specialized cells in higher vertebrates. They are made in response to molecules which are recognized by the organism as being of external, 'foreign' origin. Their role is to bind to such molecules (antigens), this initial binding activating a variety of mechanisms which lead to the destruction of the alien molecule and its removal from the system.

Immunoglobulins fall into five classes, but share some structural features. The basic Ig struture is a multimer containing two types of polypeptide, a light and a heavy chain. These are combined in the general formula $(H_2L_2)_n$, and in all natural antibodies all the heavy chains are identical and all the light chains are identical. The structure of IgG, the most commonly encountered Ig, is shown in Figure 1. There are two identical antigen binding sites at the amino terminus of the molecule, the specificity of the binding site deriving from the amino acid sequences of both the light and the heavy chains. The amino acid sequences which form the binding site for antigen are highly variable between different antibodies, while the remainder of the molecule has an amino acid sequence that is constant in all the antibodies of a given class (or subclass, in the case of IgG). The carboxy terminal half of the molecule is responsible for activating various biological effector systems leading to the elimination or destruction of the antigen. There are several subclasses of IgG with different biological properties, these being derived from differences in the amino acid sequences of the heavy chain constant regions.

Immunoglobulins are synthesized by specialized cells called B lymphocytes. These are produced in the bone marrow throughout the life of the animal, and circulate in the blood and lymph. Each B cell is programmed to synthesize Ig of a single amino acid sequence, and hence a single antigen binding specificity. This is determined by random variations introduced in the amino acid sequence of amino terminal regions contributing to the antigen binding site, the so-called *hypervariable* or *complementarity determining regions* (CDRs), these variations occurring during the maturation of each B cell. This cell-specific Ig is expressed on the membrane of the B lymphocyte, which is a small, quiescent, non-dividing cell.

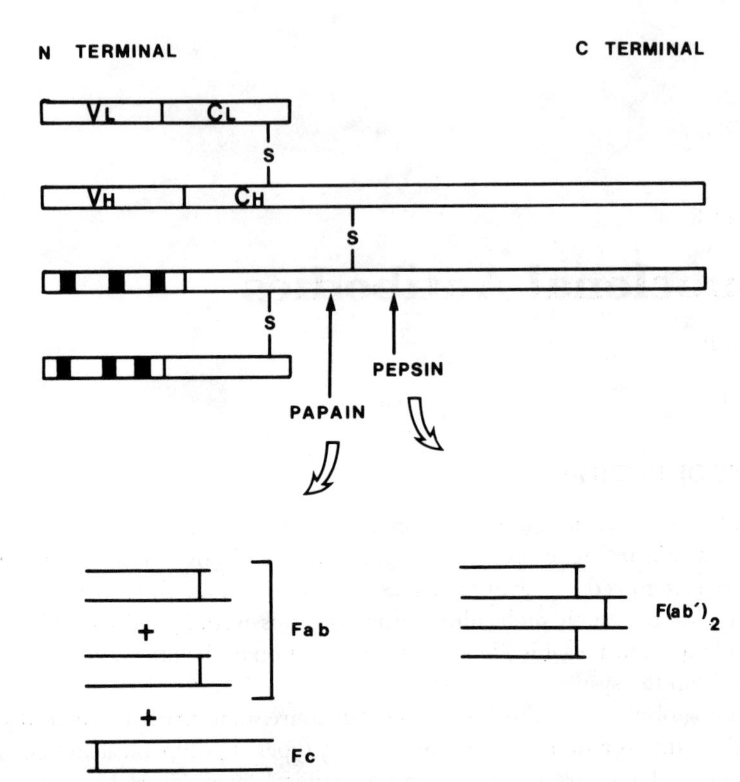

Figure 1 *The structure of Immunoglobulin G (IgG). The molecule is composed of two identical heavy chains and two identical light chains. Association between these chains is maintained by disulfide bonds, whose position is indicated schematically, and by non-covalent forces. The carboxy terminal regions of both heavy and light chains have relatively constant amino acid sequences when IgGs of the same species and subclass are compared (C_H and C_L). Three regions of extreme sequence variability indicated by black bars in the amino terminal regions (V_H and V_L) form the antigen binding site. Two proteolytic enzymes, papain and pepsin, cleave the molecule as shown to yield either monovalent or bivalent antigen-binding fragments*

The vast majority of Igs are destined never to encounter an antigen to which they can bind, and the lymphocytes which make them will eventually die and be lost. If such an encounter should occur, it acts as a trigger, (with assistance from a second type of lymphocyte, the T cell) which causes the B lymphocyte to divide, and to secrete antibody into the body fluids. All the daughter cells will make antibody identical to that made by the original parental cell, and the result is a rise in the concentration of the specific immunoglobulin, leading finally to the elimination of the antigen. This process of *clonal selection* is illustrated schematically in Figure 2.

Antibodies are potentially very valuable tools in the study of biological molecules, as they can be produced against virtually any antigen of interest. However, the antibodies in the serum of animals deliberately exposed to foreign antigens (immunized) are a complex mixture of many different specificities, and

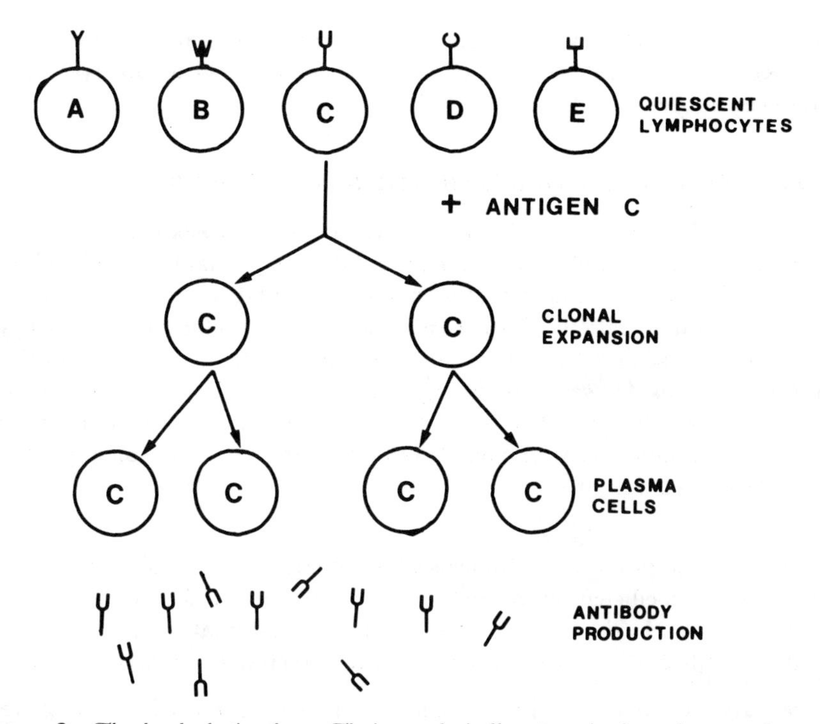

Figure 2 *The clonal selection theory. The immunologically naïve animal contains many clones of B lymphocytes, each synthesizing only a single specific antibody. Encounter with an appropriate antigen drives the cell into proliferation and maturation. The result is a clone of very many mature plasma cells, all secreting identical antibody of the same specificity as the original lymphocyte*

at best only 10% of the Ig will bind the antigen of interest. This 10% will itself constitute a range of antigen binding affinities, antibody classes, and specific combining sites, and this heterogeneity limited the usefulness of such preparations. The monoclonal antibody technique overcame all these limitations, and allowed the preparation of unlimited quantities of homogeneous antibody. It is based on the finding, mentioned above, that *each B lymphocyte synthesizes and expresses only a single specific antibody.*

The essence of the technique is to immortalize the mature antibody-producing cell (the plasma cell), so that it can proliferate indefinitely. The daughter cells of a single antibody-secreting cell, being genetically identical with the parental B lymphocyte, will synthesize and secrete Ig of exactly the same specificity. These immortalized cells can be cloned to produce cell lines (hybridomas) which produce homogeneous, or monoclonal, antibody.

In this chapter, I will discuss the background to the technique, and some practical aspects of making monoclonal antibodies (McAbs). I will also briefly review some of the areas in which they have made a major impact, and some of the recent extensions of the original concept. These have employed the techiques of molecular genetics to produce reagents of even greater potential. In the space available, I cannot give a comprehensive guide to practical techniques, and I refer the reader to Harlow and Lane[1] for an excellent laboratory manual. Similarly, Roitt[2] gives a full account of the basic immunological concepts briefly introduced above.

2 THE CONCEPTUAL BASIS OF THE McAb TECHNIQUE

The essence of the monoclonal antibody technique was to exploit the methods developed in somatic cell genetics for the production of stable hybrid cell-lines which retain the desirable features of both parent cell types. The mature antibody-secreting cell (the plasma cell, the mature product of B lymphocyte activation) has a limited life span, and will not survive or proliferate in tissue culture conditions. However, by fusing such a cell to an immortal tumour cell-line of the same lineage, a hybrid cell may be produced which retains the capacity for unlimited growth, and also secretes antibody of the same specificity as the plasma cell parent.

The key steps in this process are:

1. finding an appropriate cell-line as a fusion partner for the plasma cells;
2. the use of an efficient means to fuse the two parental cell types;
3. the use of a selective system to remove unfused parental cells;
4. the identification of those hybrid cells which secrete the desired antibody.

Interspecific mammalian cell hybrids, in which the parental cells are of different species, tend to lose chromosomes of one or the other parent. This loss is most rapid in the early generations after fusion, and the result is usually a moderately stable hybrid in which a few chromosomes of one parent are carried in a genetic background of the other parent. In contrast, *intraspecific* hybrids have less tendency to lose the chromosomes of either parent. However, when differentiated cells of different lineages are fused, it is often found that the differentiated characteristics of one or the other parent are lost in the hybrid cell, a phenomenon known as *extinction*.

These considerations are relevant to the choice of the partner cell used to immortalize the antibody-forming cells of interest. To avoid both genetic instability and loss of the differentiated characteristics of the plasma cell (in this case, the secretion of antibody is the differentiated characteristic of interest) the partner cell should be of the same species and cell lineage as the antibody-forming cells. Myelomas are tumours of B cell origin, and these are the fusion partners used in the production of McAbs.

The first monoclonal antibodies to be deliberately produced were made in mice, and the mouse has remained by far the most commonly used species for this work. The reasons for this are partly historical. The earliest McAbs were made by immunologists as part of an effort to understand the molecular mechanisms of antibody production. This work had occurred in mice, where it was supported by the availability of a great deal of genetic information, and the existence of many

[1] E. Harlow and D. Lane, 'Antibodies; A Laboratory Manual', Cold Spring Harbor Laboratory Press, New York, 1988.

[2] I. Roitt, 'Essential Immunology, 7th Edn.' Blackwell Scientific Press, Oxford, 1991.

inbred strains. Apart from this historical bias, there are sound practical reasons for the continuing use of mice in preference to other species. They are easy to breed and handle, and smaller quantities of immunogen are required than is the case for larger species.

There are several mouse myelomas in common use, and most of them are derived from those produced originally in Balb/c mice by Potter,[3] who induced them by injecting mineral oil intraperitoneally. These myelomas were established as permanent cell-lines, with the generic name of MOPC (mineral oil plasmacytoma). From these cells, mouse myelomas for use as fusion partners in McAb production have been derived by selection of variants with useful properties.

One of the properties required is the provision of a genetic marker which permits the removal of unfused parental myelomas from the mixed population of cells after fusion. A selective system originally used in somatic cell genetics is employed for this purpose.[4] Two biosynthetic pathways are available to normal cells for the synthesis of nucleotides, the *de novo* pathway and the salvage pathway. By selection with the drug 8-azaguanine, cells defective in one of the enzymes of the salvage pathway, hypoxanthine guanosine phosphoribosyl transferase (HGPRT), may be selected. These cells are unable to use the salvage pathway, but survive in normal medium through use of the *de novo* pathway. Addition of either aminopterin or methotrexate inhibits the *de novo* pathway, and under these conditions cells lacking a functional HGPRT will die, since they are unable to employ the salvage pathway. Hybrid cells between a wild type parent and an $HGPRT^-$ parent will survive in selective medium because they have a copy of the functional gene from the wild type parent. From the original MOPC myelomas, selection for variant cells lacking HGPRT lead to the first useful fusion partners for McAb production, amongst which are the cell-lines P3-X63Ag8[5] and NS1/1-Ag4-1.[6]

The original MOPC tumour cells secreted an antibody of their own. Consequently, several variants have been selected which either do not secrete or do not synthesize immunoglobulin of their own. The myelomas of choice as fusion partners in McAb production include NS0/1[7] and FO,[8] which do not express any endogenous heavy or light chains, and NS1/1-Ag4-1,[9] which synthesizes the kappa light chain, but does not secrete it.

Several methods are available to fuse the two parental cell types, including Sendai virus treatment, fusion with polyethyleneglycol (PEG),[10] and fusion mediated by a strong electrical field.[11] For the majority of purposes, PEG fusion is the method of choice, as it is easy and reproducible. PEG treatment of a

[3] M. Potter, *Physiol. Rev.*, 1972, **52**, 631.
[4] J. Littlefield, *Science*, 1964, **145**, 709.
[5] G. Kohler and C. Milstein, *Nature (London)*, 1974, **256**, 495.
[6] G. Kohler and C. Milstein, *Eur. J. Immunol.*, 1976, **6**, 511.
[7] G. Galfre and C. Milstein, *Methods Enzymol.*, 1981, **73**, 3.
[8] F. de St. Groth and D. Scheidegger, *J. Immunol. Methods*, 1980, **35**, 1.
[9] G. Kohler, S. Howe, and C. Milstein, *Eur. J. Immunol.*, 1976, **6**, 292.
[10] G. Pontecorvo, *Somat. Cell Genet.*, 1975, **1**, 397.
[11] M. Lo, T. Tsong, M. Conrad, S. Strittmatter, L. Hester, and S. Snyder, *Nature (London)*, 1984, **310**, 792.

densely packed pellet of cell results in a fusion between the membranes of opposed cells. The immediate result of this is a *heterokaryon*, a cell containing the two separate nuclei derived from the parent cells. When the nuclei break down during the first mitotic division post-fusion, the chromosomes derived from both parents segregate to the daughter cells, and the nucleus of each daughter cell reforms around both sets of chromosomes, to yield a true hybrid cell. Antibody-secreting hybrids derived by fusing antibody-secreting primary cells from the immunized animal with myeloma cells are referred to as *hybridomas*.

The third requirement for McAb production is the use of a selective system to remove unfused parental cells. Even the most efficient fusions rarely result in the formation of viable hybrids at a frequency of more than 10^{-5}, and these will exist in a background of both parental cell types. The primary cells derived from the mouse are not a problem, as they do not survive for more than a few days in tissue culture. Unfused myeloma cells (and the products of myeloma–myeloma fusions) are removed by using a selective medium which exploits their lack of a functional HGPRT gene. A selective medium (HAT) is applied, containing either methotrexate or aminopterin to block the *de novo* nucleotide synthesis pathway, and supplemented with hypoxanthine and thymidine to allow the salvage pathway to operate. The salvage pathway in the myelomas is non-functional due to the enzyme defect, and so they die in this medium. The only cells to survive are those hybrids which are able to divide continuously (a property derived from the myeloma parent) and which contain a functional HGPRT gene (derived from the mouse wild type parent[4]).

These are the key steps involved in producing antibody-secreting hybridomas. In addition, it is essential to be able to identify populations of hybridomas which are secreting antibody of interest, and also to be able to clone these to produce true monoclonal antibodies. Screening assays are dealt with in a later section.

The initial product of the cell fusion procedure is a highly heterogeneous population. In addition to stable hybrids secreting useful antibody, this population also contains hybrids secreting antibody of no interest to the experimenter, and hybrids which have an unstable karyotype, whose antibody secretion may be lost as chromosomes segregate. The minority population of interest to the experimenter may be overgrown by these fusion products. In addition, one can only guarantee the homogeneity of the antibody product of a hybridoma cell-line when that cell-line itself is the genetically homogeneous clonal descendant of a single founder cell.

For these reasons, the final requirement for the production of truly monoclonal antibodies is the ability to produce clones of hybridomas from the mixed (polyclonal) hybridoma cell populations that are the products of the original fusion. There are several methods for achieving this, but in essence they all depend on isolating single cells from a mixed population under conditions in which these can divide and produce a clonal population. This can be achieved by simple dilution of the mixed population to very low densities, followed by plating this diluted suspension in aliquots in separate tissue culture wells. At an appropriate dilution, only a fraction of the wells will receive a cell, and the chance that each well receives a given number of cells will follow the Poisson distribution.

Repeating this procedure several times, recloning each time from the previously cloned population, will eventually produce a population which is monoclonal with a high degree of probability.

An alternative is to carry out the plating of diluted cell suspensions in a semi-solid medium, in which the colonies resulting from division of the founder cells remain physically isolated, and can be recovered free of contamination with independently derived colonies. Both this method and the dilution cloning in wells method suffer from the problem that cells may tend to stick to each other, and one can never be certain that a colony arising in these situations really came from a single cell. For this reason, these methods are usually performed several times reiteratively to give a high probability of monoclonality. A practical test of whether a population is monoclonal is that at this point, 100% of the subclones produce the same antibody.

3 PRACTICAL ASPECTS OF McAb PRODUCTION

3.1 Immunization

The majority of monoclonal antibodies that have been produced are of mouse origin, with a small minority derived from the rat. This is partly because of the historical reasons mentioned above, especially the availability of good mouse myeloma cell-lines as fusion partners. In addition, mice are easy to keep, and fairly small quantities of immunogen are sufficient to induce a good immune response. Of the many mouse strains available, Balb/c is the strain of choice, as all the mouse myelomas commonly used as fusion partners are immunologically compatible with it. A certain degree of strain variation in the response to some antigens is found, and occasionally it may be necessary to use a strain other than Balb/c. If another strain is used, the hybridomas must be grown in a F1 hybrid mouse (first generation cross between Balb/c and the strain used for immunization) to produce ascitic fluid.

An outline of the procedure is shown in Figure 3. The key stages in the procedure are the immunization regime, an adequate test to check that the immunized mice are producing antibody of the required specificity, the fusion and selection procedure itself, and finally the screening of the hybridoma products and their cloning. There are various points at which alternative courses are open to the experimenter, and the procedures employed for immunization and screening will vary depending on the nature of the antigen employed.

The purpose of immunizing the animals is to induce a good immune response to the antigen of interest, this immune response being indicated by the presence of a high titre of serum antibodies. Generally, immunization involves the injection of a suitable form of the antigen, together with an immunological *adjuvant*. Adjuvants are non-specific potentiators of specific immune responses. The most widely used adjuvant in the past has been Freund's, which is a mineral oil used either alone (incomplete adjuvant) or supplemented with heat-killed *Mycobacterium* (complete adjuvant). Although it is highly potent, Freund's adjuvant suffers from two disadvantages. It can induce the formation of granulomas at the site of

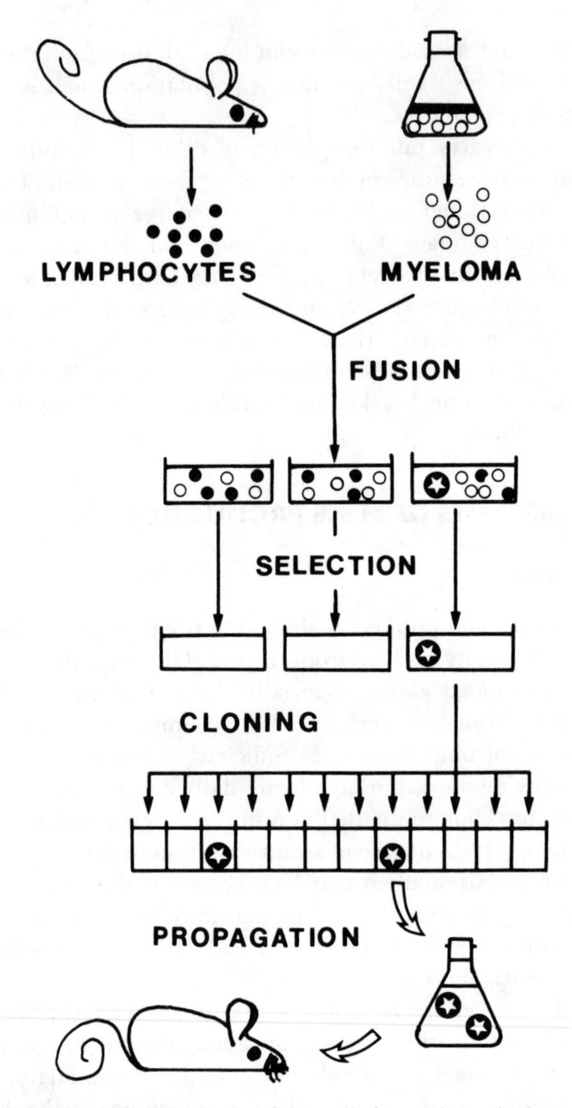

Figure 3 *Principle of McAb production. Spleen cells from an immunized mouse serve as the source of lymphocytes, some of which are secreting antibody of the required specificity. These are fused with a permanently growing myeloma partner cell, and the products are plated in many tissue culture wells in the presence of the selective agent HAT. Unfused myeloma cells die in this medium, and unfused lymphocytes are unable to survive for more than a few days in tissue culture. The wells are screened to identify those wells containing clones which secrete useful antibody. These are freshly plated at low dilution in many tissue culture wells, so that each well contains on average fewer than one cell. After a period of growth, these wells are again screened to identify those which contain antibody-producing cells. If the plating density was sufficiently low, these cells will have arisen from a single original cell,* i.e. *they are monoclonal. After cloning, the cells may be propagated in tissue culture or in mice to produce antibody*

injection, and in addition it has the potential to induce severe and long term arthritis-like symptoms if accidently injected into the experimenter. There are now a variety of oil-based commercially available substitutes for Freund's, which are claimed to be equally effective, but to have fewer undesirable side effects.

The variables in the immunization regime are: (1) the nature of the antigen, (2) the adjuvant, (3) the route of injection, and (4) the number and distribution of injections. The optimum regime varies according to the antigen of interest and the type of antibodies (*e.g.* IgM or IgG; high or moderate affinity?) required. It is always advisable to begin immunizing a group of 4–6 mice with the same antigen, since different individuals may have differing levels of responsiveness to the antigen, and it is helpful to select for fusion those mice with the highest titre of serum antibodies.

Proteins may be used either in soluble form or as particulate suspensions. If the protein is a pure species, injections of as little as 1 μg can yield good responses, although 10–20 μg are more commonly injected. Impure preparations or mixtures of many different proteins may also be used, in which case doses of 100–200 μg are usually employed. Particulate proteins are usually very good antigens, as they are readily phagocytosed by the antigen-presenting cells of the immune system. Soluble proteins may be converted to particulate preparations by binding them to a solid carrier such as agarose, or to nitrocellulose, or they may be precipitated. For example, membrane proteins prepared in sodium deoxycholate make very good immunogens after precipitations in ethanol, which removes the detergent.

Synthetic peptides are normally coupled to a larger carrier protein to convert them into effective immunogens. The carrier protein provides the accessory antigenic sites that must be recognized by helper T lymphocytes to produce a good response. Proteins such as serum albumin or thyroglobulin are well suited for this purpose. The peptide of interest can be coupled to the carrier, usually in molar ratios of 25–100 peptides/carrier using bifunctional cross-linkers such as carbodiimide or glutaraldehyde. These conjugations couple the amino terminus of the peptide to the carrier, leaving the carboxy terminus exposed, and immunizations with such conjugates usually produce antibodies preferentially recognizing the carboxy terminus. Detailed protocols for these procedures can be found in Harlow and Lane.[1]

Other classes of biological molecules, such as carbohydrates and nucleic acids, can also function as antigens, but generally they function most effectively as small haptens bound to carrier proteins. Large complex carbohydrates may induce an immune response, but frequently only a primary response (fairly low titre, predominantly IgM antibodies) is found, and in addition such molecules in large doses may induce immunological unresponsiveness (tolerance). For these classes of molecules it is therefore best to construct an appropriate hapten–carrier conjugate.

There are cases where the biologist does not have recourse to a purified preparation of the molecule of interest. This may be because the methods for such purification have not been worked out, or the molecule of interest is present in very low amounts. This situation also occurs when the biologist has no *a priori* knowledge of the molecules of interest, and is using the monoclonal antibody technique to identify previously unknown molecules. An example of such 'shot-gunning' was shown by Williams *et al.*[12] who discovered a series of rat T lymphocyte cell surface antigens by generating McAbs against impure membrane glycoproteins. The monoclonal antibody approach is the only practical way of using the immune system to identify minor antigenic species in such a complex mixture, because the cloning of the hybridomas generated allows the dissection of the complex response into its component parts.

In these types of experiments, the immunogen may consist of fractions of the tissue of interest, for example soluble extracts, organelle preparations, membrane preparations, or partially purified protein preparations derived from these sources. Whole cells can provide a useful immunogen when the molecules of interest are exposed at the cell surface. In this case, the cells (2×10^6–5×10^7) are usually innoculated intraperitoneally without adjuvant. A drawback of this method is that the antibodies produced are frequently of low affinity.

If the antibodies will be required to recognize their antigen in fixed preparations, for example in histological sections, it may be useful to fix the antigen prior to injection. Fixation with glutaraldehyde or alcohol may modify some antigenic determinants, and monoclonal antibodies produced against unfixed tissue may fail to recognize such modified determinants.

The routes most commonly used for innoculation of mice are (1) intraperitoneal, (2) subcutaneous, and (3) intravenous. Intraperitoneal injections may be used for soluble or particulate protein preparations emulsified in Freund's adjuvant or one of the more recent alternatives. Emulsification is carried out by mixing an equal volume of adjuvant and aqueous protein preparation until a stiff emulsion is formed. It is difficult to achieve this by vortexing, and the best method is to pass the mixture repeatedly between two glass syringes connected by a Luer Lock. The emulsion is ready for use when it does not disperse in water. When Freund's adjuvant is used, the first injection employs the complete version, and subsequent preparations are made with the incomplete version.

Proteins immobilized on nitrocellulose or contained in strips cut from polyacrylamide gels may also be used in conjunction with these oil-based adjuvants. In this case, the nitrocellulose or gel strip should be ground into small pieces (freezing with liquid nitrogen assists with this) and then added to the adjuvant. We find that it is best to make an emulsion with adjuvant and saline first, and then add the powdered gel/nitrocellulose, dispersing it evenly by a few additional passages between syringes. The emulsified preparation can then be injected by a worker with suitable experience into the peritoneal cavity of the mouse. Up to 500 μl can be injected at this site, but the experimenter should aim at about 200 μl of suspension, containing the required dose of antigen.

Up to 150 μl of antigen can be injected subcutaneously in an oil-based adjuvant. This route is suitable for soluble or particulate antigens emulsified in an oil-based adjuvant. Antigen is prepared in exactly the same way as described

[12] A. Williams, G. Galfre, and C. Milstein, *Cell*, 1977, **12**, 663.

above. Repeated subcutaneous injections at the same site can cause painful granulomas when Freund's adjuvant is employed, and should be avoided. An effective regime is to use an initial innoculation subcutaneously, and to carry out subsequent innoculations intraperitoneally. Subcutaneous injections can also be carried out using live cells without adjuvant, for example in the production of antibody against tumour cells. Subcutaneous implantation of strips of nitrocellulose with bound protein antigen has also been found to be effective in some cases, but the efficiency of this method must be determined empirically for each antigen.

Intravenous injections into the tail vein are used only in the case of soluble antigens, as injection of particulate matter can cause rapid death through embolism. This route is not used in primary immunization, but it is useful as the route for the final injection prior to fusion. Antigen introduced directly into the circulation via the tail vein is quickly localized in the spleen, where its presence seems to lead to a temporary localization of antibody-forming cells to this organ, which will be removed as the source of stimulated lymphocytes for cell fusion. About 100 μl of soluble antigen without adjuvant is usually employed in these injections.

The immunization regime (form of the antigen, amount injected, and number of injections) chosen will vary according to the type of antibodies required. IgM antibodies are generated early in the immune response, whereas IgGs predominate later. If IgM antibodies are required, a single dose of antigen (given intravenously if soluble, intraperitoneally if not) may be given four days prior to fusion. For IgG antibodies, a series of injections are given at 1–3 week intervals. After 6–8 such injections, most antibodies raised will be of the IgG class. If very high affinity antibodies are required, it may be advantageous to use lower amounts of antigen per injection. The reason for this is that the immune response is antigen driven, and different clones of lymphocytes are in competition for the available antigen. If the antigen concentration is low, those cells with high affinity antigen receptors (*i.e.* surface Igs) are more likely to be driven into clonal expansion than those clones expressing low affinity antibody.

The most critical injection of antigen, irrespective of the regime that precedes it, is the final boost prior to fusion. This is given 2–4 days before fusion. In the case of soluble antigens, this boost is given intravenously without adjuvant. Aggregated or particulate antigens are given intraperitoneally, again without adjuvant. Both of these routes of immunization will result in the localization of antibody-forming cells in the spleen, but only for a limited duration.

McAb of the required specificity cannot be produced from mice which are not making such antibody in the first place. Since the downstream techniques of McAb production are time-consuming, one should never proceed to cell fusion until one has verified that the mice are producing reasonable titres of antibody of the desired specificity. This may be done using the same assay that will be used for screening the products of cell fusion. Two or three drops of blood are collected after nicking the tail vein, and after clotting the serum is withdrawn and serially diluted in phosphate buffered saline. The titre of the serum is the dilution at which antibody activity above control levels (control = serum from unimmunized mice at the same dilution) can be detected. As an empirical guide, it is not usually worth proceeding to fusion unless antibody activity can be detected at dilutions of greater than 1/50. When the antibody activity is low, the immunization programme should continue until higher titres are found. If several mice have been immunized in parallel, those with the highest titre of antibody are selected for fusion. Mice selected for fusion should be rested for three weeks before their final boost prior to fusion.

An alternative to animal immunization is *in vitro immunization*. In this technique, spleen cells are prepared from a non-immunized animal, and exposed to antigen in culture medium supplemented with various growth factors.[13] After several days, the cells are used as fusion partners in the usual way. This method has the virtue of eliminating the need to immunize animals, but it tends to produce mainly IgM class antibodies, and it is difficult to see how the high affinity antibodies characteristic of hyperimmunized mice could ever be produced. Few reports of its use have appeared since it was first described.

3.2 Cell Fusion and Selection of Hybridomas

The central procedures of McAb production – cell fusion, selection of hybrid cells, and cell cloning – all rely on good tissue culture technique. Space does not permit a detailed consideration of basic cell culture techniques, and the reader is referred to Adams[14] for a good introduction. Here we will merely note that myeloma cells and hybridomas grow in suspension in various tissue culture media, which are usually supplemented with 10–20% foetal calf serum as a source of growth factors. Various medium formulations are available for the growth of hybridomas for those cases in which the absence of serum components is required.

Myeloma cells divide continuously in culture, and are generally maintained at densities of 5×10^4–1×10^6 cm^{-3}. High density cultures are subcultured by dilution of an appropriate volume of cell suspension in fresh medium. Healthy myelomas grow rapidly, and must be subcultured every 2–3 days. The cells do not tolerate high densities for long before dying, and should not be allowed to get into this condition. Stocks of myeloma cells may be maintained indefinitely in liquid nitrogen, but it is advisable to grow the cells for at least a week before using them for fusion. Like other cultured cell-lines, myelomas may become infected with intracellular parasites called mycoplasma. Such infection does not kill the cells, but results in impaired growth rates, fusion efficiency, and cloning efficiency. Myeloma cell-lines should be checked periodically for the absence of mycoplasma by either fluorescent or radioisotope tests,[15] and the presence of these contaminants should be suspected if the cells grow slowly or are in unaccountably bad condition.

The myeloma cell lines should also be checked periodically to ensure that they have retained their drug sensitivity. This can be achieved by either growing the

[13] R. Pardue, R. Brady, and J. Dedman, *J. Cell Biol.*, 1983, **96**, 1149.
[14] R. Adams, 'Cell Culture For Biochemists', Elsevier (Netherlands), Amsterdam, 1980.
[15] J. Golding, 'Monoclonal Antibodies; Principles and Practice', Academic Press, New York, 1983.

cells in 6-thioguanine (6TG; the mutation conferring aminopterin sensitivity also confers resistance to this cytotoxic drug), or by checking that test aliquots of the cells die in HAT medium. However, if the cells are routinely grown in 6TG, they should be changed to a drug-free medium a week prior to fusion to avoid toxic effects of the drug on nascent hybridomas. Successful fusions only result when healthy myelomas are used. The highest efficiencies of fusion result if the myelomas are used when they are in the log phase of growth. To achieve this, the cultures are usually diluted to $1.0–2.0 \times 10^5$ cells cm^{-3} on the day prior to fusion.

The decision to proceed to cell fusion will only be taken when the following conditions are satisfied:

1. Immunized mice are available, and have been shown to make antibody of the required specificity.
2. An assay suitable for screening the products of the fusion is available. (Assay methods are discussed in Section 3.3. The assay can be checked and optimized on the trial sera taken from the mice.)
3. Healthy myeloma cells have been in culture for at least a week.
4. The experimenter has available a continuous period of at least four weeks to devote to the care and maintenance of the fusion. Any valuable products of the fusion cannot be made secure in less time than this.

If these conditions are satisfied, the immunized mouse (or mice) destined to provide the stimulated lymphocytes are given their final innoculation with antigen. This final boost is given intravenously into the tail vein in the case of soluble antigens, or intraperitoneally in the case of aggregated antigens. The fusion should be performed 2 to 4 days after this final boost. The spleen is removed aseptically from the mouse, and gently dissociated into a single cell suspension. This suspension contains the antibody-forming cells in a mixture of other cell types. It is not necessary to attempt enrichment of the antibody-secreting cells from this mixture prior to fusion. The myeloma cells should be subcultured on the day before fusion to ensure that they will be in log phase growth on the day of fusion.

The vast majority of fusions employ polyethylene glycol (PEG) as the fusogen, although electrically mediated fusion has been employed in very special circumstances. In outline, the basic procedure used in PEG-mediated fusion is similar in most published studies. The myeloma cells are washed in serum-free medium or buffer, and pooled with a ten-fold excess of dissociated spleen cells from the immunized mouse. The pooled cells are gently centrifuged to form a mixed pellet, and the supernatant is withdrawn. 1–2 cm^3 of a 50% solution of polyethylene glycol in serum-free medium or buffer is added. After 1–2 minutes of gentle stirring, the PEG solution is slowly diluted by the addition of 20–50 cm^3 of serum-free medium, this addition taking place over about 2–4 minutes. The cells are gently centrifuged, and finally resuspended in growth medium containing serum and the selective agents (usually HAT). The resuspended cells are plated in appropriate culture vessels, sometimes with the addition of extra feeder cells. These are usually spleen or thymus cells from a non-immunized animal, and they encourage the growth of nascent hybridomas by keeping the cell density high.

There are numerous minor variations on this theme, including such variables as the PEG concentration, the addition of other agents such as lectins to promote cell agglutination prior to fusion, whether or not feeder cells are required, and whether the selective medium is added immediately, or on the day after fusion. The best course for the novice is to start by finding a procedure that works, and then explore systematically variations from this procedure.

There are several points to note about the technique. Fusion will not occur in the presence of serum proteins, which is the reason for washing the myelomas in serum-free medium. When the PEG solution is added to the mixed cell pellet and stirred gently, the formation of large aggregates of red cells is observed. This is a good sign, as it means that cell membranes are indeed fusing. All stirring operations are carried out extremely gently, as it is in this phase of the procedure that cells are contacting each other and fusing, and excessive agitation may physically break apart nascent hybrids. The time over which the PEG is diluted is also critical for the procedure to work optimally. When the pellet of cells is resuspended in growth medium after centrifuging through the diluted PEG, extreme care should be taken, as the recently fused cells are fragile and easily damaged. If a sample of the resuspended fusion mixture is examined microscopically, it is usually possible to see large myeloma cells with smaller cells adhering to them, or in various stages of merging with them.

Some workers prefer to resuspend the fusion mixture in growth medium lacking HAT, and to add an equal volume of medium containing double the normal HAT concentration on the day after fusion. In our hands, there appears to be no advantage in this procedure. There is also a variety of practices regarding feeder cells. Some workers avoid them altogether. Others resuspend the fusion mixture in a cell-free medium and plate it in tissue culture vessels to which feeders have already been added. The advantage of resuspending the fusion mixture in medium containing feeder cells is that only one set of plating operations is necessary.

The fusion mixture may be plated out in tissue culture vessels, or in the wells of tissue culture plates. We generally plate the products of one cell fusion in five or six microtitre tissue culture plates. Each plate contains 96 wells, each having a volume of about 0.3 cm^3. The total fusion is thus spread out over 500–600 wells. This increases the chance that any hybridomas arising will be uncontaminated by others arising in the same well. Alternatively, the fusion may be plated in two to four plates, each containing 24×2 cm^3 wells. In such cases, one may frequently observe several separate colonies, each presumably derived from independent parental cells, in the same well. Plating in a few large-well plates may be useful in those cases where the experimenter anticipates only a low number of colonies. Irrespective of the number of wells into which the fusion was originally plated, recloning procedures are necessary to ensure the monoclonality, and hence stability, of the hybridoma population.

During the fusion process, random associations occur between cells of all types. The approximately 10:1 ratio of spleen cells to myeloma cells means that myeloma–myeloma fusions are relatively rare, and most myelomas will fuse with spleen cells. These are the only productive fusions, and their products are the only cells to survive in the selective medium. It was found in the early days of McAb research that more successful antibody-secreting hybrids were obtained than would be expected on the basis of the fusion efficiency, the total number of spleen cells available, and the proportion of these that were secreting the desired immunoglobulins (the spleen cells include many cell types in addition to antibody-secreting cells). It appears that the activated, antibody-secreting (and perhaps dividing) plasma cell is actually more likely to fuse with the myeloma partner than the unactivated quiescent small lymphocyte. This fortuitous finding greatly increases the overall efficiency of the procedure.

In the first few days after fusion, a great deal of cell death occurs in the culture wells. Unfused lymphocytes and myeloma cells die, and the debris from these and from the red blood cells which always contaminate the spleen cell preparations are removed by macrophages (which are also components of the spleen cell mixture). Four days after fusion, the cultures are fed by withdrawing about half the medium in the well, and replacing it with fresh medium. Small colonies of dividing cells may be seen with the microscope at any time after about four days, and in the second week after fusion, these become large enough to be seen with the naked eye. The cultures are fed again at eight days after fusion, and they are usually ready to assay between ten and fourteen days post-fusion. By this time, the colonies may cover up to half of the surface area of the well containing them. The medium above such colonies is frequently yellower in colour than that in adjacent wells lacking hybridomas, as the medium becomes more acidic owing to the lactate produced by the metabolism of the rapidly growing cell. At this point, the fusion is ready to be screened to identify those wells containing hybridomas secreting useful antibody.

3.3 Screening the Products of Hybridoma Fusions

The purpose of the screening assay is to identify those wells which contain hybridoma colonies which are secreting antibodies of the required specificity. The best assay to employ will depend on the properties of the antibodies required, and also on the availability of antigen in an appropriate form to use as a target. In principle, there are three ways of screening fusions:

1. The antigen may be immobilized on a solid support, such as a nitrocellulose sheet or a PVC microtitre well, and the hybridoma supernatants applied to the immobilized antigen. After an incubation period, the antibodies bound to the support are identified by the use of a labelled secondary reagent. Screening McAbs by immunoblotting or by immunofluorescence on cultured cells or tissue sections are versions of this 'antibody capture' technique.
2. The antibodies present in the supernatants are immobilized on a solid support, and labelled antigen is applied to the support. After incubation and washing, the location of the labelled antigen is revealed by a suitable detection method. When the support has been adequately washed, labelled antigen will only be detected at positions corresponding to culture supernatants which contained active antibody.

3. It is possible in principle to screen for McAbs by using a functional test, such as inhibition of an enzyme activity or of a measurable response of cultured cells. If the antibodies are required for such blocking activities, this method has the obvious advantage of only revealing antibodies which will be immediately useful. However, such assays are difficult to set up, and may not detect potentially useful antibodies which are present only at low concentration (because of slow cell growth, for example). The assay must also be set up in such a way that the presence of the components of tissue culture medium (especially foetal calf serum) do not interfere. These disadvantages limit the applicability of this type of assay and the more robust antibody or antigen capture assays described above are safer as primary screens.

At the time when fusion products become ready for screening, usually 10–12 days post-fusion, the investigator may have between 500 and several thousand wells to screen. The cultures within these wells will be fairly dense by this time, and subculture is required to save those fusion products secreting antibody of interest. This is the period when the tissue culture burden of the technique is greatest, and it is important to identify useful cultures rapidly, to avoid wasting time and expense on maintaining and propagating hybridoma populations which are not producing useful antibody. Part of the function of the screening assay is to reduce the number of cultures that must be handled to a manageable level. As a rough guide, one operator can expect to handle up to fifty cultures beyond the initial screening stage, although the work involved in this is considerable, and it is to the operator's advantage to reduce this number further.

These considerations have several implications for the choice of screening assay. First, it should be able to detect antibodies present at only a few μg per ml. Hybridoma cultures grow at different rates, and insensitive assays may result in the loss of useful antibodies produced by 'slow' cultures. Second, it should be applicable to large numbers of wells, since the average fusion may be spread over 500–1000 wells. Third, it should produce results rapidly, preferably within 24 hours, to allow the rescue of useful cultures before they overgrow. Accurate quantitation of the amount of antibody produced is not required for screening purposes.

The first two categories of assay are the most generally useful, and can be adapted for a wide range of antigens and antibody characteristics. Immunoblotting or immunostaining of biological samples may be useful in some restricted circumstances, but these are quite labour intensive for large numbers of samples. Figure 4 illustrates commonly used versions of these assays, using 96-well microtitre plates as solid supports to immobilize either target antigen or antibody from the cultures. Detailed protocols for these assays may be found in Harlow and Lane.[1]

The first type of assay exploits the finding that proteins will adhere non-specifically to polyvinyl chloride.[16] Protein or peptide antigens (the latter may be used directly or coupled to a carrier, but if a carrier protein is used it should be

[16] T. Tsu and L. Herzenberg, in 'Selected Methods in Cellular Immunology', ed. B. Mishel and S. Shiigi, Freeman, San Francisco, 1980, pp. 373–397.

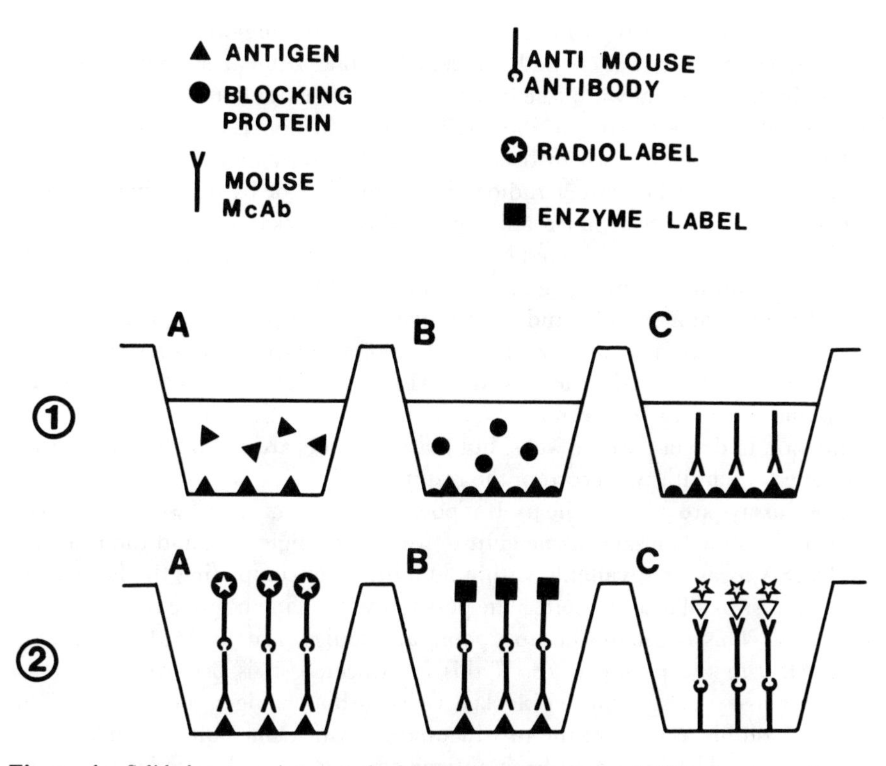

Figure 4 *Solid phase screening assay for McAbs using antigen-coated PVC microtitre wells.*

1. *Procedure. (A) PVC microtitre wells bind antigen non-specifically. (B) Residual non-specific binding sites are blocked using an irrelevant protein. (C) Antibody from positive supernatants will bind to the immobilized antibody and can be detected using an appropriate second reagent.*
2. *Detection methods. Bound McAb can be detected using either a radiolabelled (A) or enzyme-labelled (B) second reagent with an affinity for mouse immunoglobulin. (C), a 'sandwich' or 'antigen capture' assay, using anti-mouse antibody to immobilize McAb from a test supernatant. Bound McAb is detected by its ability to bind (capture) radiolabelled pure antigen*

different from that used in the initial immunization) are applied to the wells of a 96-well PVC microtitre plate in a suitable buffer. An appropriate buffer must be determined empirically for purified antigens, but 0.1 M $NaHCO_3$ pH 9.0 works well for most antigens, including tissue homogenates. Nanogram amounts of the antigen will adsorb to the plastic. The coating solution, usually containing antigen at about 1 mg ml^{-1} can be removed and reused several times. Remaining non-specific protein binding sites on the plastic are blocked by incubation in 2% albumin. Supernatants from the culture plates are sampled by the withdrawal (sterilely!) of 100 μl aliquots, which are added to the culture wells. This operation is easily performed if the fusion has been plated in 96-well plates, as the assay plate replicates the arrangement of supernatants present on the culture plate. Multichannel pipettors are available, which allow the transfer of a row of 12 wells simultaneously, and 12 plates (1152 independent wells) can be sampled in less than an hour. After an incubation, typically for 1–3 hrs, to allow the antibody in the supernatant to bind to the immobilized antigen, the plates are washed in buffer, and a labelled second reagent is applied. After a further incubation and wash, the bound second reagent is detected, and thus the location of wells containing antibody-secreting hybridomas is revealed.

The second reagent used in these assays will usually be an anti-mouse antibody, which is labelled either radioactively with ^{125}I or by its conjugation to a suitable enzyme such as horseradish peroxidase or alkaline phosphatase. ^{125}I-labelled reagents can be detected by cutting out the soft plastic wells of the assay plate and counting them in a gamma counter. Alternatively, the plates may be placed against an *X*-ray film and radioactive wells detected by autoradiography. Enzyme labels are detected by the addition of a chromogenic substrate to the wells after the last wash, the so-called ELISA (enzyme-linked immunoassay) technique. The plate can usually be scanned by eye after about 20 minutes of incubation to detect positive wells, but plate scanners are available to produce a permanent quantitative record of the results.

These assays are fast and cheap. It is possible for an unassisted worker to screen of the order of a thousand tissue culture wells in a single day, and the results of the ELISA assay are available within 30 minutes of completing the last step of the assay. As in all assays, there is the possibility that false positive results may be generated. This is usually because some antibodies (usually IgMs) stick non-specifically to the plastic wells. If this is suspected, it is best to carry out a duplicate assay using albumin blocked wells without antigen, or else to add an excess of soluble antigen of the first incubation with tissue culture supernatant. No signal should be detected from antigen-specific antibodies under these conditions. In our experience, this assay is robust and reliable, and we have used it successfully to screen for antibodies to proteins, peptides, and also to membrane-bound antigens. In the latter case, a homogenate of the tissue of interest is used to coat the plates.

If the antigen is available in radioactive form, but in limited amounts, the PVC plate assay can be adapted by immobilizing the McAb to the plastic, and then adding radioactive antigen.[17] This is usually achieved by incubating the wells first with an anti-mouse Ig antibody. The test samples are then incubated in the wells, and finally radioactive antigen is added. After washing, the presence of bound antigen is detected by counting or autoradiography.

I have described PVC plate assays at some length, because they are the most generally applicable assays available. Very similar procedures can be carried out using nitrocellulose sheets as the solid support, usually in conjunction with a manifold to ensure the regular, ordered application of the samples. However, the assay system of choice will always depend upon the nature and availability of the antigen, and the types of antibody that are required. Thus, if a specific class of antibody is required, the screening assay will employ a class-specific second reagent to detect the first antibody. Protein A binding or complement-activating antibodies are screened using protein A binding or cytotoxic assays respectively (complement-activating antibodies are able to lyse and kill target cells bearing

[17] Z. Eshlar, R. Ben-Izhak, and R. Arnon, in 'Protides of the Biological Fluids', Vol. 31, ed. H. Peeters, Pergamon Press, 1983, pp. 929–932.

the cognate antigen in the presence of complement). Even if the eventual aim is to find antibodies which have some functional effect, it may still be the best strategy to screen initially for simple antigen binding, using a suitable version of the assays described. More subtle secondary screens may be employed at a later stage, when the hybridoma cultures have been rescued and the antibody can be used at higher concentrations.

3.4 Growth and Cloning

12–14 days after fusion, the hybridoma colonies identified by the screening assay as producers of useful antibody must be transferred from their small culture wells into larger vessels. By this stage, the colonies will cover a substantial proportion of the small wells. In order to maintain cell density, we add feeder cells at each transfer of the hybridomas to a larger vessel, and typically we move from 0.3 cm^3 wells to 2 cm^3 wells, and then to culture flasks with a surface area of 25 cm^2. Fast growing hybridomas will be at a high density in such flasks within a week of their transfer from 0.3 cm^3 wells. At this point, it is important both to clone from these populations and to freeze samples of the cultures.

Two methods of cloning are commonly employed, but the essential aim of each is to isolate single cells from the hybridoma culture and to allow these cells to form a colony (clone) by division. Being the progeny of a single original cell, the daughter cells will be genetically identical with it and with each other, and they will all secrete an identical immunoglobulin. If the cloning procedure has been carried out properly, they will also be uncontaminated by other cells producing either no antibody or antibody of a different specificity.

In the first method, the hybridomas are spread in soft agar, and colonies developing are picked with a Pasteur pipette and transferred to liquid culture. These cultures are rescreened for production of antibody. 2% agar is melted in tissue culture medium, and after the addition of serum, is maintained at 46°C. 1.2 ml of this mixture is added to 0.8 cm^3 of suspended hybridomas at 10^3, 5×10^3, and 10^4 cells cm^{-3}. After mixing, the suspension is immediately plated in petri dishes and allowed to set. Colonies growing in this agar (which immobilizes the clones and prevents them mixing) are transferred to liquid cultures in 96-well plates using sterile glass pipettes about 10 days after plating. When the cultures are sufficiently dense, they are rescreened using the original assay method.

The alternative method is to clone by limiting dilution. In this case, a suspension of the hybridoma is serially diluted, and aliquots of each dilution are plated into microculture wells to which feeder cells are then added. When the density of cells in the plating suspension is sufficiently low, colonies growing in the microculture wells can be presumed to have originated from a single cell. The number of cells per well will follow the Poisson distribution, so where 63% or fewer of the wells contain colonies, these probably contain only a single clone per well. In practice, the procedure is usually repeated serially at least twice to ensure monoclonality. A useful operational criterion of monoclonality is that all the subclones on repeated recloning secrete the appropriate antibody. A more stringent criterion is to examine the antibody secreted by the cells on two dimensional gels after biosynthetic labelling with a radioactive amino acid. Spots corresponding to only a single light and a single heavy Ig chain will be found for truly monoclonal antibodies.

Once the parental myelomas have died, it is no longer necessary to maintain the hybridomas in HAT medium. However, they should be grown for a few subcultures in HT medium (containing hypoxanthine and thymidine) before growth in drug free medium, since residual aminopterin may persist in the cells after the removal of HAT, and this will poison the hybridomas in the absence of hypoxanthine and thymidine. We usually maintain our fusions in HAT until they are screened and then transfer them to larger vessels in HT. We maintain them in HT until their first subcloning, and thereafter grow them in drug-free medium.

It is important to freeze cells from the earliest possible stages of a hybridoma's life. This is a precaution against the loss of the cells through contamination, and provides a reserve population from which to attempt cloning if the antibody-secreting cells in the main population become overgrown by non-secretors early in the hybridoma's life. We usually freeze the cells left over from the first cloning in one or two vials. The cells are centrifuged, and the medium withdrawn. They are resuspended in 1–2 cm^3 of freezing mixture (culture medium supplemented with 10% foetal calf serum and 10% dimethyl sulfoxide, added to prevent freezing damage to the cells) and added to sterile freezing vials. Although some workers use various means to freeze the cells at a controlled rate, we find that it is quite adequate to leave them in a $-80\,°C$ freezer overnight, or to place them directly on solid carbon dioxide for an hour. After either of these procedures, the cells may be transferred to liquid nitrogen, where they are stable indefinitely. In general, it is best to freeze cells at a high density ($\sim 10^7$ cm^{-3}), although this may not always be possible early in a hybridoma's life. Once the cells have been cloned, the selected representative clone(s) of a particular antibody should be grown to produce enough cells to allow the freezing of 10–20 vials, which will then act as reserve stock. We always freeze a set of vials of our hybridomas after each round of cloning.

Antibody is secreted into the medium by healthy hybridomas, where it may reach concentrations of the order of 2–10 $\mu g\ cm^{-3}$. Much higher concentrations, in the region of 1–10 mg cm^{-3}, are found in the ascitic fluid produced by growing the cells as an ascitic tumour in mice. Methods for preparing larger quantities of antibody from these sources are given in Harlow and Lane.[1]

4 APPLICATIONS OF McAbs

From the time of the first description of the technique, monoclonal antibodies have shown their value in a variety of biological research applications. They did not merely represent an improvement over previously used polyclonal antisera, but offered the possibility of types of research that were simply impossible using conventional reagents. This type of activity, starting in the mid-seventies, has continued unabated into the mid-nineties, and in this time, monoclonal antibody technology has become the method of choice for tackling a wide range of research problems in biology and biochemistry. I will describe first this type of application.

There were many hopes in the 1980s that monoclonal antibodies would offer potent new tools for the diagnosis and therapy of human disease. These hopes were not fully realized at that time. The introduction of new second and third generation antibodies by technologies relying on the methods of molecular biology now seems likely to allow the full achievement of this potential, and I discuss this in the second section.

An exciting recent development with possible industrial applications is the discovery that monoclonal antibodies may possess enzyme-like catalytic activities. By exploiting the diversity of the immune system, it may be possible to make 'abzymes' of predetermined enzyme activity, and this possibility is discussed in the third section.

4.1 Antibodies in Biomedical Research

The ability to 'dissect out' single, and possibly minor, specificities from a complex immune response conferred by the McAb technique allowed its use to identify previously unknown biological molecules. This approach was first demonstrated by Williams *et al.*,[12] who carried out a 'shotgun' immunization with membrane glycoproteins from rat thymocytes. They described several antigens localized on the surface of thymocytes and the T lymphocytes to which they give rise, none of which were previously known. The patterns of antigen binding were studied by fluorescent cell sorting, and it was shown that one antibody, W3/25, recognized a subset of T cells. Quantitative binding of radioactive W3/25 antibody showed that it was a minor constituent of the cell surface, with only about 20 000 molecules per cell. The antibody was subsequently used to purify the subset of cells bearing it, and these were assayed for immunological function[18] and shown to be T helper cells. In *in vitro* assays, Webb *et al.*[19] showed that the antibody was a potent inhibitor of the activation of these cells by histoincompatible lymphoid cells, thereby implicating the molecule recognized by the antibody in some aspect of T cell activation. The antibody was later used to identify and purify its antigen by affinity chromatography.

I have described these studies in detail because they were the first to use an approach which has since become very common in studies on diverse biological systems. Previous work on T cell subsets had been carried out in the mouse, and had relied on the very weak antisera generated by intraspecific immunizations of mice with lymphocytes from different strains. These studies were beset with problems due to the poor quality of the reagents. These problems were removed, and the field was immediately clarified by the introduction of the superior technology. The application of the McAb technique to the human system followed on from the rat studies rapidly, and lead to the generation of many useful specific reagents, some of which are in use as diagnostic and therapeutic

[18] R. White, D. Mason, A. Williams, G. Galfre, and C. Milstein, *J. Exp. Med.*, 1978, **148**, 664.
[19] M. Webb, D. Mason, and A. Williams, *Nature (London)*, 1979, **282**, 841.

agents. This development would have been impossible with conventional serological reagents.

This 'shotgun' approach can be defined as the immunization of mice with impure, complex mixtures of antigens, and the use of the resultant antibodies to identify previously unknown antigens. It has been of particular value in the study of molecules which may have functions for which no easy assay is available, and which may be expressed only at low levels, or in small subpopulations of cells. The approach has been applied to the nervous systems of various organisms, including those of simple animals such as the leech, as well as mammalian systems.[20–22] It has revealed a rich diversity of previously unknown molecules, some of which are specific to cell types, synapses, or other specialized architectural features.[23,24] In some cases, the antibodies have been used to probe the functions of the molecules that they recognize, an approach that has been most successful with cell-surface antigens which mediate interactions between neural cell types.[25,26] The pattern of these studies has thus been similar to the prototypical analysis of the rat T cell surface pioneered by Williams and coworkers.[12,18,19]

The inherent advantages of monoclonal over conventional antibodies have also been exploited in cases where the antigen was already known. Antibodies have been used in extremely sensitive assays for biological molecules since 1959, when the first practical radioimmunoassay for insulin was introduced. In such assays, the concentration of the molecule of interest can be measured by the degree to which it displaces radioactive antigen from specific antibody. Such radioimmunoassays (RIAs) are both highly specific and very sensitive, such that nano- or pico-molar concentrations of the molecule of interest can be measured in small volumes (< 1 cm^3) of body fluids such as blood, urine, or cerebrospinal fluid. Such assays have been established for many peptide and steroid hormones, for amino acid derivatives such as thyroxine and triiodothyronine, and for important intracellular metabolites such as cyclic nucleotides.

The two requirements for the antibodies used in such assays are specificity and high affinity. It has been reported that the affinities of McAbs are low when compared with the effective affinities (avidities) of their polyclonal counterparts. This partly reflects the distribution of affinities found *in vivo*, and partly the bonus effect of avidity when the same antigen can be bound by different serum antibodies recognizing different epitopes on the antigen. A McAb is only able to recognize a single epitope, and such a bonus effect cannot be produced unless mixtures of antibodies are used. Nevertheless, McAbs recognizing peptide antigens with affinity constants of up to 10^{11} mol^{-1} have been produced, and the general assumption that McAbs have rather low affinities is unjustified. In practice, it may be necessary to screen many different McAbs against the same

[20] R. Hawkes, E. Niday, and A. Matus, *Proc. Natl. Acad. Sci. USA*, 1982, **79**, 2410.
[21] R. McKay and S. Hockfield, *Proc. Natl. Acad. Sci. USA*, 1982, **79**, 6747.
[22] M. Webb and P. Woodhams, *J. Neuroimmunol.*, 1984, **6**, 283.
[23] I. Sommer and M. Schachner, *Dev. Biol.*, 1981, **83**, 311.
[24] P. Woodhams, H. Kawano, P. Seeley, D. Atkinson, and W. Webb, *Neuroscience*, 1992, **46**, 57.
[25] J. Lindner, F. Rathjen, and M. Schachner, *Nature (London)*, 1983, **305**, 427.
[26] S. Chang, F. Rathjen, and J. Raper, *J. Cell Biol.*, 1987, **104**, 335.

antigen in order to find one suitable for a specific purpose such as RIA. Once such a reagent is found, however, the general advantages of McAbs apply. The high specificity of the reagents may enable the ready distinction between the ligand of interest and any chemically similar metabolites. Thus, we[27] described a McAb which recognizes the peptide hormone bradykinin (Bk) with a sub-nanomolar K_D. This antibody fails to recognize the bradykinin metabolites Bk (1–8) and Bk (1–7) which differ from the parent peptide by one and two amino acids respectively.

Many RIA kits are now marketed commercially, and it is a particular advantage to a commercial supplier to be able to guarantee that the antibody at the heart of the assay will not be subject to biological variation as a result of bleed or animal changes over time. It seems likely that McAbs will gradually replace conventional antibodies in these assays.

Conventional antibodies had been used in the affinity purification of their cognate antigens with only a limited degree of success. Frequently, the antibody affinity column purification method was less effective than the conventional purification techniques for the antigen. The main reason for this lack of success was the fact that a conventional antibody, raised against purified antigen, will still contain only a minority of antigen-specific immunoglobulins. Monoclonal antibody preparations, in contrast, contain 100% antigen-specific, identical immunoglobulin, and monoclonal antibodies immobilized on solid supports have been used to affinity purify a wide variety of different antigens.

A detailed discussion of the methods used in affinity purification of antigens on McAb affinity columns is given in Arvieux and Williams,[28] and the following discussion is necessarily brief. Soluble antigens can be purified directly from biological fluids, aqueous extracts, or fermentation broths. Membrane molecules must first be solubilized in a suitable detergent, such as Triton X-100 or deoxycholate. These preparations are then passed over the column of immobilized antibody. Such columns typically contain 5–10 mg of monoclonal Ig cm^{-3} of bed volume, usually cross-linked to an insoluble support of dextran of agarose. Many McAbs retain the ability to bind antigen in the presence of such mild detergents. After a suitable washing procedure, the specifically bound antigen is eluted from the column. This is frequently achieved by the use of denaturing conditions such as high or low pH buffers, which disrupt the bonds between antigen and antibody. The ability of the immobilized antibody to bind antigen is frequently recovered when the antibody is returned to non-denaturing conditions. Successful affinity purification of antigen on McAb columns requires antibodies of moderately high affinity to extract antigen and retain it through the washing procedure. In practice, IgM antibodies have not been found to be useful in such procedures, probably because these antibodies tend to have low affinities.

Affinity purifications using McAbs have become a routine part of academic studies on newly defined antigens. Such purifications provide material for further

[27] E. Phillips and M. Webb, *J. Neuroimmunol.*, 1989, **23**, 179.
[28] J. Arvieux and A. Williams, 'Immunoaffinity Chromatography', in 'Antibodies; A Practical Approach', IRL Press, Oxford, 1988.

biochemical analysis, including amino acid sequence determination as a preliminary to cloning the antigen.

The recognition of a single epitope on the antigen has limited the application of McAbs in the direct screening of expression libraries for expressed proteins. If the epitope is lost because of incorrect disulfide bond formation or lack of proper protein folding when the protein is expressed in a prokaryote system, the McAb will fail to detect the cloned product. Polyclonal antibodies, recognizing multiple determinants, stand a better chance of picking out clones in such expression screening studies. However, for cell surface antigens, cloned in a mammalian expression vector and expressed in Cos cells, a selection protocol based on the adhesion of antigen-expressing cells to dishes coated with the McAb has been very successful.[29]

4.2 McAbs as Therapeutic Agents

In the 1970s, there was a widespread hope that the advent of McAbs would revolutionize the treatment of human disease. The ability to make these highly specific reagents directed against any required biological target raised the possibility of attacking tumour cells via specific cell surface antigens, of modulating the response of the immune system in transplantation and autoimmune disease by directing antibodies against appropriate lymphocyte surface molecules, and of targeting specific pathogenic molecules. It appeared that McAbs might be the 'magic bullets' looked forward to by Paul Ehrlich at the turn of the century.[30] The first generation of conventional McAbs largely failed in this promise, and only one McAb, OKT3, has been licensed for clinical use.

There were three reasons for this disappointing failure. First, murine McAbs are themselves immunogenic in man, and the human immune system mounts an immunological attack upon them, leading to a rapid clearance from the system. Second, the antibodies were not always directed at key cell surface structures on the target cells, such as growth factor receptors required for tumour proliferation or lymphocyte activation. Finally, murine McAbs are frequently ineffective in recruiting the human complement mediated and antibody-dependent cell mediated cytotoxicity mechanisms which would destroy the target cells subsequent to antibody binding.

Research in the 1980s has therefore focused on the definition of better target antigens on neoplastic and lymphoid cells, on the reduction of the immune response to the therapeutic McAb, and finally on 'arming' the antibody with toxins or radioactive atoms to facilitate their destruction of target cells.

The antibody OKT3 is licensed for use in the treatment of acute renal allograft rejection, where it has been shown to be more effective than conventional broad spectrum immunosuppressants.[31] The antibody, directed against an antigen expressed on all circulating T cells, produces various side effects such as fever, vomiting, and diarrhoea, probably because it activates T cells and causes them to

[29] B. Seed and A. Aruffo, *Proc. Natl. Acad. Sci. USA*, 1987, **84**, 3365.
[30] P. Ehrlich, *Proc. Roy. Soc.*, 1900, **66**, 424.
[31] Ortho Multicentre Transplantation Group, *N. Engl. J. Med.*, 1985, **313**, 337.

release mediators such as tumour necrosis factor and Interleukin 2. It would be an improvement to target T cell surface structures actually involved in antigen recognition, or in the recognition of signalling molecules responsible for activating the cells. One antibody specific for the constant region of the T cell antigen receptor was shown in pilot studies to be successful in the treatment of acute graft rejection.[32]

A further step in this direction is to prepare McAbs reactive with the variable region on the T cell receptor which is actually responsible for binding to the antigen, since such reagents would only inhibit the function of T cells with the unwanted activity. This approach has been shown to be effective in two areas: (1) in the treatment of leukaemias, where all the leukaemic cells bear the same receptor variable region (the approach can be used for malignancies of both T and B cell origin)[33,34]; (2) immunosuppression has also been achieved by the use of McAbs which recognize cell surface molecules involved in cell–cell interactions. Antibodies against either of the components of an adhesion system, ICAM-1 or LFA-1, are effective in T cell inhibition, and inhibit allograft rejection.[35] The cell surface receptor for Interleukin 2, an important signalling molecule, is expressed on T cells only after their activation. Waldmann[36] used a McAb which blocks the binding of Il-2 to its receptor in the treatment of patients with adult T cell leukaemia with some success.

The second major area of advance has been in the reduction of the immunological rejection of the mouse McAb by the human immune system. This is most strongly mediated through the recognition of the constant regions of the mouse antibody. These have therefore been replaced by the techniques of genetic engineering to yield 'humanized' antibodies, which retain the mouse variable regions containing the antigen binding site, on a backbone of human constant region.[37–39] The complementarity-determining regions (CDRs) of the IgG molecule, responsible for forming the antigen binding site, are held in a framework region within the variable part of the molecule. The limit of the 'humanizing' approach was to construct antibodies that retained only the mouse CDRs in a framework region derived from human IgG.[38,40] Unfortunately, in some cases, such antibodies had reduced binding affinities for their antigens when compared with the original mouse McAbs, suggesting that some residues in the framework region were important for maintaining the CDRs in the correct orientation. Computer modelling is being used to try to predict those amino acids in the framework region which may be most important in this function, so they can be retained in the engineered stucture. In addition to reducing the immunogenicity of the therapeutic antibody, antibody engineering can be used to improve its

[32] H. Schlit *Transplant. Proc.*, 1988, **20**, 103.
[33] C. Janson *et al.*, *Cancer Immunol. Immunother.*, 1989, **28**, 225.
[34] R. Miller, D. Maloney, R. Warnke, and R. Levy, *N. Engl. J. Med.*, 1982, **306**, 517.
[35] A. Fisher *et al.*, *Lancet*, 1986, **11**, 1058.
[36] H. Waldmann, *Ann. Rev. Immunol.* 1989, **7**, 407.
[37] S. Rudikoff *et al.*, *Proc. Natl. Acad. Sci. USA*, 1982, **79**, 1979.
[38] P. Jones, P. Dear, J. Foote, M. Neuberger, and G. Winter, *Nature (London)*, 1986, **321**, 522.
[39] C. Queen, W. Schneider, H. Selick, P. Payne, N. Landolfi, J. Duncan, N. Audalovic, M. Levitt, R. Junghaus, and J. Waldmann, *Proc. Natl. Acad. Sci. USA*, 1989, **86**, 10029.
[40] L. Riechmann, M. Clark, H. Waldmann, and G. Winter, *Nature (London)*, 1988, **332**, 323.

effector functions. Thus, murine McAbs directed against tumour antigens were more effective in mediating the destruction of the target cells via antibody-dependent cell mediated cytotoxicity when the mouse constant regions were replaced with human IgG1 regions.[40]

The ideal McAb for human therapy would be a completely human antibody. However, humans cannot be immunized to provide antibody synthesizing cells, and hybrids between mouse myelomas and human antibody synthesizing cells are unstable for reasons discussed in Section 2. One approach to the problem of immunization is to reconstitute the immunodeficient SCID mouse with a source of human immunoprogenitor cells, such as bone marrow. These mice may then be immunized and used as a source of antibody secreting cells, the antibodies being wholly human.[41,42] To avoid the problems encountered in mouse-human interspecific cell fusions, it may be possible to immortalize such cells by infecting them with Epstein-Barr virus. No reports have yet appeared on the clinical use of antibodies derived by this method. Transgenic mice, bearing human immunoglobulin heavy chain variable and constant regions have been made, which should also allow human McAbs to be produced from hyperimmunized mice.[43]

There is a further approach which avoids the use of animals altogether. The polymerase chain reaction can be used to clone the variable regions of immunoglobulins, which are then expressed in prokaryotic systems. The libraries thus made can then be screened to find clones of antibody derivatives (single-chain domain antibodies, DABs) able to bind antigen.[44–46] Heavy and light chain variable region libraries can be screened separately, but these can also be combined randomly to generate fab-like molecules for screening. Mullinax *et al.*[46] found human fab-like clones with binding activity for tetanus toxoid in such a combinatorial library derived from human peripheral blood lymphocytes.

All of these approaches are in their infancy, and although it is not yet clear what impact they will make on the therapeutic application of McAbs, it should at least be clear that there are several possible approaches to the problems encountered in the use of rodent antibodies in human therapy.

The final major area of research has been directed at 'arming' antibodies so that they carry a cytotoxic moiety which will efficiently kill target cells. The cytotoxic moiety may be a chemical toxin or a radioactive nuclide. Although simple in concept, there are several difficulties which must be overcome for this strategy to be effective. The conjugate must retain the capacity to recognize the target efficiently, it must gain access to the cell by endocytosis (since most of the toxic moieties which have been used act by inhibiting protein synthesis within the cell), and the linkage between the toxin and the antibody must be stable

[41] J. McCune, R. Namikawa, H. Kaneshima, L. Schultz, M. Liebermann, I. Weissmann, R. Gulizia, S. Baird, and D. Wilson, *Science*, 1988, **241**, 1632.

[42] D. Mosier, R. Gulizia, S. Baird, and D. Wilson, *Nature (London)*, 1988, **335**, 256.

[43] M. Bruggemann, H. Caskey, C. Teale, H. Waldmann, G. Williams, M. Surani, and M. Neuberger, *Proc. Natl. Acad. Sci. USA*, 1989, **86**, 6709.

[44] E. Ward, D. Gussow, A. Griffiths, P. Jones, and G. Winter, *Nature (London)*, 1989, **341**, 544.

[45] W. Huse, L. Sastry, A. Kang, M. Alting-Mees, D. Burton, S. Benkovic, and R. Lerner, *Science*, 1989, **246**, 1275.

[46] B. Mullinax, E. Gross, J. Amberg, B. Hay, H. Hogrefe, M. Kubitz, A. Greener, M. Alting-Mees, D. Ardourel, J. Short, S. Sorge, and B. Shopes, *Proc. Natl. Acad. Sci. USA*, 1990, **87**, 8095.

enough to withstand passage through the body until it reaches the target cells. These difficulties are increased by the fact that the antibody–toxin conjugate itself is immunogenic, and will provoke an immune response leading to its elimination. Thus far, only modest success has attended this approach.[47,48]

One way of 'arming' an antibody is to increase its capacity to trigger the natural cytotoxic mechanisms of the immune system. This was mentioned above in connection with the increased antibody-dependent cell mediated cytotoxicity shown by murine McAbs when the mouse heavy chain constant regions were replaced with human IgG1 regions.[40] A different method of achieving this is to construct *bispecific* McAbs, in which the two binding sites recognize different antigens. One site is specific for the target antigen on the cellular target, while the other is directed toward a cell surface structure on the cytotoxic cells with the potential to kill target cells. The antibody facilitates the interaction between the target and the effector cell, with a resultant increase in the efficiency of cell mediated destruction of the target.

Bispecific antibodies can be produced by biological or chemical methods. Biological production involves the fusion of two McAb producing hybridomas, or of an immunized spleen cell with the hybridoma.[49,50] The random association between light and heavy chains in the resultant *quadroma* would be predicted to lead to secretion of a mixture of all possible combinations, amongst which will be the required bispecific antibody. However, the association between light and heavy chains, and between light and heavy chain pairs, may not be random in practice; Suresh *et al.*[50] found both parental antibodies and bispecific antibodies were secreted from a hybrid–hybrid hybridoma, but inactive antibodies resulting from heterologous chain associations were not found. The bispecific antibodies may nevertheless require purification by affinity methods to remove the unwanted monospecific parental antibodies. An alternative to this biological method of producing bispecific antibodies is to recombine light and heavy chains from appropriate sources *in vitro*.

This technique has applications beyond those of cancer chemotherapy, including enzyme immobilization and novel immunoassay and immunolocalization techniques. Recent developments in this technology were reviewed in Nolan and Kennedy.[51]

An alternative to the use of antibody–toxin conjugates is to tag the antibody with a radionuclide to kill target cells by irradiation. This is similar in principle to the established technique of using a radiolabelled antibody to assist in the localization of tumours. One advantage over the toxin conjugates is that the antibody–nuclide does not need to be internalized to inflict damage on the target cell. There has been encouraging progress in this area. Patients with hepatoma have been successfully treated with ^{131}I labelled antibodies to ferritin,[52] a combination of marrow transplantation and ^{90}Y labelled antibodies against

[47] L. Nadler, Proc. Second Intl. Symp. Immunotoxins, 1990, p. 58.
[48] V. Byers *et al.*, *Blood*, 1990, **75**, 1426.
[49] C. Milstein and A. Cuello, *Nature (London)*, 1983, **305**, 537.
[50] M. Suresh, A. Cuello, and C. Milstein, *Proc. Natl. Acad. Sci. USA*, 1986, **83**, 7978.
[51] O. Nolan and R. Kennedy, *Biochim. Biophys. Acta*, 1990, **1040**, 1.
[52] S. Order *et al.*, *J. Clin. Oncol.*, 1985, **3**, 1573.

ferritin lead to complete remission in four out of ten patients with Hodgkin's disease, and a ^{90}Y labelled McAb against a cell surface antigen expressed by malignant T cells gave partial or complete remission in five out of six patients suffering from adult T cell leukaemia.

The problems associated with the therapeutic use of McAbs are greater than they appeared at first. Will these new developments allow the achievement of the therapeutic goals which were elusive to the first generation of McAbs? Recent progress in all of the three key areas for this success would suggest grounds for optimism. It is not yet clear whether a unified technology will emerge, with a standard methodology applied to a variety of clinical problems, or whether different combinations of the methods discussed will be found to be appropriate in different cases. It is clear, however, that our increasing ability to manipulate the immune system and its products has provided a powerful weapon which may yet fulfil Ehrlich's dream of a 'magic bullet'.

4.3 Antibody Catalysis (Abzymes)

The immune system can be regarded as a huge reservoir of potential complementary surfaces, within which binding sites to essentially any epitope within the appropriate size range could be found. Linus Pauling pointed out many years ago that the essential difference between antibody recognition and substrate recognition by enzymes is that antibodies commonly recognize the 'ground state' of their cognate antigens, while enzymes have evolved to bind 'transition states' of their substrates. This stabilization of the transition state lowers the energy barriers between substrate and product, and results in the catalytic activity of the enzyme. It follows that antibodies with binding affinity for transition state intermediates may possess catalytic activities similar to those of enzymes. By exploiting the reservoir of binding sites within the immune repertoire, it may be possible to develop antibody enzymes—'abzymes'—which catalyse reactions for which no natural enzyme exists. Catalytic antibodies may be developed which cleave proteins or carbohydrates at specific residues, the protein equivalent of restriction endonucleases. Such reagents might be used as therapeutic agents, attacking specific coat proteins or sugars on viruses or tumour cells. They might also be used as selective catalysts in the synthesis of new pharmaceutical agents.

The first studies[53,54] demonstrated that antibodies raised against stable analogues of the transition states in the hydrolysis of esters and carbonates accelerated the hydrolysis of their respective 'substrates'. The antibodies were found to have substantially higher binding affinities from the transition state analogues than for the reaction substrates, although this affinity difference was not great enough to account for the 10^3–10^6 fold acceleration in reaction rates observed. In such cases, it may be that a catalytic group on an amino acid side chain exists by chance in an appropriate position within the binding site.

In the short time since these reports, abzyme catalysis of a range of reactions

[53] A. Tramontano, K. Janda, and R. Lerner, *Science*, 1986, **234**, 1566.
[54] S. Pollack, J. Jacobs, and P. Schultz, *Science*, 1986, **234**, 1570.

has been reported, including amide bond cleavage,[55] and bimolecular chemical reactions.[56] A library of mouse heavy chain variable region genes has been cloned as an alternative way of accessing the immunological repertoire for the production of abzymes.[57]

Many enzymes depend on cofactors, and antibody enzymes incorporating metal ion cofactors have been reported. Antibodies were raised against a stable complex of triethylene tetra-amine and cobalt. The antigen included a peptide co-ordinated to the cobalt, and was designed to select a combining site in the antibody that would bring a labile octahedral metal cofactor complex and a peptide into a chemically reactive relationship. Two of the antibodies produced were able to hydrolyse the peptide bond between glycine and phenylalanine with a turnover number of 6×10^4 sec^{-1}.[58]

As yet, these studies are at a very preliminary stage. However, this area seems likely to be yet another in which the ability to understand and manipulate the immune system, which led to the first generation of McAbs, will lead to the employment of these astonishing molecules in ever more diverse ways.

Dedication. This chapter is dedicated to the memory of Alan F. Williams, teacher and friend.

[55] K. Janda, D. Schloeder, S. Benkovic, and R. Lerner, *Science*, 1988, **241**, 1188.
[56] S. Benkovic, A. Napper, and R. Lerner, *Proc. Natl. Acad. Sci. USA*, 1988, **85**, 5355.
[57] L. Sastry, M. Alting-Mees, W. Huse, J. Short, J. Sorge, B. Hay, K. Janda, S. Benkovic, and R. Lerner, *Proc. Natl. Acad. Sci. USA*, 1989, **86**, 5728.
[58] B. Iverson and R. Lerner, *Science*, 1989, **243**, 1184.

CHAPTER 19

Biosensors

M. F. CHAPLIN

1 INTRODUCTION

Biosensors are analytical devices which convert biological actions into electrical signals in order to quantify them.[1,2] Biosensors make use of the specificity of biological processes; enzymes for their substrates and other ligands, antibodies for their antigens, lectins for carbohydrates, and nucleic acids for their complementary sequences. A typical biosensor has a number of connected parts (Figure 1). The biological reaction usually takes place in close contact with the electrical transducer, here shown as a 'black-box'. This intimate arrangement ensures that most of the biological reaction is detected. The resultant electrical signal is compared with a reference signal, usually produced by a similar system without the biologically active material, and the difference between these two signals is amplified, processed, and displayed or recorded.

The primary advantage of using biologically active molecules as part of a biosensor is due to their high specificity and, hence, high discriminatory power. Thus they are generally able to detect particular molecular species within complex mixtures of other materials with similar structures, which may be present at comparable or substantially higher concentrations. Often, samples can be analysed without any prior clean-up. In this aspect they show distinct advantages over most 'traditional' analytical methods; for example, colorimetric assays like the Lowry assay for proteins.

Biosensors may serve a number of analytical purposes. In some applications, for example in clinical diagnosis, it is important only to determine whether the analyte is above or below some pre-determined threshold, whereas in process control there often needs to be a continual and precise feed-back of the level of analyte present. In the former case, the biosensor must be designed to give the minimum number of false positives and, more importantly, false negatives, whereas in the latter case it is generally the response to changes in the analyte that is more important than its absolute accuracy. Minimizing false negatives is often more important in clinical analyses than minimizing false positives as in the former case a diagnosis may be missed and therapy delayed, whilst in the latter

[1] E. A. H. Hall, 'Biosensors', Open University Press, Buckingham, 1990.
[2] A. E. G. Cass, 'Biosensors, A Practical Approach', Oxford University Press, Oxford, 1990.

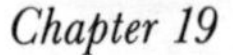

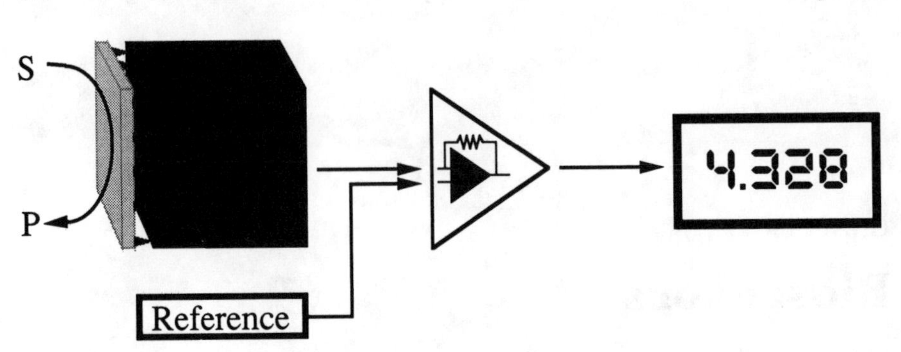

Figure 1 *The functional units of a biosensor*

case further tests may show the error. Other biosensor analytical applications may require accuracy over a wide analytical range.

Biosensors must show advantages over the use of the free biocatalyst, which possess equal specificity and discriminatory powers, if they are to be acceptable. Their main advantage usually involves their re-usability. Repetitive re-use of the same biologically active sensing material generally ensures that similar samples give similar responses as any need for accurate aliquoting of such biological materials with precisely similar activity is obviated. This avoids the possibility of introducing errors by inaccurate pipetting or on dilution. Repetitive and reagent-less methodology offers considerable savings in terms of reagent costs, so reducing the cost per assay. In addition the increased operator time per assay and the associated higher skill required for 'traditional' assay methods also involves a cost penalty which in many cases may be greater than that due to the reagents. These advantages must be sufficient to encourage the high investment necessary for the development of a biosensor and the purchase price to the end-user. Table 1 lists a number of important attributes that a successful biosensor may be expected to possess. In any particular case only some of these may be achievable.

The different types of biosensors have their own advantages and disadvantages[3] which may be summarized (Table 2). Not all types of biosensor have been used in commercial products. Apart from the important related area of colorimetric test strips, the more important commercial biosensors are all currently electrochemical. There are potentiometric biosensors available for glucose, lactate, glycerol, alcohol, lactose, L-amino acids, and cholesterol, some of which involve field effect transistors (FET). Amperometric devices have been marketed for glucose, other low molecular weight carbohydrates, and alcohol. There are strong indications that optical biosensors may be entering the market more strongly in the near future as is evidenced by the launch of BIAcore (see Section 7.2), a surface plasmon resonance device.

The market share of the application areas for biosensors is shown in Table 3. From this it can be seen that the area is generally expanding at an impressively fast pace. A note of caution should be entered here as many market research projections for the growth of the biosensor industry have, in the past, been grossly

[3] J. H. T. Luong, C. A. Groom, and K. B. Male, *Biosens. Bioelectron.*, 1991, **6**, 547.

Table 1 *The properties required of a successful biosensor*

Required property	*Achievable with ease*
Specificity	yes
Discrimination	yes
Repeatability	yes
Precision	yes
Safe	yes
Accuracy	yes, as easily calibrated
Appropriate sensitivity	yes, except in trace analysis
Fast response	yes, usually
Miniaturizable	yes, generally
Small sample volumes	yes, generally
Temperature independence	yes, may be electronically compensated
Low production costs	yes, if mass produced
Reliability	yes, generally
Marketable	difficult, due to competing methodology
Drift free	difficult but possible
Robust	no, generally need careful handling
Stability	no, except on storage or in the short term
Sterilizable	no, except on initial storage
Autoclavable	not presently achievable

Table 2 *The properties of the various biosensor configurations*

Biosensor	*Cost*	*Reliability*	*Complexity*	*Sensitivity*	*Speed of response*	*General utility*	*Present usage*	*Future prospects*
Amperometric	low	high	high	medium	medium	small	high	high
Conductimetric	very low	medium	low	medium	medium	small	small	medium
Potentiometric	low	medium	medium	medium	medium	medium	medium	high
Piezoelectric	low	low	low	low	fast	narrow	small	small
Thermometric	high	high	very high	high	slow	wide	small	medium
Optical	low	medium	medium	medium	medium	small	small	high

Table 3 *The market share and growth of biosensor application areas*

Application area	*Market share (%)*	*Annual growth rate (%)*
Clinical diagnostics	53	25
Industrial process control	11	50
Medical devices	11	30
Agriculture/veterinary	8	60
Defence	6	45
Environmental	5	35
Research	3	50
Robotics	2	30
Other	1	30

over-optimistic, and the industry sales as a whole were worth well under £50M in 1992 but are expected to take off in the near future. It is now accepted that there are substantial and investment-intensive difficulties involved in producing such robust and reliable analytical devices, which are able to operate under authentic real-life conditions, even where a novel and highly promising research device has been produced.

By far the largest biosensor application area is in clinical diagnostics. This includes monitoring of critical metabolites during surgery. The major target markets are concerned with use within the physician's office, the casualty department of a hospital, and in home diagnosis. These application areas are potentially very wide. The use of rapid biosensor techniques in Doctors' surgeries obviates the need for expensive and, most importantly, time-consuming testing at central clinical laboratories; thus diagnosis and treatment may start during the first visit of a patient. This removes the need to wait for a return visit after the clinical tests have been completed elsewhere and allowing time, perhaps, for the patient's condition to deteriorate somewhat. Also there is less likelihood of the sample being mishandled or contaminated. Centralized clinical analytical facilities remain a necessity due to the need, in many cases, for multiple different analyses on the same sample and difficulties such as regulatory compliance and quality assurance. Recent legislation imposes stringent quality assurance and control standards on clinical analyses. This law will make it much more expensive in the future to bring novel clinical biosensors to the market place, but allows the use of those biosensors that are already established, giving them a distinct competitive edge.

Home diagnosis is an area which is being opened up by, for example, pregnancy and ovulation test kits. Clearly there are risks and problems involved with their more widespread use but many people prefer to use them as indications for whether a trip to the Doctor's surgery is really necessary or not. As conselling may be necessary in some potential home diagnosis applications (for example, cancer and AIDS), controversy exists over their development.

One of the major potential uses for biosensors is for *in vivo* applications. The purpose here is to continuously monitor the levels of metabolites so that corrective action can be employed when necessary. Clearly such biosensors must be biocompatible and miniaturized so that they are implantable. In addition they should be reagentless, the reaction being controlled only by the presence of the metabolite and the stabilized bioreagent. The signal generated must be stable over the period of interest. At the present time such biosensors have a relatively short lifespan of a few days at most due mainly to problems which occur due to the body's response. The cost of the sensor is often of importance and considerable effort has been expended in the production of disposable devices using cheap integrated chip technology, disposable technology offering opportunities for continued and increased revenue. Such methodology has opened up the possibility for having a number of different sensors on one device, allowing multiparameter assays.

Industrial analysis involves food, cosmetics, and fermentation process control, and quality control and monitoring. The defence industry is interested in

Table 4 *Examples of biosensor immobilization methods*

Physical entrapment	Biologically active material held next to the sensing surface by a semi-permeable membrane which prevents it from escaping to the bulk phase but allows the analyte in. Sometimes the membrane can be made such that it increases the specificity of the sensing or reduces unwanted side-reactions. This is a simple and inexpensive method.
Non-covalent binding	Adsorption of enzymes to a porous carbon electrode. This may suffer from a gradual leaching of the enzyme to the bulk phase.
Covalent binding	Treatment of the biosensor surface with 3-aminopropyltriethoxysilane followed by coupling of biologically active material to the reactive amino groups remaining on the cross-linked siloxane surface. Proteins may be attached by use of carbodiimides which form amide links between amines and carboxylic acids. Such methods permanently attach the biological material but are difficult to reproduce exactly and often cause a large reduction in activity.
Membrane entrapment	Cross-link proteins with glutaraldehyde within a cellulose or nylon supporting net.

detectors for explosives, nerve gases, and microbial toxins. Environmental uses of biosensors are mainly in areas of pollution control. A typical application might be to detect parts per million of particular molecular species such as an industrial toxin within the highly complex mixtures produced as process effluent.

2 THE BIOLOGICAL REACTION

An important factor in most biosensor configurations is the sensing surface. This normally consists of a thin layer of biologically active material in intimate contact with the electronic transducer. In some cases the biological material may be covalently or non-covalently directly attached to the surface but usually it forms part of a thin membrane covering the sensing surface. Generally, the conversion of the biological process into an electronic signal is most efficient where the distance between the place where the biological reaction takes place and the place where the electronic transduction takes place is minimal. In addition, it is important for retention of biological activity that the biological material is not lost into analyte solutions. The immobilization technology for holding the biocatalyst in place is extensive.[4] The various methods of immobilization are summarized in Table 4.

3 THEORY

In the absence of diffusion effects (see later), most biological reactions can be described in terms of saturation kinetics:

[4] M. F. Chaplin and C. Bucke, 'Enzyme Technology', Cambridge University Press, Cambridge, 1990.

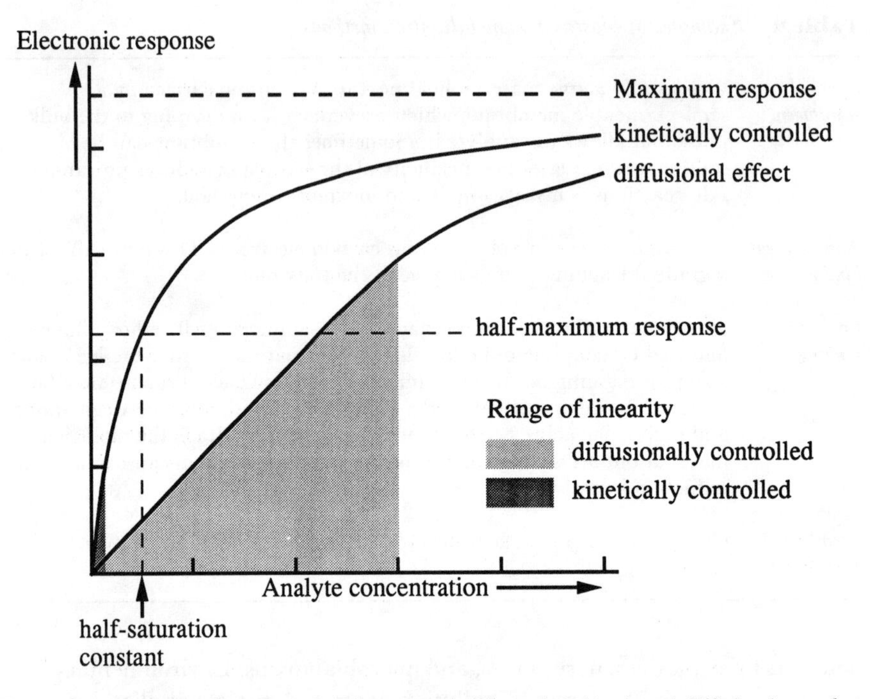

Figure 2 *The range of response of a biosensor under biocatalytic kinetic and diffusional control*

$$\text{Biological material} + \text{Analyte} \rightleftharpoons \text{Bound Analyte}$$

The bound analyte then gives rise to the biological response, so generating the electronic response. This electronic response varies with the biological response which, in turn, varies with the concentration of the bound analyte. Apart from the logarithmic relationship in potentiometric biosensors, the biological and electronic responses are often proportional.

$$\text{Bound analyte} \rightarrow \text{Biological response} \rightarrow \text{Electronic response}$$

$$\text{Electronic response} = \frac{(\text{Maximum electronic response possible}) \times (\text{Analyte concentration})}{(\text{Half saturation constant}) + (\text{Analyte concentration})}$$

where the half-saturation constant is equal to the analyte concentration which gives rise to half the maximum electronic response possible (Figure 2). The response is linear, to within 95%, at analyte concentrations up to about a twentieth of the half-saturation constant. A biosensor obeying these kinetics may be used over a wider, non-linear range if it has compensatory electronics.

Where the analyte reacts as part of the biological response (*i.e.* during a biocatalytic reaction utilizing enzyme(s) and/or microbial cells), an additional factor is the diffusion of the analyte from the bulk of the solution to the reactive surface.[4] If this rate of diffusion is less than the rate at which the analyte would otherwise react, this will reduce the concentration of analyte undergoing reaction. The rate of diffusion increases as the concentration gradient increases.

$$\text{Rate of diffusion} = (\text{Diffusivity constant}) \times (\text{Analyte concentration gradient})$$

where the analyte concentration gradient is given by the difference between the bulk analyte concentration and the analyte concentration on the sensing surface of the biosensor, divided by the distance through which the analyte must diffuse.

As most biocatalytic biosensor configurations utilize membrane-entrapped biocatalyst, this concentration gradient depends not only on the analyte concentration in the bulk and membrane but also on the membrane's thickness; the thicker the membrane, the greater the diffusive distance from the bulk of the solution to the distal sensing surface of the biosensor, and the greater the amount of biocatalyst present. Both effects increase the likelihood that the overall reaction will be controlled by diffusion. Thus, such biosensors can be designed to be under diffusional or kinetic control by varying the membrane thickness. When the rate of analyte diffusion is slower than the rate at which the biocatalyst can react, the electronic response decreases due to the lower level of analyte available for reaction. A steady-state is rapidly established when the rate of arrival of the analyte equals its rate of reaction. This steady-state condition may be determined wholly by the rate of diffusion, wholly by the rate of reaction, or by an intermediate dependency. Where it depends solely on the rate of diffusive flux of the analyte, this determines the electronic response.

$$\text{Electronic response} \propto \text{Analyte concentration}$$

As the rate of diffusion depends on the bulk concentration of the analyte, this electronic response is linearly related to the bulk analyte concentration and, most importantly and intriguingly, is independent of the properties of the enzyme. Thus, the biosensor is linear over a much wider range of substrate concentrations (see Figure 2) and relatively independent of changes in the pH and temperature of the biocatalytic membrane, so long as the system remains diffusion controlled. It should be noticed, however, that under these conditions the response is reduced relative to a system containing the same amount of biocatalyst but not diffusionally limited. Maximum sensitivity to analyte concentration would be accomplished by the utilization of thin membranes containing a high biocatalyst activity and a well-stirred analyte solution. The overall kinetics of most biosensor configurations are difficult to predict as they depend on diffusivities in the bulk phase and within the biocatalytic volume; the nature, porosity, and physical properties of any membrane; the intrinisic biocatalytic kinetics; the electronic transduction process and kinetics; the way in which the analyte is presented; and on other non-specific factors. Generally, such overall kinetics are determined experimentally using the complete biosensor and, hence, it is very important that the biosensor configuration is reproducible.

4 ELECTROCHEMICAL METHODS

Electrochemical biosensors are generally fairly simple devices. They include amperometric biosensors which determine the electric current associated with

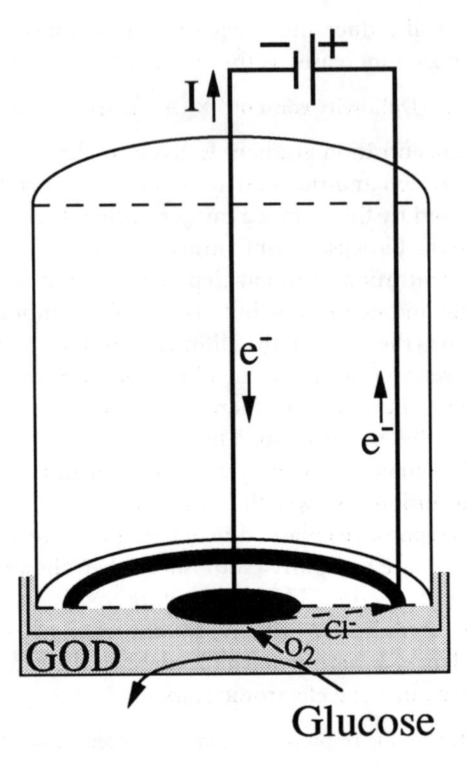

Figure 3 *Amperometric glucose biosensor based on the oxygen electrode utilizing glucose oxidase (GOD)*

the electrons involved in redox processes, potentiometric biosensors which use ion selective electrodes to determine changes in the concentration of chosen ions (*e.g.* hydrogen ions), and conductimetric biosensors which determine conductance changes associated with changes in the overall ionic environment.

4.1 Amperometric Biosensors

Enzyme-catalysed redox reactions can form the basis of a major class of biosensors if the flux of redox electrons can be determined. Normally, a constant voltage is applied between two electrodes and the current, due to the electrode reaction, determined. The first and simplest biosensor was based on this principle. It was for the determination of glucose and made use of the Clark oxygen electrode.[5] Figure 3 shows a section through such a simple amperometric biosensor. A potential of 0.6 V is applied between the central platinum cathode and the surrounding silver/silver chloride anode. Dissolved molecular oxygen at the platinum cathode is reduced and the circuit is completed by means of the saturated KCl solution. The electrode compartment is separated from the biocatalyst by a thin plastic membrane, permeable only to oxygen, and often

[5] L. C. Clark and C. H. Lyons, *Ann. N.Y. Acad. Sci.*, 1962, **102**, 29.

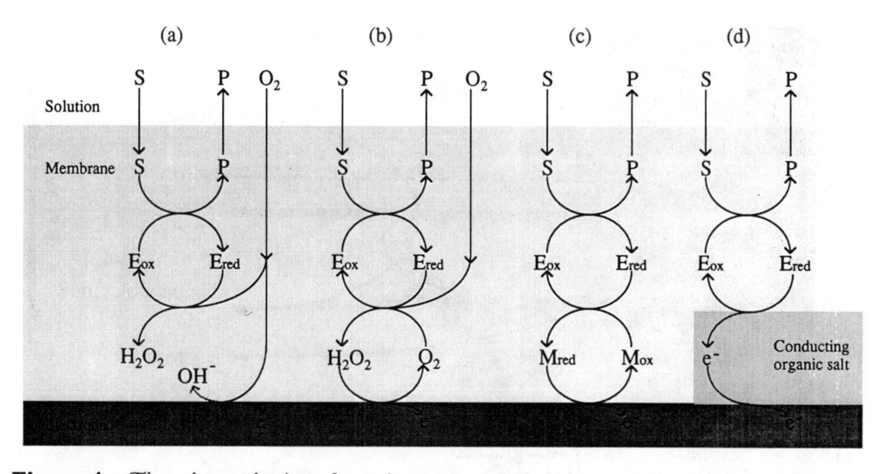

Figure 4 *The redox mechanisms for various amperometric biosensor configurations*

made of Teflon. The biocatalyst is retained next to the electrode by means of a membrane which is permeable only to low molecular weight molecules including the reactants and products.

$$\text{Cathode reaction } (-0.6\text{ V}) \quad O_2 + 4H^+ + 4e^- \longrightarrow 2H_2O$$
$$\text{Anode reaction} \quad 4Ag^0 + 4Cl^- \longrightarrow 4AgCl + 4e^-$$

Glucose may be determined by the reduction in the dissolved oxygen concentration when the redox reaction, catalysed by glucose oxidase, occurs

$$\text{Glucose} + O_2 \xrightarrow{\text{glucose oxidase}} \delta\text{-Gluconolactone} + H_2O_2$$

Conditions can be chosen such that the rate at which oxygen is lost from the biocatalyst-containing compartment is proportional to the bulk glucose concentration. Other oxidases can be used in this biosensor configuration and may be immobilized as part of a membrane by treatment of the dissolved enzyme(s), together with a diluent protein, with glutaraldehyde on a cellulose or nylon support. An alternative method of determining the rate of reaction is to detect the hydrogen peroxide produced directly, by reversing the polarity of the electrodes and using a cellulose acetate membrane in place of the Teflon membrane to retard anions and allow passage of the hydrogen peroxide (Figure 4b).

$$\text{Cathode reaction} \quad 2AgCl + 2e^- \longrightarrow 2Ag^0 + 2Cl^-$$
$$\text{Pt Anode reaction } (+0.6\text{ V}) \quad H_2O_2 \longrightarrow O_2 + 2H^+ + 2e^-$$

These electrodes can be developed further for the determination of substrates for which no direct oxidase enzyme exists. Thus sucrose can be determined by placing an invertase layer over the top of the glucose oxidase membrane in order to produce glucose which can then be determined. Interference from glucose in the sample can be minimized by including a thin anti-interference layer of glucose oxidase and peroxidase over the top of both layers which removes the glucose without significantly reducing the oxygen diffusion to the electrode. An

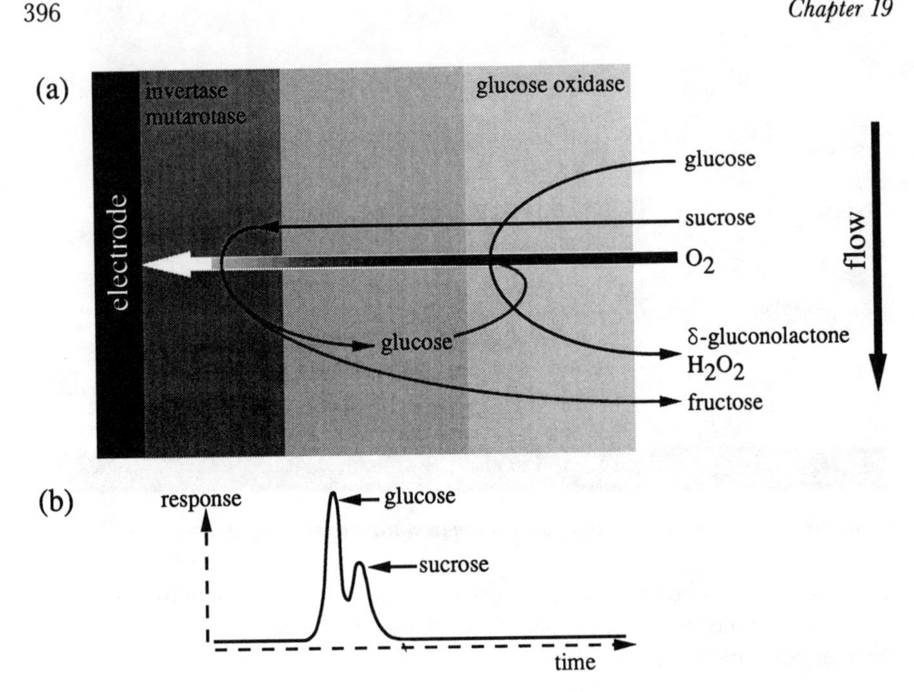

Figure 5 *A kinetically controlled anti-interference membrane for sucrose. The sample is presented as a rapid pulse of material in the flowing stream*

alternative approach to assay sucrose and glucose together[6] makes use of the lag period in the response due to the necessary inversion of the sucrose delaying its response relative to the glucose (Figure 5).

Fish freshness can be determined using a similar concept; the nucleotides in fish changing due to a series of reactions after death. Fish freshness can be quantified in terms of its **K** value, where

$$\mathbf{K} = \frac{(\mathrm{HxR} + \mathrm{Hx}) \times 100}{(\mathrm{ATP} + \mathrm{ADP} + \mathrm{AMP} + \mathrm{IMP} + \mathrm{HxR} + \mathrm{Hx})}$$

HxR, IMP, and Hx represent inosine, inosine-5′-monophosphate, and hypoxanthine respectively. After fish die their ATP undergoes catabolic degradation through a series of reactions outlined below:

$$\mathrm{ATP} \rightarrow \mathrm{ADP} \rightarrow \mathrm{AMP} \rightarrow \mathrm{IMP} \rightarrow \mathrm{HxR} \rightarrow \mathrm{Hx} \rightarrow \mathrm{Xanthine} \rightarrow \mathrm{Uric\ acid}$$

The intermediary accumulation of inosine and hypoxanthine relative to the nucleotides is an indicator of how long the fish has been dead and its storage conditions, and hence its freshness. A commercialized fish freshness biosensor has been devised which utilizes a triacyl cellulose membrane containing immobilized nucleoside phosphorylase and xanthine oxidase over an oxygen electrode.

$$\mathrm{Inosine} + \mathrm{Phosphate} \xrightarrow[\text{phosphorylase}]{\text{nucleoside}} \mathrm{Hypoxanthine} + \mathrm{Ribose\text{-}phosphate}$$

[6] E. Watanabe, M. Takagi, S. Takei, M. Hoshi, and C. Shu-gui, *Biotech. Bioeng.*, 1991, **38**, 99.

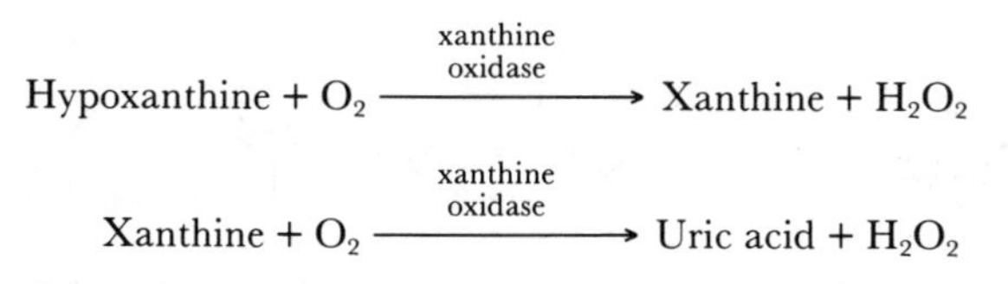

The electrode may be used to determine the reduction of oxygen,[7] due to its reaction given above, or the increase in hydrogen peroxide. The nucleotides can be determined using the same electrode and sample, subsequent to the addition of nucleotidase and adenosine deaminase.

K values below 20 show the fish is very fresh and may be eaten raw. Fish with a **K** value between 20 and 40 must be cooked but those with a **K** value above 40 are not fit for human consumption. Clearly a relatively simple probe to accurately and reproducibly determine fish freshness has significant economic importance to the fish industry. In its absence freshness is determined completely subjectively by inspection.

Although such biosensors are easy to produce they do suffer from some significant drawbacks. The reaction is dependent on the concentration of molecular oxygen which precludes its use in oxygen-deprived environments such as *in vivo*. Also the potential used is sufficient to cause other redox processes to occur, such as vitamin C oxidation/reduction, which may interfere with the analyses. Much research has been undertaken on the development of substances that can replace oxygen in these reactions.[8] Generally oxidases are far more specific for the oxidizable reactant than they are for molecular oxygen itself, as the oxidant, and many materials can act as the oxidant. The optimal properties of such materials include fast electron transfer rates, the ability to be easily regenerated by an electrode reaction, and retention within the biocatalytic membrane. In addition, they should not react with other molecules present, including molecular oxygen. Many such molecules, now called mediators, have been developed. Their redox reactions are summarized in Figure 6.

The mediated biosensor reaction consists of three redox processes:

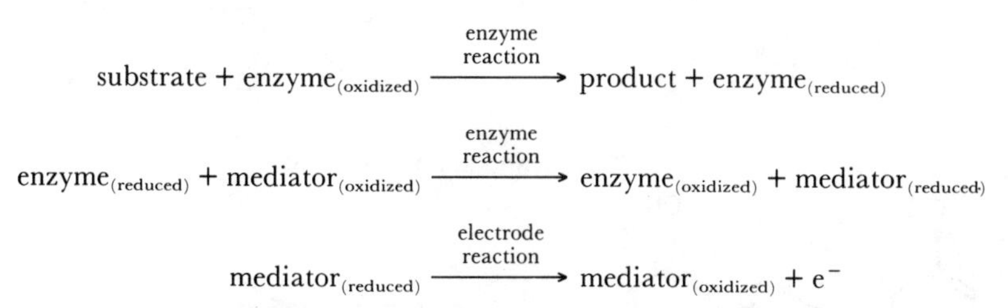

When a steady-state response has been obtained the rates of all of these processes and the rate of the diffusive flux in must be equal (see Figure 4c). Any of these, or a combination, may be the controlling factor. The overall response is, therefore, difficult to predict.

[7] E. Watanabe, K. Toyama, I. Karube, H. Matsuoka, and S. Suzuki, *Appl. Microbiol. Biotechnol.*, 1984, **19**, 18.

[8] F. W. Scheller, F. Schubert, B. Neumann, D. Pfeiffer, R. Hintsche, I. Drasfeld, U. Wollenberger, R. Renneberg, A. Warsinke, G. Johansson, M. Skoog, X. Yang, V. Bogdanovskaya, A. Bückmann, and S. Y. Zaitsev, *Biosens. Bioelectron.*, 1991, **6**, 245.

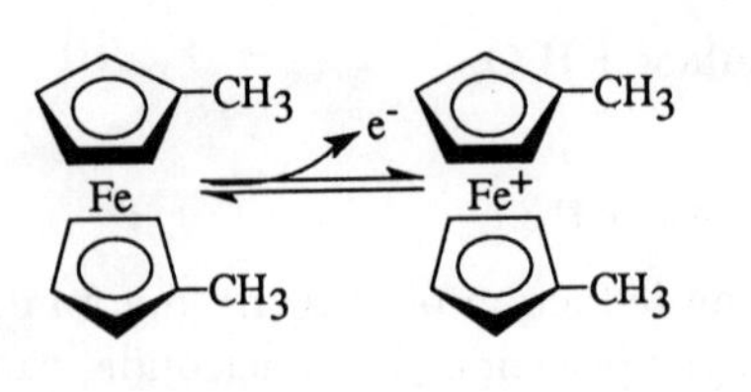

Ferrocene

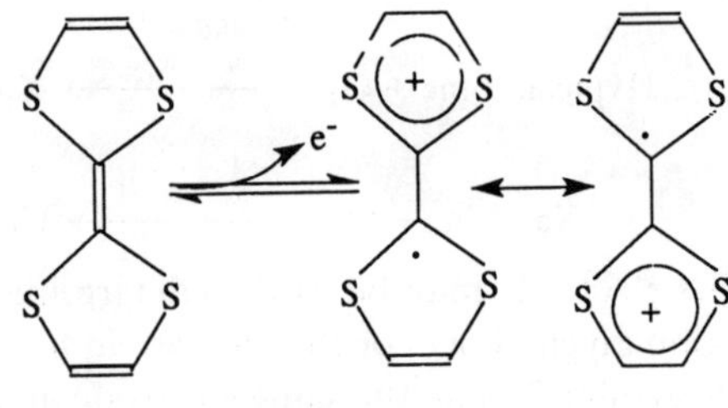

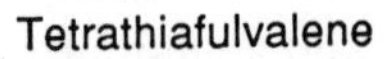

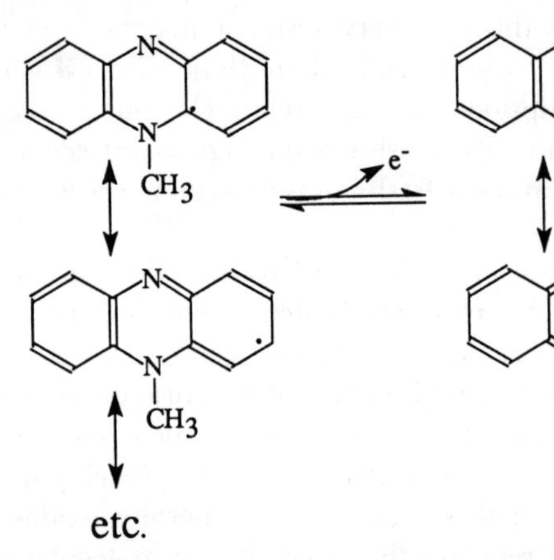

N-Methyl phenazinium cation

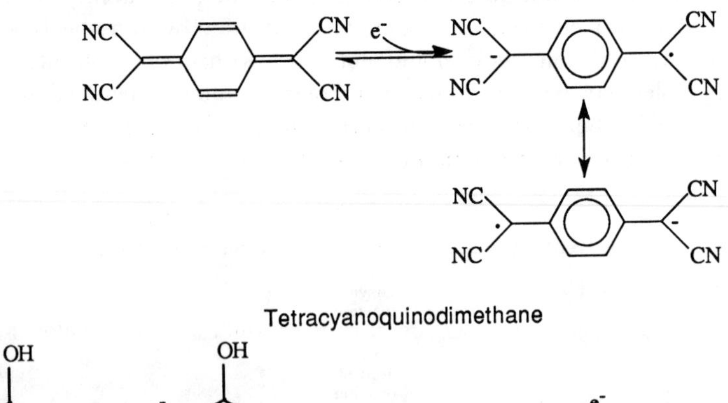

OH

OH

e-

OH

•OH
+

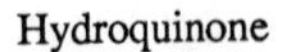

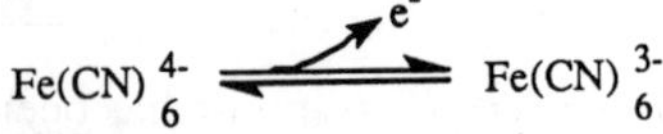

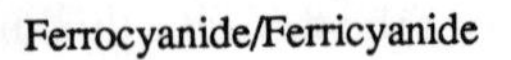

Figure 6 *The redox reactions of amperometric biosensor mediators. Tetracyanoquinodimethane acts as a partial electron acceptor whereas ferrocene, tetrathiafulvalene, and* N*-methyl phenazinium can all act as partial electron donors. Hydroquinone and ferricyanide are soluble mediators*

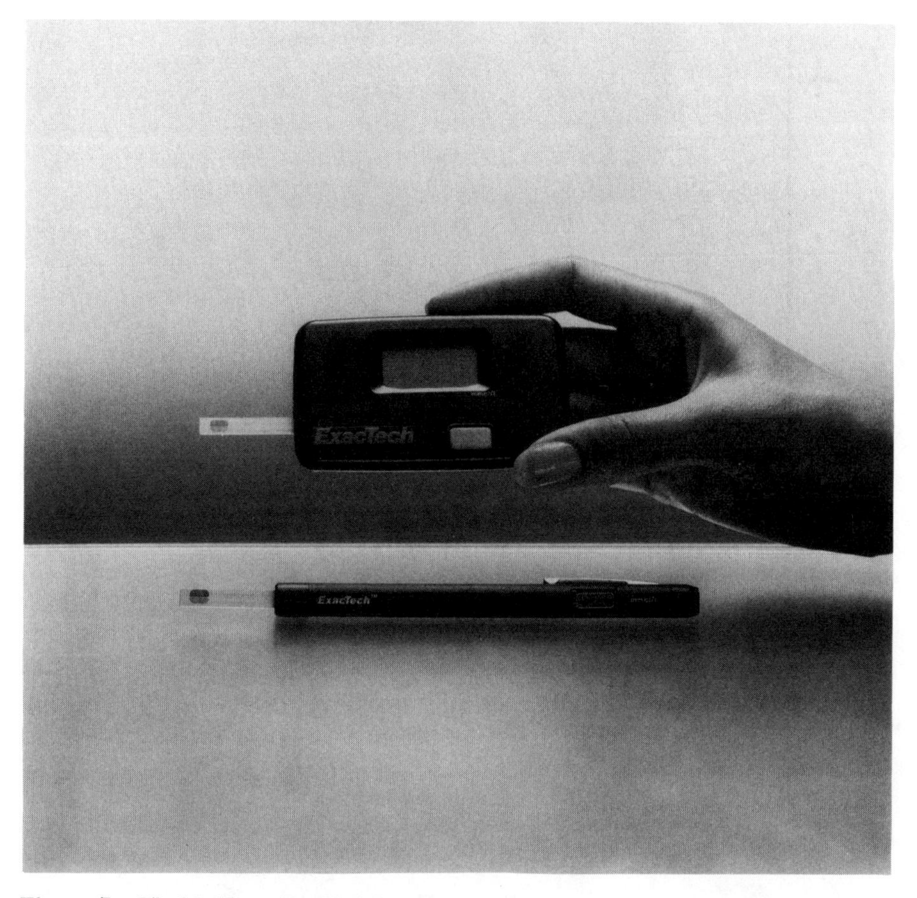

Figure 7 *The MediSense ExacTech Companion and Pen Sensors, both with disposable test strips. They are manufactured by MediSense Contract Manufacturing Ltd., Abingdon, Oxfordshire, UK*

A blood-glucose biosensor, for the control of diabetes, has been built and marketed based on this mediated system[9] with an appearance very similar to a watch-pen (Figure 7). The sensing area is a single-use disposable electrode arrangement and clips into what would be the nib end of the device. It is produced by screen printing onto a plastic strip and consists of an Ag/AgCl reference electrode and a carbon working electrode containing glucose oxidase and a derivatized ferrocene mediator. Both electrodes are covered with a wide mesh hydrophilic gauze to enable even spreading of a blood drop sample and to prevent localized cooling effects due to uneven evaporation. The electrodes are kept dry until use and have a shelf life of six months when sealed in aluminium foil. They can detect glucose concentrations of 2–25 mM in a single drop of blood and display the result within 30 seconds. Such biosensors are currently used by

[9] D. R. Matthews, R. R. Holman, E. Brown, J. Steemson, A. Watson, S. Hughes, and D. Scott, *Lancet*, 1987, **i**, 778.

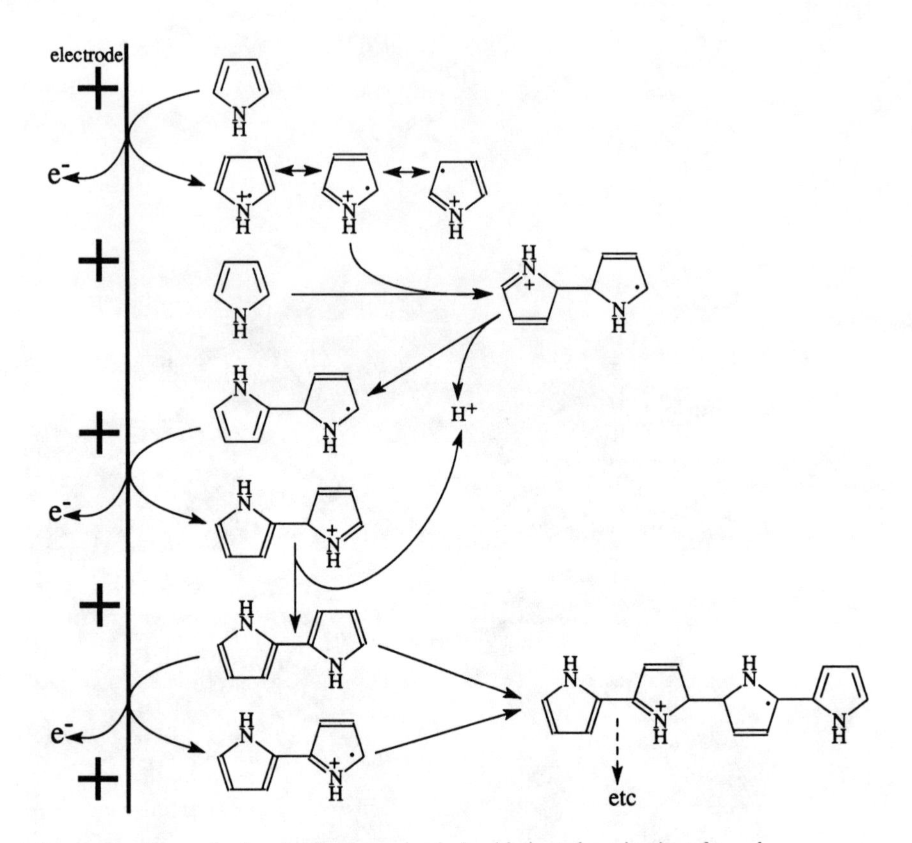

Figure 8 *The mechanism for the electrochemical oxidative polymerization of pyrrole*

hundreds of thousands of people with diabetes in more than 50 countries worldwide.

When an oxidase is unable to react rapidly enough with available mediators, horseradish peroxidase, which rapidly reacts with ferrocene mediators, can be included with the enzyme. This catalyses the reduction of the hydrogen peroxide produced by the oxidase, with consequent oxidation of the mediator. In this case the mediator is acting as an electron donor rather than acceptor. The oxidized mediator then can be rapidly reduced at the electrode at moderate redox potential.

$$\text{mediator}_{(\text{reduced})} + \tfrac{1}{2}H_2O_2 + H^+ \xrightarrow{\text{peroxidase}} \text{mediator}_{(\text{oxidized})} + H_2O$$

$$\text{mediator}_{(\text{oxidized})} + e^- \xrightarrow{\text{electrode reaction}} \text{mediator}_{(\text{reduced})}$$

A major advance in the development of micro-amperometric biosensors came with the discovery that pyrrole can undergo electrochemical oxidative polymerization (Figure 8) under conditions mild enough to entrap enzymes and mediators at the electrode surface without denaturation. A membrane, entrapping the

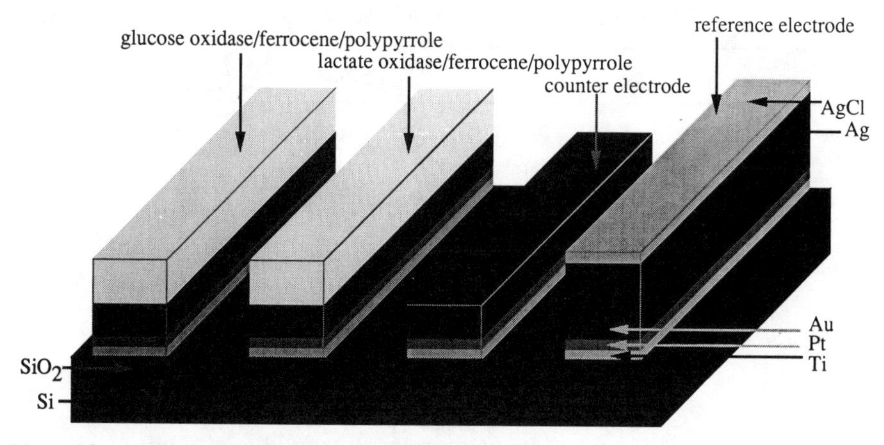

Figure 9 *A combined micro-electrode for glucose and lactate*

biocatalyst and mediator, can be formed at the surface of even extremely small electrodes by polymerizing pyrrole in the presence of biocatalyst. This allows silicon chip microfabrication methods to be used and for many different sensors to be laid down on the same chip (Figure 9).

Another advance has been the use of conducting organic salts on the electrode. These may allow the direct transfer of electrons from the reduced enzyme to the electrode without the use of any (other) mediator (Figure 4d). Conducting organic salts consist of a mixture of two types of planar aromatic molecules, electron donors and electron acceptors (see Figure 6), which partially exchange their electrons. These molecules form segregated stacks, containing either the donor or acceptor molecules, with some of the electrons from the donors being transferred to the acceptors. The electrons which have been partially transferred are mobile up and down the stacks giving the organic crystals a high conductivity. There must not be a total electron transfer between the donor and acceptor molecules or the crystal becomes an insulator through lack of electron mobility. Although these electrodes give the appearance of direct electron transfer to the electrode, as both the components of the organic salts, in the appropriate redox state, may be able to mediate the reaction in a similar manner to the ferrocene derivatives by themselves, it is highly probable that these electrodes are behaving as a highly insoluble mediator prevented from large scale leakage by electrostatic effects.

4.2 Potentiometric Biosensors

Changes in ionic concentrations are easily determined by use of ion-selective electrodes. This forms the basis of potentiometric biosensors. Many biocatalysed reactions involve charged species each of which will absorb or release hydrogen ions according to their pK_a and the pH of the environment. This allows a relatively simple electronic transduction using the commonest ion-selective electrode, the pH electrode. Table 5 shows some biocatalytic reactions that can be

Table 5 *Biocatalytic reactions that can be used with ion-selective electrode biosensors*

Electrode	*Reactions*
Hydrogen ion	
Penicillin	$\text{penicillin} \xrightarrow{\text{penicillinase}} \text{penicilloic acid} + H^+$
Lipid	$\text{triacylglycerol} \xrightarrow{\text{lipase}} \text{glycerol} + \text{fatty acids} + H^+$
Urea	$H_2NCONH_2 + H_2O + 2H^+ \xrightarrow{\text{urease (pH 6)}} 2NH_4^+ + CO_2$
Ammonia	
L-Phenylalanine	$\text{L-phenylalanine} \xrightarrow{\text{phenylalanine ammonia-lyase}} NH_4^+ + \textit{trans}\text{-cinnamate}$
L-Asparagine	$\text{L-asparagine} + H_2O \xrightarrow{\text{asparaginase}} NH_4^+ + \text{L-aspartate}$
Adenosine	$\text{adenosine} + H_2O + H^+ \xrightarrow{\text{adenosine deaminase}} NH_4^+ + \text{inosine}$
Iodide	
Peroxide	$H_2O_2 + 2I^- + 2H^+ \xrightarrow{\text{peroxidase}} 2H_2O + I_2$

utilized in potentiometric biosensors. Potentiometric biosensors can be miniaturized by the use of field effect transistors (FET).

Ion-selective field effect transistors (ISFET) are low cost devices that are in mass production. Figure 10 shows a diagrammatic cross-section through an npn hydrogen ion responsive ISFET with a biocatalytic membrane covering the approximately 0.025 mm^2 ion-selective membrane. The build-up of positive charge on this surface (the gate) repels the positive holes in the p-type silicon causing a depletion layer and allowing the current to flow. The reference electrode is likely to be an identical ISFET without any biocatalytic membrane. A major practical problem with the manufacture of such enzyme-linked FETs (ENFETs) is protection of the silicon from contamination by the solution, hence the covering of waterproof encapsulant. Because of their small size they only require minute amounts of biological material and can be fabricated in a form whereby they can determine several analytes simultaneously. A further advantage is that they have a more rapid response rate when compared with the larger ion-selective electrode devices. The enzyme can be immobilized to the silicon nitride gate using polyvinyl butyral deposited by solvent evaporation and cross-linked with glutaraldehyde. Some fabrication problems still exist, however, and these are currently being addressed. In particular, they need on-chip temperature compensation.

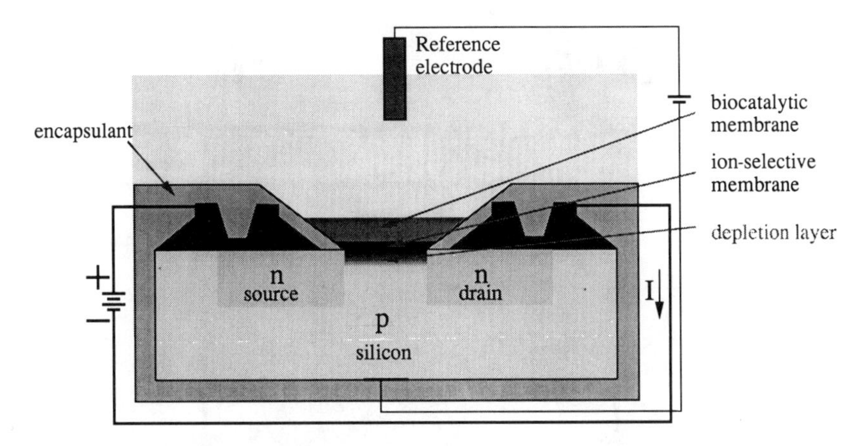

Figure 10 *An FET-based potentiometric biosensor*

A potentiometric biosensor for fish freshness similar in biochemical principle to that described earlier has been developed, which determines the pH charges associated with the reactions, utilizing an amorphous silicon field effect transistor.[10]

4.3 Conductimetric Biosensors

Many biological processes involve changes in the concentrations of ionic species. Such changes can be utilized by biosensors which detect changes in electrical conductivity. A typical example of such a biosensor is the urea sensor, utilizing immobilized urease,[11] and used as a monitor during renal surgery and dialysis (Figure 11). The reaction gives rise to a large change in ionic concentration making this type of biosensor particularly attractive for monitoring urea concentrations.

$$NH_2CONH_2 + 3H_2O \xrightarrow{\text{urease}} 2NH_4^+ + HCO_3^- + OH^-$$

An alternating field between the two electrodes allows the conductivity changes to be determined whilst minimizing undesirable electrochemical processes. The electrodes are interdigitated to give a relatively long track length (~ 1 cm) within a small sensing area (0.2 mm^2). A steady-state response can be achieved in a few seconds allowing urea to be determined within the range 0.1–10 mM. The output is corrected for non-specific changes in pH and other factors by comparison with the output of a non-enzymic reference electrode pair on the same chip. The method can easily be extended to use other enzymes and enzyme combinations which produce ionic species, for example amidases, decarboxylases, esterases, phosphatases, and nucleases.

[10] M. Gotoh, E. Tamiya, A. Seki, I. Shimizu, and I. Karube, *Anal. Lett.*, 1988, **21**, 1785.
[11] P. Vadgama, *Analyst (London)*, 1986, **111**, 875.

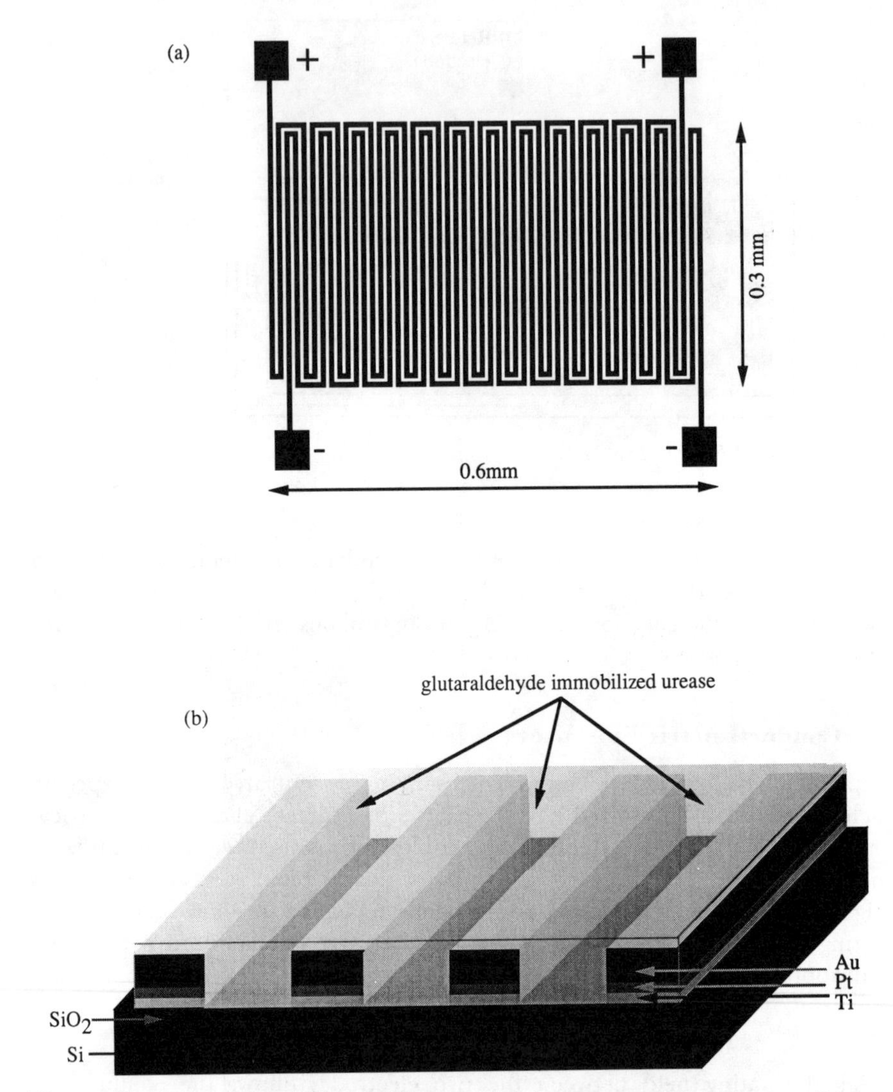

Figure 11 *Parts of a conductimetric biosensor electrode arrangement: (a) top view, (b) cross-sectional view. The tracts are about* 5000 nm *wide and the thickness of the various layers are approximately:* SiO_2 550 nm, *Ti* 100 nm, *Pt* 100 nm, *Au* 2000 nm

5 THERMOMETRIC BIOSENSORS

A general property of many enzyme reactions is the production of heat (Table 6). This forms the basis of thermometric biosensors (Figure 12),[12,13] sometimes also called calorimetric or thermal biosensors. An important factor in the manufacture of such biosensors is assuring that the resulting heat changes are not affected

[12] G. G. Guilbault, B. Danielsson, C. F. Mendenlus, and K. Mosbach, *Anal. Chem.*, 1983, **55**, 1582.
[13] K. Mosbach, *Biosens. Bioelectron.*, 1991, **6**, 179.

Table 6 *Exothermic reactions used in thermometric biosensors*

Analyte	*Reaction*	*Biocatalyst*
Antigens	ELISA	catalase/antibody
Ascorbic acid	oxidation	ascorbate oxidase
Cholesterol	oxidation	cholesterol oxidase
Ethanol	oxidation	alcohol oxidase
Glucose	oxidation	glucose oxidase
Glycerol	catabolism	*Gluconobacter oxydans* cells
Hydrogen peroxide	redox	catalase
Lactate	oxidation	lactate oxidase
Penicillin G	hydrolysis	β-lactamase
Pyruvate	reduction	yeast lactate dehydrogenase
Oxalic acid	oxidation	oxalate oxidase
Urea	hydrolysis	urease
Uric acid	oxidation	uricase

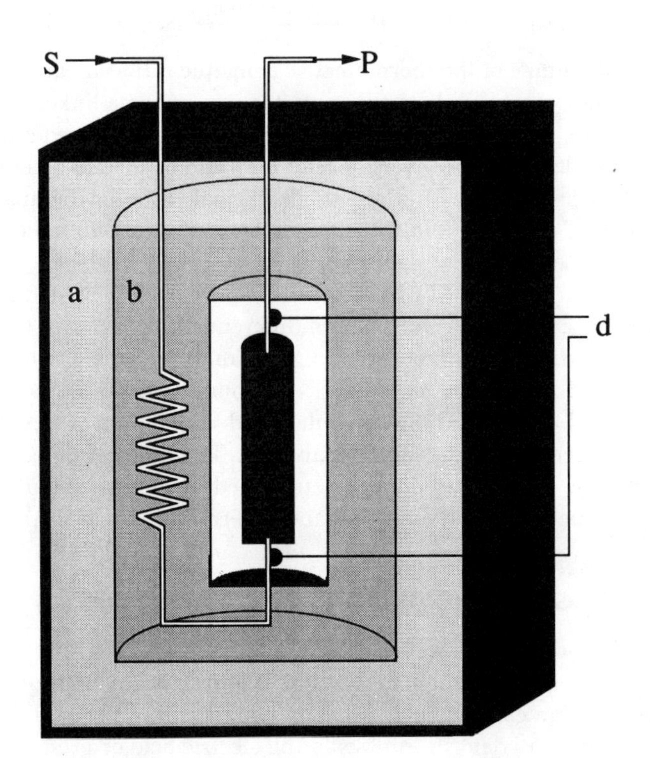

Figure 12 *Sectional view through a thermometric biosensor: (a) insulated box; (b) heat exchanger, aluminium cylinder; (c) biocatalytic packed bed reactor; (d) matched thermistors*

by environmental fluctuations. For this reason, the reaction is confined within a heat-insulated box and the analyte stream is passed through a heat exchanger. The reaction takes place within a small packed bed reactor and the difference in temperature between the incoming analyte and the product stream is determined by matched thermistors. This difference is only a fraction of a degree centigrade but temperatures can be resolved down to 0.0001 °C.

Thermometric devices have had only limited commercial success when compared to the other types of biosensors but they have proved to have a wide utility. One advantage that they have over other biosensor configurations is the ease with which a number of reactions can be linked together within the reactor. This allows the possibility of utilizing recycling reactions by co-immobilizing other enzymes and introducing other reactants into the analyte stream. Thus the sensitivity to lactate can be increased by more than an order of magnitude by co-immobilizing lactate dehydrogenase with the lactate oxidase, allowing substrate recycling and effectively reacting the lactate analyte hundreds of times.

$$\text{lactate} + O_2 \xrightarrow{\text{lactate oxidase}} \text{pyruvate} + H_2O_2$$

$$\text{pyruvate} + \text{NADH} + H^+ \xrightarrow{\text{lactate dehydrogenase}} \text{lactate} + \text{NAD}^+$$

A major advantage of the thermometric principle is that it can be extended to the use of whole viable cells and as part of an enzyme-linked immunoassay (ELISA) system. Immobilized viable cells within the packed bed reactor can not only be used to achieve bioconversions but may also be used as an environmental monitor when presented with a metabolizable substrate in the analyte stream. The presence of toxic materials in this substrate stream will affect the general metabolic rate, so indicating their presence.

Thermometric ELISA (TELISA) systems, in a similar manner to other ELISA methods, may have a number of different configurations. One method is to apply a mixture of unlabelled antigen (analyte) and a fixed amount of enzyme-labelled antigen to the packed bed column containing an immunosorbent. Increased concentrations of unlabelled antigen increases the amount bound in competition to the labelled antigen. The amount of labelled antigen remaining in the column can then be determined by pulsing the substrate for the labelling enzyme through the column and determining the heat produced.

6 PIEZOELECTRIC BIOSENSORS

The piezoelectric effect is due to some crystals containing positive and negative charges which separate when the crystal is subjected to a stress, causing the establishment of an electric field. As a consequence, if this crystal is subjected to an electric field it will deform. An oscillating electric field of a resonant frequency will cause the crystal to vibrate with a characteristic frequency dependent on its composition and thickness as well as the way it has been cut. As this resonant frequency varies when molecules adsorb to the crystal surface, a piezoelectric crystal may form the basis of a biosensor. Even small changes in resonant frequencies are easy to determine accurately by modern electronic techniques. Differences in mass, adsorbed to the sensing surface, even as small as a nanogram per square centimetre can be measured. Changes in frequency are generally determined relative to a similarly treated reference crystal without the active biological material. A biosensor for cocaine in the gas phase may be made by attaching cocaine antibodies to the surface of a piezoelectric crystal. This biosensor changes frequency by about 50 Hz for one part per billion cocaine in the sample atmosphere and can be re-used on flushing for a few seconds with clean air. The relative humidity of the air is important as if it is too low the response is less sensitive, and if it is too high the piezoelectric effect may disappear altogether.

Enzymes with gaseous substrates or inhibitors can also be attached to such crystals, as has been proved by the production of a biosensor for formaldehyde incorporating formaldehyde dehydrogenase and organophosphorus insecticides incorporating acetylcholinesterase respectively.

One of the drawbacks preventing the more widespread use of piezoelectric biosensors is the difficulty in using them to determine analytes in solution. The frequency of a piezoelectric crystal depends on the liquid's viscosity, density, and specific conductivity and, under unfavourable conditions, the crystal may cease to oscillate completely. There is also a marked effect of temperature due to its effect on viscosity. The binding of material to the crystal surface may be masked by other intermolecular effects at the surface and bulk viscosity changes consequent upon even quite small concentration differences.

Antibody–antigen binding can be determined by measuring the frequency changes in air after drying the crystal but such procedures are difficult to reproduce on repetitive use of the same crystal. A biosensor has been developed, using this principle, for the detection of enterobacteria.[14]

7 OPTICAL BIOSENSORS

Optical biosensors are currently generating considerable interest, particularly with respect to the use of fibre optics and optoelectronic transducers. These allow the safe non-electrical remote sensing of materials in hazardous or sensitive (*i.e. in vivo*) environments. An advantage of optical biosensors is that no reference sensor is needed: a comparative signal is generally easily generated using the same light source as the sampling sensor. A simple example of this is the fibre optic lactate sensor (Figure 13) which senses changes in molecular oxygen concentrations by determining its quenching of a fluorescent dye.

$$O_2 + \text{lactate} \xrightarrow[]{\text{lactate monooxygenase}} CO_2 + \text{acetate} + H_2O$$

The presence of oxygen quenches (reduces) the amount of fluorescence generated by the dyed film. An increase in lactate concentration reduces the oxygen concentration reaching the dyed film, so alleviating the quenching and consequentially causing an increase in the fluorescence output.

Simple colorimetric changes can be monitored in some biosensor configurations (Figure 14). Bromocresol green changes colour from yellow to blue-green

[14] M. Plomer, G. G. Guilbault, and B. Hock, *Enzyme Microb. Technol.*, 1992, **14**, 230.

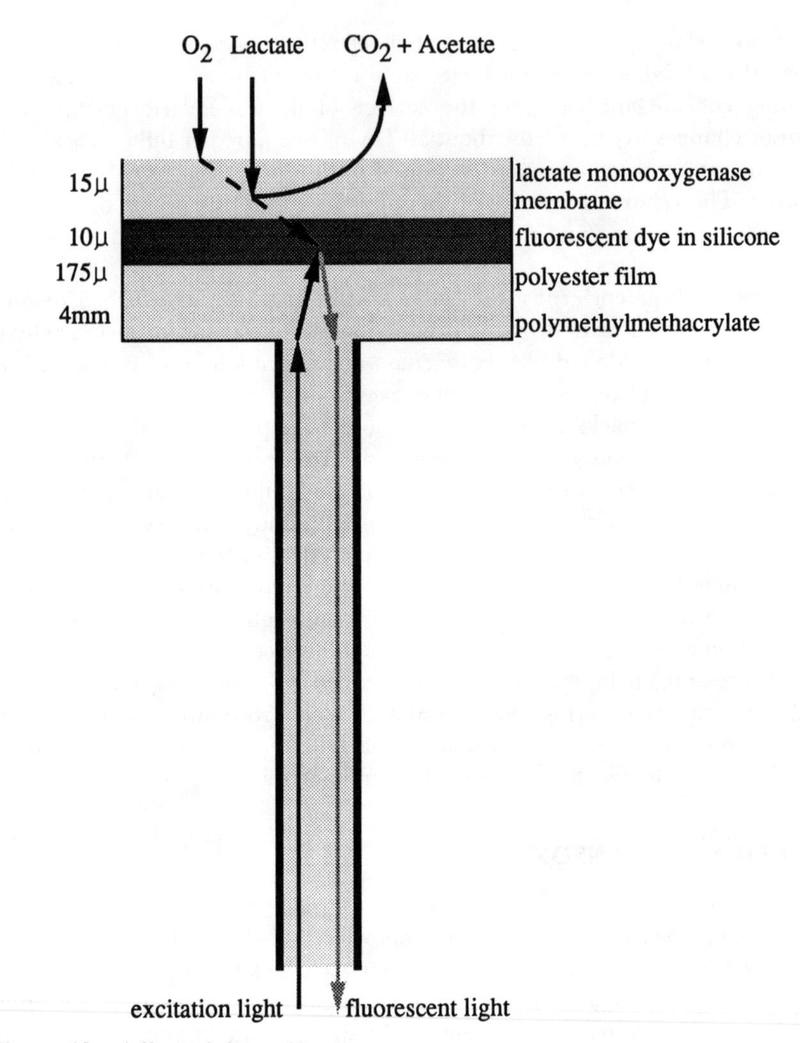

Figure 13 *A fibre optic lactate biosensor*

on binding to serum albumin at pH 3.8 and this change, which is maximal at 630 nm, may be simply detected by the absorption of light passing through the biosensing cell. The biosensor[15] shows a linear response to albumin over the range 5–35 mg cm^{-3}. A solid colorimetric gas assay for alcohol vapour has been described where solid alcohol oxidase, peroxidase, and 2,6-dichloroindophenol are dispersed on thin transparent microcrystalline cellulose TLC plates which undergo a colour change detected by transmission densitometry.[16]

One of the most widely established biosensor technologies is the low technology single-use colorimetric assays based on a paper pad impregnated with reagents. This industry revolves mainly round blood and urine analysis and is

[15] M. J. Goldfinch and C. R. Lowe, *Anal. Biochem.*, 1980, **109**, 216.
[16] E. Barzana, A. M. Klibanov, and M. Karel, *Anal. Biochem.*, 1989, **182**, 109.

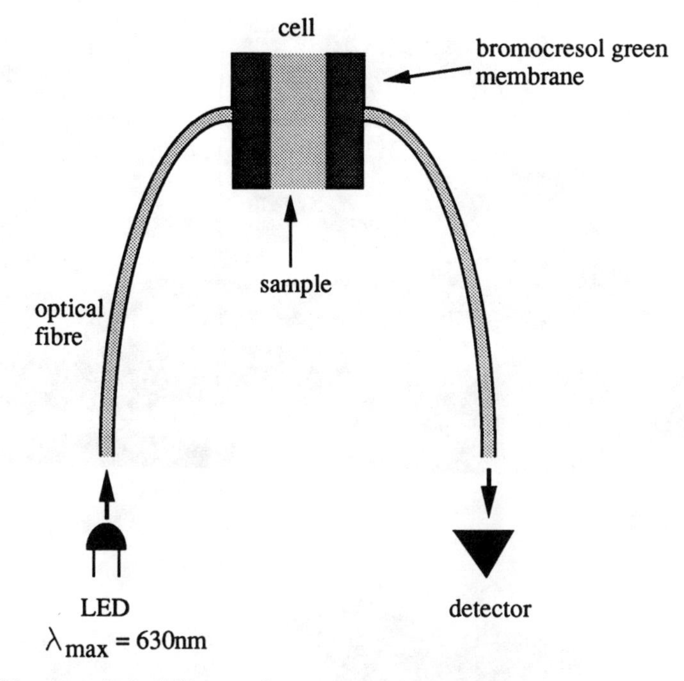

Figure 14 *An optical cell biosensor for serum albumin*

currently worth about $100M in the USA with test strips costing 5–6 ¢ for urine analysis and 40–60 ¢ for blood analysis. A particularly important use for these colorimetric test strips is in the monitoring of whole-blood glucose in the control of diabetes. In this case the strips contain glucose oxidase and horseradish peroxidase together with a chromogen (*e.g.* *o*-toluidine) which changes colour when oxidized in the peroxidase-catalysed reaction with the hydrogen peroxide produced by the aerobic oxidation of glucose.

$$\text{chromogen(2H)} + H_2O_2 \xrightarrow{\text{peroxidase}} \text{dye} + 2H_2O$$

The colour produced can be determined by visual comparison to a test chart or by the use of a portable reflectance meter. Advances in this area have been concerned with increasing the variety of test strips available and incorporating anti-interference layers to produce more reproducible and accurate assays.

It is possible to link up luminescent reactions to biosensors as light output is a relatively easy phenomena to transduce to an electronic output. Thus the reaction involving immobilized (or free) luciferase can be used to detect the ATP released by the lysis of micro-organisms.[17]

$$\text{Luciferin} + \text{ATP} + O_2 \xrightarrow{\text{luciferase}} \text{oxyluciferin} + CO_2 + \text{AMP} + \text{Pyrophosphate} + \text{light}$$

This allows the rapid detection of urinary infections by detecting the microbial content of urine samples.

[17] M. F. Chaplin, in 'Physical Methods for Microorganisms Detection', ed. W. M. Nelson, CRC Press, Boca Raton, USA, 1991, p. 81.

(a)

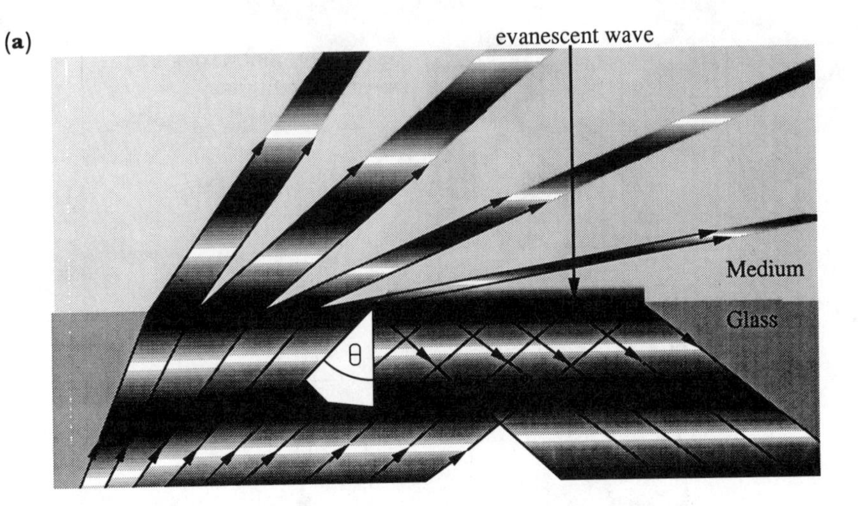

(b)

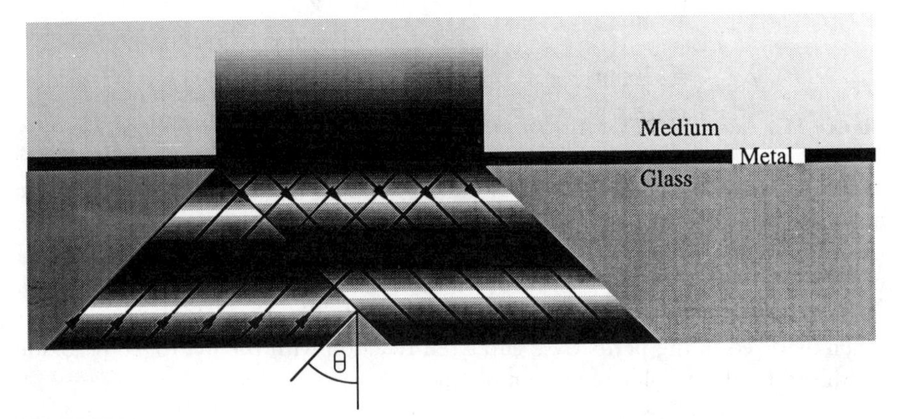

Figure 15 *Production of (a) an evanescent wave and (b) surface plasmon resonance. At acute enough angles of incidence the light is totally internally reflected at the glass surface. In (a), an evanescent wave extends from this surface into the air or water medium. This process is amplified in (b) by the presence of a thin metal film*

7.1 Evanescent Wave Biosensors

A light beam will be totally reflected when it strikes an interface between two transparent media from the side with the higher refractive index at angles of incidence (θ) greater than the critical angle (Figure 15a). This is the principle that allows transparent fibres to be used as optical waveguides.[18] At the point of reflection an electromagnetic field is induced which penetrates into the medium with the lower refractive index; usually air or water. This field is called the evanescent wave and it rapidly decays exponentially with the penetration distance and generally has effectively disappeared within a few hundred nanometres. The exact depth of penetration depends on the refractive index and

[18] B. I. Bluestein, I. M. Walczak, and S-Y. Chen, *Trends Biotechnol.*, 1990, **8**, 161.

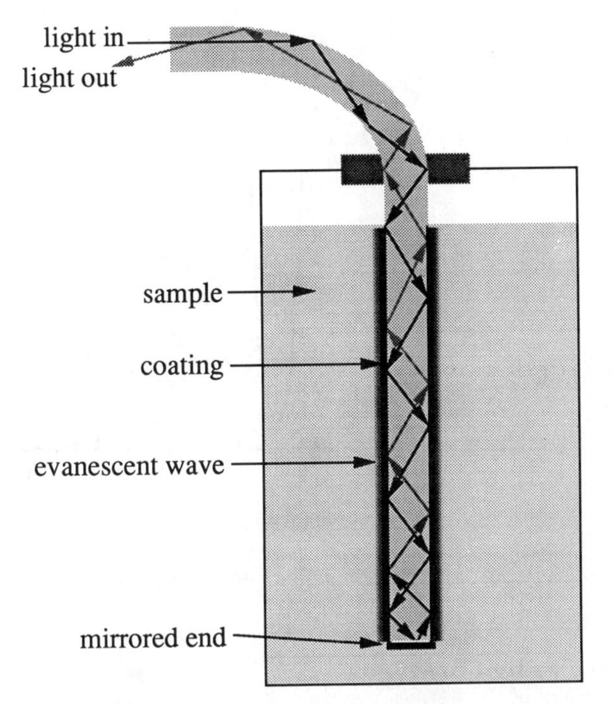

Figure 16 *The principle behind the evanescent wave immunosensor. The light output is reduced by absorption within the evanescent wave*

the wavelength of the light and can be controlled by the angle of incidence. The evanescent wave may interact with the medium. The resultant electromagnetic field may be coupled back into the higher refractive index medium (usually glass) by essentially the reverse process, giving rise to changes in the light emitted down the wave guide. In this process it can be used to detect changes occurring in the liquid medium.

Various effects, due to biological sensing processes, can be determined including changes in absorption, optical activity, fluorescence, and luminescence. Because of the small degree of penetration, this system is particularly sensitive to biological processes in the immediate vicinity of the surface and independent of any bulk processes or changes, and can even be used for the continuous monitoring of apparently opaque solutions.

This biosensor configuration is particularly suitable for immunoassays as there is no need to separate bulk components as the wave only penetrates as far as the antibody–antigen complex. Surface bound fluorophores may be excited by the evanescent wave and the excited light output detected after it is coupled back into the fibre (Figure 16), effectively by the reverse of the process causing the evanescent wave. Sensors can be fabricated which measure oxidase substrates using the principle of quenching of fluorescence by molecular oxygen as described earlier. Another advantage of only sensing a surface reaction less than a micron thick is that the volume of analyte needed may be very small indeed.

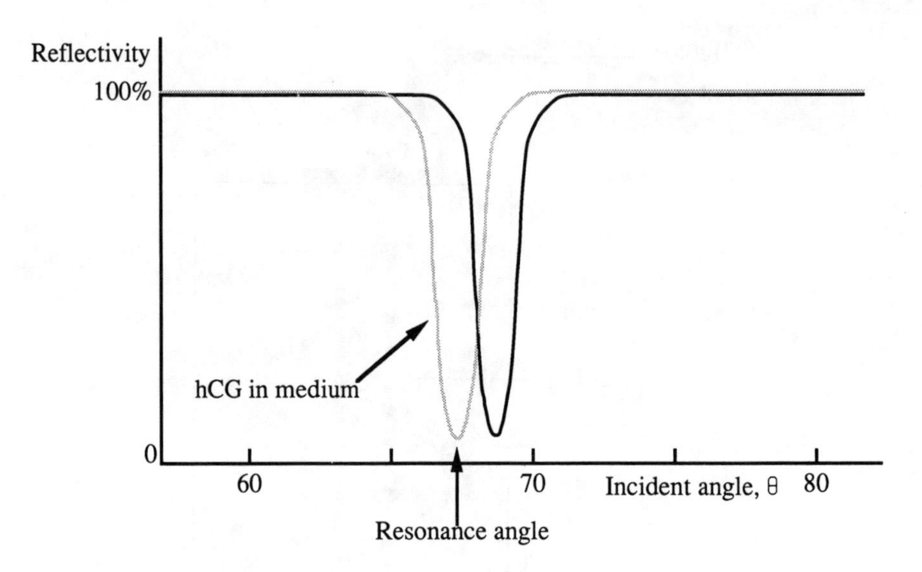

Figure 17 *The change in absorption due to surface plasmon resonance*

7.2 Surface Plasmon Resonance

The evanescent field generated by the total internal reflection of light within a fibre optic or prism may be utilized in a different type of optical biosensor by means of the phenomena of surface plasmon resonance (SPR). If the surface of the glass is covered with a layer of metal (usually pure gold, silver, or palladium) then the electrons at the surface may be caused to oscillate in resonance with the photons generating a surface plasmon wave and amplifying the evanescent field on the far side of the metal (Figure 15b). If the metal layer is thin enough to allow penetration of the evanescent field to the opposite surface, the effect is critically dependent on the medium adjacent to the metal. This effect occurs only when the light is at a specific angle of incidence dependent on its frequency, the thickness of the metal layer, and the refractive index of the medium immediately above the metal surface within the evanescent field. The generation of this surface plasmon resonance absorbs some of the energy of the light so reducing the intensity of the internally reflected light (Figure 16). Changes occurring in the medium caused by biological interactions may be followed by noting the changes in the intensity of the reflected light or the resonance angle. Figure 17 shows the change in the resonance angle of a human chorionic gonadotrophin (hCG) biosensor on binding hCG to surface-bound hCG antibody.[19] The sensitivity in such devices is limited by the degree of uniformity of the surface and the bound layer and the more sensitive devices minimize light scattering.

The biological sensing can be achieved by attaching the bioactive molecule to the medium-side of the metal film. Physical adsorption may be used but, because

[19] J. W. Attridge, P. B. Daniels, J. K. Deacon, G. A. Robinson, and G. P. Davidson, *Biosens. Bioelectron.*, 1991, **6**, 201.

Table 7 *Whole cell biosensors*

Analyte	*Organism*	*Biosensor*
Ammonia	*Nitrosomonas* sp.	amperometric (O_2)
Biological oxygen demand (BOD)	many	amperometric (O_2/mediated) or potentiometric (FET/H_2)
Cysteine	*Proteus morganii*	potentiometric (H_2S)
Glutamate	*Escherichia coli*	potentiometric (CO_2)
Glutamine	*Sarcina flava*	potentiometric (NH_3)
Herbicides	Cyanobacteria	thermometric or amperometric (mediated)
Nicotinic acid	*Lactobacillus arabinosus*	potentiometric (H^+)
Sulfate	*Desulfovibrio desulfuricans*	potentiometric (SO_3^-)
Thiamine	*Lactobacillus fermenti*	amperometric (mediated)

this may lead to undesired denaturation and weak binding, covalent binding is often preferred. Gold films can be coated with a monolayer of long-chain 1,ω-hydroxyalkyl thiols which are copolymerized to a flexible un-crosslinked carboxymethylated dextran gel enabling the subsequent binding of bioactive molecules. This flat plate system, marketed as BIAcore by Pharmacia Biosensor AB,[20] allows the detection of parts per million of protein antigen where the appropriate antibody is bound to the gel. Typical analyses need less than 50 μl and take times of 5–10 minutes. Similarly, biosensors for DNA detection can be constructed by attaching a DNA or RNA probe to the metal surface when as little as a few femtograms of complementary DNA or RNA can be detected[21] and, as a bonus, the rate of hybridization may be determined. Such biosensors retain the advantages of the use of evanescent fields as described earlier.

8 WHOLE CELL BIOSENSORS

As biocatalysts, whole microbial cells can offer some advantages over pure enzymes when used in biosensors.[22] Generally, microbial cells are cheaper, have longer active lifetimes, and are less sensitive to inhibition, pH, and temperature variations than the isolated enzymes. Against these advantages, such devices usually offer longer response and recovery times, a lower selectivity, and they are prone to biocatalytic leakage. They are particularly useful where multistep or coenzyme-requiring reactions are necessary. The microbial cells may be viable or dead. The advantage of using viable cells is that the sensor may possess a self-repair capability but this must be balanced against the gentler conditions necessary for use and problems that might occur due to membrane permeability. Different types of whole cell biosensors are shown in Table 7.

[20] S. Löfås and B. Johnsson, *J. Chem. Soc., Chem. Commun.*, 1990, 1526.

[21] T. Schwarz, D. Yeung, E. Hawkins, P. Heaney, and A. McDougall, *Trends Biotechnol.*, 1991, **9**, 339.

[22] D. M. Rawson, *Int. Ind. Biotech.*, 1988, **8(2)**, 18.

Table 8 *A selection of immunosensors*

Analyte	*Sensing method*	*Biosensor*
Human chorionic gonadotrophin	antibody/catalase	amperometric (O_2)
	antibody	SPR
Hepatitis B surface antigen	antibody/peroxidase	potentiometric (I^-)
Insulin	antibody/catalase	amperometric (O_2)
T2 toxin	antibody	evanescent wave

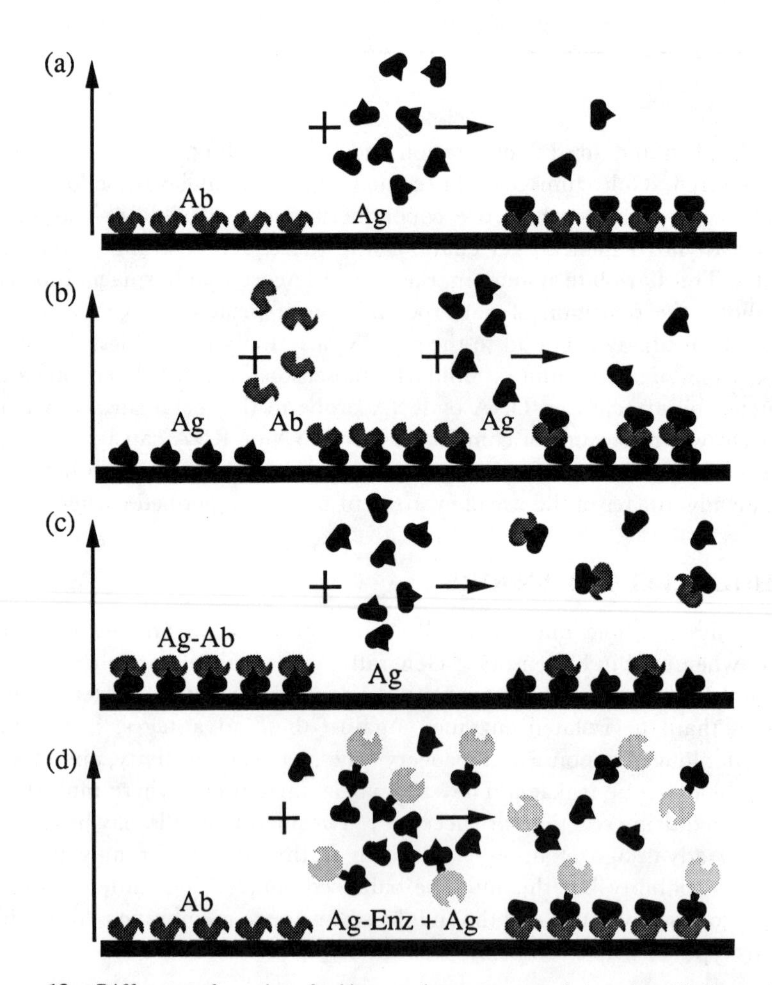

Figure 18 *Different configurations for biosensor immunoassays: (a) antigen binding to immobilized antibody, (b) immobilized antigen binding antibody which binds free second antigen, (c) antibody bound to immobilized antigen partially released by competing free antigen, (d) immobilized antibody binding free antigen and enzyme-labelled antigen in competition*

9 IMMUNOSENSORS

The immunodiagnostic market has been estimated as almost 3 billion dollars in the USA alone.[18] At the present time only an insignificant proportion of this is contributed by immunosensors. However, this area is expected to grow in importance over the next few years.

Most biosensor configurations may be used as immunosensors (Table 8) and some of these have been mentioned earlier. Figure 18 shows some of the configurations possible. Direct binding of the antigen to immobilized antibody (Figure 18a) or antigen–antibody sandwiches (Figure 18b) may be detected using piezoelectric or SPR devices,[23] as can antibody release due to free antigen (Figure 18c). Binding of enzyme-linked antigen (Figure 18d) or antibody can form the basis of all types of immunosensors but has proved particularly useful in thermometric and amperometric devices. The amount of enzyme activity bound in these immunosensors is dependent on the relative concentrations of the competing labelled and unlabelled ligands and so it can be used to determine the concentration of unknown antigen concentrations.

The main problems involved in developing immunosensors centre on non-specific binding and incomplete reversibility of the antigen–antibody reaction both of which reduce the active area, and hence sensitivity, on repetitive assay. Single-use biosensing membranes are a way round this but they necessitate strict quality control over production.

10 CONCLUSION

Biosensors form an interesting and varied part of biotechnology. They have been applied to solve a number of analytical problems and some biosensors have achieved notable commercial success. They have not yet reached their full potential with many more commercial products being expected over the next few years.

[23] J. R. North, *Trends Biotechnol.*, 1985, **3**, 180.

Subject Index